New Methodologies for Understanding Radar Data

Related titles on radar:

Advances in Bistatic Radar Willis and Griffiths
Airborne Early Warning System Concepts, 3rd Edition Long
Bistatic Radar, 2nd Edition Willis
Design of Multi-Frequency CW Radars Jankiraman
Digital Techniques for Wideband Receivers, 2nd Edition Tsui
Electronic Warfare Pocket Guide Adamy
Foliage Penetration Radar: Detection and characterisation of objects under trees Davis
Fundamentals of Ground Radar for ATC Engineers and Technicians Bouwman
Fundamentals of Systems Engineering and Defense Systems Applications Jeffrey
Introduction to Electronic Warfare Modeling and Simulation Adamy
Introduction to Electronic Defense Systems Neri
Introduction to Sensors for Ranging and Imaging Brooker
Microwave Passive Direction Finding Lipsky
Microwave Receivers with Electronic Warfare Applications Tsui
Phased-Array Radar Design: Application of radar fundamentals Jeffrey
Pocket Radar Guide: Key facts, equations, and data Curry
Principles of Modern Radar, Volume 1: Basic principles Richards, Scheer and Holm
Principles of Modern Radar, Volume 2: Advanced techniques Melvin and Scheer
Principles of Modern Radar, Volume 3: Applications Scheer and Melvin
Principles of Waveform Diversity and Design Wicks *et al.*
Pulse Doppler Radar Alabaster
Radar Cross Section Measurements Knott
Radar Cross Section, 2nd Edition Knott *et al.*
Radar Design Principles: Signal processing and the environment, 2nd Edition Nathanson *et al.*
Radar Detection DiFranco and Rubin
Radar Essentials: A concise handbook for radar design and performance Curry
Radar Foundations for Imaging and Advanced Concepts Sullivan
Radar Principles for the Non-Specialist, 3rd Edition Toomay and Hannan
Test and Evaluation of Aircraft Avionics and Weapons Systems McShea
Understanding Radar Systems Kingsley and Quegan
Understanding Synthetic Aperture Radar Images Oliver and Quegan
Radar and Electronic Warfare Principles for the Non-specialist, 4th Edition Hannen
Inverse Synthetic Aperture Radar Imaging: Principles, algorithms and applications Chen and Martorella
Stimson's Introduction to Airborne Radar, 3rd Edition Baker, Griffiths and Adamy
Test and Evaluation of Avionics and Weapon Sytems, 2nd Edition McShea
Angle-of-Arrival Estimation Using Radar Interferometry: Methods and applications Holder
Biologically-Inspired Radar and Sonar: Lessons from Nature Balleri, Griffiths and Baker
The Impact of Cognition on Radar Technology Farina, De Maio and Haykin
Novel Radar Techniques and Applications, Volume 1: Real Aperture Array Radar, Imaging Radar, and Passive and Multistatic Radar Klemm, Nickel, Gierull, Lombardo, Griffiths and Koch
Novel Radar Techniques and Applications, Volume 2: Waveform Diversity and Cognitive Radar, and Target Tracking and Data Fusion Klemm, Nickel, Gierull, Lombardo, Griffiths and Koch
Radar and Communication Spectrum Sharing Blunt and Perrins
Systems Engineering for Ethical Autonomous Systems Gillespie
Shadowing Function from Randomly Rough Surfaces: Derivation and applications Bourlier and Li
Photo for Radar Networks and Electronic Warfare Systems Bogoni, Laghezza and Ghelfi
Multidimensional Radar Imaging Martorella
Radar Waveform Design Based on Optimization Theory Cui, De Maio, Farina and Li
Micro-Doppler Radar and its Applications Fioranelli, Griffiths, Ritchie and Balleri
Maritime Surveillance with Synthetic Aperture Radar Di Martino and Antonio Iodice
Electronic Scanned Array Design Williams
Advanced Sparsity-Driven Models and Methods for Radar Applications Gang Li
Deep Neural Network Design for Radar Applications Gurbuz

New Methodologies for Understanding Radar Data

Edited by
Amit Kumar Mishra and Stefan Brüggenwirth

The Institution of Engineering and Technology

Published by SciTech Publishing, an imprint of The Institution of Engineering and Technology, London, United Kingdom

The Institution of Engineering and Technology is registered as a Charity in England & Wales (no. 211014) and Scotland (no. SC038698).

First published 2021

The Institution of Engineering and Technology
Michael Faraday House
Six Hills Way, Stevenage
Herts, SG1 2AY, United Kingdom

www.theiet.org

British Library Cataloguing in Publication Data
A catalogue record for this product is available from the British Library

ISBN 978-1-83953-188-0 (hardback)
ISBN 978-1-83953-189-7 (PDF)

Typeset in India by Exeter Premedia Services Private Limited

Contents

About the editors

Amit Kumar Mishra is a professor at the Radar Remote Sensing Group, University of Cape Town, South Africa. Amit has over 15 years of experience in radar and radio system design and signal processing. His current interests are around the use of cognitive architectures and advanced signal processing and machine learning to understand radar signals. He has published more than 35 papers in ISI listed journals and over 120 papers in peer reviewed conference proceedings.

Stefan Brüggenwirth is head of the Cognitive Radar Department at the Fraunhofer FHR, Germany and NATO SET panel member-at-large for machine learning and artificial intelligence for RF-systems. Stefan studied aerospace engineering in Stuttgart and Munich, where he did his PhD in the development of a cognitive system architecture for UAV mission management. He spent 2.5 years at NASA Ames Research Center in Mountain View. His research focuses on radar target identification, signal processing architectures and self-learning radar systems.

Chapter 1

New methods for radar signal understanding

Stefan Brüggenwirth[1] and Amit Kumar Mishra[2]

1.1 A philosophical introduction

The questions of understanding their environment, the correct interpretation of sensations, the principles behind the material world and also the limits of perception have always concerned mankind. Western philosophy reaches back to Plato's (428–348 BC) *theory of forms* or equivalently *theory of ideas* which states that every (visual) sensory experience is just a perceivable projection from the higher realm of ideas as shown in the *Allegory of the Cave* (Figure 1.1).

These ideas represent the real, eternal, unchangeable and immaterial properties of the *thing-in-itself*, accessible by the human mind. The Greek word ιδέα (idea) is derived from the word seeing (*idein*) and reflects the human capability to infer and abstract the a priori existing category and ideal *archetype* of things from raw sensory input. Logic and mathematical formulation can be used to capture and approximate this realm of ideas.

Epistemology (or *theory of knowledge*) is a main branch of philosophy that is concerned with the nature of knowledge, its justification and the rationality of belief. With the modern era philosophy in the seventeenth century, a dispute arose whether knowledge comes primarily from sensory experience (*empiricism*) or our knowledge can be derived entirely from reasoning (*rationalism*). The dispute was resolved by Immanuel Kant by synthesis as expressed in his *criticism of pure knowledge.*

Thus, all human knowledge begins with intuitions, proceeds from thence to concepts and ends with ideas.

As opposed to Plato, the essential idea of an object is a *metaphysical* concept and hence cannot be grasped rationally. The concept of *logical positivism* discussed in the *Vienna Circle* (including mathematicians and physicists such as Mach, Hertz and Gödel) in the early twentieth century constraints knowledge and valid theories

[1]Fraunhofer FHR, Germany
[2]University of Cape Town, South Africa

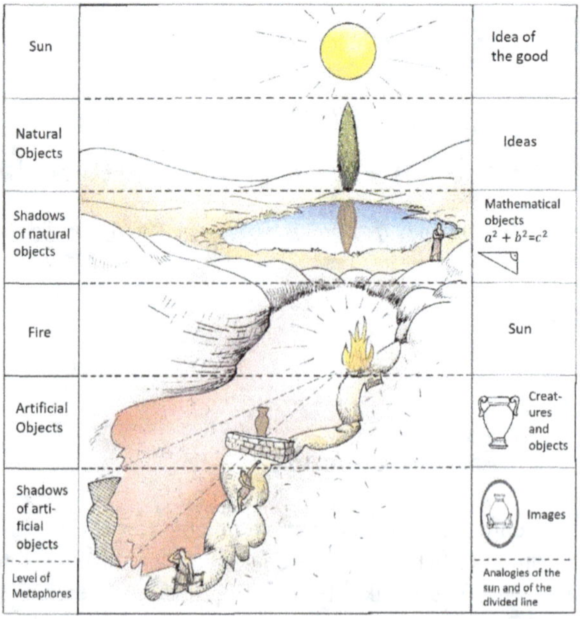

Figure 1.1 Platon's Allegory of the Cave Source: [1]

to empirical perceptions, *verifiable* by experiments. In modern theory of science, according to Popper, this concept of verification is replaced by *falsifiability*. This means that an established scientific hypothesis is valid until a better one is found that approximates reality more closely. Hence scientific theories are often refined over time as models of different complexity, accuracy and preconditions. For instance, different interpretations of electromagnetism as a light ray, wave, particle or from quantum physics are perfect examples to show this abstraction from reality and the specific context of every model.

Another important philosophical approach developed in the twentieth century by Edmund Husserl and his scholars is known as *Phenomenology*. It stresses subjective aspects when humans interpret physically perceived phenomena, by considering the structures of consciousness and the influence of individual experiences, intuitions and emotions. It is important to recognise this potential bias when seeking objective statements. On the other hand, human perception heavily depends on individual factors that are clearly hard to automate.

1.2 Automated reasoning

The term *ontology* was introduced in the sixteenth century for the philosophical study of being and is also used today in computer science to represent knowledge, concepts and categories and relationships between them in a digital form, e.g. a

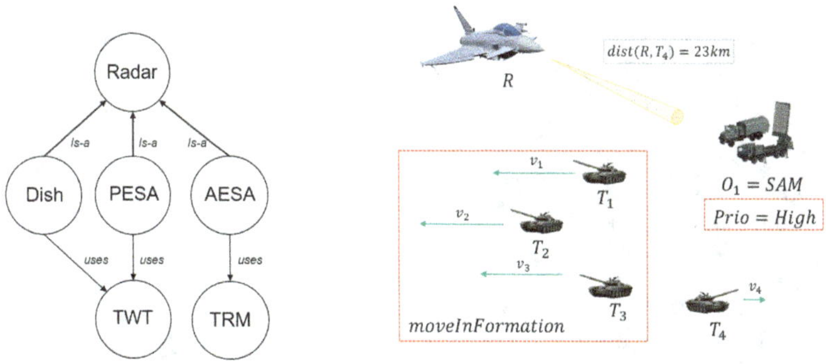

$$\forall\, O: Is - A = Radar \Rightarrow \{Dish, PESA, AESA\}$$

$$\forall\, O: uses = TWT \Rightarrow \{Dish, PESA\}$$

$$\forall\, T_i, T_j: v_i = v_j \wedge dist(T_i, T_j) < 100m \Rightarrow moveInFormation(T_i, T_j)$$

$$\forall\, O_j: O_i = SAM \wedge dist(R, O_i) < 25km \Rightarrow Prio(O_j) = High$$

Figure 1.2 Examples of symbolic reasoning systems. Left: Radar-ontology, Right: Rule-based situation assessment.

formal logic, data structure or programming language. *Semantics* (from ancient Greek: *semantikos*, 'significant') is the study of meaning in these language structures (as opposed to *syntax*, which refers to the combinatorics of language units). Both are key to automated understanding of data using *formal reasoning* or *knowledge-based* methods. Symbol-transforming-knowledge-based systems (Figure 1.2) are a key foundation of *artificial intelligence* and have been successfully used in *expert systems, computer algebra software, theorem proving* or *formal verification* of safety-critical systems.

In the last decade, *subsymbolic* or *connectionist* approaches using *data-driven machine learning approaches* have seen a major breakthrough due to the availability of large sets of training data and hardware acceleration. *Deep learning methods* have revolutionised image perception (Figure 1.3) or speech recognition. However, they are *black-box approaches* and *explainable neural nets* are still a topic of research. A fundamental drawback in these pattern-recognition algorithms lies in the lack of *contextual knowledge*, e.g. the human-level skill of attaching higher-level meaning (e.g. semantics) to a received signal from personal experience with similar data, relationships to other patterns in the broader scene or a thorough scientific understanding of the signal model.

Since natural human language is more expressive than any formal computer language used for automated reasoning, human input (e.g. as programmer, ontology-engineer or for annotating data) is always required to capture and formalise human-level understanding and close the *semantic gap,* as shown in Figure 1.4. In the AI community, this is well known as the symbolic vs non-symbolic AI. One of the challenges for current AI research is the interface between these two.

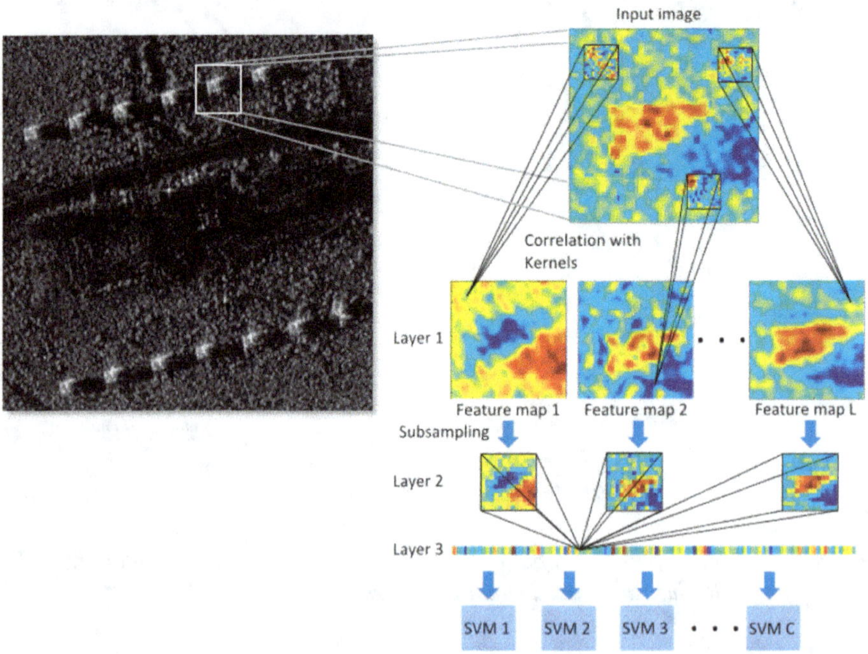

Figure 1.3 SAR Scene (left) with a CNN for automated target recognition

Due to this principle shortcoming, humans have been considered to outperform machines in perception tasks since the early days of human-computer-interaction. Even though deep learning approaches can excel human in certain pattern recognition and object classification tasks todays, humans are still essential for complex scene understanding.

In particular, radar data is notoriously hard to understand and interpret. It is well known that analysts of satellite Synthetic Aperture Radar (SAR) data go through lengthy education and training period to understand all the physical effects in the real scene that show up in the processed SAR image. Often enough data from electro-optical sensors is still preferred as it resembles more naturally our everyday visual perception.

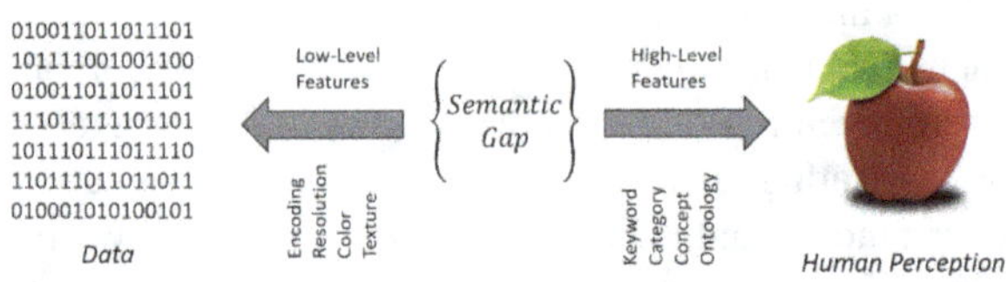

Figure 1.4 Semantic gap

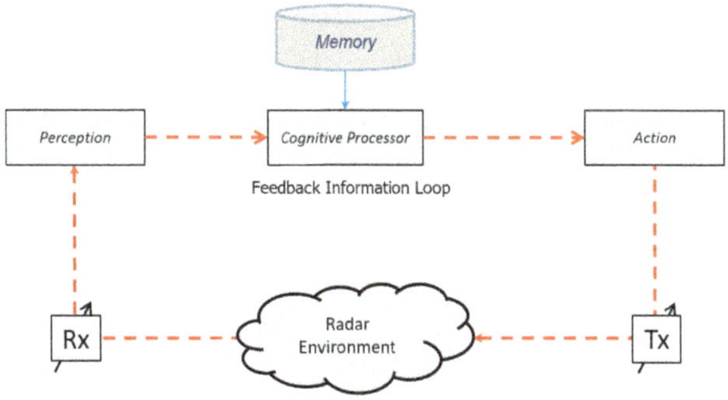

Figure 1.5 Perception–action cycle of a cognitive radar system with memory

1.3 The role of perception in cognitive radar

The concept of cognitive radar has recently received increased attention. The current IEEE definition (IEEE Std P686-2017) emphasises the perception–action cycle (Figure 1.5) and the self-learning capability:

> A radar system that automatically forms an on-going perception of the target scene and takes actions accordingly. It may use short and long-term memory to improve performance of a given function. In comparison to adaptive radar, cognitive radar learns to adapt operating parameters as well as processing parameters and may do so over extended time periods.

Artificial intelligence is commonly used as a tool for implementing both the perception and action parts of a cognitive radar system. It is also generally acknowledged that a cognitive radar system operates on a high automation and abstraction level, e.g. on high-level goals or tasks supplied by a human operator. Accurate, robust and semantically rich perception skills are hence key to a cognitive radar system. Though the action part is generally not discussed in this book, it is interesting to note that radar as an active sensor has the ability to systematically collect missing bits of information from the environment and focus attention (in terms of radar resources) according to its situational awareness.

Different cognitive radar architectures have been derived. Some root in human cognitive frameworks, some in computer science and intelligent agent theory, some in data fusion and some in knowledge-based radar signal processing. The role and concrete implementation of the perception block in Figure 1.5 depends very much on the philosophy and cognitive radar framework used. Some concepts that the authors are familiar with are presented in Sections 1.3.1–1.3.4.

1.3.1 Fuster's model of cognition

One of the comprehensive works on modelling human cognition has been done by Fuster [2]. In a series of papers, he has analysed the roles of perception–action

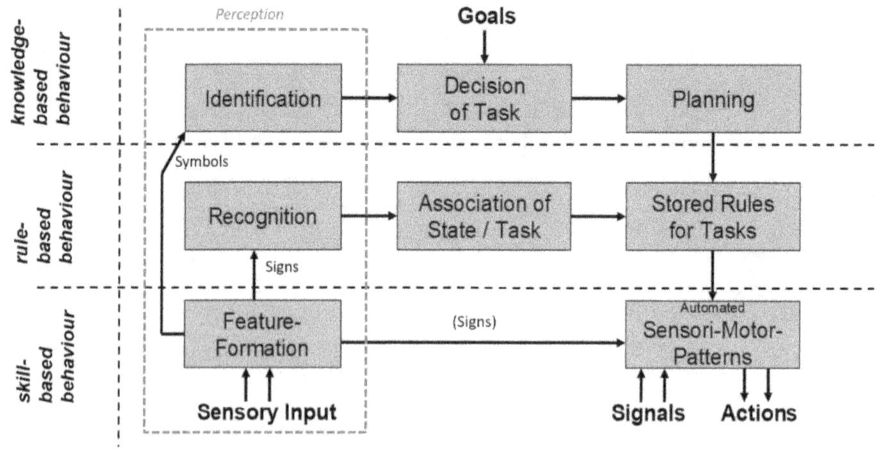

Figure 1.6 Perception–action cycle of a cognitive radar system with memory

feedback and layered architecture of prefrontal cortex in the emergence of cognition. This has been a major inspiration behind Haykin's proposal on cognitive radar [3].

1.3.2 Three-layer cognitive radar architecture

The cognitive radar architecture developed at Fraunhofer FHR is based on the three-layer model of human cognitive performance by Rasmussen. As shown in Figure 1.6, the complex human cognition process is broken down into eight simplified cognitive subfunctions and these are arranged in three behavioural layers of increasing abstraction. The skill-based behaviour represents subconscious subsymbolic processing of input sensory data inside the grey perception block and execution of sensorimotor patterns on the action side. Rule-based behaviour allows a human to quickly and consciously react to a known situation by learned procedures. Knowledge-based behaviour provides the highest flexibility by a semantically rich, symbolic identification of the environment and deliberate task planning. The three layers of behaviour are associated with different types of perceptions.

Signals are sensory data representing time-space variables from dynamic spatial configuration in the environment, and they can be processed by the organism as continuous variables. *Signs* indicate a state in the environment with reference to certain conventions for acts. Signs are related to certain features in the environment and the connected conditions for action. Signs cannot be processed directly, they serve to activate stored patterns of behaviour. *Symbols* represent other information, variables, relations and properties and can be formally processed. Symbols are abstract constructs related to and defined by a formal structure of relations and processes, which by conventions can be related to features of the external world.

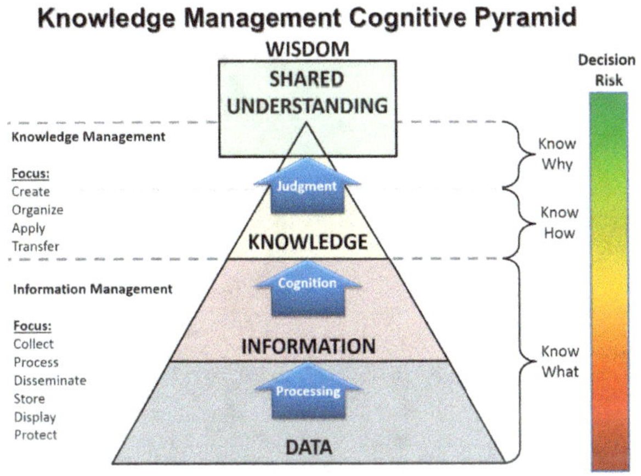

Figure 1.7 *DIKW hierarchy, wisdom hierarchy, knowledge hierarchy, information hierarchy and the data pyramid*

1.3.3 DIKW pyramid

As we have discussed so far, 'understanding' is a complex and complicated process. In modelling this process, the data-information-knowledge-wisdom (DIKW) hierarchy has been used often. The origin of this model is uncertain. However, the use of DIKW has mostly been seen in the latter half of the twentieth century. For signal or data understanding, this is an efficient model for cognitive processes. As a model, it is intuitive to understand; shows action items as how to implement; shows a long-term vision and how that is connected to what can be achieved now; and is not too abstract while not being restrictive (and thereby leaving scopes for disruptive innovations and discoveries).

Figure 1.7 shows the structure of the DIKW pyramid. The DIKW pyramid gives a good model to understand the synthesis of wisdom from data. This represents the perception-leg of cognitive engineering. In addition to this, if we can abstract an action-leg, then it can be used as a framework to implement cognitive systems. Figure 1.8 shows the cognitive model where the perception-leg is modelled using the DIKW framework. Details of the model can be found in [4].

1.3.4 Bio-inspired approaches

Cognitive algorithms are mostly inspired by human brain. In addition to humans, other members of biosphere can act as great inspiration as well. Especially in radar, there are species (like bats and dolphins) that use echolocation [5] and hence are sources of inspiration. In an interesting review paper [6], the team from University College London (UCL) has discussed a range of things that radar engineers can learn from bats starting from signal processing to waveform design [7]. Another interesting animal that uses electric signal echolocation is the weakly electric fishes found in

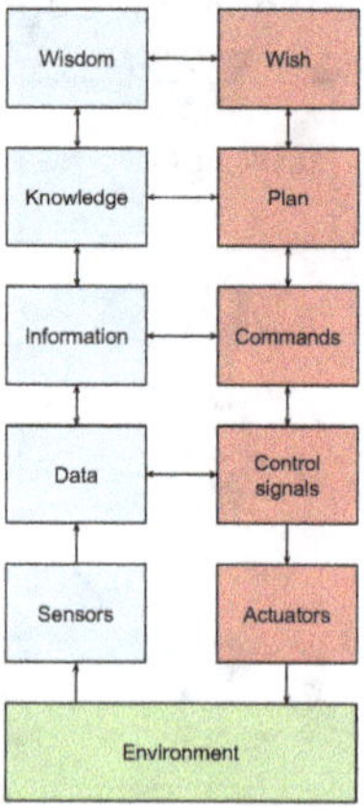

Figure 1.8 *The DIKW model of cognitive processing. Note that each layer can be regarded as a layer in the prefrontal cortex. A layer above another represents a more abstract way of processing.*

deep rivers [8]. The layered structure of their signal-processing neurons is an inspiration in itself for neural network designers. Besides that, these fishes use the same set of electric sensors to achieve three tasks, viz. echolocation, target identification and communication!

The benefit of bio-inspired approaches inspired by smaller animals is the fact that most of these animals use a very small set of neurons to achieve interesting feats. Second, most of them use a range of steps starting from signal design, agile signal transmission, target/situation-specific processing and diversity gains to work efficiently in their ecosystem. We expect more bio-inspired radar architectures to be tried and tested in future.

1.4 A brief history of endeavours to understand signal coming from radar

1.4.1 What is signal understanding?

In the last decade, a few universities have been working on a multistatic radar system called Nextrad [9]. Figure 1.9 shows the range-time-intensity (RTI) image of one of the recordings collected using Nextrad in 2019.[a]

[a]The authors thank the Nextrad team, especially Simon Lewis, for sharing this image.

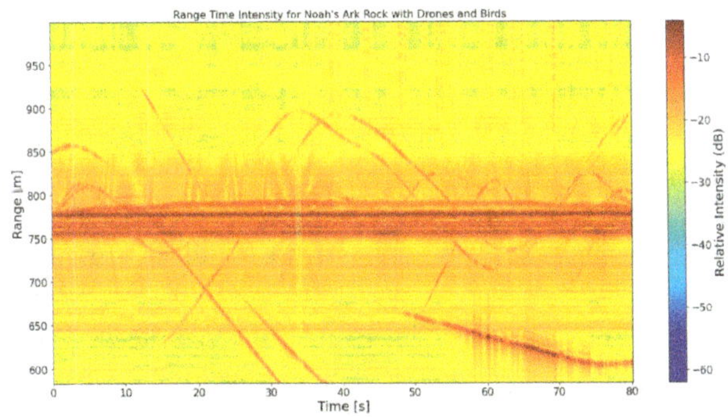

Figure 1.9 *RTI plot of a dataset collected using Nextrad in 2019 in Simons Town (South Africa). Noah's Arc Rock creates a strong return at around 760 m range. There are two drones flying which create the periodic signature between 700 and 850 m range bins. The strong target between bins 600 and 650 m from about 60–70 seconds is a small speedboat passing into the beam.*

This is a scene filled with a number of events going on. We have been involved during the trials. Hence, we could tag some of the stronger signals in the plot. However, imagine we are given this RTI plot and asked to interpret it without any background knowledge. 'Radar signal understanding' is hence a challenging task and a key functionality in modern radar system. In this book, we hope to summarise some of the efforts in various domains and provide new perspectives that appear on the horizon.

1.4.2 A brief history

Original radar engineers were more focused on detecting the 'target'. Detecting a ship or an aircraft is not an easy task. However, the environment of operation is mostly simple enough to be modelled into two classes, viz. target and clutter. This is no more an acceptable assumption when we operate in complex environments such as looking at a battlefield with multiple types of objects, targets and clutters. The situation becomes even more complicated when we consider most of the civilian applications (e.g. vehicular radar or disaster management radar). If we think of the example given in the previous section using Nextrad data, we can see where the problem lies. There are too many objects in the scene and the mathematical models we have are not rich enough to account for them all. The situation is very similar to the interpretation of shadows of artificial objects in Plato's *Allegory of the Cave* (Figure 1.1).

Understanding the shadows in the allegory and understanding the radar signal are two very similar challenges. However, we are a little better off in radar research, since we can take a peek at the scene and use the extra information to help us!

The most powerful step will be to make up our mind about what do we want to find from the haystack. In this case, we are not just interested in signal or maximising signal-to-noise ratio (SNR). Rather, we want to extract what is important to us. We have to redefine what is 'information' for us and design our system to maximise information-to-noise ratio.

Conventional radar signal processing relies heavily on matched filter which is an optimal signal-processing step if we are only interested in maximising SNR. Bell, in his seminal work [10], gave one of the earliest analysis on waveform design to maximise information-to-noise ratio. Typically, information-theoretic bounds (e.g. Cramer–Rao or Weiss–Weinstein bound) define a maximum of information that can be extracted from the stochastic environment – and a good engineering approach comes close to that!

The research around signal design to enhance the information content has been enabled by two current trends. First, there has been a consistent effort towards having software-defined radars where the signal transmitted can be changed digitally. Second, there has been a series of innovations in the investigation of more interesting waveform like nonlinear chirp [11] and Orthogonal Frequency Division Multiplexing (OFDM) waveform.

In addition to the active ways to understand the scene, there is a wealth of research on the development of phenomenological procedure to model radar signal based on physical phenomena. One of the earliest efforts can be ascribed to Huynen [12], who gave an elegant method to decompose fully polarimetric radar signal into components with physical significance.

Similar approach using single polarised radar signal can be designed where the radar signal can be decomposed into canonical scattering mechanisms [13]. If these scattering mechanisms are determined by the physical shape of the object then this will give us an understanding of the physical shape of the target. However, the challenge with real-life objects and scenes is that they are much richer than canonical objects. Hence, the signal shows a lot more richness. One of the interesting and rigorous analyses on this has been done by Rihaczek [14].

The recent approach of multi-dimensional radar imaging [15] stresses the importance of having 'many eyes' to obtain an improved, richer and more effective picture of targets. In this regard, additional information provided by multi-channel, multi-frequency, multi-polarisation and multi-static radar (or combinations thereof) are used to 'understand' more about targets and the environment.

Automated target recognition (ATR) [15] or non-cooperative target identification (NCTI) is today's standard technology in military and civilian radar systems. They support the human operator in ground-target recognition in SAR scenes or threat assessment for missile defence. Micro-Doppler features are typically used in automotive radar for pedestrian detection or even automated emergency braking.

Referring again to Platon's *Allegory of the Cave*, we have come a long way and certainly increased our level of insight. We have tried to (metaphorically) change the type of light that is creating the shadows. We have interpreted the shadows using our previous experiences and mathematical models. The above review is by no means complete. Signal understanding is and shall continue to be a major field of innovation in radar engineering.

1.5 The scope and a brief outline of this monograph

In the previous section, we discussed the significance of the efforts towards understanding radar signal. Different parts of this endeavour have been termed using different phrases such as signal classification, scene classification, target recognition, parameter extraction. However, we can identify the common thread. All these efforts have one thing in common. They all help radar engineers interpreting the shadows in Plato's *Allegory of the Cave*!

The research in the domain of radar signal understanding has seen some interesting new developments. This has been mainly due to the developments around cognitive radar and the use of modern machine learning algorithms. It was the editors' belief that an edited book to consolidate such efforts will be extremely beneficial. Maybe it is time to put all these pieces of research under one umbrella of 'radar signal understanding' to give a coherent and holistic picture. It was exciting to note that the IET appreciated this vision and many of our well-known colleagues consented to contribute chapters to this humble endeavour.

1.5.1 Brief outline of the book

Part 1 focuses on statistical and conventional methods for interpreting radar data. Chapters in this part are as follows.

- The chapter entitled 'Understanding FM Radar Signal through Parameter Estimation' by Abratkiewicz and Semantikos briefly reviews signal-processing techniques for frequency-modulated waveform parameter estimation.
- The chapter 'Non-Linear Estimation Algorithms to Analyse Radar Signals' by Singh *et al.* discusses some of the modern nonlinear signal processing algorithms to process radar signals.
- The chapter 'Understanding Radar Signals in a Highly Heterogeneous Clutter Environment' by Braeu and Rial provides a mathematical overview and signal models for highly heterogeneous clutter environment encountered in ground-penetrating radar.
- In the next chapter, Jahangir, Baker and Oswald describe the concept of staring radar and the analysis of its signal in discriminating small airborne targets (like drones) in their chapter titled 'Signature Characteristics and Discrimination of Small Airborne Targets with a Staring Radar'
- The contribution 'SAR Coherent Change Detection' by Blacknell, Andre and Bennett gives an overview of modern methods in change detection in SAR images using physics-based derivations.

Part 2 focuses on compressed sensing and cognitive methods for understanding radar data. Chapters in this part are as follows.

- The chapter 'Modern GPR Target Recognition Methods' by Giovanneschi, Mishra and Gonzalez introduced novel dictionary-learning methods for detection of mines in ground-penetrating radar applications.
- In the chapter 'Jet Engine Recognition via Sparse Decomposition of ISAR Images' by Wagner and Ender, a novel compressed sensing-based approach is introduced based on the mathematical models of jet engine inlets for NCTI.
- In their chapter titled 'Waveform Design for Efficient Radar Signal Recovery', Alaee-Kerahroodi, Shankar and Ottersten discuss transmitter signal design schemes to enable better signal recovery.

Part 3 focuses on machine learning methods for understanding radar and remote-sensing data. Chapters in this part are as follows.

- In the first chapter in this part of the book, Dominicos, Mishra and Rautenbach discuss deep-learning algorithms applied to predict ice concentrations using remote-sensed SAR images in the chapter titled 'Sea Ice Concentration Estimation Techniques Using Machine Learning: An End-To-End Workflow for Estimating Concentration Maps from SAR Images'.
- In the next chapter titled 'Cognitive Radar for Ground Moving Target Imaging', Gelli, Giusti and Martorella present a cognitive architecture to form SAR images of moving ground targets.
- The work 'GAN4SAR: Generative Adversarial Networks for Synthetic Aperture Radar Imaging of Targets Signature' by Vignaud describes the generation of SAR/ISAR images of man-made targets using a novel deep learning method called Generative Adversarial Network.
- In the chapter 'Micro Doppler Analysis: Theory and Applications' by Neumann, an introduction to micro-Doppler processing for target identification is given from an industry point of view.
- The contribution 'Kinematic and Linguistic Interpretation of Human Motion via RF Signal Analysis' by Gurbuz and Malaia shows how deep learning methods can be applied for human activity recognition.

References

[1] Burkard F.-P., Kunzmann P. 'dtv-Atlas Philosophie, Deutscher Taschenbuch Verlag with friendly permission of dtv Verlagsgesellschaft mbH & Co. KG, Translation by Stefan Brüggenwirth'. 1991, p. 40.

[2] Fuster J. *The Prefrontal Cortex*. Academic Press; 2015.

[3] Haykin S. 'Cognitive radar: a way of the future'. *IEEE Signal Processing Magazine*. 2006, vol. 23(1), pp. 30–40.

[4] Mishra A.K. 'A DIKW architecture for cognitive engineering'. *Procedia Computer Science*. 2018, vol. 123(1), pp. 285–9.

[5] Au W.W., Simmons J.A. 'Echolocation in dolphins and bats'. *Physics today*. 2007, vol. 60(9),40.
[6] Vespe M., Jones G., Baker C.J. 'Lessons for radar'. *IEEE Signal Processing Magazine*. 2009, vol. 26(1), pp. 65–75.
[7] Baker C.J., Inggs M., Mishra A.K. 'Waveform processing-domain diversity and ATR'. 19th European Signal Processing Conference; 2011. pp. 431–5.
[8] Kumar Mishra A., Mishra A.K. 'WEBCA: Weakly-Electric-Fish Bioinspired cognitive architecture'. *Procedia Computer Science*. 2018, vol. 145(1), pp. 365–9.
[9] Inggs M., Lewis S., Palama R., Ritchie M., Griffiths H. 'Report on the 2018 trials of the multistatic NeXtRAD dual band polarimetric radar'. IEEE Radar Conference (RadarConf). IEEE; 2019. pp. 1–6.
[10] Bell M.R. 'Information theory and radar waveform design'. *IEEE Transactions on Information Theory*. 1993, vol. 39(5), pp. 1578–97.
[11] Doerry A.W. 'Generating nonlinear FM chirp waveforms for radar'. *Sandia Report SAND 2006-5856; andia National Laboratories: Albuquerque, NM, USA*. Livermore, CA, USA; 2006.
[12] Huynen J.R. 'A revisitation of the phenomenological approach with applications to radar target decomposition'. *Illinois Univ At Chicago Circle Communications Lab*. 1982.
[13] Potter L.C., Moses R.L. 'Attributed scattering centers for SAR ATR'. *IEEE Transactions on Image Processing*. 1997, vol. 6(1), pp. 79–91.
[14] Rihaczek A.W., Hershkowitz S.J. 'Radar resolution and complex-image analysis (book)'. *Norwood*. MA: Artech House; 1996.
[15] Martorella M. 'Multidimensional radar imaging'. *Institution of Engineering and Technology*. 2019.

Chapter 2

Understanding FM radar signal through parameter estimation

Karol Abratkiewicz[1] *and Piotr Jerzy Samczyński*[1]

This chapter is a short review through modern digital signal processing techniques used for frequency-modulated radar waveform parameter estimation. The theory related to novel algorithms developed in recent years is briefly described, with particular emphasis on their usefulness in radar, ER, and spectrum sensing applications. The main goal of the chapter is to present promising algorithms that serve to estimate parameters of radar signals with the assumption that the waveforms are of a completely unknown nature and the phase terms are not given. Primarily, the chapter is devoted to time-frequency (TF) signal processing methods which are usually dedicated to non-stationary waveforms, which radar signals are. However, the chapter also presents the matched filtering-based approach, which precisely estimates radar signal parameters enabling the understanding of data contained in the waveform. All the methods provided are supported with real-life signal analysis, and the outcomes are illustrated and described.

2.1 Problem statement

There are various definitions of the term *radar signal*. Depending on the type of radar, the transmitted waveforms can be divided into two main categories: continuous wave (CW) and pulse wave (PW). Each of these two groups can be consistently divided depending on the type of modulation. The most popular modulations used in modern radars are frequency modulations, e.g. linear, non-linear and shift keying. Besides classical signals with frequency modulation, in modern applications, there are also more advanced waveform diversity methods applied to design various waveforms in order to perform better detection of targets, improve signal interference robustness and noise suppression [1]. The usage of waveform diversity techniques allows the radar signal cross-ambiguity function to be optimised.

[1]Radar Research Laboratory, Institute of Electronic Systems, Warsaw University of Technology, Warsaw, Poland

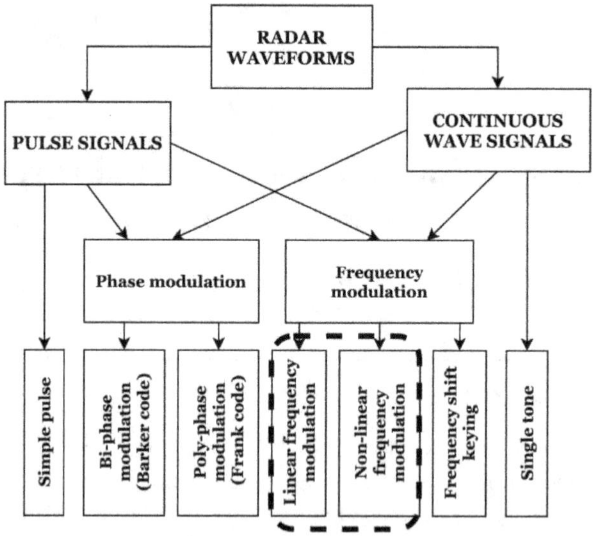

Figure 2.1 A simple graph illustrating radar signal diversity. The marked group is the focus of this chapter.

Therefore, the signal is designed in order to match the target echo response and/or to distinguish the target in the interfered environment. Typically, the radar waveforms are designed to maximise detection capabilities for assumed types of objects. A simplified diagram presenting radar signal diversity is depicted in Figure 2.1; however, this topic itself is very wide and in this chapter the authors have focused mostly on the understanding of radar signal parameter estimation of the signals with frequency modulation, which is widely used in most radars in service.

Radar waveform parameter estimation is a process which is used in many new applications. In the beginning, the radar signature was interesting mostly for electronic warfare (EW) purposes and used for the identification and classification of the radar transmitter by electronic intelligence (ELINT) receivers. Recently, there has been rapid development in passive radar technology. In many cases, this approach uses commercial CW transmitters as an illuminator of opportunity, e.g. FM radio, DAB and DVB-T. However, there is also a class of passive coherent location (PCL) systems which uses other radars as transmitters of opportunity. In such a case, the estimation of the radar waveform parameters is crucial for the processing in passive radars which is based on such illuminators. Other rapidly developing applications where the radar waveform parameters have to be estimated are spectrum monitoring and spectrum sharing. In recent years, there has been huge interest from telecommunication companies in occupying more and more of the spectrum for new mobile and Internet access purposes. The developers of 4G and 5G technologies have been working on the topic of spectrum sharing for their mobile providers, and are more and more interested in spectrum sharing with radar systems. For these applications,

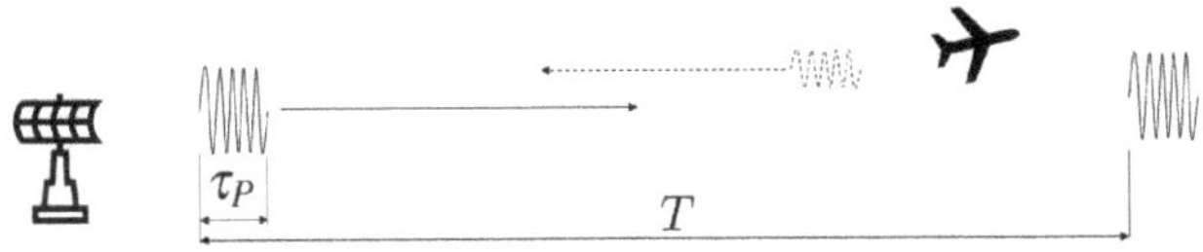

Figure 2.2 *An illustration of the main idea of radar operation. Solid line – transmitted signal, dotted line – reflected signal.*

the topic of radar parameter estimation is also crucial to find the best way for such spectrum sharing, where cognitive techniques might additionally be applied [2–4].

It should be underlined that in all the aforementioned applications, the receiver used for analysing the radar signature is in most cases located some distance from the active radar. Therefore, signals reaching the receiver are corrupted due to several reasons, like multi-path propagation, Doppler shift, etc. In such a case it is important to apply proper techniques to estimate the waveform parameters which are robust to all the phenomena present in a real-life measurement scenario.

The principle of radar operation and basic parameters are presented in (Figure 2.2).

The pulses of duration τ_P are transmitted by the radar with the *pulse repetition frequency* (PRF), given by

$$f_{\mathrm{PRF}} = \frac{1}{T} \tag{2.1}$$

where T denotes the *pulse repetition interval* (PRI), which is the time between consecutive pulses. Another main parameter defined in radar nomenclature is the duty factor, given as

$$D = \frac{\tau_P}{T}.100\% \tag{2.2}$$

In pulse radars, $D < 100\%$ (in most applications, D is in the range from a fraction of a per cent to single or even tens of percentage points), but for CW systems the wave is transmitted all the time and in such a case $D = 100\%$, and in CW considerations it is assumed that the signal duration is equal to PRI.

The signals received by the radar in most cases have a more complex nature than the wave emitted by the transmitting antenna. In a real scenario, the radar echo is corrupted by additional multi-path effects, caused by reflections from different terrain obstacles, such as trees, buildings, among others. Additionally, the real targets are built from many reflecting components (many scatterers) and the receiving echo is a superposition of the wave reflected from different scatterers with varied amplitude and phase, depending on each scatterer's reflectivity. Moreover, some scatterers might have additional movement not only related to the target's trajectory but also relevant to the target's centre of gravity, e.g. the rotation of a helicopter target's blades, or the motion of the legs and arms of a walking or running human target. These movements of components impact with additional Doppler shift, and as these components are much smaller in relation to the main movement echo related to the target distance changes, thus they are described as micro-Doppler components of the target. Therefore, for the purposes of the description a general signal model is presented, and further modifications depending on the application will be considered individually in this chapter.

2.1.1 Radar waveform definition

In this section, the definitions of the radar waveform used for further analyses in this chapter are defined. In principle, by way of its antenna the radar emits specially designed waveforms to detect, identify and characterise targets of interest [5–7] based on their echo. In the simplest case, a radar sends a simple pulse on a constant carrier frequency. In practice, the pulses may have the character of an amplitude-keyed harmonic wave and may additionally be modulated. As already mentioned, the most common modulations in radar applications are linear or non-linear frequency ones, described by the signal phase function. The former case is usually called a linear frequency-modulated (LFM) pulse, and the latter is known in radar nomenclature as a non-linear frequency-modulated (NLFM) pulse.

Let us return to the unified definition of a modulated radar waveform. In general, the model of such a signal can be defined as follows:

$$x(t) = A(t)\exp\left(j\Phi(t)\right) \tag{2.3}$$

where j is the imaginary unit $j = \sqrt{-1}$ and $A(t)$ is the amplitude in pulse radar systems defined by the rectangular function as follows:

$$A(\mathrm{t}){=}\Pi(t) = \begin{cases} 0, & \text{if } |t| > \frac{\tau_P}{2} \\ \frac{1}{2}, & \text{if } |t| = \frac{\tau_P}{2} \\ 1, & \text{if } |t| < \frac{\tau_P}{2} \end{cases} \tag{2.4}$$

where τ_P is the pulse duration (see Figure 2.2). In (2.3), the key component is the phase $\Phi(t)$ that defines the signal's instantaneous character. In most radars for air surveillance applications, the pulsed signals with a frequency modulation are used. In such a case, the phase can be defined by the $\Phi(t)$-th order polynomial as follows:

$$\Phi(t) = a_0 + \sum_{i=1}^{N} \frac{a_n}{n} t^n \tag{2.5}$$

where $\{a_0, a_1, \ldots, a_N\}_N$ constitutes the set of polynomial phase coefficients. This is where the name *polynomial phase signal* (PPS) comes from for such a phase definition. In fact, this set is usually limited to a few components and, besides one case, is of an even order. Namely, for $n = 1$ the waveform is a harmonic signal with a constant frequency $f = \frac{a_1}{2\pi}$ Hz. The effect of linear frequency modulation occurs for n=2, which gives the quadratic polynomial phase. The NLFM signal is obtained for n=4, 6, 8, $n = 4, 6, 8$, etc. Moreover, these components define the rate of the phase change, and in consistency with (2.3) and (2.5) are expressed in $\frac{\text{rad}}{\text{s}}$ for $n = 2$, $\frac{\text{rad}}{\text{s}^3}$ for $n = 4$ and $\frac{\text{rad}}{\text{s}^5}$ for $n = 6$, etc., and after normalisation with respect to 2π the radians are unified to Hz. This model will be used in the further part of this chapter to define radar pulses of a different nature generated by the transmitter.

As already stated, in a real-life scenario the signal received by a radar is disturbed by the echo from stationary targets, often called *ground clutter*, and the echo from moving objects (introducing additional Doppler shift). Moreover, in some systems additional factors can appear, namely direct leakage between the transmitting and receiving channels or direct signal interference (in multistatic or passive radars). In such cases, each particular component may be considered separately with individual signal attributes that can be additionally shifted in the frequency domain due to the Doppler effect, and in the time domain which corresponds to the range (bistatic or monostatic). Thus, the received signal can be written as

$$y(t) = \underbrace{x(t)}_{\text{direct signal}} + \underbrace{\sum_{n=1}^{N} \mathcal{A}_n(t-\tau_n)x(t-\tau_n)}_{\text{clutter echo}} + \underbrace{\sum_{m=1}^{M} \mathcal{A}_m(t-\tau_m)x(t-\tau_m)\cdot\exp\left(-j2\pi f_{D,m}t\right)}_{\text{echo from moving objects}} + \underbrace{w(t)}_{\text{AWGN}} \tag{2.6}$$

where $\mathcal{A}_n(t)$ and $\mathcal{A}_m(t)$ are, respectively, the amplitude of a particular component, τ_m and τ_n are given shift in time related to the particular component, $f_{D,m}$ is the Doppler shift resulting from the mth object movement and $w(t)$ is an additive white Gaussian noise (AWGN). A perfect illustration of such a signal can be a micro-Doppler signature obtained for a flying helicopter, where each rotating part produces an individual echo component with an instantaneous phase related to the momentary velocity vector and position; additionally, the signal is disturbed by an echo from stationary objects and the ground, as well as the direct leakage between the antennas (for an active monostatic radar working with a CW mode). For the ideal scenario, the active radar sends a pulse and receives its echo which is delayed and shifted in the Doppler frequency, caused by the target movement with respect to the radar. In practice, the receiving channel of the active radar is corrupted by the ground clutter, which introduces additional interferences from stationary targets to the echo signal. The same interferences are present in electronic devices used for radar waveform parameter estimation, such as ELINT receivers or passive radars using other pulse radars as a source of illumination (see Figure 2.3).

Thus, in the further part of this chapter novel methods allowing such signal analysis to be performed are presented.

2.2 Modern methods for analysis of radar signals

The most frequently used methods for signal analysis are TF methods. However, in many radar applications these techniques might be insufficient. As already presented, in a real-life scenario the signal received by the radar is composed of many different echoes. This can result in the analysis of its spectrum or time dependencies being too cursory. In practice, in ELINT and passive radar receivers utilising other radars as a source of illumination a few main signal parameters are estimated, namely pulse (signal) duration, bandwidth, and the coarse type of the waveform (modulation type). For this purpose, a few different methods are usually applied which involve time, frequency, and TF analysis in order to achieve a complete set of information and classify and/or reconstruct the signal. In the literature, many flourished methods and algorithms have been proposed with various capabilities but also

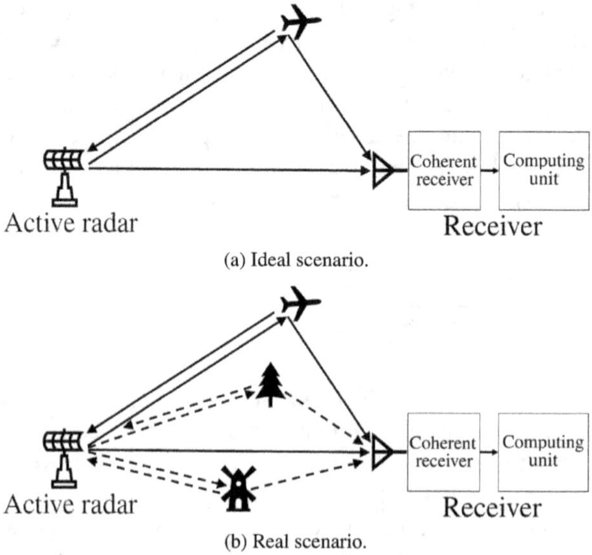

(a) Ideal scenario.

(b) Real scenario.

Figure 2.3 The problem of multi-path propagation in ER systems

limitations. For example, the wavelet transform is characterised by heterogeneous resolution on the time-scale plane, which may result in badly estimated signal attributes. Among the many different techniques for the TF signal representation, a few of them are usually used interchangeably – namely the short-time Fourier transform (STFT) and the Wigner-Ville (WV) distribution – that allow relatively fast and precise signal representation in the TF domain to be obtained and consequently enable signal parameters to be estimated. The STFT (deeply studied in the further part of this chapter) is based on the Fourier transform computed on the part of the signal multiplied by the analysing window. The window is shifted along the waveform and the FT is computed in each iteration, and consequently the TF representation of the signal is obtained. The WV distribution of the signal $x(t)$ is defined using the following relationship [8]:

$$\mathcal{W}_x(t,\omega) = \int_{\mathbb{R}} x\left(t+\frac{\tau}{2}\right) x^*\left(t-\frac{\tau}{2}\right) e^{-j\omega\tau} \mathrm{d}\tau \tag{2.7}$$

where $\omega = 2\pi f$ is the angular frequency and $(\cdot)^*$ is the symbol of the complex conjugate. The WV distribution function is optimal when analysing the LFM pulse, but the method suffers from so-called cross terms resulting from the non-linear operation. Therefore, if a pure LFM signal is processed, the distribution is precise with the optimal resolution on the TF plane (the resolution defines the capability of the energy accumulation in the vicinity of the main signal component). If the waveform has non-linear components, e.g. the NLFM pulse or additional multi-path reflections, the distribution is degraded by cross terms. Taking into account that in real scenarios it is almost impossible to eliminate the multi-path, as this phenomenon is

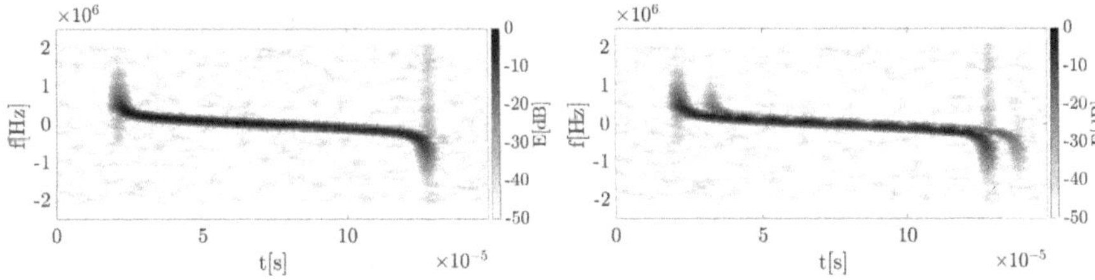

(a) Spectrogram of the NLFM pulse without (left) and with (right) multi-path effect.

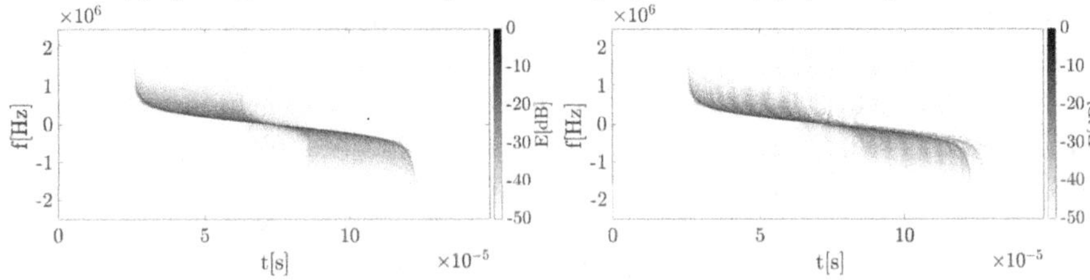

(b) WV distribution of the NLFM pulse without (left) and with (right) multi-path effect.

Figure 2.4 Exemplary distributions obtained using the STFT and the WV methods

caused by the surrounding radar environment, the classical WV method usually fails also for NLFM pulses.

The comparison of these two most used techniques is illustrated in Figure 2.4 using a real-life NLFM pulse with and without the multi-path effect in the form of energy distribution (energy distribution is defined as a squared absolute value of the transform).

As can be seen, the main disadvantage of the STFT distribution is a limited resolution which is, however, constant and is strongly related to the analysing window (this problem is described in the Sections 2.2.2.3–2.2.2.5). For the WV distribution, the energy concentration near the main component is strong but due to the non-linear terms in the phase of the signal additional, spurious components appear. This effect disqualifies the simple WV method from potential algorithms that may be used in various radar signal analysis and reconstruction. In the literature, many different modifications of this method have been proposed, e.g. a smoothed WV distribution that reduces the effect of cross-terms, but these methods usually spoil the final resolution of the distribution, making them similar to other, less computationally complex methods.

In recent years, a few novel algorithms have been proposed of particular usefulness in the radar domain, which has opened new possibilities in understanding complex waveforms. In the following sections of this chapter, recent techniques for radar signal analysis and understanding are described.

2.2.1 Extended generalised chirp transform

The extended generalised chirp transform (EGCT) is the algorithm proposed in 2013 in [9] to be used in passive bistatic radar (PBR) applications utilising pulse radars as a source of illumination; however, the technique is not limited to radar signals. In general, the method can be freely applied for other phase models which is its valuable advantage.

As presented in Figure 2.3, the signal reaching the receiver propagates a particular range from the transmitter to the receiving device. The receiver, in general, can be a PBR, ELINT, EW, or electronic reconnaissance (ER) system. The waveform is usually disturbed by the multi-path effect being a sum of reflected copies of the pulse from stationary and moving objects, therefore the receiving signal can be defined by Guerci [6]. In order to obtain a precise reconstruction of the waveform, its parameters have to be estimated and then the signal can be reconstructed (if needed).

The EGCT method is very precise due to the matched filtering principle applied in this technique. The matched filter is a special type of linear filter that maximises the signal-to-noise ratio (SNR) at its output [10]. This approach is commonly used in radars, therefore is a natural consequence of typical methods in the considered domain. Generally speaking, the matched filter is the product of the convolution of an unknown signal and the filter response, which is a conjugated time-reversed reference signal. In active radar applications, the reference waveform is the out-going pulse and this pattern which exists in copies that are received, delayed and disturbed by the AWGN is searched for. Simply put, after the matched filtering procedure the reflected copies of the transmitted pulse are distinguished from the received signal with maximised SNR. The matched filtering of the signal $x(t)$ is defined as [10] follows:

$$y(\tau) = \int_{\mathbb{R}} x(t)g(\tau - t)\mathrm{d}t \tag{2.8}$$

where $g(t)$ is the matched filter impulse response that satisfies the relationship $g(t) = x^*(-t)$. In ER, ELINT, EW and PBR applications, this problem comes down to the convolution of the signal model and the gathered waveform; however, the received signal is always disturbed.

A simple comparison of the matched filtering with optimal and not optimal parameters is presented in Figure 2.5 using an exemplary simulated linear chirp.

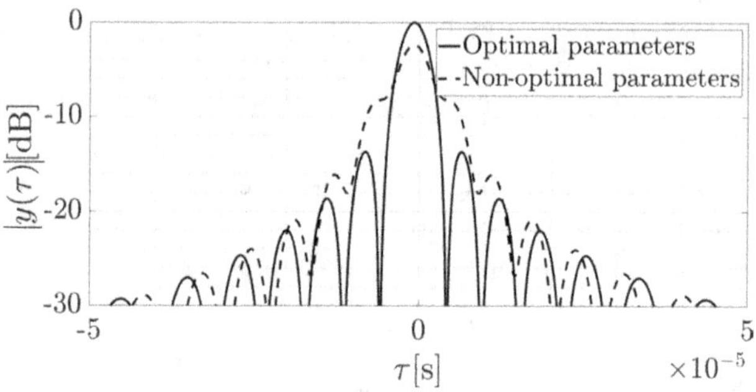

Figure 2.5 A comparison of the matched filtering for the filter with optimal and non-optimal parameters

The more precise parameters the filter has, the higher the peak and the lower the side-lobes are obtained after matched filtering (which ensures better detection capability – the filtered echo is stronger and more focused around its estimated position/delay). The property was adopted in the EGCT method that gives the optimal signal estimate in the thought of this concept by finding the maximum value of $y(t)$ in [8] among the analysed set of values.

Assuming that $x(t)$ is the signal under consideration given in (2.3), the matched filtering process is performed using (2.8), where $g(t) = \exp\left(-j\Phi_{\boldsymbol{P}}(t)\right)$ is the matched filter impulse response. The phase coefficients vector is composed of the terms $\boldsymbol{P} = [a_0, a_1, ..., a_N]$. Thus, the phase is expressed as polynomial functions, in accordance with (2.5):

$$\Phi_{\boldsymbol{P}}(t) = a_0 + \sum_{n=1}^{N} \frac{a_n}{n} t^n. \tag{2.9}$$

In order to obtain the estimated phase model, a maximum likelihood estimator (MLE) can be applied for the phase coefficients vector, which can be defined as

$$\hat{\boldsymbol{P}} = \underset{n}{\operatorname{argmax}} \left|y(\tau)\right| \tag{2.10}$$

where $|\cdot|$ denotes the absolute value operator. In practice, this problem comes down to the formulation of a K – points set to search the particular component value. The major disadvantage of this method is that this set may be completely undefined when analysing an unknown signal, thus the processing time can be significant. The simplified pseudo code is presented in Algorithm 2.1.

Input: $x(t)$
Result: $\hat{\boldsymbol{P}}$

```
for each phase coefficient a_n do
    for each value from K – point search area do
        calculate y(τ) using (1.8);
    end
    calculate P̂ using (1.10);
end
```

***Algorithm 2.1**: EGCT algorithm pseudocode*

The matched filtering can be performed in both time (as a convolution of the signal) and frequency domains (as a multiplication of their spectrum and then a calculation of the inverse FT), therefore the processing time may be reduced by using Fast Fourier Transform (FFT) algorithm for the frequency domain implementation. Examples of using this technique can be found in [9, 11, 12], where NLFM radar pulses were investigated and their parameters were successfully estimated.

In order to present the outcomes obtained for the real-life signal, a single NLFM pulse with unknown parameters originating from a Raytheon ASR-10SS air traffic control (ATC) radar was processed [13]. Since the phase attributes as well as the order

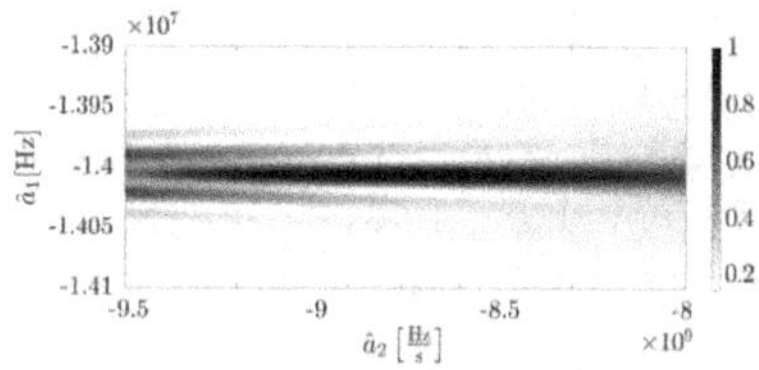

(a) EGCT maps for the coefficients a_1 and a_2.

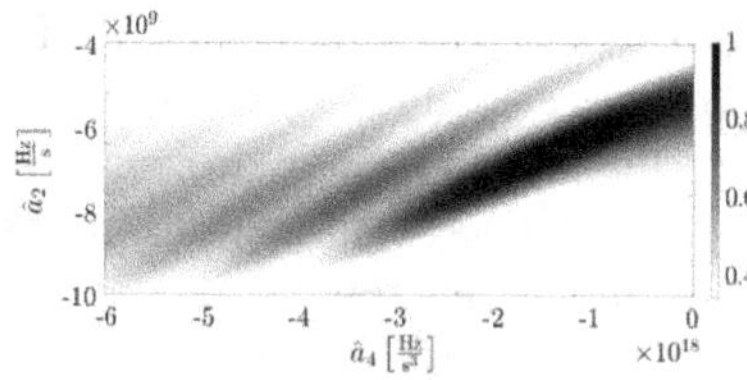

(b) EGCT maps for the coefficients a_2 and a_4.

Figure 2.6 EGCT maps for the coefficients a_1, a_2 *and* a_4

defining the phase were not given *a priori*, such conditions allowed perfect validation of the method to be performed. The presented algorithm was applied, and the results in the form of EGCT maps are illustrated in Figure 2.6. During the test the signal was assumed to be described by the 4th order polynomial phase as follows (the initial phase was assumed to be and the amplitude was normalised):

$$x(t) = \exp\left(j2\pi\left(a_1 t + \frac{a_2 t^2}{2} + \frac{a_4 t^4}{4}\right)\right),\ t \in (t_1, t_2) \tag{2.11}$$

In the presented results (see Figure 2.6), the maximum of the EGCT maps (black colour) indicates the value of estimated parameters $\hat{a}_1$, $\hat{a}_2$ and $\hat{a}_4$, respectively. In the example carried out, the estimated values are as follows: $\hat{a}_1 = -1.3994 \cdot 10^7$ Hz, $\hat{a}_2 = -7.4771 \cdot 10^9 \frac{\text{Hz}}{\text{s}}$ and $\hat{a}_4 = -2.0461 \cdot 10^{18} \frac{\text{Hz}}{\text{s}^3}$. In fact, the actual values of these parameters are not known, since the transmitter was non-cooperating. Therefore, the assumption of the non-linear nature of the pulse can be confirmed through the matched filtering of the received pulse and the reconstructed one, obtained using the estimated components. According to the matched filtering principle that is illustrated in Figure 2.5, the narrower and the higher peak, the more accurate the pulse estimate achieved. Thus, it can be assumed that the received pulse is an LFM or an NLFM waveform, and the matched filtering results can be compared. Such a comparison is presented in Figure 2.7, where these outcomes for both cases are illustrated.

As shown, for the assumed NLFM phase model the matched filtering result is characterised by both a higher and narrower peak, which confirms the appropriate

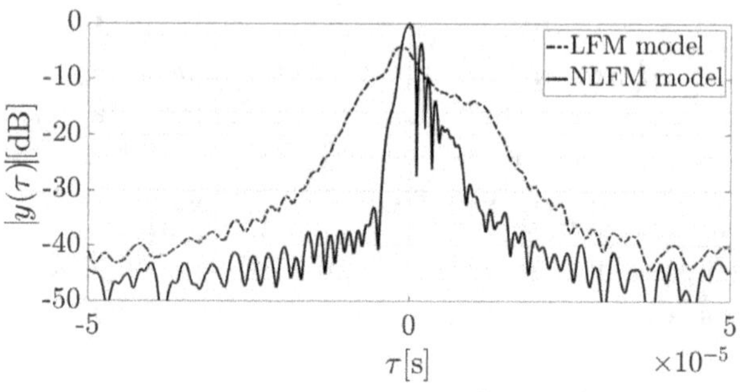

Figure 2.7 A comparison of the matched filtering for the assumed signal models LFM and NLFM, while the actual signal model was unknown

assumption related to the phase model. In addition, the actual phase polynomial degree is not known, hence the operation can be evaluated for different orders since the best results are obtained (the phase model can be established to be 6th, 8th, 10th, etc. order polynomial).

The EGCT method can be used as a reference algorithm in many applications due to the precise outcomes resulting from the MLE [10]. Typically, this technique is implemented in post-processing realisations due to the required computational effort disqualifying a real-time operation. If signal components are entirely unknown, the matched filtering may be time-consuming; however, the precision of this algorithm makes it useful and robust.

2.2.2 Time-frequency methods

The waveform analysis method in the TF domain described in the further part of the chapter is based on the STFT principle and its modifications, which is one of the most intuitive approaches to non-stationary signal processing. Additionally, in the literature it has been stated that the STFT has some interesting properties, allowing not only the frequency of signal components to be estimated but also additional waveform parameters to be extracted. In the following, selected methods are described with particular emphasis on their usefulness in radar applications.

2.2.2.1 STFT background

In the literature, the STFT of the signal $x(t)$ is defined as follows [8]:

$$F_x^g(t,\omega) = \int_{\mathbb{R}} x(\tau) g^*(t-\tau) e^{-j\omega\tau} \mathrm{d}\tau = A_x^g(t,\omega) e^{j\phi_x^g(t,\omega)} \tag{2.12}$$

where $(\cdot)^*$ is a complex conjugate operator, $A_x^g(t,\omega)$ is the amplitude of the distribution, and $\phi_x^g(t,\omega)$ is its phase. The STFT can be interpreted and implemented in two ways according to (2.12). In the first one, $g(t)$ is a narrow-band filter impulse response and the convolution product of the signal with the filter centred at the output angular frequency ω, thus the output can be thought of as the frequency content of the waveform in each moment t and in the vicinity of the angular frequency ω. In the second interpretation, $g(t)$ is an even real window that moves along the signal and is multiplied by a short excerpt of the waveform, and the FT is computed on each such product (this is where the common name *Moving Window Method* (MWM) comes from). Adjacent waveform parts (frames) of the signal extracted and multiplied by the analysis window may overlap, which leads to a more accurate distribution at the cost of the number of operations that have to be performed. Due to the real character of the window, the complex conjugation is not necessary, and to distinguish the window from the impulse response notation, the window is denoted as $h(t)$, which by substituting to (2.12) leads to

$$F_x^h(t,\omega) = \int_{\mathbb{R}} x(\tau) h(t-\tau) e^{-j\omega\tau} \mathrm{d}\tau = A_x^h(t,\omega) e^{j\phi_x^h(t,\omega)}. \tag{2.13}$$

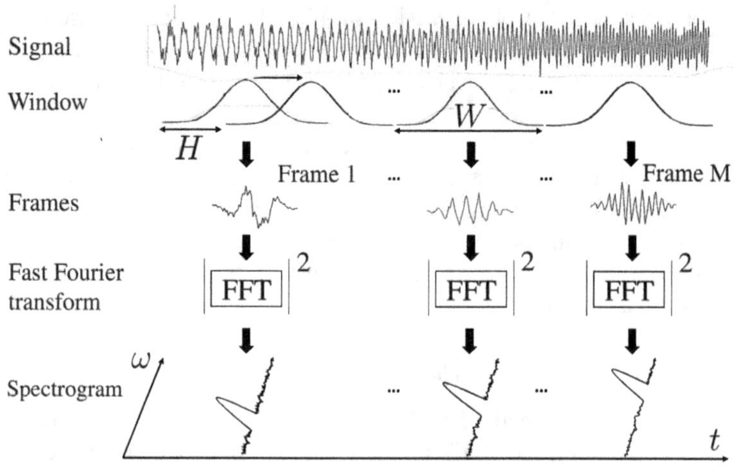

Figure 2.8 A simplified diagram presenting the STFT computation

Both forms are equivalent, and thus the easier one to understand (in the authors' opinion) defined by (2.13) is considered in this chapter.

The STFT is a complex-valued distribution that contains both the real and imaginary part for each (t, ω) coordinate. A squared absolute value of the STFT gives the energy distribution

$$E_x^h(t,\omega) = \left|F_x^h(t,\omega)\right|^2 \tag{2.14}$$

usually called the spectrogram. A simplistic diagram presenting how the STFT is actually computed is illustrated in Figure 2.8, and an exemplary spectrogram of the radar chirp is depicted in Figure 2.4. Consecutive parts (frames) of the input signal are multiplied by the analysing window (of a width W samples) moving along the waveform by H samples in each iteration. For each such product, the FFT algorithm is applied which results in temporary spectral information. An entire distribution is obtained whilst the window reaches the last sample of the considered signal.

Considerations presented in [14] showed that phase information is also contained in the STFT amplitude, which has important consequences. Let us start with a small modification of the STFT output. The exponential form of the transformation may be 'inconvenient' in some algorithms, thus the following operation is admissible:

$$\Phi_x^h(t,\omega) = \ln\left(F_x^h(t,\omega)\right) = \Lambda_x^h(t,\omega) + j\phi_x^h(t,\omega) \tag{2.15}$$

where $\Lambda_x^h(t,\omega) = \ln\left(A_x^h(t,\omega)\right)$. Equation (2.15) is known as a *complex phase* of the STFT and is used for the estimation of many signal parameters in the TF domain. Namely, Barnes in [15] proposed the instantaneous bandwidth of the signal component distributed in the TF domain. This concept was extended in [16] and defined as a *signed channelised instantaneous bandwidth* defined by

$$\Sigma_x^h(t,\omega) = \frac{\partial \Lambda_x^h(t,\omega)}{\partial t} = \Re\left[\frac{\partial \ln\left(F_x^h(t,\omega)\right)}{\partial t}\right], \quad \Sigma_x^h(t,\omega) \in \mathbb{R} \tag{2.16}$$

for $\Re$ being the real part. Also, in [16] a novel signal distribution was presented, namely the *signed local group duration* – a derivative with respect to the angular frequency of the real part of the complex phase of STFT, expressed as

$$\mho_x^h(t,\omega) = -\frac{\partial \Lambda_x^h(t,\omega)}{\partial \omega} = -\Re\left[\frac{\partial \ln\left(F_x^h(t,\omega)\right)}{\partial \omega}\right], \quad \mho_x^h(t,\omega) \in \mathbb{R}. \tag{2.17}$$

Both (2.16) and (2.17) deliver a local estimate of the signal parameters on the TF plane for each (t, ω) point. The former $(\Sigma_x^h(t,\omega))$ defines the instantaneous bandwidth of the component on the TF plane, energy dominates in a given (frequency) channel and at the time t, and is concentrated around the angular frequency ω. The latter $(\mho_x^h(t,\omega))$ is related to the duration of the signal component in a given (frequency) channel, delayed by t with respect to this channel, and centred on the angular frequency ω.

Equivalent representations of (2.16) and (2.17) were presented, e.g. in [17]. As shown, these distributions can be computed more efficiently through using modified analysing windows. Hence, the signed channelised instantaneous bandwidth can be expressed as

$$\sum_x^h(t,\omega) = \Re\left[\frac{\partial F_x^h(t,\omega)}{\partial t}\frac{1}{F_x^h(t,\omega)}\right] = \Re\left[\frac{F_x^{\mathcal{D}h}(t,\omega)}{F_x^h(t,\omega)}\right] \tag{2.18}$$

and the signed local group duration may be defined by

$$\sum_x^h(t,\omega) = -\Re\left[\frac{\partial F_x^h(t,\omega)}{\partial t}\frac{1}{F_x^h(t,\omega)}\right] = \Re\left[\frac{F_x^{\mathcal{T}h}(t,\omega)}{F_x^h(t,\omega)}\right]. \tag{2.19}$$

In (2.18), the expression $\mathcal{D}h$ stands for the first-order derivative of the analysing window $\mathcal{D}h = \frac{\mathrm{d}h(t)}{\mathrm{d}t}$ whereas in (2.19) $\mathcal{T}h = th(t)$ is a window multiplied by the linear time ramp with a root in 0 as presented in Figure 2.9. Equations (2.18) and (2.19) result from

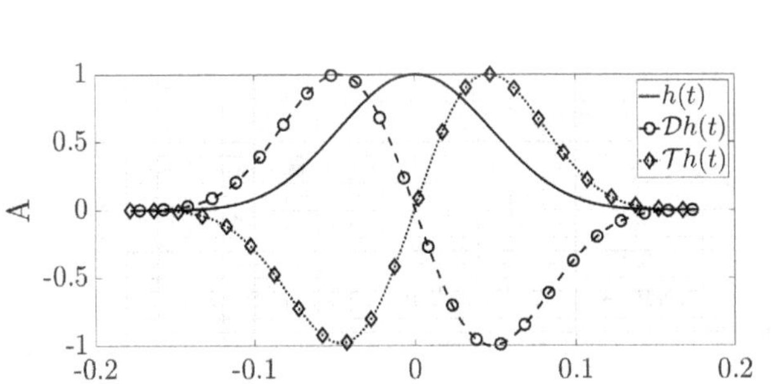

Figure 2.9 Fifth term Blackman-Harris window with its (normalised) modifications

$$\frac{\partial F_x^h(t,\ \omega)}{\partial t} = \int_{\mathbb{R}} x(\tau)\frac{\partial h(t-\tau)e^{-j\omega(t-\tau)}}{\partial t}\mathrm{d}\tau = F_x^{\mathcal{D}h}(t,\ \omega) \tag{2.20}$$

and

$$\frac{\partial F_x^h(t,\omega)}{\partial \omega} = \int_{\mathbb{R}} x(t-\tau)h(\tau)\frac{\partial e^{-j\omega\tau}}{\partial \omega}\mathrm{d}\tau = \int_{\mathbb{R}} x(t-\tau)h(\tau)j\tau e^{j\omega\tau}\mathrm{d}\tau = -jF_x^{\mathcal{T}h}(t,\omega). \tag{2.21}$$

Thus, (2.16) and (2.17) can be rewritten as

$$\Sigma_x^h(t,\omega) = \frac{\partial \Lambda_x^h(t,\omega)}{\partial t} = \Re\left[\frac{\partial \ln(F_x^h(t,\omega))}{\partial t}\right] = \Re\left[\frac{F_x^{\mathcal{D}h}(t,\omega)}{F_x^h(t,\omega)}\right] \tag{2.22}$$

$$\mho_x^h(t,\omega) = -\frac{\partial \Lambda_x^h(t,\omega)}{\partial \omega} = -\Re\left[\frac{\partial \ln(F_x^h(t,\omega))}{\partial \omega}\right] = \Im\left[\frac{F_x^{\mathcal{T}h}(t,\omega)}{F_x^h(t,\omega)}\right]. \tag{2.23}$$

The provided set of distributions is fundamental for further considerations regarding radar signal parameter estimation in the TF domain and consequently their understanding. The following techniques are strongly based on the presented background, thus its adoption is necessary for the rest of this chapter.

2.2.2.2 Chirp rate estimation

The chirp rate (CR) is a parameter which can be effectively used to estimate the NLFM radar waveform parameters. The classical STFT spectrogram allows particular components to be assessed in terms of the instantaneous frequency and the energy. According to a literature survey conducted by the author of this chapter, higher-order phase components were not estimated locally and directly from the STFT amplitude and/or phase until 2013. The authors did not identify official publications on this topic. A novel interesting method was revealed by Czarnecki and Moszyński [16] and deals with the second-order component of the PPS signal, thus the reduced (for the sake of the explanation of this method) signal model can be defined as

$$x(t) = A(t)\exp\left(j2\pi\left(f_0 t + \frac{\alpha t^2}{2}\right) + j\phi_0\right) \tag{2.24}$$

with $A(t)$ being a signal amplitude, ϕ_0 is the initial phase, f_0 stands for the carrier frequency and α is the CR (in the literature this phase coefficient is also called frequency rate or frequency slope) – the parameter that defines how the signal's instantaneous frequency changes in time. For instance, an exemplary signal is defined by

$$x(t) = \exp\left(j2\pi\left(100t + \frac{25t^2}{2}\right)\right),\ t \in (0, 4\ \mathrm{s}) \tag{2.25}$$

where the carrier frequency $f_0 = 100$ Hz and additionally the instantaneous frequency linearly increases by 25 Hz each second. For example, in $t = 0$ s the waveform instantaneous frequency is $f_i|_{t=0\ \mathrm{s}} = 100$ Hz, after 0.2 s the instantaneous frequency is equal to $f_i|_{t=0.2\ \mathrm{s}} = 105$ Hz and after $t = 4$ s, 0.2 Hz.

The CR value can be relatively simply computed by differentiating twice the signal phase with respect to time:

$$\hat{\alpha}(t) = \frac{1}{2\pi}\frac{\mathrm{d}^2}{\mathrm{d}t^2}\mathrm{W}^{-1}(\angle \mathrm{x(t)}) \tag{2.26}$$

where W^{-1} operator (known as a phase unwrap) serves to prevent jumps between consecutive angles that are greater than or equal to π rad of the phase given by $\angle x(t)$. However, in practical cases the signal is usually disturbed by noise, weak or interfered with by the multi-path phenomenon, thus this operation fails. In turn, the classical spectrum analysis suffers from excluded time dependencies of the signal, therefore the frequency changes in time are neglected.

Completely different properties are observable in the TF domain. Namely, as previously stated, (2.16) is associated with the instantaneous bandwidth of the component, whilst (2.17) expresses a group delay of the component. As observed in [16–18], the ratio of these two quantities gives the previously mentioned CR estimate (sometimes called $\mathfrak{K}$ estimator in honour of its inventor – Krzysztof)

$$\mathfrak{K}(t,\omega) = \frac{\Sigma_x^h(t,\omega)}{\mho_x^h(t,\omega)} = -\frac{\partial \Lambda_x^h(t,\omega)}{\partial t} \Big/ \frac{\partial \Lambda_x^h(t,\omega)}{\partial \omega} = \frac{\Re\left[\frac{F_x^{\mathcal{D}h}(t,\omega)}{F_x^h(t,\omega)}\right]}{\Im\left[\frac{F_x^{\mathcal{T}h}(t,\omega)}{F_x^h(t,\omega)}\right]}, \forall_t, \forall_\omega\ \mathfrak{K}(t,\omega) \in \mathbb{R}. \tag{2.27}$$

The distribution obtained through (2.27) is called a *phase accelerogram* or just *accelerogram*. As illustrated in Figure 2.10, the CR can be determined by the local slope of the component over the 2D distribution. An interesting property of the method is that the signal CR is estimated using the information contained in the STFT amplitude, which is rarely considered in the time domain analysis. A real-life NLFM radar pulse in the form of the energy distribution and the phase accelerogram is presented in Figure 2.11.

As can be observed, the frequency slope is computed for each point over the plane, which is of particular importance when analysing multi-component waveforms. What is meaningful in the considered case, the instantaneous CR is computed along the pulse, allowing its parameter estimation to be performed. Thanks

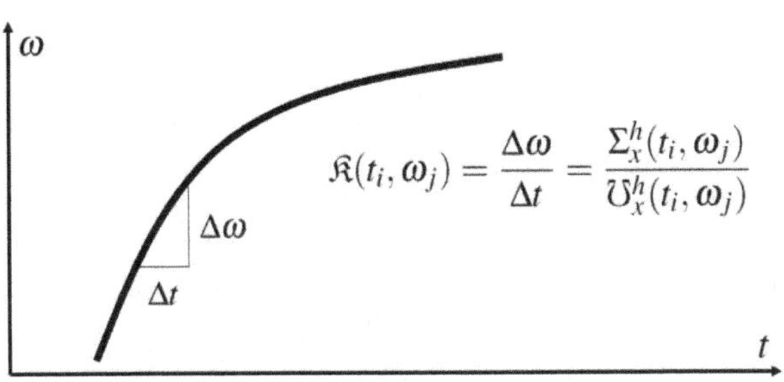

Figure 2.10 A graphical interpretation of CR estimation in the TF domain

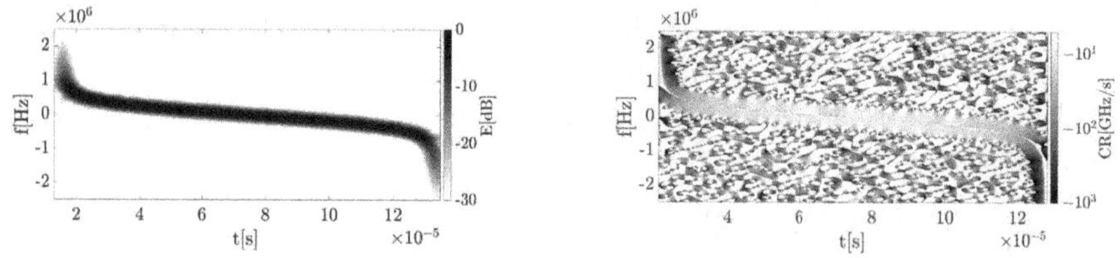

(a) Spectrogram of the NLFM signal.

(b) Phase accelerogram obtained using the K estimator.

Figure 2.11 The results for the real-life NLFM radar pulse – the spectrogram and the accelerogram

to preserved time dependencies, the signal can be precisely defined, recognised and classified regardless of the number of phase components (usefulness in multi-path conditions).

The method suffers from the uncertainty principle that spoils the resolution. This phenomenon results from an unidentified value at the energy ridge (attractor – the point in which the energy is the highest) which can be observed on the accelerogram presented in Figure 2.11 in the centre of the chirp. In each signal frame (each spectrogram column) at the point where the pulse has the maximum energy, the CR estimate is undefined. The reason for this can be explained by analysing Figure 2.9, where modified windows are illustrated. Both $\mathcal{D}h(t)$ and $\mathcal{T}h(t)$ have a root in $t = 0$, which manifests itself in numerical instabilities near the attractor, and consequently a badly conditioned CR estimate at this $(t, \boldsymbol{\omega})$ coordinate. Therefore, the method was robustified by Fourer *et al.* in [17].

The uncertainty effect resulting from dividing 0 by 0 was figured out by using L'Hôpital's rule [17]. Finally, the new estimators were revealed with a conserved principle of operation, however, with improved numerical stability near the energy ridge. This property can be observed by analysing Figure 2.12.

The window derivative is still characterised by the root in $t = 0$, but the roots of $\mathcal{T}^2h(t)$ in this point were dismissed. By applying L'Hôpital's rule in (2.27) and

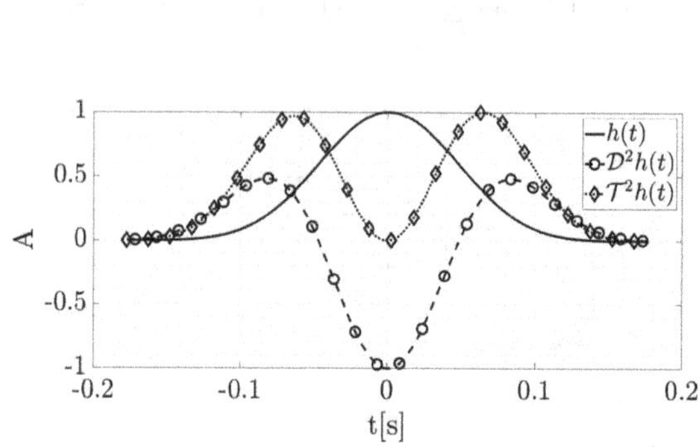

Figure 2.12 Higher-order Blackman-Harris window derivative and the same window multiplied by the second-order time ramp

differentiating the numerator and the denominator once with respect to time, the $\mathfrak{D}$ estimator can be defined as follows ($\mathfrak{D}$ comes from its author – Dominique):

$$\mathfrak{D}(t,\omega) = \frac{\partial}{\partial t}\Re\frac{\partial \ln F_x^h(t,\omega)}{\partial t} \Big/ \frac{\partial}{\partial t}\Re\frac{\partial \ln F_x^h(t,\omega)}{\partial \omega} = \frac{\frac{\partial}{\partial t}\Re\left[\frac{F_x^{\mathcal{D}h}(t,\omega)}{F_x^h(t,\omega)}\right]}{\frac{\partial}{\partial t}\Im\left[\frac{F_x^{\mathcal{T}h}(t,\omega)}{F_x^h(t,=\omega)}\right]} =$$

$$\frac{\Re\left[\frac{F_x^{\mathcal{D}^2h}(t,\omega)}{F_x^h(t,\omega)} - \left[\frac{F_x^{\mathcal{D}h}(t,\omega)}{F_x^h(t,\omega)}\right]^2\right]}{\Im\left[\frac{F_x^{\mathcal{D}\mathcal{T}^2h}(t,\omega)}{F_x^h(t,\omega)} - \frac{F_x^{\mathcal{T}h}(t,\omega)F_x^{\mathcal{D}h}(t,\omega)}{\left[F_x^h(t,\omega)\right]^2}\right]},\quad \forall_t, \forall_\omega\ \mathfrak{D}(t,\omega) \in \mathbb{R}. \tag{2.28}$$

The symbols applied in (2.28) are the same as for the $\mathfrak{K}$ estimator, namely, with analogy to Figure 2.12 $\mathcal{D}^2h(t) = \frac{d^2h(t)}{dt^2}$ and $\mathcal{T}^2h(t) = t^2h(t)$. In general, the estimator can be extended to any order of the window derivative and the time ramp; however, it does not bring benefits related to the accuracy or computational complexity [17], and therefore it is usually not considered in practical implementations.

The second estimator can be obtained through differentiating the numerator and the denominator once with respect to frequency in the thought of L'Hôpital's rule, which leads to the $\mathfrak{F}$ estimator ($\mathfrak{F}$ in honour of the author – François):

$$\mathfrak{F}(t,\omega) = \frac{\partial}{\partial \omega}\Re\frac{\partial \ln F_x^h(t,\omega)}{\partial t} \Big/ \frac{\partial}{\partial \omega}\Re\frac{\partial \ln F_x^h(t,\omega)}{\partial \omega} = \frac{\frac{\partial}{\partial \omega}\Re\left[\frac{F_x^{\mathcal{D}h}(t,\omega)}{F_x^h(t,\omega)}\right]}{\frac{\partial}{\partial \omega}\Im\left[\frac{F_x^{\mathcal{T}h}(t,\omega)}{F_x^h(t,\omega)}\right]} =$$

$$\frac{\Im\left[\frac{F_x^{\mathcal{D}\mathcal{T}h}(t,\omega)}{F_x^h(t,\omega)} - \frac{F_x^{\mathcal{T}h}(t,\omega)F_x^{\mathcal{D}h}(t,\omega)}{\left[F_x^h(t,\omega)\right]^2}\right]}{\Re\left[\frac{F_x^{\mathcal{T}^2h}(t,\omega)}{F_x^h(t,\omega)} - \left[\frac{F_x^{\mathcal{T}h}(t,\omega)}{F_x^h(t,\omega)}\right]^2\right]},\quad \forall_t, \forall_\omega\ \mathfrak{F}(t,\omega) \in \mathbb{R}. \tag{2.29}$$

Additionally, the fourth estimator (called the $\mathfrak{M}$ estimator) was proposed which is the geometrical mean of the $\mathfrak{D}$ and $\mathfrak{F}$ estimators and is defined as

$$\pm\mathfrak{M}(t,\omega) = \sqrt{\mathfrak{D}(t,\omega)\mathfrak{F}(t,\omega)} = \sqrt{-\frac{\Re\left[\frac{F_x^{\mathcal{D}^2h}(t,\omega)}{F_x^h(t,\omega)} - \left[\frac{F_x^{\mathcal{D}h}(t,\omega)}{F_x^h(t,\omega)}\right]^2\right]}{\Re\left[\frac{F_x^{\mathcal{T}^2h}(t,\omega)}{F_x^h(t,\omega)} - \left[\frac{F_x^{\mathcal{T}h}(t,\omega)}{F_x^h(t,\omega)}\right]^2\right]}}, \tag{2.30}$$

$\forall_t, \forall_\omega\ \mathfrak{M}(t,\omega) \in \mathbb{R}^+$.

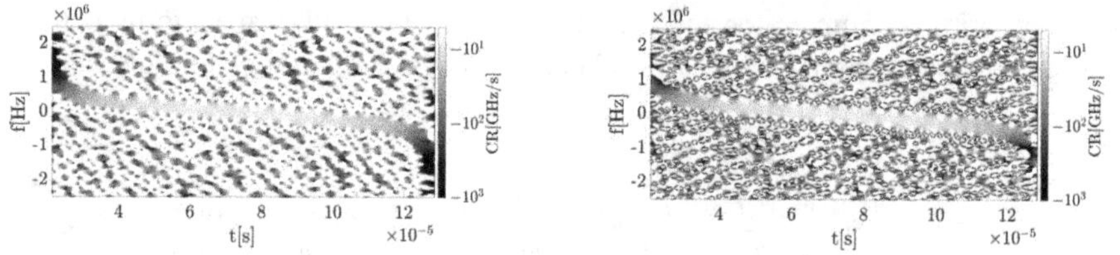

(a) Accelerogram of the NLFM signal obtained using the D estimator (1.28).

(b) Accelerogram of the NLFM signal obtained using the F estimator (1.29).

Figure 2.13 Phase accelerograms obtained according to (2.28) and (2.29)

The advantage of this estimator is that it does not use mixed derivatives. However, due to the geometrical mean operation the sign is lost. In consequence, its usefulness is limited to these applications in which the CR is suspected to be constant with a known sign. Such an assumption cannot always be made in radars or EW and ELINT systems, since the signal impinging the receiver may be completely unknown. For this reason, (2.30) is not considered in the further part of this chapter.

Indeed, despite the diversity in mathematical definitions of the considered CR estimators, their principle of operation is identical as in Figure 2.10. The main difference between them is the computational complexity (higher for $\mathfrak{D}$ and $\mathfrak{F}$ estimators) and the accuracy in the vicinity of the attractor (the point in which the CR is supposed to be the most accurate). The results in the form of accelerograms for the same NLFM waveform are shown in Figure 2.13.

As can be seen, all the estimators provided comparable outcomes, allowing CR to be estimated around the main component. The primary advantage of the presented TF methods is their ability to separate components in the TF domain which is unobtainable for the analysis in the time or the frequency domain. This is particularly important for multi-component signal analysis.

2.2.2.3 Adaptive short-time Fourier transforms

One of the main disadvantages existing in STFT signal analysis is a fixed window width that results in a constant TF resolution. In order to satisfy the high resolution of the signal distribution, the knowledge on the waveform parameters has to be given *a priori*. This results from the Heisenberg-Gabor uncertainty principle [19] which states that the simultaneous and precise determination of the signal components in both the time and the frequency domain is impossible. Simply put, using a single, fixed window, one can obtain a high resolution in time at the cost of a spoiled frequency resolution (for such a case a 'narrow' window in the time domain is applied), and the frequency resolution can be improved making the window wider at the cost of decreasing the time resolution. A graphical illustration of this dependency is depicted in Figure 2.14.

As can be noticed, for the narrow window the time dependencies can be easily extracted for the distribution; however, along the frequency axis the energy is 'blurred', which prevents a precise instantaneous frequency extraction, even if the

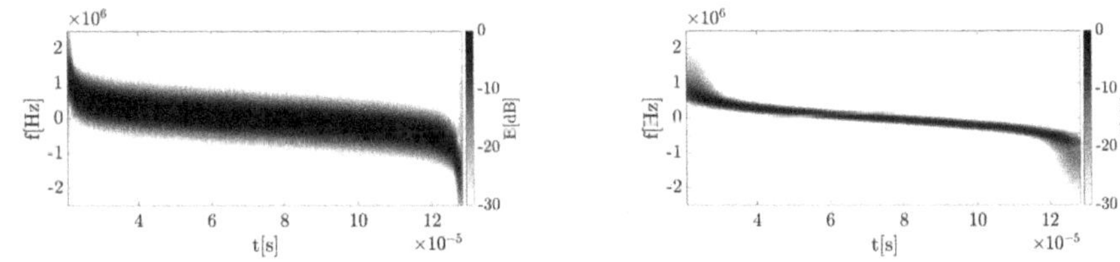

(a) Spectrogram for the "narrow" window.

(b) Spectrogram for the "wide" window.

Figure 2.14 A comparison of the spectrogram resolution obtained using different windows

variability in the time domain may be thoroughly extracted. Otherwise the conditions are noticeable for the second case when the wide time-varying window is applied. Then, the instantaneous frequency may be relatively precisely estimated, but information about time dependencies is difficult to distinguish. It can be concluded that the window of constant width is usually insufficient when analysing multi-component or fast varying waveforms, which, e.g. radar signals are. The second problem is that the signal character defines the optimal analysing window width thus this parameter has to be adjusted adaptively.

Cohen in [20] provided a formula describing the optimal time-varying window width based on the CR for LFM waveforms. This optimal window width allows good resolution in both time and frequency domains to be obtained. For FM signals, this issue comes down to finding the window width, for which the ratio of the frequency spread to time spread is approximately equal to the instantaneous CR of the signal, which can be approximated as [20]

$$\frac{B}{A} \approx \frac{\Delta f}{\Delta t} \tag{2.31}$$

where B is the window frequency spread given by [20]

$$B = \sqrt{\frac{\mathcal{H}_2 - \mathcal{H}_1^2}{\mathcal{H}_0}} \tag{2.32}$$

where $\mathcal{H}_i = \int_{\mathbb{R}} f^i \left|H(f)\right|^2 \mathrm{d}f$ and $H(f)$ is a FT of the window $h(t)$. The window time spread is described as follows:

$$A = \sqrt{\frac{h_2 - h_1^2}{h_0}} \tag{2.33}$$

where $h_i = \int_{\mathbb{R}} t^i \left|h(t)\right|^2 \mathrm{d}t$. For LFM signals, the optimal window width is given by [20]

$$\mathscr{R} = \frac{B}{A} \approx \sqrt{\frac{1}{4\pi|\alpha|}} \tag{2.34}$$

where α is the CR included in the signal phase given by (2.24). Equation (2.34) is valid for any analysis window and, e.g. in the Gaussian window case given by

$$h(t) = \frac{1}{\sqrt{2\pi}\sigma} e^{\frac{t^2}{2\sigma^2}} \tag{2.35}$$

$B = \sqrt{\frac{1}{8\pi^2\sigma^2}}$, $A = \sqrt{\frac{\sigma^2}{2}}$ and the standard deviation of the window (its time spread) σ is related to the $\mathscr{R}$ by $\mathscr{R}^2 = \frac{\sigma^2}{2}$, which leads to

$$\sigma = \sqrt{\frac{1}{2\pi\ |\alpha|}}. \tag{2.36}$$

This relationship is satisfied for the linear chirp in which the frequency slope is constant for the entire pulse duration. However, in many radar applications different LFM signals are often used, thus it is obvious that this condition can be inefficient. Such waveforms impose instantaneous (local) conditioning of the analysis parameters related directly to the signal's CR that can be estimated using the previously described CR estimation method. Therefore, if α can be treated as a function of time as $\alpha \to \alpha(t)$, then the instantaneous window width can be modified for each frame of the signal (for each column of the spectrogram/accelerogram), therefore the instantaneous standard deviation becomes a function:

$$\sigma(t) = \sqrt{\frac{1}{2\pi\ \left|\alpha(t)\right|}}. \tag{2.37}$$

Equation (2.37) satisfies the local time spread of the Gaussian window and allows optimal instantaneous character to be obtained in the term of the energy concentration on the TF plane [21].

The information contained in the spectrogram and accelerogram creates the possibility of adaptive window width control [22]. Namely, if the energy distribution is optimised using the local CR of the signal, the dominant value of this parameter can be distinguished from the distribution. Therefore, it can be applied:

$$\tilde{\alpha}(t) = \mathcal{R}_x^h(t, \arg\max_{\omega}(E_x^h(t,\omega)) \tag{2.38}$$

where $E_x^h(t,\omega) = \left|F_x^h(t,\omega)\right|^2$ as stated in (2.14), and $\mathcal{R}_x^g(t,\omega)$ stands for an accelerogram obtained for one of the $\mathfrak{K}$, $\mathfrak{D}$ or $\mathfrak{F}$ estimators ($\mathfrak{D}$ and $\mathfrak{F}$ are recommended due to the numerical instabilities near the energy ridge for the $\mathfrak{K}$ estimator). Finally, the adaptive short-time Fourier transform (ASTFT) can be written as

$$\grave{F}_x^h(t,\omega) = \int_{\mathbb{R}} x(\tau)\grave{h}(t-\tau)e^{-j\omega\tau}\mathrm{d}\tau \tag{2.39}$$

where

$$\grave{h}(t) = \frac{1}{\sqrt{2\pi}\sigma(t)} e^{-\frac{t^2}{2(\sigma(t))^2}} \tag{2.40}$$

and $\sigma(t)$ is defined by (2.37) and α determined by (2.38). In general, in the considered case the Gaussian window is applied, but any differentiable window is admissible, and the results are supposed to be consistent. As can be expected, besides the optimisation of the energy distribution, the CR estimate is also computed more precisely using the adaptive window. The results are presented in the following subchapter.

2.2.2.4 Adaptive chirplet transform

The chirplet technique is strongly related to the STFT, but the main difference is the analysis window. For the STFT case the window is even and real, e.g. the Gaussian or the Blackman-Harris window as presented previously. For the chirplet transform (CT), the window has an additional degree of freedom – its own chirp rate. For TF analysis, the chirplet window can be interpreted as a rotation of the TF cell of an angle proportional to its CR, which has significant consequences. Namely, if the CR of the chirplet window is the same as the CR of the signal, the energy is more concentrated near the attractor. The technique assumes, however, the appropriately chosen width of the analysis window. Such an approach was proposed in [23], where the adaptive chirplet transform (ACT) based on the CR estimation was revealed.

The CT can be defined as a convolution product of a signal $x(t)$ with a function g_α defined as [23]

$$C_x^h(t,\omega) = \int_{\mathbb{R}} x(\tau)h(t-\tau)e^{-j\alpha\frac{(t-\tau)^2}{2}}e^{-j\omega\tau}\,\mathrm{d}\tau \tag{2.41}$$

where $h(t)$ is a differentiable analysis window with a frequency-modulated component providing the transformation kernel:

$$g_\alpha(t,\omega) = h(t)e^{-j\alpha\frac{t^2}{2}}e^{-j\omega t} = h_\alpha(t)e^{-j\omega t} \tag{2.42}$$

with α being the CR of the window. For the Gaussian case, the chirplet window can be defined by

$$h_\alpha(t) = \frac{1}{\sqrt{2\pi}\sigma}e^{-\frac{t^2}{2\sigma^2}}e^{-j\alpha\frac{t^2}{2}}. \tag{2.43}$$

In the optimal case, the window CR and the signal CR are consistent, therefore the energy distribution is enhanced around such components. As presented in [23], the CT can be defined as self-tuning by combining (2.41) with the CR estimation process presented in this chapter. This allows extraction of the dominant component (and the related CR) and consequently adaptive changeability of the chirplet window is admissible the same way as for the ASTFT defined previously. In the thought of this technique, the TF signal distribution is computed using (2.41) with the assumption that the initial window CR is zero, which means that in this step, the CT is treated as the STFT defined by (2.13). Next, the same as in the defined ASTFT algorithm, the phase accelerogram is computed using one of the CR estimators in the TF domain ($\mathfrak{D}$ and $\mathfrak{F}$ estimators are preferred) which is used for the computation of the local phase acceleration (using the same window with $\alpha = 0$), and in the next step to obtain the dominant CR defined as

$$\tilde{\alpha}(t) = \mathcal{R}_x^g(t, \arg\max_\omega(E_x^g(t,\omega)) \tag{2.44}$$

for the energy distribution for a CT given by $E_x^g(t,\omega) = \left|C_x^g(t,\omega)\right|^2$, and $\mathcal{R}_x^g(t,\omega)$ being an accelerogram obtained for one of the CR estimators. As can be seen, this is the same relationship as (2.38). Finally, the ACT is given in the following manner:

$$\acute{C}_x^h(t,\omega) = \int_{\mathbb{R}} x(\tau)h(t-\tau)e^{-j\tilde{\alpha}(t-\tau)\frac{(t-\tau)^2}{2}}e^{-j\omega\tau}\mathrm{d}\tau. \tag{2.45}$$

It can be expected that the CR estimate is also more accurate as presented in [23]. Due to the limited scope of this chapter, the readers interested in this topic are kindly referred to the literature [23, 24]. The results for this technique in the form of energy distribution are presented in the next section as a comparison of all the techniques considered in this chapter.

2.2.2.5 Double-adaptive chirplet transform

The presented adaptive methods are still limited in their adaptivity, therefore in the ASTFT and ACT cases there are conditions in which the energy concentration is insufficient. As can be noted, both the ASTFT and the self-tuning CT have influence on the analysis window parameters. Taking this feature into account, the merging of these techniques as well as providing a double-adaptive window with the width and the CR adjusted simultaneously has been proposed in [24]. The technique is called double-adaptive chirplet transform (DACT). In fact, the principle of operation is the same as in the previously described techniques, but it takes advantages from both, allowing the adaptive width and the CR to be obtained as presented in Figure 2.15 and Table 2.1.

The window applied in DACT is characterised by two degrees of freedom – the first one is its width (consistent with a Cohen definition for the LFM signal, but applied locally), and the second is its CR that additionally focuses the energy near the energy ridge.

As in the two previously defined techniques, the dominant CR is extracted in each frame of the signal using (2.38), and both window parameters can be optimised

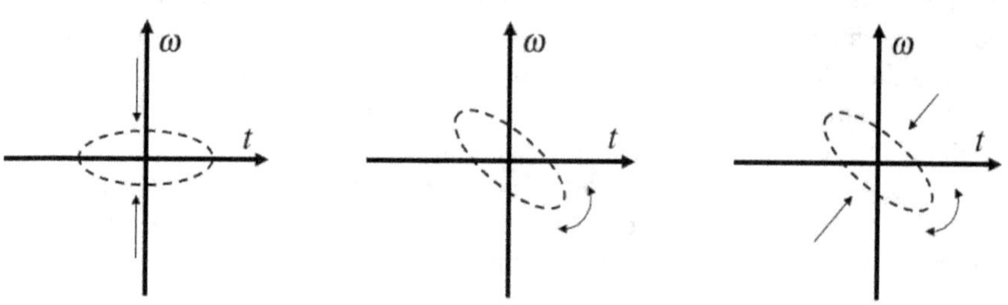

(a) Adaptive STFT window. (b) Adaptive CT window. (c) Adaptive DACT window.

Figure 2.15 A comparison of the TF representation of the adaptive windows useful in radar signal analysis

Table 2.1 Comparison of the analysing windows used in the described techniques

Method	Analysing window (Gaussian case)
STFT	$h(t) = \underbrace{\frac{1}{\sqrt{2\pi}\sigma} e^{\frac{-t^2}{2(\sigma)^2}}}_{\text{fixed window}}$
ASTFT	$\grave{h}(t) = \underbrace{\frac{1}{\sqrt{2\pi}\sigma(t)} e^{\frac{-t^2}{2(\sigma(t))^2}}}_{\text{adaptive width}}$
ACT	$\acute{h}(t) = \underbrace{\frac{1}{\sqrt{2\pi}\sigma} e^{\frac{-t^2}{2(\sigma)^2}}}_{\text{fixed window}} \overbrace{e^{-j\tilde{\alpha}(t)\frac{t^2}{2}}}^{\text{adaptive rotation}}$
DACT	$\acute{\grave{h}}(t) = \underbrace{\frac{1}{\sqrt{2\pi}\sigma(t)} e^{\frac{-t^2}{2(\sigma(t))^2}}}_{\text{adaptive width}} \overbrace{e^{-j\tilde{\alpha}(t)\frac{t^2}{2}}}^{\text{adaptive rotation}}$

using the knowledge about the local signal's character. Therefore, the DACT can be defined as

$$\acute{\grave{C}}_x^h(t,\omega) = \int_{\mathbb{R}} x(\tau) \frac{1}{\sqrt{2\pi}\sigma(t-\tau)} e^{-\frac{(t-\tau)^2}{2(\sigma(t-\tau))^2}} e^{-j\tilde{\alpha}(t-\tau)\frac{(t-\tau)^2}{2}} e^{-j\omega\tau} \mathrm{d}\tau \tag{2.46}$$

where

$$\sigma(t) = \sqrt{\frac{1}{2\pi\left|\tilde{\alpha}(t)\right|}}, \quad \tilde{\alpha}(t) \neq 0 \tag{2.47}$$

expresses the instantaneous standard deviation of the Gaussian window (its time spread), and $\tilde{\alpha}(t)$ is an estimated dominant CR as stated in (2.38) expressed in $\left(\frac{\mathrm{Hz}}{\mathrm{s}}\right)$. This example is provided for the Gaussian window case, but the method can be freely modified using any differentiable window. In this particular situation, the double-adaptive Gaussian window is defined as

$$\acute{\grave{h}}(t) = \grave{h}(t) \cdot e^{-j\tilde{\alpha}(t)\frac{t^2}{2}} = \frac{1}{\sqrt{2\pi}\sigma(t)} e^{-\frac{t^2}{2(\sigma(t))^2}} e^{-j\tilde{\alpha}(t)\frac{t^2}{2}}. \tag{2.48}$$

In (2.47), there is a statement $\tilde{\alpha}(t) \neq 0$ that exists in the particular Gaussian window case. However, when analysing LFM and NLFM radar pulses this condition is inherently fulfilled, but it has to be kept in mind if other waveforms are investigated.

Contrary to other analysed techniques, the window is more complex (requiring definition of the instantaneous standard deviation and the chirp rate in each moment); however, its computation does not require additional processing since its

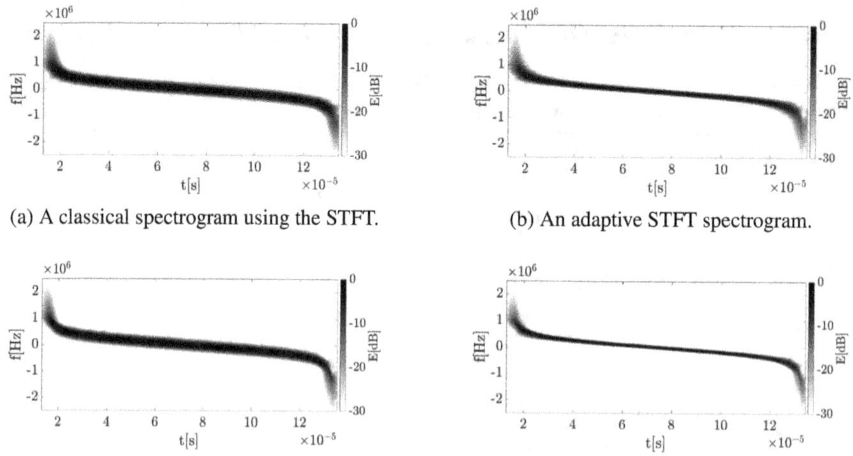

(a) A classical spectrogram using the STFT.

(b) An adaptive STFT spectrogram.

(c) An adaptive CT spectrogram.

(d) A DACT spectrogram.

Figure 2.16 A comparison of the TF energy distributions using the classical and adaptive transforms

parameters are dependant only on the local CR estimate. Therefore, as presented in [24, 25], the computational complexity and the processing time are the same for the DACT method as in the ASTFT and the ACT.

Figure 2.16 presents the juxtaposition of all of the techniques described. As can be observed, the best energy concentration is obtained for the DACT technique for the entire pulse duration – for the linear part of the chirp, as well as for its ends when the frequency rate grows rapidly. This is thanks to the combined approach that tailors the analysis parameters depending on the pulse character. As shown, the ACT did not provide a significant improvement in the energy concentration itself. Such a phenomenon exists if the initial window is characterised by an imprudently selected window width. Therefore, for the precise signal analysis the window has to be defined by the appropriately conditioned width not only its CR. This is the reason why the authors of [23] proposed the multi-window technique; however, this solution is more computationally complex than the DACT.

As can be expected, besides the improved energy distribution the phase accelerograms are also more precise. The results for all investigated methods are illustrated in Figure 2.17.

The accelerograms present the distribution of the same signal; however, their character is dictated by the applied window which is apparent especially for the transforms in which the window width is adjusted (ASTFT and DACT). The influence of this parameter manifests in the distribution of noise, and its decomposition over the TF plane follows the pattern of the chirp.

The methods presented are a subset of the techniques allowing signal parameter estimation to be performed. Essentially, the algorithms described in this chapter may contribute to deeper signal understanding and the assessment of its components. Not only these methods are limited to NLFM pulse analysis but their versatility

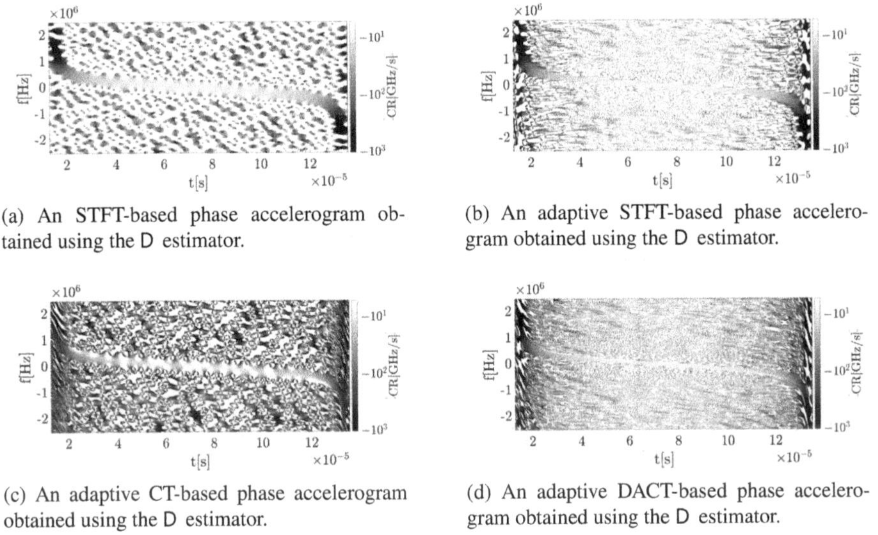

(a) An STFT-based phase accelerogram obtained using the D estimator.

(b) An adaptive STFT-based phase accelerogram obtained using the D estimator.

(c) An adaptive CT-based phase accelerogram obtained using the D estimator.

(d) An adaptive DACT-based phase accelerogram obtained using the D estimator.

Figure 2.17 A comparison of the CR estimation results using classical and adaptive transforms

has allowed their application in other fields of radar signal processing and completely different types of signals. As can be found in the literature, CR estimation can be successfully applied in NLFM [12] or micro-Doppler analysis [26], in passive forward-scattering radar [27] or linear chirp in strongly disturbed multi-path propagation conditions [24]. Therefore, such methods constitute a novel perspective to analysing radar signals, which enables additional information of the waveform phase to be distinguished.

2.2.2.6 Time-frequency reassignment

The reassignment technique was proposed by Kodera *et al.* in [19] due to the finite TF representation resolution resulting from the Heisenberg-Gabor uncertainty principle. In general, the method is used for sharpening the distribution that improves its readability, component separation and the separability of particular terms. In [28], the method was generalised by Auger and Flandrin, therefore, it is applicable in any TF representation belonging to the Cohen class. In this chapter, only the STFT is considered as an illustrative example; however, in the literature other representations can be found in which the TF reassignment was applied, e.g. [29–31].

In fact, the TF reassignment is the non-linear transformation that maps the energy to new coordinates on the plane, such as $(t, \omega) \mapsto (\hat{t}(t, \omega), \hat{\omega}(t, \omega))$, with $\hat{t}$ and $\hat{\omega}$ being expected to be closer to the true support of the analysed signal. Therefore, the energy surrounding the component with the highest energy locally attracts this energy, providing its high concentration as presented in Figure 2.18.

Basically, the reassignment technique allows thorough understanding of radar signal parameters and its principle of operation can be illustrated using the STFT

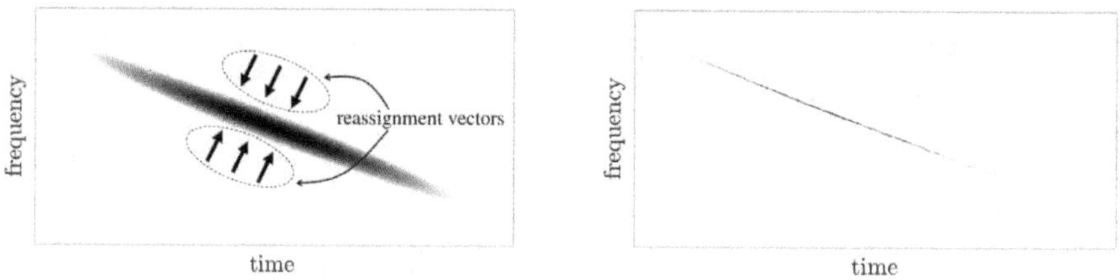

(a) An exemplary pulse processed using the STFT (a description of the term reassignment vectors is provided further).

(b) Concentrated pulse processed using the TF reassignment.

Figure 2.18 A graphical illustration of the TF reassignment

as an exemplary transform. The reassignment focuses signal components in the TF domain by moving energy from each (t, ω) coordinate to its *group delay* and the *instantaneous frequency* which define the reassignment vectors (see Figure 2.18), and are directly related to the STFT attributes resulting from its complex phase (see (2.22) and (2.23)). In their nature, the distributions may be considered as gradients along the time and frequency axes, and this property was straightforwardly used in the spectrogram concentration. In the investigated STFT approach, the reassignment operators may be estimated along the time and frequency directions, respectively [28]:

$$\hat{t}(t,\omega) = -\Im\left(\frac{\partial\Phi_x^h(t,\omega)}{\partial\omega}\right) = t - \Re\left(\frac{\partial\Phi_x^h(t,\omega)}{\partial\omega}\right) = t - \Re\left(\frac{F_x^{\mathcal{T}h}(t,\omega)}{F_x^h(t,\omega)}\right) \tag{2.49}$$

$$\hat{\omega}(t,\omega) = \omega + \Im\left(\frac{\partial\Phi_x^h(t,\omega)}{\partial t}\right) = \omega + \Im\left(\frac{F_x^{\mathcal{D}h}(t,\omega)}{F_x^h(t,\omega)}\right) \tag{2.50}$$

where $\Re$ is the real and $\Im$ is the imaginary part. $\hat{t}(t,\omega)$ denotes the relocation along the t axis, and $\hat{\omega}(t,\omega)$ expresses the reassignment vector along the frequency axis. As can be observed, the relationships used in (2.49) and (2.50) are known from the previously illustrated properties of the STFT and a meaningful conclusion arises herein. The amplitude of the STFT is useful for CR estimation in the TF domain and can be successfully applied for the concentration of the spectrogram.

Finally, the energy relocation using the reassignment method can be expressed as [28]

$$R_x^h(t,\omega) = \iint_{\mathbb{R}^2} E_x^h(\tau,\Omega)\delta(t-\hat{t}(\tau,\Omega))\delta(\omega-\hat{\Omega}(\tau,\Omega))\mathrm{d}\tau\mathrm{d}\Omega \tag{2.51}$$

where $\delta(\cdot)$ denotes the Dirac delta function, and $E_x^h(\tau,\omega)$ is the energy in the form of a spectrogram expressed by (2.14). The main disadvantage of the reassignment method is that is not reversible, thus only the distribution computation is obtainable without the possibility of inverse transformation. Thus, in the literature other methods are proposed, e.g. the synchrosqueezing that is reversible and allows mode extraction and the signal reconstruction to be performed [32, 33]. Contrary to the reassignment method, the synchrosqueezing relies on the relocation of the energy along only one axis: t or ω, which results in worse concentration capabilities in some applications, but the reversibility may be a key merit, especially when applying a higher-order synchrosqueezing transform, as presented in [34]. The technique, however, has potential in radar applications as presented, e.g. in [35, 36], but its description is beyond the scope of this chapter. The authors encourage the readers to explore the theory and applications related to the synchrosqueezing transform.

The following question might arise when talking about the reassignment method: How can one utilise this technique to understand radar waveforms? As can be expected, the energy concentration may provide sharpening of the TF ridges and consequently improved readability of signal components. This can be especially meaningful for multi-component signals, which, e.g. micro-Doppler waveforms are. The micro-Doppler signature is a signal which reaches the radar receiver when a complex moving object exists in the radar beam. The main condition is that the resolution of the radar (understood as the smallest distance between two targets, allowing for their clear distinction) has to be significantly smaller than the size of the object (the smaller the resolution, the better the radar's capabilities to differentiate spatially distributed targets). The radar resolution is given as $\Delta R = \frac{c}{2B}$, where c is the speed of light and B stands for signal bandwidth. There is no direct relation imposing the required radar resolution for the analysis of a particular object, therefore the phrase *significantly smaller* was used. For instance, for the radar with a bandwidth $B = 100$ MHz the resolution is $\Delta R = 1.5$ m, and for such conditions the observation of drone propellers is rather impossible but allows helicopter blades to be observed individually.

As an example, the micro-Doppler signature of a walking human has been presented in Figure 2.19 for the STFT and the reassignment technique [37]. The signal was obtained using the XY-DemoRad System by XY Sensing Ltd [38] and provided the resolution $\Delta R = 15$ cm, which is sufficient for human walk observation. At frequencies from -0.2 to -0.3 kHz, the echo from a torso is clearly visible, and additionally waving legs and arms produced an echo spread in the entire observed frequency range. The STFT-based distribution (the spectrogram in Figure 2.19 on the left) illustrates united components that cannot be easily determined. After the

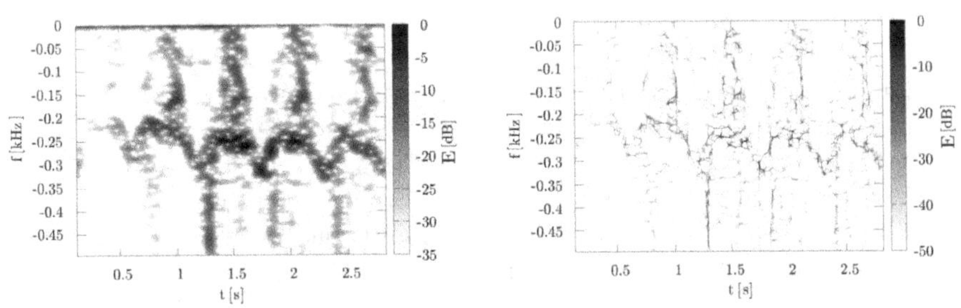

(a) A micro-Doppler signature of a walking human processed using the STFT.

(b) A reassigned micro-Doppler signature of a walking human

Figure 2.19 A micro-Doppler signature obtained using the STFT and concentrated using the reassignment technique energy concentration (the right-hand image in Figure 2.19), the signals corresponding to particular moving parts of the body are clearly visible and can be directly differentiated and distinguished. Moreover, for the sake of the clarity of the concentration capabilities of the reassignment method, the Rényi entropy was computed as a classical measure of the TF distribution concentration [39]. The γ order Rényi entropy of any TF signal representation $\mathrm{TF}_x(t,\omega)$ is given as

$$H_R^\gamma\left(\mathrm{TF}_x\right) = \frac{1}{1-\gamma}\log_2\left(\frac{\iint_{\mathbb{R}^2}\left(\left|\mathrm{TF}_x(t,\omega)\right|\right)^\gamma dt d\omega}{\iint_{\mathbb{R}^2}\left(\left|\mathrm{TF}_x(t,\omega)\right|\right) dt d\omega}\right). \tag{2.52}$$

and is expressed in bits. The lower value expresses the better concentration properties. For this particular case $\gamma = 3$, and the outcomes for the STFT and the reassigned STFT are as follows: $H_R^\gamma\left(\left|F_x^h(t,\omega)\right|\right) = 20.373$ and $H_R^\gamma\left(R_x^h(t,\omega)\right) = 17.040$, which confirms robustness in the separability of components in the concentrated spectrogram.

The illustrated example highlights the main advantages of the reassignment technique which can be successfully used in various areas of radar signal processing, e.g. in forward-scattering radar as presented in [40]. It can be supposed that analogously to the CR estimation process, the usage of the appropriate analysis window may still enhance the readability of the TF distribution. As in the phase accelerogram computation, the window width and its CR resulted in better energy concentration and a more precise estimate [24]. The same properties are still valid for the reassignment method, and the optimised window parameters have a meaningful influence on the reassigned spectrogram quality as presented, e.g. in [23].

For this purpose, the CR estimation method is useful when assessing the instantaneous frequency rate of the signal. Hence, if the signal frequency slope in each particular moment is known, the double adaptive window may be applied, which is given by (2.48). Since the reassigned distribution quality depends on the spectrogram, it has to be computed as precisely as possible, applying optimal processing parameters. This method is iterative, therefore if the initial window width is conditioned imprudently, the adaptive estimate can be still disturbed and the process might require additional repetitions. Basically, the adaptive reassignment can be defined by the following steps:

1. Compute the classical spectrogram with an initial window using (2.14).
2. Compute the phase accelerogram using (2.27), (2.28) or (2.29).
3. Distinguish the dominant CR for each frame using (2.38).
4. Create the set of optimal windows for each particular frame using (2.48).
5. Compute the group delay and the instantaneous frequency by (2.49) and (2.50).
6. Perform the adaptive reassignment process using (2.51).

The classical reassignment method and its adaptive version were applied for analysis of the NLFM pulse with the multi-path propagation effect and the results are illustrated in Figure 2.20. The signal was depicted in the initial part of this chapter in Figure 2.4. The TF reassignment technique without optimisation does not

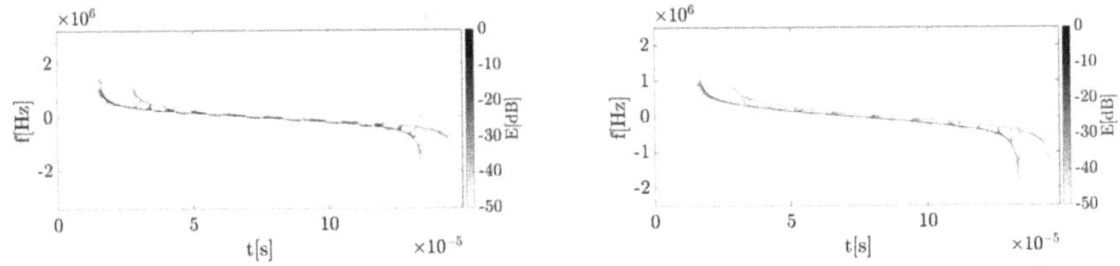

(a) A TF reassigned NLFM pulse with the multi-path propagation effect.

(b) An adaptively reassigned NLFM pulse with the multi-path propagation effect.

Figure 2.20 A comparison of the classical and adaptive reassignment technique in the context of NLFM pulse parameter estimation

satisfy the expected differentiation between the direct signal and the multi-path copy of the signal. In the middle part of the pulse, both components are united despite the energy relocation. After the adaptive processing, the components are better distinguishable thanks to the energy mapping resulting from the local estimate of the signal parameters.

2.3 Conclusions

In this section, the reassignment technique was briefly described and illustrated using real-life signals. This technique may be freely used for all transforms belonging to the Cohen class; however, in this chapter the focus was on the STFT for the sake of consistency with the previously considered methods. To the authors' observation, the method is gaining popularity in the radar society, but due to its huge potential in the near future the reassignment method may be found in many systems, especially in classification tasks, where a spectrogram of improved readability can deliver more information about the target than the classical method.

2.4 Summary

This chapter described recent algorithms dedicated to the analysis of non-stationary radar signals, which can be effectively used for different purposes, such as passive radar applications utilising other radars as a source of illumination, ELINT, EW or ER systems, among others. It is also a review of current research conducted by the authors in the field of signal recognition and classification. New techniques which were developed over the past few decades were summarised and illustrated using real-life radar waveforms such as NLFM pulses or micro-Doppler signatures. An interesting property (not mentioned in this chapter due to its limited scope) of the reassignment method, but also of the CR estimation in the TF domain, is that both can be implemented using the recursive filter bank approach, as presented in [41, 42]. Such a technique allows reduced computational complexity to be obtained, with preserved capabilities of the distribution computation. The authors believe that the techniques described in this chapter will be involved in practical applications where the radar signals are present.

References

[1] Pillai S.U., Li K.Y., Selesnick I., Himed B. *Waveform Diversity: Theory & Applications*. McGraw-Hill Education; 2011.

[2] Khawar A., Abdelhadi A., Clancy T.C. *Spectrum Sharing between Radars and Communication Systems: A Matlab Based Approach*. 1st ed. Springer Publishing Company, Incorporated; 2017.

[3] Dimas A., Li B., Clark M., Psounis K., Petropulu A. 'Spectrum sharing between radar and communication systems: can the privacy of the radar be preserved?'. 2017 51st Asilomar Conference on Signals, Systems, and Computers; Pacific Grove, CA; 2017. pp. 1285–1289.

[4] Shajaiah H., Abdelhadi A., Clancy C. 'Spectrum sharing approach between radar and communication systems and its impact on radar's detectable target parameters'. 2015 IEEE 81st Vehicular Technology Conference (VTC Spring); Glasgow; 2015. pp. 1–6.

[5] Bell M.R. 'Information theory and radar waveform design'. *IEEE Transactions on Information Theory*. 1993, vol. 39(5), pp. 1578–1597.

[6] Guerci J.R. 'Optimal radar waveform design'. Academic Press Library in Signal Processing; 2014. pp. 729–757.

[7] Khobahi S., Bose A., Soltanalian M. 'Deep radar waveform design for efficient automotive radar sensing'. 2020 IEEE 11th Sensor Array and Multichannel Signal Processing Workshop (SAM); China, Hangzhou; 2020. pp. 1–5.

[8] Boashash B., Analysis T.F. *Time Frequency Analysis*. San Diego, CA, USA: Elsevier Science; 2003.

[9] Samczyński P. 'Extended generalized chirp transform for signal parameter estimation in bistatic passive pulse radars'. 2013 14th International Radar Symposium (IRS); Dresden; 2013. pp. 155–160.

[10] Mahafza B.R. *Radar Systems Analysis and Design Using MATLAB*. CRC Press, Inc; 2000.

[11] Świercz E. 'Estimation of polynomial frequency modulation law for FM signals based on modified extended generalized chirp transform'. 2017 Signal Processing Symposium (SPSympo); Jachranka; 2017. pp. 1–5.

[12] Abratkiewicz K., Samczyński P., Czarnecki K. 'Radar signal parameters estimation using phase accelerogram in the time-frequency domain'. *IEEE Sensors Journal*. 2019, vol. 19(13), pp. 5078–5085.

[13] Raytheon Website. *ASR-10SS solid-state primary surveillance radar [online]*. Available from https://www.raytheon.com/capabilities/products/asr10ss [Accessed 06 Jun 2020].

[14] Auger F., Chassande-Mottin Éric., Flandrin P. 'On phase-magnitude relationships in the short-time Fourier transform'. *IEEE Signal Processing Letters*. 2012, vol. 19(5), pp. 267–270.

[15] Barnes A.E. 'The calculation of instantaneous frequency and instantaneous bandwidth'. *GEOPHYSICS*. 1992, vol. 57(11), pp. 1520–1524.

[16] Czarnecki K., Moszyński M. 'A novel method of local chirp-rate estimation of LFM chirp signals in the time-frequency domain'. 36th International Conference on Telecommunications and Signal Processing (TSP); Rome; 2013. pp. 704–708.

[17] Fourer D., Auger F., Czarnecki K., Meignen S., Flandrin P. 'Chirp rate and instantaneous frequency estimation: application to recursive vertical synchrosqueezing'. *IEEE Signal Processing Letters*. 2017, vol. 24(11), pp. 1724–1728.

[18] Czarnecki K. 'The instantaneous frequency rate spectrogram'. *Mechanical Systems and Signal Processing*. 2016, vol. 66-67(2), pp. 361–373.

[19] Kodera K., De Villedary C., Gendrin R. 'A new method for the numerical analysis of non-stationary signals'. *Physics of the Earth and Planetary Interiors*. 1976, vol. 12(2–3), pp. 142–150.

[20] Cohen L. *Time-Frequency Analysis: Theory and Applications*. Prentice-Hall; 1995.

[21] Pei S.-C., Huang S.-G. 'STFT with adaptive window width based on the chirp rate'. *IEEE Transactions on Signal Processing*. 2012, vol. 60(8), pp. 4065–4080.

[22] Abratkiewicz K., Samczyński P. 'An adaptive spectrogram and accelerogram algorithm for electronic warfare applications'. 2020 IEEE International Radar Conference (RADAR); Washington, DC, USA; 2020. pp. 530–535.

[23] Czarnecki K., Fourer D., Auger F., Rojewski M. 'A fast time-frequency multiwindow analysis using a tuning directional kernel'. *Signal Processing*. 2018, vol. 147(5), pp. 110–119.

[24] Abratkiewicz K. 'Double-adaptive chirplet transform for radar signature extraction'. *IET Radar, Sonar & Navigation*. 2020, vol. 14(10), pp. 1463–1474.

[25] Abratkiewicz K., Samczyński P. 'A block method using the chirp rate estimation for NLFM radar pulse reconstruction'. *Sensors*. 2019, vol. 19(22), p. 5015.

[26] Abratkiewicz K., Gromek D., Samczynski P. 'Chirp rate estimation and micro-doppler signatures for pedestrian security radar systems'. 2019 Signal Processing Symposium (SPSympo); Krakow, Poland; 2019. pp. 212–215.

[27] Abratkiewicz K., Krysik P., Gajo Z., Samczyński P. 'Target Doppler rate estimation based on the complex phase of STFT in passive forward scattering radar'. *Sensors*. 2019, vol. 19(16), p. 3627.

[28] Auger F., Flandrin P. 'Improving the readability of time-frequency and timescale representations by the reassignment method'. *IEEE Transactions on Signal Processing*. 1995, vol. 43(5), pp. 1068–1089.

[29] Fourer D., Auger F. 'Recursive versions of the reassigned scalogram anc of the synchrosqueezed wavelet transform'. 2017 22nd International Conference on Digital Signal Processing (DSP); London; 2017. pp. 1–5.

[30] Fourer D., Auger F., Hu J. *Reassigning and synchrosqueezing the Stockwell Transform: complementary proofs [online]*. 2015. Available from https://hal.archives-ouvertes.fr/hal-01467244/document [Accessed 5 May 2020].

[31] Auger F., Flandrin P., Lin Y.-T., *et al.* 'Time-frequency reassignment and synchrosqueezing: an overview'. *IEEE Signal Processing Magazine*. 2013, vol. 30(6), pp. 32–41.

[32] Meignen S., Pham D.-H. 'Retrieval of the modes of multicomponent signals from downsampled short-time Fourier transform'. *IEEE Transactions on Signal Processing*. 2018, vol. 66(23), pp. 6204–6215.

[33] Oberlin T., Meignen S., Perrier V. 'On the mode synthesis in the synchrosqueezing method'. Proceedings of the 20th European Signal Processing Conference (EUSIPCO), Bucharest; 2012. pp. 1865–1869.

[34] Oberlin T., Meignen S., Perrier V. 'Second-Order synchrosqueezing transform or invertible reassignment? towards ideal time-frequency representations'. *IEEE Transactions on Signal Processing*. 2015, vol. 63(5), pp. 1335–1344.

[35] Zhu M., Feng Z., Zhou X., Xiao R., Qi Y., Zhang X. 'Specific emitter identification based on Synchrosqueezing transform for civil radar'. *Electronics*. 2020, vol. 9(4), p. 658.

[36] Behera R., Meignen S., Oberlin T. 'Theoretical analysis of the second-order synchrosqueezing transform'. *Applied and Computational Harmonic Analysis*. 2018, vol. 45(2), pp. 379–404.

[37] Abratkiewicz K., Gromek D., Stasiak K., Samczynski P. 'Time-Frequency Reassigned Micro-Doppler Signature Analysis Using the XY-DemoRad System'. Proceedings of Signal Processing Symposium (SPSympo), Krakow, Poland; 2019. pp. 331–334.

[38] Samczynski P., Stasiak K., Gromek D., Kulpa K., Misiurewicz J. 'Xy-demorad – novel k- and mm-band radar demo kit for educational and commercial applications'. 2019 20th International Radar Symposium (IRS); Ulm; 2019. pp. 1–11.

[39] Baraniuk R.G., Flandrin P., Janssen A.J.E.M., Michel O.J.J. 'Measuring time-frequency information content using the Renyi entropies'. *IEEE Transactions on Information Theory*. 2001, vol. 47(4), pp. 1391–1409.

[40] Płotka M., Abratkiewicz K., Malanowski M., Samczyński P., Kulpa K. 'The use of the reassignment technique in the time-frequency analysis applied in VHF-Based passive forward scattering radar'. *Sensors*. 2020, vol. 20(12), p. 3434.

[41] Fourer D., Auger F., Flandrin P. 'Recursive versions of the Levenberg-Marquardt reassigned spectrogram and of the synchrosqueezed STFT'. IEEE International Conference on Acoustics, Speech and Signal Processing (ICASSP); Shanghai; 2016. pp. 4880–4884.

[42] Abratkiewicz K., Samczyński P., Fourer D. 'A comparison of the recursive and FFT-based reassignment methods in micro-doppler analysis'. 2020 International Radar Conference (RadarConf); Florence, 21–25 Sept. 2020; 2020. pp. 1–6.

Chapter 3

Non-linear estimation algorithms to analyse radar signals

U K Singh[1], A K Mishra[2], V Bhatia[3], and R Mitra[4]

3.1 Introduction

With the advent of digital signal processing and the rapid increase in digitisation, radar systems have evolved in a variety of ways to yield better target localisation. The main performance evaluation criterion for the radar system is based on detection (of target or event of interest) and/or estimation (of parameters of interest).

For many years estimators based on Fourier transform (FT) method have been used in the radar systems. An estimator based on FT is in fact a maximum likelihood (ML) estimator [1–3] and, hence, is based on batch processing, i.e. the radar observations are stored and then processed simultaneously. This makes the existing estimator time-consuming and, at times, unsuitable for radar applications demanding fast processing of data. Hence, there is a scope of developing adaptive estimation techniques suitable for tracking and monitoring applications as they provide an online estimate of required parameters [4, 5]. Furthermore, in practice, the radar return is affected by clutter, which is unknown a priori. This makes the conventional estimator unsuitable for practical deployment as the traditional estimation techniques are developed assuming the absence of clutter. Hence, there is a scope of developing adaptive estimation techniques capable of providing online estimates and also deals with the harmful effects of non-Gaussian clutter [6].

In the wake of improving the estimation performance of the conventional radar systems, recently kernel adaptive filters (KAFs)-based non-linear estimators are developed [4, 5]. The estimators based on KAFs utilise representer theorem and yield an initial non-linear estimate of the unknown parameters. After that, the initial estimate is adaptively updated using the convex cost function in high-dimensional

[1]University of Luxembourg, Luxembourg
[2]Department of Electrical Engineering, University of Cape Town, South Africa
[3]Department of Electrical Engineering, Indian Institute of Technology, India
[4]École de Technologie Supérieure (ÉTS), Université du Québec, Canada

reproducing kernel Hilbert space (RKHS). Optimisation of the convex cost function in RKHS results in an accurate estimation of the unknown parameters in comparison to conventional estimation techniques based on FT. Later, the non-linear adaptive estimators based on KAFs are also developed for estimation in the presence of non-Gaussian clutter [6]. Further, with the motive of estimating the parameters of a non-static target, Kalman filter-based non-linear estimators are explored, which finds application in tracking.

Therefore, in this chapter, a description of different non-linear estimation techniques, to enhance the estimation performance of the radar system, is given. We discuss some non-linear adaptive estimation techniques along with the non-linear extension of Kalman filter-based estimation techniques. The chapter proceeds with a description of the general radar signal model. Subsequently, the conventional batch processing estimation method is discussed. Next, the adaptive online non-linear estimators and non-linear estimators based on the Kalman filter are discussed. The chapter ends with a summary drawn from the discussion of the various adaptive online estimation techniques.

3.2 General radar signal model

For a radar system, the generalised radar return, $r(m, l)$, for the mth pulse in a group of M pulses and the lth frequency/time sample can mathematically be expressed as

$$r(m, l) = \exp(j2\pi m\theta_1)\exp(j2\pi l\theta_2) + c(m, l) + w(m, l), \tag{3.1}$$

where θ_1 and θ_2 are the unknown parameters corresponding to the target's range and velocity, respectively, please note that the range and velocity can be equivalently estimated by estimating delay and Doppler shift. In the above expression, $c(m, l)$ is the clutter sample, generally modelled as a non-Gaussian distribution and $w(m, l)$ is the thermal noise usually modelled as a Gaussian distribution.

In the forthcoming Sections 3.4.2 and 3.4.3, the estimators based on KAFs are applied over linear frequency modulated (LFM), orthogonal frequency-division multiplexed (OFDM) and stepped frequency (SF) radar systems in the presence and absence of clutter. Therefore, next, the signal model for LFM, OFDM and SF radar systems is discussed in detail.

3.2.1 Radar signal model in Gaussian noise

In a given coherent pulse interval (CPI), let the radar system transmit M radar pulses [7, 8]. The transmitted waveform is the pulse of a finite width which repeats after a certain PRI. Processing of the waveform is considered on PRI basis, thus for CPI of M transmitted pulses, any arbitrary mth transmitted pass-band pulse at central frequency f_c is

$$s_m(t) = \{s(t - mT_{\mathrm{PRI}})\}\exp(j2\pi f_c t), \tag{3.2}$$

where $s(t)$ is the baseband transmitted pulse for LFM or OFDM radar and T_{PRI} is the PRI.

The echo from the target is the delayed version of $s_m(t)$. If echo is from a single target located at an unknown range R and approaching towards the radar with an unknown velocity v, then the time delay by which an arbitrary mth pulse in CPI gets delayed is

$$\tau_m = \tau_o - \frac{2}{c}\{vmT_{\text{PRI}}\}, \tag{3.3}$$

where c is the velocity of light and τ_o is the unknown delay introduced in the pulse because of the initial position of a target, given by $\tau_o = \frac{2R}{c}$.

The received baseband radar return for any arbitrary mth pulse is

$$r_m(t) = \zeta\{s_m(t - \tau_m)\} + w(t), \tag{3.4}$$

where ζ[a] represents the complex attenuation factor which is proportional to the target RCS and $w(t)$ denotes the measurement and thermal noise process modelled as complex zero mean additive white Gaussian noise (AWGN).

3.2.2 LFM radar signal model

Let the transmitted LFM pulse be denoted as

$$s_{LFM}(t) = a\exp(j\pi\gamma t^2), \text{ for } 0 \leq \text{t} \leq \text{T}_o, \tag{3.5}$$

where a and γ are the amplitude and frequency sweep rate, respectively. T_0 is the duration of the transmitted baseband LFM pulse. The instantaneous frequency of $S_{LFM}(t)$ is

$$f_i(t) = \gamma t, \tag{3.6}$$

where $\gamma = \frac{L\Delta f}{T_o}$ as $L\Delta f$ is the bandwidth of $S_{LFM}(t)$.

Substituting (3.5) in (3.4) and omitting ζ yield the LFM radar return for the mth transmitted pulse as

$$r^m_{\text{LFM}}(t) = \{s_{\text{LFM}}(t - mT_{\text{PRI}} - \tau_m)\}\exp(j2\pi f_c(t - \tau_m)) + w(t). \tag{3.7}$$

The detailed derivation of the LFM radar return (r_{LFM}) from a single scatterer is given in [2]. Following the steps given in [2], radar return for LFM pulse is given by

$$r_{\text{LFM}}(m, l) = \exp(-j2\pi mf_c\tau_m)\exp(-j2\pi l\Delta f\tau_m) + w(m, l), \tag{3.8}$$

where τ_m is the delay defined in (3.3).

Substituting τ_m from (3.3) yields

$$r_{\text{LFM}}(m, l) = \exp(j2\pi mf_dT_{\text{PRI}})\exp(-j2\pi l\Delta f\tau_o)\exp(j2\pi f_d ml\left(\frac{T_{\text{PRI}}\Delta f}{f_c}\right)) + w(m, l), \tag{3.9}$$

[a]The effect of ζ is not considered for estimation as the target is assumed to be perfect reflector i.e, $|\zeta|^2$.

where $f_d = \frac{2vf_c}{c}$ is the unknown Doppler shift introduced by the velocity of the moving target.

3.2.3 OFDM radar signal model

Let the transmitted baseband OFDM pulse $s_{\text{OFDM}}(t)$ comprise of L-sub-carriers with a frequency separation of Δf. Then to maintain orthogonality between sub-carriers, T_o satisfies $T_0 = \frac{1}{\Delta f}$. Hence, the OFDM pulse is given by

$$s_{\text{OFDM}}(t) = \sum_{l=0}^{L-1} a_l \exp(j2\pi l \Delta f t), \text{ for } 0 \leq t \leq T_o, \tag{3.10}$$

where a_l is the modulation symbol which takes values from the set of constant modulus PSK constellation set such that $|a_l|^2 = 1$.

The corresponding OFDM radar return is obtained by substituting (3.10) in (3.4) and given as

$$r^m_{\text{OFDM}}(t) = \sum_{l=0}^{L-1} a_l \exp(j2\pi l \Delta f(t - mT_{\text{PRI}} - \tau_m)) \exp(j2\pi f_c(t - \tau_m)) + w(t). \tag{3.11}$$

Removing carrier frequency yields

$$r^m_{\text{OFDM}}(t) = \sum_{l=0}^{L-1} a_l \exp(j2\pi l \Delta f(t - mT_{\text{PRI}} - \tau_m)) \exp(-j2\pi f_c \tau_m) + w(t). \tag{3.12}$$

After sampling at $t_s = mT_{\text{PRI}} + \frac{n}{L\Delta f}$, the discrete form $(r^m_{\text{OFDM}}(n))$ of $r^m_{\text{OFDM}}(t)$ is given by [9]

$$\begin{aligned} r^m_{\text{OFDM}}(n) = &\sum_{l=0}^{L-1} a_l \exp(j2\pi l \frac{n}{L}) \exp(-j2\pi l \Delta f \tau_o) \exp(j2\pi (f_c + l\Delta f)\{\frac{2vmT_{\text{PRI}}}{c}\}) \times \\ &\exp(-j2\pi f_c \tau_o) + w(n). \end{aligned} \tag{3.13}$$

Assuming $f_c + l\Delta f \simeq f_c$ and utilising the steps given in [4], the OFDM radar return (r_{OFDM}) is given by

$$r_{\text{OFDM}}(m, l) = \exp(j2\pi m f_d T_{\text{PRI}}) \exp(-j2\pi l \Delta f \tau_o) + w(m, l). \tag{3.14}$$

3.2.4 Radar signal model in non-Gaussian clutter

Performance of the KMC-based estimators is analysed over both the LFM and SF radar system, which, in a given CPI transmits M radar pulses [7, 8, 10–12]. The transmitted waveform is a pulse of finite width and repeats after a certain PRI. We consider processing the waveform on PRI basis. Then, from (3.2), the transmitted pass-band pulse is given by

$$s_m(t) = \left(s(t - mT_{\text{PRI}})\right) \exp(j2\pi f_c t), \tag{3.15}$$

where $s(t)$ is either the LFM or SF baseband transmitted pulse.

Without loss of generality, let us consider that the target is located at an unknown range R_o and S scatterers are located at range $R_s \forall s = [1, \cdots, S]$ from the radar. The received pass-band radar return for an arbitrary mth pulse is given by

$$r_m(t) = \left(s_m(t - \tau_m)\right) + \sum_{s=1}^{S} s_m(t - \tau^s_m) + w(t), \tag{3.16}$$

where τ_m is defined in Section 3.2 and $\tau_m^s = \frac{2R_s}{c}$.

The second term in (3.16) represents the combined effect of all the unwanted reflections from the clutter henceforth, denoted as a random process $c(t)$. From (3.15) and (3.16), yield

$$r_m(t) = \left(s(t - mT_{\text{PRI}} - \tau_m)\right) \exp(j2\pi f_c(t - \tau_m)) + c(t) + w(t). \quad (3.17)$$

3.2.5 LFM radar signal model in non-Gaussian clutter

After substituting (3.5) into (3.17) and following the steps given in [2], the radar return of LFM radar for *m*th pulse and *l*th frequency sample is given by

$$r_{\text{LFM}}(m,l) = x_{\text{LFM}}(m,l) + c(m,l) + w(m,l), \quad (3.18)$$

where $x_{\text{LFM}}(m,l) = \exp(j2\pi m f_d T_{\text{P}}) \exp(-j2\pi l \Delta f \tau_o) \exp(j2\pi f_d m l(\frac{T_{\text{P}}\Delta f}{f_c}))$, $c(m,l)$ are the sample from clutter follows the K distribution. For a detailed description of K distributed clutter, readers are advised to refer to [6].

For $m = 0, \cdots, M-1$ and $l = 0, \cdots, L-1$, concatenating (3.18) in a vector yields

$$\mathbf{r}_{\text{LFM}} = \mathbf{x}_{\text{LFM}} + \mathbf{c} + \mathbf{w}, \quad (3.19)$$

where

$$r_{LFM} = [r_{LFM}(0,0), r_{LFM}(0,1), \cdots, r_{LFM}(M-1, L-1)]^T, \mathbf{c} = [c(0,0), c(0,1), \cdots, c(M-1, L-1)]^T,$$
$$\mathbf{w} = [w(0,0), w(0,1), \cdots, w(M-1, L-1)]^T, \quad (3.20)$$

and

$$\mathbf{x}_{\text{LFM}} = [x_{\text{LFM}}(0,0), x_{\text{LFM}}(0,1), \cdots, x_{\text{LFM}}(M-1, L-1)]^T.$$

3.2.6 SF radar signal model in non-Gaussian clutter

For SF radar, $s(t)$ for *m*th pulse is given by

$$s(t) = b \exp(j2\pi m \Delta f t), \quad \text{for} \quad 0 \le t \le T_o \quad (3.22)$$

where b is the amplitude and Δf is the frequency increment of SF waveform.

Substituting (3.22) into (3.17) and sampling at an interval of $lT_o + mT_{\text{P}}$, the radar return for SF radar for *m*th pulse and *l*th fast time index is given by

$$r_{\text{SF}}(m,l) = x_{\text{SF}}(m,l) + c(m,l) + w(m,l), \quad (3.23)$$

where

$$x_{\text{SF}}(m,l) = \exp(j2\pi f_d(mT_{\text{P}} + lT_o)) \exp(-j2\pi l \Delta f \tau_o)$$

Concatenating (3.23) in a vector yields

$$\mathbf{r}_{\text{SF}} = \mathbf{x}_{\text{SF}} + \mathbf{c} + \mathbf{w}, \quad (3.24)$$

where

$$\mathbf{r}_{\text{SF}} = [r_{\text{SF}}(0,0), r_{\text{SF}}(0,1), \cdots, r_{\text{SF}}(M-1, L-1)]^T. \tag{2.25}$$

and

$$\mathbf{x}_{\text{SF}} = [x_{\text{SF}}(0,0), x_{\text{SF}}(0,1), \cdots, x_{\text{SF}}(M-1, L-1)]^T. \tag{2.26}$$

For better estimation of delay and Doppler shift in the presence of Gaussian noise, the KLMS-based estimators are discussed in Section 3.4.2. However, since KLMS-based estimator works on the principle of minimisation of MSE criterion in RKHS, they are prone to yield inaccurate estimates in the presence of non-Gaussian clutter. Therefore, to improve the estimation of delay and Doppler shift in the presence of non-Gaussian clutter, new KMC-based sparse estimators are discussed in Section 3.4.3.

3.3 Batch processing estimation method

In the radar signal model presented in (3.1) the unknown parameters θ_1 and θ_2 are usually considered to be deterministic. Subsequently, under consideration of perturbation due to thermal noise, the classical estimation theory suggests the estimation of θ_1 and θ_2 with the non-Bayesian estimation approach. In radar, the popular non-Bayesian estimation approaches include the ML estimation and least squares (LS) estimation approaches. The estimation approaches based on ML and LS are in fact the batch processing estimation approach. Specifically, this implies that for estimation, the observations parametrised by the unknown parameters are first stored for some specific time duration, which in radar terminology called CPI. The stored observations are then simultaneously processed for the estimation. In simple words, the classical ML and LS estimator follows the offline estimation approach and subject to processing delays in the estimation of parameters. For some radar applications where the fast processing of observations is required, with a security perspective, this delay in the processing of radar observations is unavoidable.

It should be noted that under the assumption of perturbation follows Gaussian distribution, the estimator based on ML and LS yields similar solution for an estimation problem. Therefore, the working of the classical estimator based on ML is discussed next.

3.3.1 Maximum likelihood estimation

In this section, the classical ML estimation approach and its limitations with radar perspectives are discussed.

In the ML-based estimation approach, the logarithm of probability density function (PDF) of the observation parametrised by the unknown parameters usually called the log-likelihood function is maximised [13]. Subsequently, the optimisation of the equivalent cost function yields the estimated value of the parameters. The radar signal model, shown in (3.1), if collected for all $m = [0, 1, \cdots, M-1]$ and $l = [0, 1, \cdots, L-1]$, the radar signal model in vector form can be written as

$$\mathbf{r} = \mathbf{s} + \mathbf{w}, \tag{3.27}$$

where $\mathbf{s} = [1, \cdots, \exp(j2\pi m\theta_1)\exp(j2\pi l\theta_2), \cdots, \exp(j2\pi(M-1)\theta_1)\exp(j2\pi(L-1)\theta_2)]^T$ and $\mathbf{w} = [w(0,0), \cdots, w(m,l), \cdots, w(M-1, L-1)]^T \in \mathcal{N}_{\mathbb{C}}(\mathbf{0}, \sigma_w^2\mathbf{I})$, i.e. the samples of $\mathbf{w}$ are additive and Gaussian distributed with zero mean and variance σ_w^2.

In ML estimation for the estimation of θ_1 and θ_2 from (3.27), the resultant ML cost function is of the form

$$L(\theta_1, \theta_2) = \frac{1}{ML\sigma_w^2} \|r - s\|^2 \tag{3.28}$$

For the estimation of unknown parameters, (3.28) is minimised with respect to unknown θ_1 and θ_2 as

$$\underbrace{\min}_{\hat{\theta}_1, \hat{\theta}_2} \mathcal{L}(\theta_1, \theta_2). \tag{3.29}$$

Using (3.1), $\mathcal{L}(\theta_1, \theta_2)$ can be further written as following

$$\mathcal{L}(\theta_1, \theta_2) = \frac{1}{ML\sigma_w^2} \sum_m \sum_l | r(m,l) - (\exp(j2\pi m\theta_1)\exp(j2\pi l\theta_2)) |^2 . \tag{3.30}$$

As shown in (3.30), the likelihood function $\mathcal{L}(\theta_1, \theta_2)$ is non-linearly dependent on θ_1 and θ_2, hence the close form solution by minimising (3.30) is difficult to obtain. Additionally, because of the exponential dependence of θ_1 and θ_2, the resulting cost function ($\mathcal{L}(\theta_1, \theta_2)$) is non-convex in the parameters θ_1 and θ_2, therefore its minimisation yields sub-optimum solution. To solve (3.28), (3.30) is approximated as

$$\mathcal{L}(\theta_1, \theta_2) = \sum_m \sum_l r(m,l)\exp(j2\pi m\theta_1)\exp(j2\pi l\theta_2). \tag{3.31}$$

For the continuous range of θ_1 and θ_2, (3.31) corresponds to the 2D discrete-time Fourier transform (DTFT) of $\mathbf{r}$ with frequency variable θ_1 and θ_2. Consequently, the estimates of θ_1 and θ_2 are obtained by maximising $\mathcal{L}(\theta_1, \theta_2)$ over the 2D plane spanned by the possible values of θ_1 and θ_2. For illustration, the obtained 2D plot after performing (3.31) is shown in Figure 3.1. Practically, DTFT is approximated as the sampled version of DTFT (discrete Fourier transform (DFT)), the sampling of DTFT makes the continuous range of possible values of θ_1 and θ_2 limited to the finite discrete range as $\{0, \frac{1}{K}\theta_{1max}, \ldots, \frac{K-1}{K}\theta_{1max}\}$ and $\{0, \frac{1}{N}\theta_{2max}, \ldots, \frac{N-1}{N}\theta_{2max}\}$, respectively. Hence, $\forall k \in [0, 1, \cdots, K-1]$ and $\forall n \in [0, 1, \cdots, N-1]$. Equation (3.31) is further approximated as

$$\mathcal{L}(k, n) = \sum_m \sum_l r(m,l)\exp(j2\pi m\frac{k}{K})\exp(j2\pi l\frac{n}{N}). \tag{3.32}$$

Subsequently, $\underbrace{\max}_{\hat{\theta}_1, \hat{\theta}_2} \mathcal{L}(\theta_1, \theta_2) \cong \underbrace{\max}_{\hat{k}, \hat{n}} \mathcal{L}(k, n)$ is solved by searching for the index k and n which maximises (3.30) over discrete 2D plane spanned by the possible index values k and n (commonly known as FT method). Therefore, in ML estimator, approximation of (3.31) to (3.32) leads to the sub-optimum estimate of θ_1 and θ_2.

In addition to the above discussion, the limitations of ML estimator are listed below:

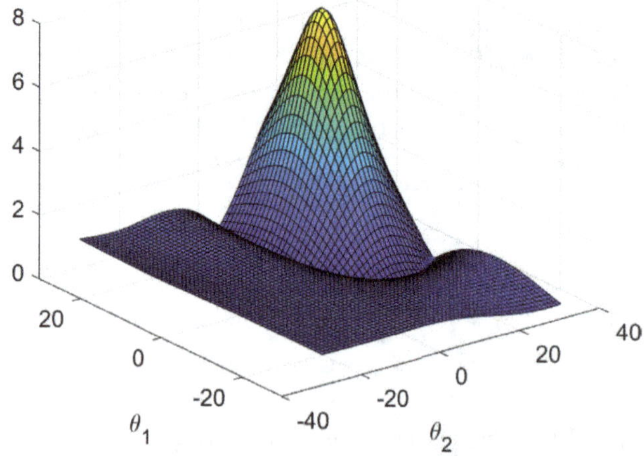

Figure 3.1 2D plot for ML estimation of θ_1 and θ_2 from (3.25)

1. For the non-linear estimation problem, the ML estimator suffers from signal-to-noise ratio (SNR) threshold effect and this causes a rapid increase in the mean square error (MSE) after a certain SNR threshold [14].
2. ML estimator is implemented by performing the 2D fast Fourier transform (FFT) over radar signal **r**. This operation results in the computational complexity of ML to be of the order of $O(K^2N^2 \log_2 KN)$. Moreover, to estimate the index at which the 2D FFT of **r** is maximised, as shown in Figure 3.1, coarse search over the 2D grid is performed, which for implementing the ML estimator, adds extra computation to $O(K^2N^2 \log_2 KN)$.

3.4 Online non-linear estimator

In this section, the new class of online non-linear estimators based on the RKHS and Kalman filter is discussed in detail. First, a detailed introduction of the RKHS-based estimation approach is given. Subsequently, a simple example of frequency estimation from discrete noisy samples of complex exponential using estimators based on RKHS is detailed. After that, the working of RKHS-based estimator with application to various radar systems, including LFM, OFDM [3, 15–17] and SF radar systems, is discussed in detail. Subsequently, a detailed discussion of the performance of the non-linear estimators based on RKHS with the help of various simulation results is given. Next, a brief introduction to the Kalman filter-based estimation approach is presented. Further, the working of the non-linear EKF- and UKF-based estimation techniques is discussed. Simulations to study the performance of EKF- and UKF-based estimators over the LFM radar system are detailed next.

3.4.1 Reproducing kernel Hilbert space-based estimation

As illustrated in (3.1), the parameters of interest (θ_1 and θ_2) are exponentially related to the radar return ($r(m, l)$). Further, in (3.1), for given θ_1 or vice versa, if the inverse relationship $\left(g(r(m, l)) = \{\theta_1, \theta_2\}\right)$ between the unknown parameter set ($\{\theta_1, \theta_2\}$) and $r(m, l)$ is known, the set of unknown parameters $\{\theta_1, \theta_2\}$ can be efficiently estimated. However, as the exact nature of $g(\cdot)$ is unknown and the estimation of $g(\cdot)$ is non-trivial and not straightforward, $g(\cdot)$ can be estimated by non-linear adaptive kernel-based function approximation algorithms commonly known as KAFs.

In wireless communication, the KAFs are applied to perform various function approximation problems, for instance, to recreate/estimate the transmitted symbols at the receiver from received symbols affected by channel non-linearity and thermal noise [18–21]. In this chapter, for the first time, the problem of estimating θ_1 and θ_2 utilising (3.1) is dealt with the perspective of solving function approximation problem. Particularly, estimators based on kernel least mean square (KLMS) and kernel maximum correntropy (KMC) are discussed, and performance is tested over LFM, OFDM and SF radar systems.

For estimation, at the kth instant, estimators based on KAF first transform the radar return ($\mathbf{r}(k)$) for all m and l and for given θ_1 and θ_2 to RKHS ($\mathbb{H}$) via an implicit mapping function $\Phi(\cdot)$. After transforming $\mathbf{r}(k)$ to $\mathbb{H}$ as $\Phi(\mathbf{r}(k))$, at kth instant for an arbitrary weight vector $\omega(k-1)$ in $\mathbb{H}$, the estimate of $g(\cdot)$ which in turn gives the estimate of desired parameter ($d(k)$) can be written as

$$\hat{d}(k) = \hat{g}(\mathbf{r}(k)) = \langle \boldsymbol{\omega}(k-1), \Phi(\mathbf{r}(k))\rangle_{\mathbb{H}}, \tag{3.33}$$

where $\langle \cdot, \cdot \rangle_{\mathbb{H}}$ represents the inner product operation in RKHS.

In (3.33), from basic adaptive filtering theory, it is explicit that for some optimum ω^*, the obtained estimate ($\hat{d}(k)$) can approximately resemble the true value ($d(k)$) s.t. $\hat{d}(k) = \langle \omega^*, \Phi(\mathbf{r}(k))\rangle_{\mathbb{H}} \simeq d(k)$. The $\omega(k-1)$ reaches the ω^* through the following iterative update.

$$\omega(k) = \omega(k-1) - \mu \nabla_{\omega(k-1)} \mathcal{J}_k(\omega), \tag{3.34}$$

where μ is the learning rate and $\mathcal{J}_k(\omega)$ is the suitable cost function.

As shown in (3.34), in order to update $\omega(k)$ iteratively (with increment in k), ($\mathcal{J}_k(\omega)$) is successively optimised in terms of $\omega(k-1)$ till the minimum of $e(k)$ ($e(k) = d(k) - \langle \omega(k-1), \Phi(\mathbf{r}(k))\rangle_{\mathbb{H}}$) is reached. For optimum working of the estimator based on KAF, the $\mathcal{J}_k(\omega)$ will be chosen according to the statistics of the observation. For instance, if the observation is perturbed by Gaussian distributed thermal noise, MSE, i.e. $\mathcal{J}_k(\omega) = \mathbb{E}[|e(k)|^2]$ is undoubtedly the suitable criterion for updating $\omega(k)$. However, in case the observation is perturbed by non-Gaussian clutter, the MSE will not yield suitable results, and, therefore, some suitable optimisation criterion like maximum correntropy criterion (MCC) is needed. Different optimisation criterion corresponds to a different type of estimators based on KAFs. In this chapter, since the estimation is performed in the presence of thermal noise and clutter, the employed optimising criterion, cost function and type of KAF-based estimators in accordance with the type of statistics of perturbation are described in Table 3.1.

Table 3.1 Types of criterion, cost function and KAF-based estimators according to different types of perturbation

Criterion	Cost function $(\mathcal{J}_k(\omega))$	Perturbation	KAF
MSE	$\mathbb{E}\left[\lvert e(k)\rvert^2\right]$	Thermal noise	KLMS
MCC	$\mathbb{E}\left[exp(-\frac{e^2(k)}{2\sigma_c^2})\right]$	Thermal noise and clutter	KMC

σ_c is the correntropy Gaussian kernel width.

Before giving a detailed description of the estimators based on KLMS and KMC, a simple example of the estimation of single-tone frequency from a discrete exponential signal perturbed by Gaussian distributed thermal noise is performed next. The standard example is presented to gauge a basic idea about the performance of the estimators based on KLMS and KMC in a non-linear estimation problem. The signal model for the considered elementary example is given by

$$r(m) = \exp(j2\pi\left(\frac{f_o}{f_s}\right)m) + w(m), \tag{3.35}$$

where f_o is the fundamental frequency to estimate and f_s is the sampling frequency.

In comparison to (3.1), the example shown in (3.35) can be considered as the estimation of a single parameter θ_1, equivalent to the estimation of f_o normalised by f_s. For simulation, it is assumed that SNR= 30 dB, $f_s = 5$ kHz, f_o can take value in the frequency range of $\{0, \cdots, \frac{f_s}{2}\}$ and the observation $r(m)$ is processed for ten samples, i.e. $M = 10$. The training and testing of estimators are done for 500 and 20 samples, respectively.

It is shown in Figure 3.2 that in the estimation of f_o from (3.35), minimum and approximately equal MSE, is achieved by the estimators based on KLMS and KMC. The MSE achieved by the estimators based on KLMS and KMC is of the order of 10^{-2}. This is because both estimators based on KLMS and KMC are capable of dealing with Gaussian noise. Nevertheless, since the estimator based on KLMS is not capable of dealing with the effects of non-Gaussian clutter, in the forthcoming Section 3.4.3, it is illustrated that higher MSE is achieved by the estimator based on KLMS. On the contrary, estimators based on KMC, due to the optimisation of MCC, handle the effect of non-Gaussian clutter and, consequently, yield the minimum MSE in the estimation of parameters.

3.4.2 KLMS-based estimators

In this subsection, to improve delay and Doppler shift estimation in presence of thermal noise, adaptive estimators based on KLMS in RKHS are discussed in detail. Since, popular estimator based on FT [2, 22] optimises non-convex ML cost function, it is prone to inaccurate estimates of delay and Doppler shift, especially in low

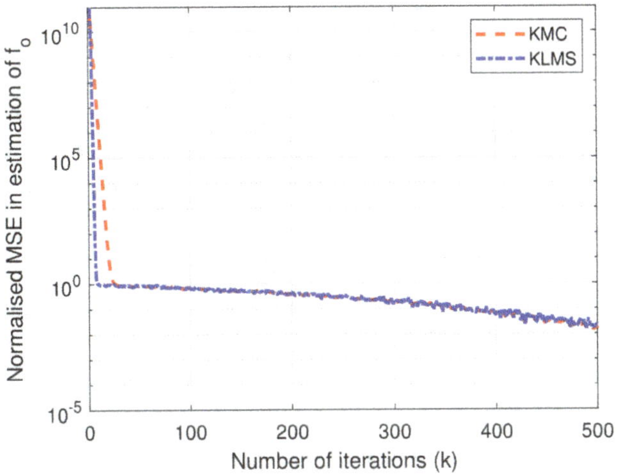

Figure 3.2 Normalised MSE in the estimation of single-tone frequency

SNR region. In comparison to the existing estimator based on FT, the KLMS-based estimation approach has a global minima, due to the optimisation of convex cost-function in RKHS [18, 19, 23]. Moreover, unlike existing estimation approaches for the aforementioned parameters, wherein the estimator's performance is typically enhanced by increasing the number of observations (thereby increasing computational complexity), the KLMS-based estimation technique selectively adapts the estimator in RKHS using Platt's novelty criterion (NC) [20, 21, 24, 25]. This yields a sparse estimator named KLMS-NC for unknown parameters in RKHS and convergence to the optimal parameter value is achieved by using a dictionary of a finite number of observations. Upon sparsification by NC, the KLMS-based estimator is found to be computationally simple for practical deployment. Another challenge with kernel-based estimator is in choosing an appropriate width of the popular Gaussian kernel. Silverman's rule yields the width of a kernel by exploiting the statistical properties of the incoming observation [26]. However, the kernel width obtained by Silverman's rule is not always optimal [27] and is prone to estimation errors. Hence, the kernel width using the incoming radar observations is learned and adapted, the resulting estimator is named KLMS-modified NC. The estimator based on KLMS-modified NC yields better convergence and smaller dictionary sizes, thereby providing a versatile and robust estimator for the estimation of delay and Doppler shift.

3.4.2.1 KLMS-based estimation of range and velocity

In the KLMS-based estimation algorithm, estimation of delay and Doppler shift is pursued independently. The assumption allows delay and Doppler shift's value space to be limited to one-dimension, i.e. $\{\tau_o, f_d\} \in \mathbb{R}$. Further, it should be noted that as the delay and Doppler shift is linearly related to the range and velocity of the target, estimation of delay and Doppler shift will invariably correspond to an estimation of range and velocity, respectively.

The estimation of τ_o is carried out by treating f_d as a known quantity. The range $\{\tau_{omin}, \tau_{omax}\}$ in which τ_o is expected to take the value is divided into K equal intervals, i.e. $[\tau^1_{omin} \dots \tau^K_{omax}]$. Following this, at any instant k, and corresponding to any arbitrary delay τ^k_o, (3.9) can be represented as the collection of M-radar return vector observations

$$\begin{aligned} \mathbf{r}^k_m &= \exp(j2\pi m f_d T_{\text{PRI}}) \exp\left(j2\pi f_d ml\left(\frac{T_{\text{PRI}}\Delta f}{f_c}\right)\right) \\ &\mathbf{d}_l + \mathbf{w}_m \ \forall\, m = [0, 1, \dots, M-1], \end{aligned} \tag{3.36}$$

for LFM radar system and as $\mathbf{r}^k_m = \exp(j2\pi m f_d T_{\text{PRI}})\mathbf{d}_l + \mathbf{w}^a_f \ \forall\, m = [0, 1, \dots, M-1]$for OFDM radar system, where

$$\mathbf{d}_l \in \mathcal{C}^{L\times 1} = [1, \exp(-j2\pi\Delta f\tau^k_o), \dots, \exp(-j2\pi(L-1)\Delta f\tau^k_o)]^T. \tag{3.37}$$

Hence, this yields

$$\mathbf{r}^k_{\tau o|fd} \in \mathbb{C}^{ML\times 1} = [\mathbf{r}^k_0, \mathbf{r}^k_1, \dots, \mathbf{r}^k_{M-1}]^T. \tag{3.38}$$

Equivalently, at instant k, the vector for the observations of radar return corresponding to any arbitrary Doppler shift f^k_d from the set $[f^1_{dmin} \dots f^K_{dmax}]$ of K-Doppler shifts is given as

$$\mathbf{r}^k_{fd|\tau o} \in \mathbb{C}^{ML\times 1} = [\mathbf{r}^k_0, \mathbf{r}^k_1, \dots, \mathbf{r}^k_{L-1}]^T, \tag{3.39}$$

where $\mathbf{r}^k_l = \exp(-j2\pi l\Delta f\tau_o)\mathbf{d}_m + \mathbf{w}_l$ and

$$\begin{aligned} \mathbf{d}_m \in \mathcal{C}^{M\times 1} = [1, \dots, &\exp(j2\pi m f^k_d T_{PRI}) \exp\left(j2\pi f^k_d ml\left(\frac{T_{PRI}\Delta f}{f_c}\right)\right), \dots, \\ &\exp(j2\pi(M-1) f^k_d T_{\{PRI\}}) \exp\left(j2\pi f^k_d (M-1) l\left(\frac{T_{PRI}\Delta f}{f_c}\right)\right)]^T, \end{aligned} \tag{3.40}$$

for LFM radar system and

$$\mathbf{d}_m \in \mathcal{C}^{M\times 1} = [1, \dots, \exp(j2\pi m f^k_d T_{\text{PRI}}), \dots, \exp(j2\pi(M-1) f^k_d T_{\text{PRI}})]^T, \tag{3.41}$$

for OFDM radar system.

After arranging the radar return in an appropriate form, next, the objective is to estimate the system with a mapping function $g(\cdot)$ which at k^{th} time instant maps the input vector $\mathbf{r}(k) \in \mathbb{C}^{ML}$ (where $\mathbf{r}(k) = \mathbf{r}^k_{\tau o|fd}, \mathbf{r}^k_{fd|\tau o}$)[b] respectively, into the corresponding output scalar $d(k) \in \mathbb{C}$ (where $d(k) = \tau_o{}^k, f_d{}^k$) as $\{g(\mathbf{r}(k)) = d(k)\}$. The KLMS algorithm proceeds with mapping the input vector $\mathbf{r}(k)\forall k$ into the high-dimensional

[b]"," is used in the relation ($r(k) = r^k_{\tau 0|fd}$ or $r^k_{fd|\tau 0}$) as a notation for 'or', which means that $\mathbf{r}(k)$ can be either $r^k_{\tau 0|fd}$ or $\mathbf{r}^k_{\text{fd}|\tau 0}$.

RKHS ($\mathbb{H}$) via complex reproducing Mercer kernel[c] $\kappa : \mathbb{X} \times \mathbb{X} \longrightarrow \mathbb{C}$, where $\mathbb{X} \subseteq \mathbb{C}^{ML}$. The kth input vector is mapped implicitly in $\mathbb{H}$ as $\Phi(\text{r(k)})$,

$$\Phi(\mathbf{r}(k)) = \kappa(\mathbf{r}(k), \cdot), \tag{3.42}$$

where κ is the mapping function used in a kernel-based adaptive algorithm, which implicitly maps the observations from finite-dimensional Euclidean space to $\mathbb{H}$.

From Mercer theorem, the reproducing kernel $\kappa(\mathbf{r}(k), \mathbf{r}(k'))$ can be written as

$$\kappa(\mathbf{r}(k), \mathbf{r}(k')) = \langle \Phi(\mathbf{r}(k)), \Phi(\mathbf{r}(k')) \rangle_{\mathbb{H}}, \tag{3.43}$$

where $\langle \cdot \rangle_{\mathbb{H}}$ represents the inner product operation in $\mathbb{H}$ and k' is the index of the vector corresponding to k'^{th} instant.

Let $\omega(k-1)$ be a weight vector defined as a function in linear space $\mathbb{H}$, then from the LMS algorithm in $\mathbb{H}$, which is equivalently called KLMS in input space, gives an estimate of the desired output $d(k)$ as $\hat{g}(k) = \langle \omega(k-1), \Phi(r(k)) \rangle_{\mathbb{H}}$. The learning rule (weight vector update rule) for this algorithm is obtained by minimising MSE, as follows:

$$\mathbb{E}[\mathcal{J}_k(\omega)] = \mathbb{E}[|d(k) - \hat{g}(k)|^2]. \tag{3.44}$$

In (3.44), MSE is estimated by the error of the instantaneous measurement, i.e. $\hat{\mathbb{E}}[\mathcal{J}_k(\omega)] = \mathcal{J}_k(\omega)$, hence

$$\mathcal{J}_k(\omega) = |d(k) - \hat{g}(k)|^2. \tag{3.45}$$

For stochastic gradient update, the gradient with respect to ω is calculated, i.e. $\nabla_\omega \mathcal{J}_k(\omega)$ is computed by applying the rule of Wirtinger's calculus given in [28, 29]. Hence, the adaptation factor by which ω is updated in$\mathcal{J}_k(\omega) = -e(k)\mu\Phi(\mathbf{r}(k))$, where $e(k) = d(k) - \langle \omega(k-1), \Phi(\mathbf{r}(k)) \rangle_{\mathbb{H}}$ and μ is the learning rate parameter. Thus, the final weight vector-update equation is $\omega(k) = \omega(k-1) + \mu e(k)\Phi(\mathbf{r}(k))$, for the initial condition $\omega(0) = 0$, repeated application of the weight vector-update yields

$$\omega(k) = \mu \textstyle\sum_{i=0}^{k} e(i)\Phi(\mathbf{r}(i)). \tag{3.46}$$

Thus, the estimated output at k^{th} time instant is given by

$$\hat{g}(k) = \langle \Phi(\mathbf{r}(k)), \omega(k-1) \rangle_{\mathbb{H}} = \langle \Phi(\mathbf{r}(k)), \mu \textstyle\sum_{i=0}^{k-1} e(i)\Phi(\mathbf{r}(i)) \rangle_{\mathbb{H}}. \tag{3.47}$$

From Mercer's theorem as defined in (3.43), the estimated output $\hat{g}(k)$ at k^{th} iteration is

$$\hat{g}(k) = \mu \textstyle\sum_{i=0}^{k-1} e(i)\kappa(\mathbf{r}(i), \mathbf{r}(k)). \tag{3.48}$$

[c]Mercer kernel can have different functional forms; however, Gaussian Mercer kernel $(k(x, y = exp(-\frac{|x-y|^2}{\sigma^2})))$ is used widely.

For kernel-based estimation, usually, the most commonly used Gaussian kernel function

$$\kappa(\mathbf{r}(i), \mathbf{r}(k)) = \exp\left(- \frac{\sum_{j=0}^{ML-1} (\mathbf{r}_j(i) - \mathbf{r}_j^*(k))^2}{\sigma^2} \right) \tag{3.49}$$

is used [28], where σ is the kernel width and $\mathbf{r}_j(\cdot)$ is the jth element of $\mathbf{r}(\cdot)$.

After replacing desired outputs (true parameters) and their corresponding inputs (observations) as $d(k) = \tau_o{}^k, f_d{}^k$ and $\mathbf{r}(k) = \mathbf{r}^k_{\tau_o|f_d}, \mathbf{r}^k_{f_d|\tau_o}$, the pseudocode for the online estimator based on KLMS is given in Algorithm 1. In Algorithm 1, the dictionary $\mathcal{D}_k$ keeps track of all the input samples used for training an estimator. The computational complexity of the estimator based on KLMS is $O(k)$.

3.4.2.2 Estimator based on KLMS-NC

At each instant k and for every input-output pair $\{\mathbf{r}(k), d(k)\}$, iterative learning and updating the estimates, unboundedly increases the size of dictionary $\mathcal{D}_k$ as given in Algorithm 1. The increasing dictionary size is due to the utilisation of every observation $\mathbf{r}(k)$ for estimation. However, not all the observations play a significant role in the estimation of the non-linear system. To address the problem of increasing sample size, Platt's novelty criterion (NC) for sparsification techniques has been introduced in [25].

Algorithm 1 KLMS algorithm to estimate τ_o and f_d

Inputs:
$\mathbf{r}(k) = \mathbf{r}^k_{\tau_o|f_d}, \mathbf{r}^k_{f_d|\tau_o} \forall\, k,\ \ d(k) = \tau_o{}^k, f_d{}^k\ \forall\, k$

Initialize:
$e(0) \leftarrow 0,\ \ \hat{g}(0) \leftarrow \kappa(\mathbf{r}(0), \cdot),\ \ \mathcal{D}_k \leftarrow \{\},\ \ \mathbf{a}_k \leftarrow \{\},$
choose μ and σ, $\mathbf{r}(0) \leftarrow \mathbf{r}^0_{\tau_o|f_d}, \mathbf{r}^0_{f_d|\tau_o}$

while $\{\mathbf{r}(k), d(k)\}$ available **do**
$\hat{g}(k) = \mu \sum_{i=0}^{k-1} e(i)\kappa(\mathbf{r}(i), \mathbf{r}(k))$
$e(k) = d(k) - \hat{g}(k),\ \mathcal{D}_k \leftarrow \mathbf{r}(k),\ \mathbf{a}_k \leftarrow e(k)$

end while

In dictionary sparsification technique, for curbing the growth of linearly increasing dictionary size, only important input observations are added as the new centre. This is accomplished by applying constraints on the Euclidean distance of an input observation. With each newly incoming input $\mathbf{r}(k)$, its Euclidean distance from the previously stored centres is calculated as $\mathbf{dis}(\mathbf{r}(k), \mathbf{r}(j)) = \|\mathbf{r}(k) - \mathbf{r}(j)\|_2\ \forall j = 0, 1, \ldots, S$ (S represents the number of elements in $\mathcal{D}_k$). If minimum of $\mathbf{dis}(\mathbf{r}(k), \mathbf{r}(j))$ is less than or equal to a pre-assigned small number δ_1, i.e. $\underbrace{min}_{0\leq j\leq S} \mathbf{dis}(\mathbf{r}(k), \mathbf{r}(j)) \leq \delta_1$ then the input observation will be discarded and not be added in the dictionary $\mathcal{D}_{k-1}$ as a new input centre. However, if $\underbrace{min}_{0\leq j\leq S} \mathbf{dis}(\mathbf{r}(k), \mathbf{r}(j)) > \delta_1$ then absolute of the error $|\, e(k)\,|$ associated with the input observation is compared against another pre-assigned small number δ_2. If $|\, e(k) \,| > \delta_2$, then the input observation is treated as an important observation and gets added in the dictionary as a new centre, thereby results in modification of $\mathcal{D}_{k-1}$ as $\mathcal{D}_k = \{\mathcal{D}_{k-1} \cup \mathbf{r}(k)\}$. Else the observation is discarded, while the contents and size of dictionary remain unchanged. By doing this, the sparsification method effectively reduces the required number of training regressors while preserving a desirable performance.

The pseudocode for the online estimator based on KLMS with sparsification is described in Algorithm 2. The computational complexity of the estimator based on KLMS-NC is $O(|\mathcal{D}_{k-1}^{NC}|)$.

3.4.2.3 Estimator based on KLMS-modified NC

The width σ is an important parameter for Gaussian kernel function. Both the learning rate and estimates are sensitive to the choice of σ. For large values of σ, all the input vectors will look similar in the RKHS (with inner product close to unity). On the contrary, low values of σ result in almost zero inner product which in turn reflect that all the input vectors are different. Hence, choosing an appropriate value of σ is vital in KLMS. There are methods to choose the finite kernel size like cross-validation, penalising function, plug-in method and Silverman's rule. The penalising function and plug-in methods are computationally expensive in the online learning system. Silverman's rule is derived for Gaussian approximation, and it is widely used to select the kernel size in kernel density estimation [26, 30, 31]. However, for online learning algorithms, Silverman's rule provides the sub-optimum value of kernel width [27, 31].

Algorithm 2 KLMS algorithm to estimate τ_o and f_d with sparsification

Inputs:
$\mathbf{r}(k) = \mathbf{r}^k_{\tau_o|f_d}, \mathbf{r}^k_{f_d|\tau_o} \ \forall k, \ d(k) = {\tau_o}^k, {f_d}^k \ \forall k$

Initialize:
$e(0) \leftarrow 0, \ \hat{g}(0) \leftarrow \kappa(\mathbf{r}(0), \cdot), \ \mathcal{D}_k \leftarrow \{\}, \ \mathbf{a}_k \leftarrow \{\},$
choose μ and σ, $\mathbf{r}(0) \leftarrow \mathbf{r}^0_{\tau_o|f_d}, \mathbf{r}^0_{f_d|\tau_o}$

while $\{\mathbf{r}(k), d(k)\}$ available **do**
 $\hat{g}(k) = \mu \sum_{i=0}^{|\mathcal{D}_{k-1}|} e(i)\kappa(\mathcal{D}^i_{k-1}, \mathbf{r}(k))$
 $e(k) = d(k) - \hat{g}(k)$
 $\mathbf{dis}(j) = \| \mathbf{r}(k) - \mathcal{D}^j_{k-1} \|$ for $0 \leq j \leq S$
 if $\underbrace{min}_{0\leq j\leq S}$ $\mathbf{dis}(j) \geq \delta_1$ and $|e(k)| \geq \delta_2$ **then**
  $\mathcal{D}_k = \{\mathcal{D}_{k-1} \cup \mathbf{r}(k)\}, \mathbf{a}_k = \{\mathbf{a}_{k-1} \cup e(k)\}$
 else
  $\mathcal{D}_k = \mathcal{D}_{k-1}, \ \mathbf{a}_k = \mathbf{a}_{k-1}$
 end if
end while

Thus, as KLMS is an online learning algorithm where the weights are updated at each iteration the kernel width σ can also be updated simultaneously by reducing the MSE at each iteration [27]. If σ_k is the kernel width at k^{th} iteration then the update equation for σ_k is given by

$$\sigma_k = \sigma_{k-1} - \eta \frac{\partial}{\partial \sigma_{k-1}} e^2(k). \tag{3.50}$$

Due to system non-linearity, $d(k)$ can be approximated as $d(k) = \hat{g}(\mathbf{r}(k)) + u(k)$, wherein $\hat{g}(\cdot)$ is the approximation of non-linear mapping which recovers the delay/Doppler spread from the radar return and $u(k)$ is the approximation error at the kth time instant.

If $g_{k-1}\{\cdot\}$ is the estimated mapping after learning upto k^{th} instant, then the error $e(k)$ associated with input $\mathbf{r}(k)$ is

$$e(k) = d(k) - g_{k-1}(\mathbf{r}(k)) = \tilde{g}_{k-1}(\mathbf{r}(k)) + u(k), \tag{3.51}$$

where $\tilde{g}_{k-1}(\mathbf{r}(k)) = \hat{g}(\mathbf{r}(k)) - g_{k-1}(\mathbf{r}(k))$.

Substituting (3.49) in (3.48) yields

$$\begin{aligned} \sigma_k &= \sigma_{k-1} - 2\eta e(k)\frac{\partial}{\partial\sigma_{k-1}}[\tilde{g}_{k-1}(\mathbf{r}(k)) + u(k)] \\ &= \sigma_{k-1} + 2\eta\mu e(k)e(k-1)\frac{\partial}{\partial\sigma_{k-1}}[\kappa_{\sigma_{k-1}}(\mathbf{r}(k-1)), \mathbf{r}(k)] \\ &= \sigma_{k-1} + \rho e(k)e(k-1)\sum_{j=0}^{ML-1}(\mathbf{r}_j(k-1) - \mathbf{r}_j^*(k))^2\frac{\kappa_{\sigma_{k-1}}(\mathbf{r}(k-1), \mathbf{r}(k))}{\sigma_{k-1}^3}, \end{aligned} \tag{3.52}$$

where $\rho = 4\eta\mu$ is the kernel width learning parameter.

Similar to μ, ρ controls the speed of convergence of kernel width to its steady-state value. Pseudocode for the online estimator based on KLMS with adaptive kernel width and sparsification is described in Algorithm 3. The computational complexity of the estimator based on KLMS-modified NC is $O(|\mathcal{D}_{k-1}^{MNC}|)$.

Algorithm 3 KLMS algorithm to estimate τ_o and f_d with adaptive kernel width and sparsification

1: **Inputs:**
$\mathbf{r}(k) = \mathbf{r}^k_{\tau_o|f_d}, \mathbf{r}^k_{f_d|\tau_o} \forall\, k,\ d(k) = \tau_o{}^k, f_d{}^k\ \forall\, k$
2: **Initialize:**
$e(0) \leftarrow 0,\ \hat{g}(0) \leftarrow \kappa(\mathbf{r}(0),\cdot),\ \mathcal{D}_k \leftarrow \{\},\ \mathbf{a}_k \leftarrow \{\},$
choose μ, σ_0, and ρ, $\mathbf{r}(0) \leftarrow \mathbf{r}^0_{\tau_o|f_d}, \mathbf{r}^0_{f_d|\tau_o}$
3: **while** $\{\mathbf{r}(k), d(k)\}$ available **do**
$\hat{g}(k) = \mu\sum_{i=0}^{|\mathcal{D}_{k-1}|} e^*(i)\kappa(\mathcal{D}^i_{k-1}, \mathbf{r}(k))$
$e(k) = d(k) - \hat{g}(k)$
$\mathbf{dis}(j) = \| \mathbf{r}(k) - \mathcal{D}^j_{k-1} \|$ for $0 \le j \le S$
4: **if** $\underbrace{min}_{0\le j\le S}\ \mathbf{dis}(j) \ge \delta_1$ and $|e(k)| \ge \delta_2$ **then**
$\mathcal{D}_k = \{\mathcal{D}_{k-1} \cup \mathbf{r}(k)\},\ \mathbf{a}_k = \{\mathbf{a}_{k-1} \cup e(k)\}$
$\sigma_k = \sigma_{k-1} + \rho e(k)e(k-1)\sum_{j=0}^{ML-1}(\mathbf{r}_j(k-1) - \mathbf{r}_j^*(k))^2\left(\frac{\kappa_{\sigma_{k-1}}(\mathbf{r}(k-1),\mathbf{r}(k))}{\sigma_{k-1}^3}\right)$
5: **else**
$\mathcal{D}_k = \mathcal{D}_{k-1},\ \mathbf{a}_k = \mathbf{a}_{k-1}$
6: **end if**
7: **end while**

3.4.2.4 Simulation results for KLMS-based estimators

In this section, simulations for the KLMS-based estimation of delay and Doppler shift are discussed in detail. The performance of the KLMS-based estimation algorithms

Table 3.2 Parameters values used for simulations for LFM radar system

Parameters	KLMS $\{\tau_0, f_d\}$	KLMS $\{\tau_0, f_d\}$	KLMS-modified NC $\{\tau_0, f_d\}$
μ	0.15, 0.2	0.53, 0.2	0.55, 0.1
σ	1.25×10^{-6}, 10^{-2}	0.4×10^{-6}, 10^{-6}	Dynamic
ρ	Nil, Nil	Nil, Nil	$10^{4.5}$, 10^{6}
σ_0	Nil, Nil	Nil, Nil	0.01, 0.01

is compared with FT and Cramer-Rao lower bound (CRLB). The detailed derivation of CRLBs is given in [4]. Simulation parameters for both LFM and OFDM radar are given in Tables 3.2 and 3.3 , respectively. Parameters of a waveform used in the simulations are described in Table 3.4. The SNR is fixed at 10 dB for estimation of delay and Doppler shift.

3.4.2.5 Estimation of delay for LFM and OFDM radar

Delay estimation is carried out by uniformly splitting the range of delay true values $\{\tau_{0min}, \tau_{0max}\}$ into 5 050 equal parts. Samples corresponding to all 5 050 delays are formed according to (3.36). From [18], out of 5 050 samples, the first 5 000 samples are used for training and the remaining 1% for testing. Comparative plot of MSE convergence versus iteration and dictionary size corresponding to LFM radar and OFDM radar is shown in Figure 3.3. From Figure 3.3, it is observed that the final MSE achieved by KLMS-NC and KLMS-modified NC is lower than KLMS. Additionally, modification in KLMS as KLMS-NC and KLMS-modified NC provides low-computational complexity and suitable kernel width, respectively. For both LFM and OFDM radar system, regressors used for the training of estimator in two modifications of KLMS are shown in Figure 3.3. In LFM radar system out of 5 000 regressors, only 160 and 100 regressors are used after incorporating NC and modified NC in KLMS, respectively. Similarly, the OFDM radar utilises 155 for NC and 135 for modified NC, out of 5 000 regressors. Therefore, sparsification of samples by NC reduces the computational complexity of an algorithm and provides lower MSE than KLMS without NC as depicted in Figure 3.3. Comparative analysis

Table 3.3 Parameters values used for simulations for OFDM radar system

Parameters	KLMS $\{\tau_0, f_d\}$	KLMS $\{\tau_0, f_d\}$	KLMS-NC $\{\tau_0, f_d\}$	KLMS-modified NC$\{\tau_0, f_d\}$
μ	0.15, 0.25	0.25, 0.25	0.35, 0.25	0.25, 0.25
σ	10^{-5}, 10^{-2}	10^{-5}, 10^{-2}	10^{-5}, 10^{-5}	Dynamic
ρ	Nil, Nil	Nil, Nil	Nil, Nil	$10^{6.5}$, $10^{6.5}$
σ_0	Nil, Nil	Nil, Nil	Nil, Nil	0.01, 0.01

Table 3.4 Values of radar parameters

Parameters	Values
Number of subcarriers (L)	8
Number of pulses (M)	8
Subcarrier spacing (Δf)	76.25 kHz
Pulse duration $(T_o = \frac{1}{\Delta f})$	0.013×10^{-3}
Pulse repetition interval $(T_{PRI} = 1.25T_o)$	0.016 Km
Bandwidth $(L\Delta f)$	610 kHz
Set of delay true value $\{\tau_{0_{min}}, \tau_{0_{max}}\} = \left\{\frac{1}{L\Delta f}, T_0\right\}$	$\{0.0016 \times 10^{-3}, 0.013 \times 10^{-3}\}$ sec
Set of Doppler true value $\{f_{d_{min}}, f_{d_{max}}\} = \left\{-\frac{1}{2T_{PRI}}, \frac{1}{2T_{PRI}}\right\}$	$\{-30.4, 30.4\}$ KHz

of the number of computations required for the convergence of estimator MSE for KLMS, KLMS-NC and KLMS-modified NC for the given LFM and OFDM radar system model and corresponding to Figure 3.3 is shown in Table 3.5.

The following relation $O(|\mathcal{D}_{k-1}^{MNC}|) < O(|\mathcal{D}_{k-1}^{NC}|) < k$ holds in which $O(|\mathcal{D}_{k-1}^{NC}|)$ and $O(|\mathcal{D}_{k-1}^{MNC}|)$ are the cardinality of the dictionary for estimators based on KLMS-NC and KLMS-modified NC, respectively.

3.4.2.6 Estimation of Doppler shift for LFM and OFDM radar

Similar to delay estimation, estimation of Doppler shift is carried out by uniformly splitting the true value range of Doppler shift $\{f_{d_{min}}, f_{d_{max}}\}$ in 5 050 equal parts. Input samples of an estimator are formed corresponding to each of 5 050 Doppler shifts utilising (3.37). To illustrate the effectiveness of KLMS-based estimation algorithms the input samples are as per [18], where 5 000 samples are used for training of an estimator and the last 50 samples are used for testing. At each kth iteration, the value of delay τ_o^k is treated as known and is chosen randomly from its available range $\{\tau_{o_{min}}, \tau_{o_{max}}\}$. Comparative plot for the convergence of MSE in Doppler shift estimation versus iterations and evolution of dictionary size for all three KLMS algorithms for LFM and OFDM radar are shown in Figure 3.3. Modification in KLMS by employing NC and modified NC reduces the final MSE. With modified NC, KLMS provides the lowest MSE as compared to KLMS. Moreover, for LFM radar, contrary to NC case where 150 samples are utilised for training of an estimator, KLMS-modified NC utilises 100 samples out of 5 000 as shown in Figure 3.3. Similar to LFM radar, as shown in Figure 3.3, in OFDM radar system too the implementation of modified NC results in utilising only 100 samples out of 5 000. Whereas, KLMS-NC results in the utilisation of 160 samples. Improvement in the performance of an estimator for both (LFM radar and OFDM radar) is because of

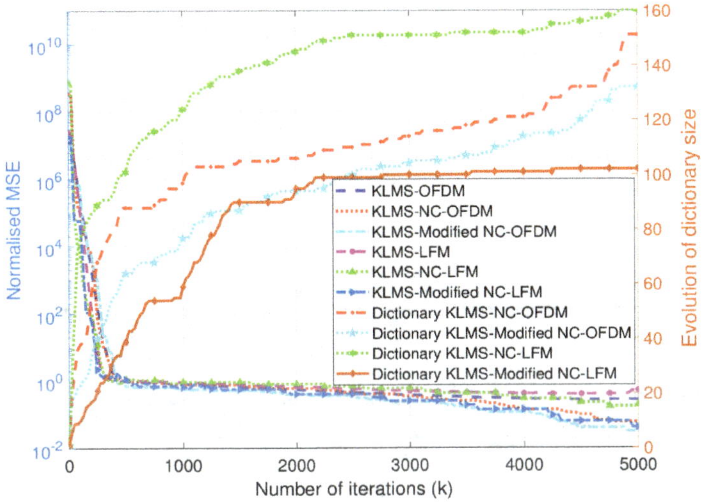

(a)

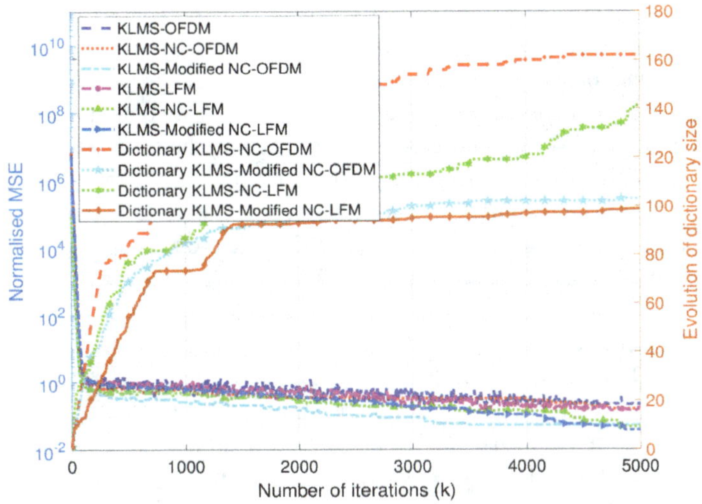

(b)

Figure 3.3 *Normalised MSE and dictionary size in the estimation of (a) delay and (b) Doppler shift for LFM and OFDM radar with estimators based on KLMS, KLMS-NC and KLMS-modified NC*

Table 3.5 Comparison of computational complexity of estimators based on KLMS, KLMS-NC and KLMS-modified NC

Estimation algorithm	Number of multiplications	Complexity order
KLMS	5 000	$O(K)$
KLMS-NC	150	$O(\lvert D_{k-1}^{NC}\rvert)$
KLMS-modified NC	100	$O(\lvert D_{k-1}^{MNC}\rvert)$

the suitable kernel width which is achieved gradually with successive iteration along with the reduction in estimator MSE.

3.4.2.7 Performance comparison of KLMS algorithms with FT method

A performance comparison between KLMS and FT is performed in this section. Since the FT method gives approximate ML estimates by minimising the non-convex cost function, the FT method is prone to produce erroneous estimates for the considered system model. Moreover, as the estimates are obtained by searching the location of the peak in the discrete 2D spectrogram, the estimates are susceptible to the resolution of spectrogram peak (finer resolution of peak yields accurate estimates). Hence, for comparison between two estimators (FT and KLMS), the variance of a 2D spectrogram is used as a metric. The metric is chosen to measure the variance in the widening of a peak in the 2D spectrogram against the increase in the noise level.

The performance of FT and KLMS algorithm using NC and modified NC for the delay and Doppler shift estimation is examined over the SNR ranges from −20 dB to 30 dB. As shown in Figure 3.4, for the whole range of SNR, KLMS-modified NC performs far better than FT. The improvement in estimation performance is considerable for LFM radar and OFDM radar, as shown in Figure 3.4, for the range of SNR typically from 0 dB to 30 dB, KLMS-modified NC yields estimate of delay with lower variance than KLMS-NC. However, for lower SNR range of −15 dB to −1 dB KLMS-NC and KLMS-modified NC have similar performance. Similar trend can also be observed for OFDM radar.

From Figure 3.4, it can be observed that KLMS-modified NC and KLMS-NC exhibit similar convergence characteristics in certain scenarios when the kernel width estimates by Silverman's rule are close to the optimal value of the kernel width. However, from Figure 3.4, KLMS-modified NC performs consistently better than KLMS-NC for OFDM radar in general. Furthermore, KLMS in both extensions (NC and modified NC) and for both LFM radar and OFDM radar yields estimates (delay and Doppler shift) closest to the CRLBs [4].

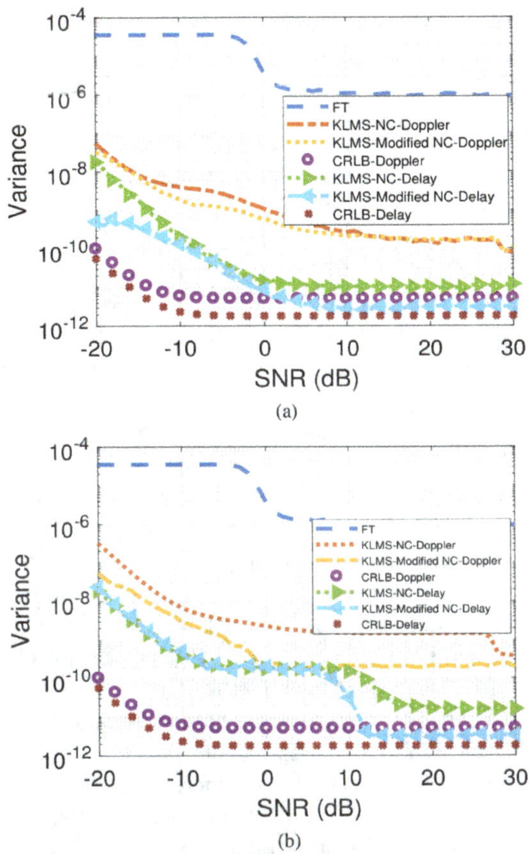

Figure 3.4 *Comparison of FT method with KLMS-NC and KLMS-modified NC algorithms for delay and Doppler shift estimation in (a) LFM and (b) OFDM radar*

3.4.3 KMC-based estimators

Previously, for efficient estimation of the target's range and velocity, KLMS-based adaptive estimators were introduced and tested for LFM and OFDM radar systems. The KLMS-based estimators are viable due to the universal approximation of an arbitrary unknown function and convexity of the cost function in RKHS [18, 19, 23, 32]. Convexity and universal representation make estimators based on KLMS better suited in comparison to conventional estimation techniques based on FT. Nevertheless, the radar return is affected by the undesired reflections from objects [33–36], these reflections are collectively called clutter and modelled by non-Gaussian distributions [37–40]. As shown previously, the KLMS-based estimator uses MSE for adaptation, which is optimal for additive Gaussian distortion [33, 34]. However, the optimality of KLMS-based estimator for additive non-Gaussian distortion is not guaranteed because MSE considers second-order statistics of error.

In this subsection, the problem of delay and Doppler shift estimation is pursued in a more realistic and practical environment, perturbed by thermal noise and clutter. Subsequently, limitations of the conventional estimators based on FT and KLMS, in the presence of non-Gaussian clutter, are overcome by the estimator based on KMC algorithm. The KMC-based estimator uses MCC in RKHS and provides improved performance both in terms of computational complexity and estimation accuracy. MCC is an information theoretic learning (ITL) criterion and can be considered as a similarity measure between the two random variables [41, 42]. Moreover, MCC is an online estimate of Renyi-α information criterion for $\alpha = 2$ [43]. Additionally, maximising the similarity between the desired parameter and the estimated parameter results in a smooth loss function and rejection of outliers. Furthermore, KMC, unlike other estimators based on KLMS criterion, considers higher-order statistics for estimation [33, 34, 41, 42, 44, 45]. Hence, maximisation of smooth loss function and consideration of higher-order statistics makes estimators based on KMC viable in the presence of clutter. However, the practical deployment of KMC-based estimator is limited by two aspects: (i) KMC-based estimator approximates the unknown function using the radar return in RKHS. Consequently, they suffer from linear temporal increase in computational complexity [21, 46] and (ii) for smooth function approximation, the KMC-based estimator employs a continuous Mercer's kernel; however, the accuracy of function approximation by KMC algorithm depends on the choice of appropriate kernel width [18, 19, 27], which makes the choice of appropriate kernel width for MCC-based estimation crucial.

To mitigate the above limitations of KMC-based estimator for range and velocity estimation in the presence of non-Gaussian clutter, the two variants of KMC-based estimator are introduced. First, for curbing the increasing computational complexity of KMC-based estimation in RKHS, similar to KLMS-NC, Platt's novelty criterion (NC) [25] with KMC is used. Second, for the selection of an appropriate kernel width, an adaptive estimator for stochastic update of the kernel width using MCC is discussed, namely the KMC-modified NC.

3.4.3.1 KMC-based estimation

Similar to estimators based on KLMS, for KMC-based estimators also, estimations of delay and Doppler shift are pursued individually, whereby, the true value of delay and Doppler shift is limited to one dimension, i.e. $\tau_o \in \mathbb{R}$, and $f_d \in \mathbb{R}$.

For estimation of delay and Doppler shift, constant attenuation factor, ζ, in (3.19) and (3.24) is considered to be unity, hence

$$\mathbf{r}_{\mathrm{LFM}} = \mathbf{x}_{\mathrm{LFM}} + \mathbf{c} + \mathbf{w}, \tag{3.53}$$

$$\mathbf{r}_{\mathrm{SF}} = \mathbf{x}_{\mathrm{SF}} + \mathbf{c} + \mathbf{w}. \tag{3.54}$$

Subsequently, for given f_d, the range $(\tau_{o_{min}}, \tau_{o_{max}})$ in which τ_o is expected to take the value is divided into K equal intervals, i.e. $[\tau^1_{o_{min}} \ldots \tau^K_{o_{max}}]$. At any instant k, (3.53) and (3.54) can be written as $\mathbf{r}^k_m = \exp(j2\pi m f_d T_{\mathrm{P}}) \exp(j2\pi f_d m l(\frac{T_{\mathrm{P}}\Delta f}{f_c}))\mathbf{d}_l + \mathbf{c} + \mathbf{w} \; \forall \, m = [0, 1, \ldots, M-1]$ for LFM radar system, where

$$\mathbf{d}_l \in \mathcal{C}^{L\times 1} = [1, \ldots, \exp(-j2\pi(L-1)\Delta f \tau_o^k)] \tag{3.55}$$

and as $\mathbf{r}_m^k = \exp(j2\pi f_d(mT_P + lT_o))\mathbf{d}_l + \mathbf{c} + \mathbf{w} \forall m = [0, 1, \ldots, M-1]$ for SF radar system, where $\mathbf{d}_l \in \mathbb{C}^{L\times 1} = [1, \cdots, \exp(-j2\pi(L-1)\Delta f \tau_o)]$. Hence, this yields

$$\mathbf{r}_{\tau_o|f_d}^k \in \mathbb{C}^{ML\times 1} = [\mathbf{r}_0^k, r_1^k, \ldots\ldots, r_{m-1}^k]^T. \tag{3.56}$$

Equivalently, at instant k, vector for the observations of radar return for LFM and SF radar system corresponding to any arbitrary Doppler shift f_d^k from the set $[f_{d_{min}}^1 \cdots f_{d_{max}}^K]$ of K-Doppler shifts is given as

$$\mathbf{r}_{f_d|\tau_o}^k \in \mathbb{C}^{ML\times 1} = [\mathbf{r}_0^k, r_1^k, \ldots\ldots, r_{L-1}^k]^T. \tag{3.57}$$

where $\mathbf{r}_l^k = \exp(-j2\pi l \Delta f \tau_o)\mathbf{d}_m + \mathbf{c} + \mathbf{w}$ and

$$\mathbf{d}_m \in \mathcal{C}^{M\times 1} = [1, \ldots, \exp(j2\pi(M-1)f_d^k T_p)\exp(j2\pi f_d^k (M-1)l(\tfrac{T_p \Delta f}{f_c}))]^T \tag{3.58}$$

for LFM radar system and

$$\mathbf{d}_m \in \mathcal{C}^{M\times 1} = [\exp(j2\pi f_d^k l T_o), \cdots, \exp(j2\pi f_d^k((M-1)T_P + lT_o))] \tag{3.59}$$

for SF radar system.

3.4.3.2 Estimator based on KMC-NC

In this section, the KMC-NC-based estimator is described in detail.

Similar to estimators based on KLMS and following (3.41) and (3.42), if $\omega(k-1)$ is a weight-vector in $\mathbb{H}$, then kernel trick, which is equivalently the inner product in $\mathbb{H}$, gives an estimate of the desired output $d(k)$ as $\hat{g}(k) = \langle \omega(k-1), \Phi(\mathbf{r}(k)) \rangle_{\mathbb{H}}$. Subsequently, the learning rule given by MCC is

$$\mathcal{J}(\omega) = \mathbb{E}[\kappa_{\sigma_c}(d(k), \hat{g}(k))], \tag{3.60}$$

where $\kappa_{\sigma_c}(d(k), \hat{g}(k)) = \exp\left(-\frac{(d(k)-\hat{g}(k))^2}{2\sigma_c^2}\right)$ is the Gaussian kernel function and is correntropy Gaussian kernel function width.

The correntropy in (3.60) is difficult to evaluate without knowing the joint probability density function of $d(k)$ and $\hat{g}(k)$ [33–35]. Therefore, (3.60) is estimated in terms of error of the instantaneous measurement at k^{th} instant as $\hat{\mathcal{J}}(\omega) = \mathcal{J}_k(\omega)$ and given by

$$\mathcal{J}_k(\omega) = \kappa_{\sigma_c}(d(k), \hat{g}(k)) = \exp\left(-\frac{(d(k)-\hat{g}(k))^2}{2\sigma_c^2}\right). \tag{3.61}$$

For stochastic gradient update of ω, the gradient of $\mathcal{J}_k(\omega)$ with respect to ω is calculated (i.e. $\nabla_\omega \mathcal{J}_k(\omega)$). Hence, the weight-vector ω is updated by the factor $\mu \exp\left(-\frac{e^2(k)}{2\sigma_c^2}\right)e(k)\Phi(\mathbf{r}(k))$, where $e(k) = d(k) - \langle \omega(k-1), \Phi(\mathbf{r}(k)) \rangle_{\mathbb{H}}$ is the instantaneous error in adaptive estimation and μ is the learning rate parameter as $\omega(k) = \omega(k-1) + \mu \exp\left(-\frac{e^2(k)}{2\sigma_c^2}\right)e(k)\Phi(\mathbf{r}(k))$. Subsequently, for the initial condition $\Omega(0) = \mathbf{0}$, repeated application of the weight-vector update yields

$$\omega(k) = \mu \sum_{i=0}^{k} \exp\left(-\frac{e^2(i)}{2\sigma_c^2}\right) e(i)\Phi(\mathbf{r}(i)). \tag{3.62}$$

Thus, the estimated output at k^{th} time instant is given by

$$= \mu \sum_{i=0}^{k-1} \exp\left(-\frac{e^2(i)}{2\sigma_c^2}\right) e(i)\langle \Phi(\mathbf{r}(i)), \Phi(\mathbf{r}(k))\rangle_{\mathbb{H}}. \tag{3.63}$$

Using Mercer's theorem, the estimated output $\hat{g}(k)$ at k^{th} iteration is given as

$$\hat{g}(k) = \mu \sum_{i=0}^{k-1} \exp\left(-\frac{e^2(i)}{2\sigma_c^2}\right) e(i)k(r(i)), r(k)) \tag{3.64}$$

As shown in (52), unlike the estimator based on KLMS where the estimate $g(k)$ is given by $\hat{g}(k) = \mu \sum_{i=0}^{k-1} e(i)\kappa(\mathbf{r}(i), \mathbf{r}(k))$ (3.48), the estimate of $g(k)$ with the estimator based on KMC has an additional factor $(\exp(-\frac{e^2(i)}{2\sigma_c^2}))$ which depends upon the instantaneous error $e(k)$. With a suitable choice of σ_c, this additional factor, while dealing with non-Gaussian clutter facilitates the estimator based on KMC to suppress the effect of heavy-tailed outliers.

Further, similar to estimator based on KLMS, at each instant k and for every input–output pair $\{\mathbf{r}(k), d(k)\}$, repeated learning and updating the estimates, increases the computational complexity of KMC-based estimator as shown in (3.64). The computational complexity increases because each incoming observations $\mathbf{r}(k)$ is used for estimation. However, not all the observations are significant for estimation. Hence, to selectively choose the observations and simultaneously reducing the complexity, the sparsification technique based on NC [25], described in Section 3.4.2 is used.

After replacing desired outputs (true parameters) and their corresponding inputs (observations) as $d(k) = {\tau_o}^k, {f_d}^k$ and $\mathbf{r}(k) = \mathbf{r}^k_{\tau_o|f_d}, \mathbf{r}^k_{f_d|\tau_o}$, respectively, the pseudocode for the online estimator based on KMC with sparsification is described in Algorithm 4.

Algorithm 4 KMC algorithm to estimate τ_o and f_d with sparsification

1: **Inputs:**
  $\mathbf{r}(k) = \mathbf{r}^k_{\tau_o|f_d}, \mathbf{r}^k_{f_d|\tau_o} \; \forall\, k, \;\; d(k) = {\tau_o}^k, {f_d}^k \; \forall\, k$

2: **Initialize:**
  choose μ and σ, $\mathbf{r}(0) \leftarrow \mathbf{r}^0_{\tau_o|f_d}, \mathbf{r}^0_{f_d|\tau_o}, e(0) \leftarrow 0,$
  $\hat{g}(0) \leftarrow \kappa(\mathbf{r}(0), \cdot), \quad \mathcal{D}_k \leftarrow \{\}, \;\; \mathbf{a}_0 \leftarrow e(0)$

3: **while** $\{\mathbf{r}(k), d(k)\}$ available **do**

4: $\hat{g}(k) = \mu \sum_{i=0}^{|\mathcal{D}_{k-1}|} \exp(-\frac{e^2(i)}{2\sigma_c^2}) e(i)\kappa(\mathcal{D}^i_{k-1}, \mathbf{r}(k))$

5: $e(k) = d(k) - \hat{g}(k)$

6: $\mathbf{dis}(j) = \left\|\mathbf{r}(k) - \mathcal{D}^j_{k-1}\right\|$ for $0 \le j \le S$

7: **if** $\underbrace{\min}_{0 \le j \le S} \mathbf{dis}(j) \ge \delta_1 \;\&\; |e(k)| \ge \delta_2$ **then**
  $\mathcal{D}_k = \{\mathcal{D}_{k-1} \cup \mathbf{r}(k)\}, \mathbf{a}_k = \{\mathbf{a}_{k-1} \cup e(k)\}$

8: **else**
  $\mathcal{D}_k = \mathcal{D}_{k-1}, \;\; \mathbf{a}_k = \mathbf{a}_{k-1}$

9: **end if**

10: **end while**

3.4.3.3 Estimator based on KMC-modified NC

As both the learning rate and parameter estimates are sensitive to the choice of σ, a suitable kernel width σ (size of the kernel) is a significant parameter for the estimator based on KMC. Consequently, similar to estimators based on KLMS, selecting a proper value of σ is vital for estimators based on KMC. Therefore, in this section, an iterative update equation to learn σ with radar observation ($\mathbf{r}(k)$) is developed.

Since KMC is an online learning algorithm in which the weights in RKHS are updated at each iteration, the kernel width σ can also be updated simultaneously by applying MCC (i.e. maximising the correntropy) [27]. If σ_k is the kernel width at k^{th} iteration then the stochastic update equation for σ_k is given by

$$\sigma_k = \sigma_{k-1} - \eta \frac{\partial}{\partial \sigma_{k-1}} \left(\kappa_{\sigma_c}(e(k))\right), \tag{3.65}$$

where $e(k)$ is the instantaneous estimation error s.t. $e(k) = d(k) - \hat{g}(k)$ and $\kappa_{\sigma_c}(\cdot)$ is the correntropy function.

Since $\kappa_{\sigma_c}(e(k)) = \exp\left(-\frac{e^2(k)}{2\sigma_c^2}\right)$, (3.65) can be written as

$$\sigma_k = \sigma_{k-1} + \frac{\eta}{\sigma_c^2} \exp\left(-\frac{e^2(k)}{2\sigma_c^2}\right) e(k) \frac{\partial}{\partial \sigma_{k-1}} (e(k)). \tag{3.66}$$

Subsequently, $d(k)$ can be written as

$$d(k) = \hat{g}(\mathbf{r}(k)) + u(k), \tag{3.67}$$

where $\hat{g}(\cdot)$ is the estimate of mapping which estimate the delay/Doppler spread from the radar return and $u(k)$ is the approximation error at the kth time instant. If $g_{k-1}(\cdot)$ is the estimated mapping after learning up to k^{th} instant, then error $e(k)$ associated with input $\mathbf{r}(k)$ is given by

$$e(k) = d(k) - g_{k-1}(\mathbf{r}(k)) = \tilde{g}_{k-1}(\mathbf{r}(k)) + u(k), \tag{3.68}$$

where $\tilde{g}_{k-1}(\mathbf{r}(k)) = \hat{g}(\mathbf{r}(k)) - g_{k-1}(\mathbf{r}(k))$.

Substituting (3.68) into (3.66) yields

$$\begin{aligned} \sigma_k &= \sigma_{k-1} + \frac{\mu\eta}{\sigma_c^2} \exp\left(-\frac{e^2(k)}{2\sigma_c^2}\right) e(k) \frac{\partial}{\partial \sigma_{k-1}} \{\tilde{g}_{k-1}(\mathbf{r}(k)) + u(k)\} \\ &= \sigma_{k-1} - \frac{\mu\eta}{\sigma_c^2} \exp\left(-\frac{e^2(k)}{2\sigma_c^2}\right) \exp\left(-\frac{e^2(k-1)}{2\sigma_c^2}\right) e(k)e(k-1) \\ &\times \frac{\partial}{\partial \sigma_{k-1}} [\kappa_{\sigma_{k-1}}(\mathbf{r}(k-1)), \mathbf{r}(k)]. \end{aligned} \tag{3.69}$$

Since $\kappa_{\sigma_{k-1}}(\mathbf{r}(k-1)), \mathbf{r}(k)) = \exp\left(-\frac{\|\mathbf{r}(k-1) - \mathbf{r}^*(k)\|^2}{\sigma_{k-1}^2}\right)$, (3.69) is simplified as

$$\begin{aligned} \sigma_k = \sigma_{k-1} - \rho \exp\left(-\frac{e^2(k)}{2\sigma_c^2}\right) \exp\left(-\frac{e^2(k-1)}{2\sigma_c^2}\right) e(k)e(k-1) \exp\left(-\frac{\|r(k-1) - r^*(k)\|^2}{\sigma_{k-1}^2}\right) \\ \times \frac{\|r(k-1) - r^*(k)\|^2}{\sigma_{k-1}^3} \end{aligned} \tag{3.70}$$

where $\rho = \frac{\mu\eta}{2\sigma_c^2}$ is the kernel width learning parameter.

As shown in (3.70), the kernel width at each iteration k is adaptively updated by the derived expression for learning σ_k and converges to an optimal kernel width. Moreover, as compared to the estimator based on KLMS-modified NC (3.52), the additional factor $(\exp\left(-\frac{e^2(k)}{2\sigma_c^2}\right))$ in (3.70), similar to (3.64), compensates for the effect of non-Gaussian clutter and provides better adaptation of kernel width for the KMC-based estimator.

Pseudocode for the online estimator based on KMC with adaptive kernel width and sparsification is described in Algorithm 5.

Algorithm 5 KMC algorithm to estimate τ_o and f_d with adaptive kernel width and sparsification

Inputs:
$\mathbf{r}(k) = \mathbf{r}^k_{\tau_o|f_d}, \mathbf{r}^k_{f_d|\tau_o} \forall k,\ d(k) = {\tau_o}^k, {f_d}^k\ \forall k$

Initialize:
$e(0) \leftarrow 0,\ \hat{g}(0) \leftarrow \kappa(\mathbf{r}(0), \cdot),\ \mathcal{D}_k \leftarrow \{\},\ \mathbf{a}_k \leftarrow \{\},$
choose $\mu,\ \sigma_0,$ and $\rho,\ \mathbf{r}(0) \leftarrow \mathbf{r}^0_{\tau_o|f_d}, \mathbf{r}^0_{f_d|\tau_o}$

while $\{\mathbf{r}(k), d(k)\}$ available **do**

$\hat{g}(k) = \mu \sum_{i=0}^{|\mathcal{D}_{k-1}|} \exp\left(-\frac{e^2(i)}{2\sigma_c^2}\right) e^*(i)\kappa(\mathcal{D}^i_{k-1}, \mathbf{r}(k))$

$e(k) = d(k) - \hat{g}(k)$

$\mathbf{dis}(j) = \left\|\mathbf{r}(k) - \mathcal{D}^j_{k-1}\right\|$ for $0 \le j \le S$

if $\underbrace{min}_{0\le j\le S}\ \mathbf{dis}(j) \ge \delta_1\ \&\ |e(k)| \ge \delta_2$ **then**

$\mathcal{D}_k = \{\mathcal{D}_{k-1} \cup \mathbf{r}(k)\},\ \mathbf{a}_k = \{\mathbf{a}_{k-1} \cup e(k)\}$

$$\sigma_k = \sigma_{k-1} - \rho \exp\left(-\frac{e^2(k)}{2\sigma_c^2}\right) \exp\left(-\frac{e^2(k-1)}{2\sigma_c^2}\right) e(k)e(k-1) \exp\left(-\frac{\|\mathbf{r}(k-1) - \mathbf{r}^*(k)\|^2}{\sigma^2}\right) \times \frac{\|\mathbf{r}(k-1) - \mathbf{r}^*(k)\|^2}{\sigma_k^3}$$

else

$\mathcal{D}_k = \mathcal{D}_{k-1},\ \mathbf{a}_k = \mathbf{a}_{k-1}$

end if

end while

3.4.3.4 Simulation results for KMC-based estimators

In this section, simulation results for delay τ_o and Doppler shift f_d estimation using the estimators based on KMC are present and discussed in detail. Subsequently, the comparative performance of the KMC-based estimators with the previously discussed KLMS-based estimators and CRLB corresponding to τ_o and f_d is discussed. The readers are advised to refer [6] to see details on the derivation of CRLBs.

The free parameter values of KMC and KLMS algorithm for LFM and SF radar are given in Tables 3.6–3.9. Parameters of LFM radar and SF radar used in simulations are described in Tables 3.10 and 3.11, respectively. Without loss of generality, for the estimation of delay and Doppler shift, the SCR=$\frac{\mathbf{x}^H\mathbf{x}}{\mathrm{tr}(\mathbb{E}(\mathbf{c}^H\mathbf{c}))}$ is fixed at 30 dB and clutter-to-noise ratio (CNR=$\frac{\mathrm{tr}(\mathbb{E}(\mathbf{c}^H\mathbf{c}))}{N\Omega_w^2}$) is fixed at 10 dB [10].

In this work, the effect of non-Gaussian clutter is modelled as described in [6]. Moreover, to consider the effect of impulsive noise and outliers the particular class of heavy-tailed spherically invariant random process (SIRP), K-distributed random process is considered. For this, the random variable α is drawn from gamma distribution with shape and size parameter, as ν and μ_c, respectively. As ν decreases, the tail of the resultant distribution function for clutter becomes heavier, which in turn introduces impulsive noise and outliers. Hence, to consider the effect of high-tailed clutter, simulations are performed for $\nu = 0.1$ as in [39, 40].

3.4.3.5 Estimation of delay with KMC-NC and KMC-modified NC-based estimator

For estimating delay, the set of possible true values of $\tau_o(\mathbb{D})$ is formed by partitioning the interval between $\tau_{o_{min}}$ and $\tau_{o_{max}}$ into K parts, i.e. $\mathbb{D} \in [\tau^1_{o_{min}} \ldots \tau^K_{o_{max}}]$. Thereafter, radar returns (learning and testing samples) corresponding to K different delays are formed as per (3.54). For learning and testing of the estimators based on KMC-NC, KLMS-NC, KMC-modified NC and KLMS-modified NC, the K is chosen as 5 500. Out of these 5 500 radar returns, 5 000 are used for the learning of the estimators while the remaining 500 are used to evaluate the estimators' performance.

In Figure 3.5, for both LFM and SF radar in the presence of clutter and thermal noise, the MSE performance of the estimators based on KMC-NC is compared with the existing adaptive estimator based on KLMS-NC and the two linear estimators based on Kalman filter MCC (MCC-KF) [44] and adaptive MCC (AMCC) [45]. From Figure 3.5, it can be observed that the use of MCC in RKHS facilitates mitigation of non-Gaussian clutter. Particularly, with suitable choice of σ_c for the estimator based on KMC-NC as given in Table 3.6, the additional factor ($\exp\left(-\frac{e^2(i)}{2\sigma_c^2}\right)$) provides robustness against non-Gaussian clutter and yields accurate estimates of delay τ_o as compared to estimator based on KLMS-NC, MCC-KF and AMCC. Also, in Figure 3.5, it can be observed that the dictionary size of the KMC-based and existing estimator achieve respective upper-bounds and consequently result in lower dictionary size for KMC-based estimator as compared to KLMS-NC-based estimator. Furthermore, from Figure 3.5, a drastic reduction in the computational complexity is observed for the KMC-NC-based estimator, which in turn, makes KMC-NC-based estimator practically deployable.

Further, Figure 3.6 compares MSE corresponding to the estimates of delay for both LFM and SF radar system obtained by the estimators based on KMC-modified NC and KLMS-modified NC, respectively. As shown in Figure 3.6, it is observed that the performance of KLMS-modified NC-based estimator is impaired in presence of clutter. On the contrary, for the estimator based on KMC-modified NC, using ITL criteria like MCC and the selection of proper kernel width by adaptive-learning

Table 3.6 Parameter values used for simulations for KMC estimation algorithms for LFM radar

Parameters	KMC-NC $\{\tau_0, f_d\}$ LFM radar	KMC-modified NC $\{\tau_0, f_d\}$ LFM radar
μ	0.55, 0.55	0.55, 0.55
σ	$10^{-5.5}$, $10^{-6.4}$	Adaptive
ρ	Nil, Nil	$10^{7.7}$,$10^{9.2}$
σ_0	Nil, Nil	$10^{-5.4}$, $10^{-6.6}$
σ_c	0.0025, 0.0017	0.0117, 0.012

at every iteration provides a better estimate of delay and, hence, results in lower MSE floor. Further, in Figure 3.6, it is observed that the computational complexity of the estimator based on KMC-modified NC is reaching analytical upper-bound and significantly reduced in comparison to the estimator based on KLMS-modified NC. As shown in Figure 3.6, due to the deleterious effect of clutter on the gradient information (since a stochastic gradient algorithm adapts the kernel width), the KLMS-modified NC converges to higher MSE floor. However, for modified kernel MCC, the said effect is less pronounced due to the robustness of the MCC criterion to non-Gaussian clutter.

Computations required for the convergence of the estimator MSE for KMC-NC and KMC-modified NC for the given LFM radar system model and corresponding to Figures 3.5a and 3.6a are shown in Table 3.12.

3.4.3.6 Estimation of Doppler shift with KMC-NC and KMC-modified NC-based estimator

For estimating Doppler shift, similar to delay estimation, the interval between $f_{d_{max}}$ and $f_{d_{min}}$ is divided in to K parts to form a set of unknown Doppler frequencies as $\mathbb{F} \in [f^1_{d_{min}} \ldots f^K_{d_{max}}]$. Thereafter, the radar return (learning and testing samples) is formed corresponding to each Doppler shift utilising (3.55). The K is chosen as the same for the delay estimation, i.e. $K = 5\,500$, with 5 000 samples used for learning

Table 3.7 Parameter values used for simulations for KMC estimation algorithms for SF radar

Parameters	KMC-NC $\{\tau_0, f_d\}$ SF radar	KMC-modified NC $\{\tau_0, f_d\}$ SF radar
μ	0.55, 0.55	0.55, 0.55
σ	10^{-55}, $10^{-6.1}$	Adaptive
ρ	Nil, Nil	10^{6}, $10^{9.2}$
σ_0	Nil, Nil	$10^{-5.2}$, $10^{-6.4}$
σ_c	0.0025, 0.0036	0.01,0.02

Table 3.8 Parameter values used for simulations for KLMS estimation algorithms for LFM radar

Parameters	KLMS-NC $\{\tau_0, f_d\}$ LFM radar	KLMS-modified NC $\{\tau_0, f_d\}$ LFM radar
μ	0.9460, 0.987	0.982, 0.983
σ	$10^{-5.5}$, $10^{-6.4}$	Adaptive
ρ	Nil, Nil	$10^{7.7}$, $10^{9.2}$
σ_0	Nil, Nil	$10^{-5.4}$, $10^{-6.6}$

Table 3.9 Parameter values used for simulations for KLMS estimation algorithms for SF radar

Parameters	KLMS-NC $\{\tau_0, f_d\}$ SF radar	KLMS-modified NC $\{\tau_0, f_d\}$ SF radar
μ	0.9581, 0.93	0.9735, 0.983
σ	$10^{-5.5}$, $10^{-6.1}$	Adaptive
ρ	Nil, Nil	10^{-6}
σ_0	Nil, Nil	$10^{-5.2}$, $10^{-6.4}$

and the rest 500 samples used for evaluating the MSE performance. For both the LFM and SF radar system, the improvement in the estimation accuracy and computational complexity of the estimator based on KMC-NC for Doppler shift estimation over estimator based on KLMS-NC is illustrated in Figure 3.7, respectively. Similar to delay estimation, improved estimation of Doppler shift is achieved with lower computational complexity, upon incorporation of ITL criteria like MCC.

Moreover, from Figure 3.8, it can be inferred that adaptive learning of the kernel width for KMC-modified NC-based estimator results in lower MSE with reasonable dictionary size. Hence, from Figure 3.8, it can be concluded that the estimator based on KMC-modified NC is viable for Doppler shift estimation as compared to

Table 3.10 Values of radar parameters for LFM radar

LFM radar parameters	Values
Number of fast index (L)	8
Number of pulses (M)	8
Frequency spacing (Δf)	76.25 kHz
Pulse duration (T_0)	0.013×10^{-3} s
Pulse repetition interval (T_p)	16 ms
Bandwidth ($L\Delta f$)	610 MHz
Set of delay true value (T_{0min}, T_{0max})	$(0.0016 \times 10^{-3}, 0.013 \times 10^{-3})$ s
Set of Doppler true value (fd_{min}, fd_{max})	(−30.4, 30.4) kHz

Table 3.11 Values of radar parameters for SF radar

SF radar parameters	Values
Number of fast index (L)	20
Number of pulses (M)	25
Frequency spacing (Δf)	5 kHz
Pulse duration (T_0)	0.1μ sec
Pulse repetition interval (T_p)	$40\mu s$
Bandwidth ($L\Delta f$)	125 MHz
Set of delay true value (T_{0min}, T_{0max})	$(0.0008 \times 10^{-6}, 40 \times 10^{-6})$ sec
Set of Doppler true value (fd_{min}, fd_{max})	(−12.5, 12.5) kHz

KLMS-modified NC for deployment in practical radar systems impaired by clutter and thermal noise.

The following relations $O(|\mathcal{D}_{k-1}^{KMC-NC}|) < O(|\mathcal{D}_{k-1}^{KLMS-NC}|)$ and $O(|\mathcal{D}_{k-1}^{KMC-MNC}|) < O(|\mathcal{D}_{k-1}^{KLMS-MNC}|)$ holds.

3.4.3.7 Comparison of estimators based on KMC and KLMS

For performance-evaluation of the estimators, variance in the estimation of delay and Doppler shift with estimators based on KMC and estimators based on KLMS for both LFM and SF radar is performed at SCR ranges from −20 dB to 30 dB and compared against the derived CRLBs [2]. For simulations, the CNR is fixed at 10 dB.

Next, the performance of estimators based on KMC and KLMS with NC and modified NC for LFM radar and for delay and Doppler shift estimation are shown in Figure 3.9. Particularly, as shown in Figure 3.9, the estimator based on KMC-NC for

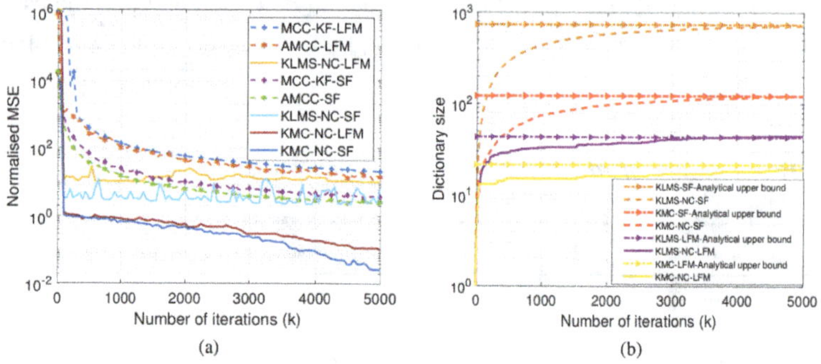

Figure 3.5 (a) Normalised MSE and (b) Dictionary size in the estimation of delay for LFM and SF radar with estimators based on KMC-NC, KLMS-NC, AMCC and MCC-KF in the presence of non-Gaussian clutter

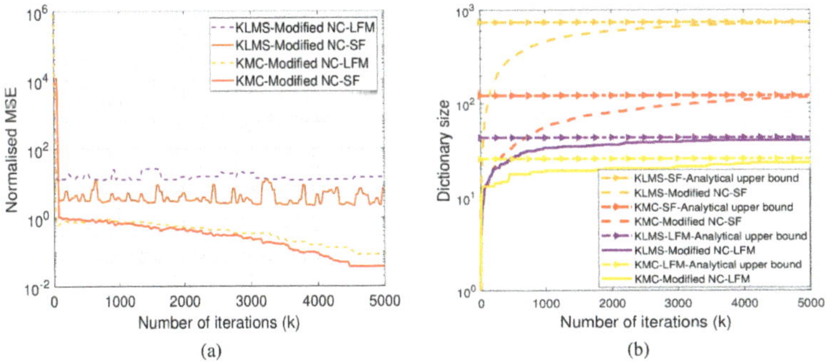

Figure 3.6 (a) Normalised MSE and (b) Dictionary size in the estimation of delay for LFM and SF radar with estimators based on KMC-modified NC and KLMS-modified NC in the presence of non-Gaussian clutter

all SCRs outperforms KLMS-NC-based estimator and achieves variance closer to CRLB. Further, KMC-modified NC-based estimator converges to a variance lower than KLMS-modified NC and closer to CRLB.

Further, in Figure 3.10, for SF radar, the variance of the KMC-NC-based estimator is compared with KLMS-NC for estimation of delay and Doppler shift. From Figure 3.10, it can be observed that the KMC-NC-based estimator converges to lower variance as compared to a corresponding KLMS-NC-based estimator. Further, variance in the estimation of delay and Doppler shift using KMC-NC-based estimator is closer to the achievable CRLB as compared to KLMS-NC. The same trend of lower estimator-variance and increased proximity of the estimator variance to the CRLB is observed for KMC-modified NC-based estimator, as compared to KLMS-modified NC-based estimator.

3.4.4 Kalman filter

Referring to radar return model $r(m, l)$, explained in (3.1), the unknown information corresponds to the range and velocity of the target, denoted by θ_1 and θ_2,

Table 3.12 Computational complexity of estimators based on KMC-NC and KMC-modified NC

Estimation algorithm	Number of multiplications	Complexity order
KLMS-NC	90	$O\left(\left\|D_{k-1}^{KLMS-NC}\right\|\right)$
KMC-NC	20	$O\left(\left\|D_{k-1}^{KMC-NC}\right\|\right)$
KLMS-modified NC	70	$O\left(\left\|D_{k-1}^{KLMS-MNC}\right\|\right)$
KMC-modified NC	10	$O\left(\left\|D_{k-1}^{KMC-MNC}\right\|\right)$

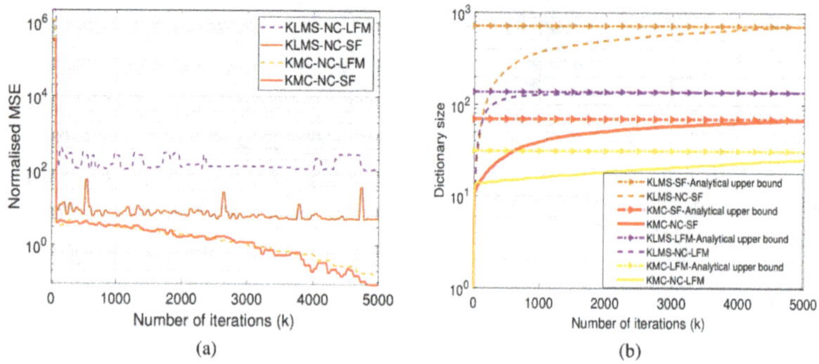

Figure 3.7 (a) Normalised MSE and (b) Dictionary size in estimation of Doppler shift for LFM and SF radar with estimators based on KMC-NC and KLMS-NC in the presence of non-Gaussian clutter

respectively, is assumed constant for processing interval of CPI of M pulses. This follows the assumption of constant target velocity $\nu T << \frac{c}{2B}$, i.e. for $T = MT_{\text{PRI}}$, the target remains in the same range gate. However, for some radar applications like target tracking, where continuous monitoring of the moving and accelerating target is necessary, the above assumption about the target is ruled out [9, 36, 47]. Therefore, for such radar applications, if the target is assumed to move with a velocity s.t. causing the target's Doppler shift to increment by $\left(\frac{1}{T_{\text{PRI}}M}\right)$ which in turn results in increment of delay by $\left(\frac{1}{2B}\right)$, the mathematical model defining the state of the target can be formulated as

$$\theta_{k+1} = f(\theta_k) + \eta_k = \theta_k + \Delta\theta + \eta_k, \quad \text{for } k = [1, 2, \ldots, K], \tag{3.71}$$

$$\theta_{k+1} = f(\theta_k) + \eta_k = \theta_k + \Delta \tag{3.72}$$

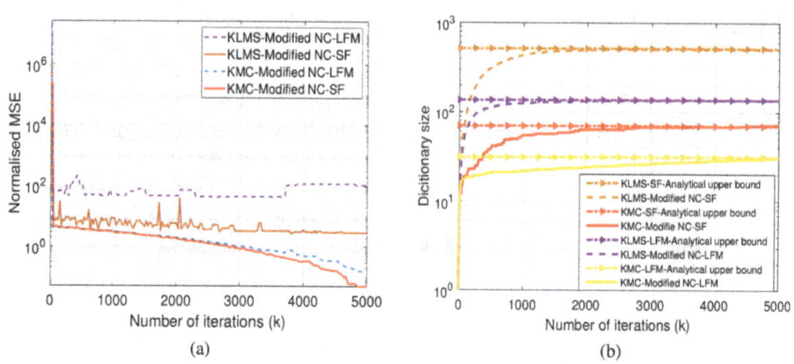

Figure 3.8 (a) Normalised MSE and (b) Dictionary size in estimation of Doppler shift for LFM and SF radar with estimators based on KMC-modified NC and KLMS-modified NC in the presence of non-Gaussian clutter

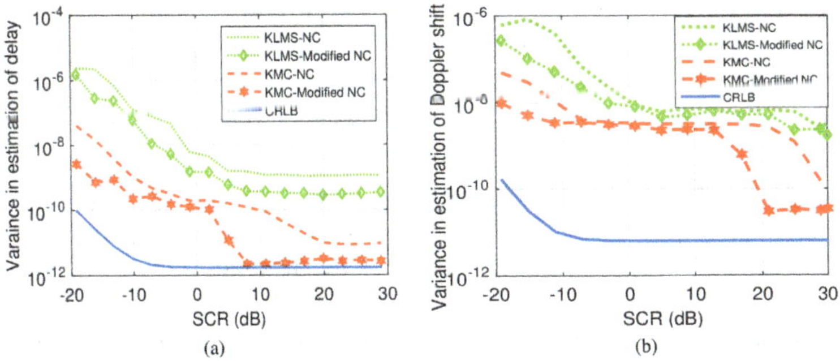

Figure 3.9 Comparative performance of the estimator based on KMC-NC, KLMS-NC, KMC-modified NC and KLMS-modified NC for estimation of (a) delay and (b) Doppler shift for LFM radar system

where $\theta = [\theta_1 \ \theta_2]^T$, $\Delta\theta = [\frac{1}{2B} \ \frac{1}{T_{PRI}M}]^T$ and η_k represents the error in modelling the dynamic state of the target.

Using (3.1) and (3.71) with an assumption of perturbation due to thermal noise, the measurement model defining the radar return corresponding to the state model (3.71) is given by

$$\mathbf{r}_{k+1} = h(\mathbf{x}_{k+1}) + \mathbf{v}_{k+1}, \tag{3.73}$$

where $h(x\_{k+1}) = \begin{bmatrix} \mathbf{Re}\left(\exp(j2\pi m\theta_{k+1}(1))\exp(j2\pi l\theta_{k+1}(2))\right) \\ \mathbf{Im}\left(\exp(j2\pi m\theta_{k+1}(1))\exp(j2\pi l\theta_{k+1}(2))\right) \end{bmatrix}$ and $\mathbf{v}_{k+1}$ represents measurement noise.

The problem of estimating θ_1 and θ_2 utilising $r(m, l)$ is formulated as state space estimation problem consists of state and measurement model as (3.69) and (3.71), respectively. In literature, the celebrated Kalman filter has been effectively used for handling the state space estimation problem [47–50]. Further, in the next Ssection

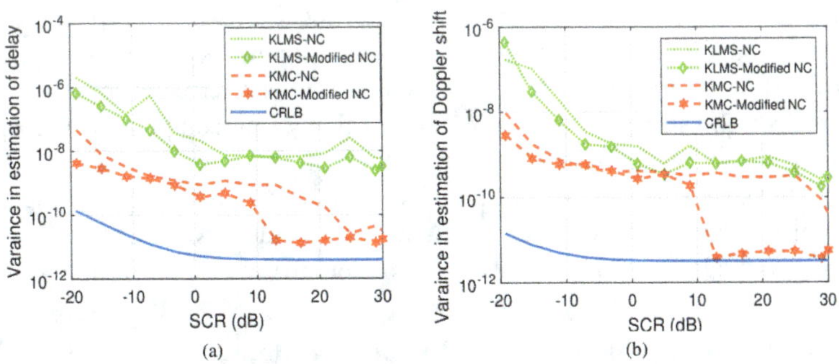

Figure 3.10 Comparative performance of the estimator based on KMC-NC, KLMS-NC, KMC-modified NC and KLMS-modified NC for estimation of (a) delay and (b) Doppler shift for SF radar system

3.4.5, non-linear versions of the Kalman filter called the extended Kalman filter (EKF) and an unscented Kalman filter (UKF) are discussed and found to outperform the conventional estimator based on FT and adaptive estimator based on KLMS. Therefore, next, a brief introduction about the basic working of the Kalman filter is given.

The Kalman filter works in two phases (a) Prediction and (b) Update, over time the filter recursively repeats the prediction and update step until the filter converges to a minimum MSE level. The two phases of the Kalman filter are described as follows:

Prediction: Starting with the initial prediction of the state vector, $\hat{\theta}_o$, and initial state error covariance matrix $\mathbf{P}_o = \mathbb{E}[\hat{\theta}_o\hat{\theta}_o^T]$, the state vector and state error covariance matrix at kth time instant are predicted by

$$\begin{aligned}\hat{\boldsymbol{\theta}}_{k|k-1} &= \mathbf{F}_{k-1}\hat{\boldsymbol{\theta}}_{k-1} + \boldsymbol{\eta}_{k-1}\\ \mathbf{P}_{k|k-1} &= \mathbf{F}_{k-1}\mathbf{P}_{k-1}\mathbf{F}_{k-1}^T + \mathbf{Q}_{k-1},\end{aligned} \tag{3.74}$$

where $\mathbf{F}_{k-1}$ is the state matrix and $\mathbf{Q}_{k-1} = \mathbb{E}[\eta_{k-1}\eta_{k-1}^T]$ is the process covariance matrix.

Update: Subsequently, following prediction, using the Kalman gain matrix, $\mathbf{K}_k = \mathbf{P}_k\mathbf{H}_k^T(\mathbf{H}_k\mathbf{P}_k\mathbf{H}_k^T + \mathbf{R}_k)^{-1}$ (where $\mathbf{H}_k$ is the measurement matrix and $\mathbf{R}_k = \mathbb{E}[\mathbf{v}_k\mathbf{v}_k^T]$ is the measurement noise covariance matrix), the predicted state vector and state error covariance matrix is updated by

$$\begin{aligned}\hat{\theta}_k &= \hat{\theta}_{k|k-1} + K_k(r_k - H_k\hat{\theta}_{k|k-1})\\ P_k &= (I - K_kH_k)P_{k|k-1}.\end{aligned} \tag{3.75}$$

3.4.5 *EKF- and UKF-based estimation of delay and Doppler shift*

In this subsection, the EKF- and UKF-based estimators for τ_o and f_d are described in details. In noisy environments (as in the case of LFM radar system), Bayesian framework-based estimators are most accepted for several decades (refer to [51–53] for detailed discussion). The Bayesian framework is based on state space formulation [51] of the system model, and it is implemented in two steps: prediction and update. Two popular simplifications of Bayesian framework are Gaussian filtering [51] and particle filtering [54]. The Gaussian filters are preferred over particle filters due to their high estimation accuracy at appreciably low computational cost. The EKF and UKF are two popular Gaussian filters.

First, the state space model for the LFM radar system is formulated and a brief discussion on Bayesian parameter estimation is provided. After that, an elaborate discussion of the EKF- and UKF-based estimation of τ_o and f_d is given.

3.4.5.1 State space model for LFM radar system

The state space model consists of state and measurement models, where the state model characterises the state dynamics while the measurement model represents the mathematical relation between the state and the measurement. Note that, the state is defined with the unknown/desired parameters (τ_o and f_d in our case), while the measurement consists of the observed quantities (the returning signal $r_{\text{LFM}}(m, l)$, given in (3.9)). Subsequently, the state and measurement variables are formulated as $\mathbf{x} = [\tau_o \quad f_d]^T$[d] and $\mathbf{r} = [\mathbf{Re}(r_{\text{LFM}}(m, l)) \quad \mathbf{Im}(r_{\text{LFM}}(m, l))]^T$, respectively. The model for targets with constant acceleration is formulated; however, other practical scenarios, like the targets with constant velocity or stationary target, can be modelled as special cases of this model. The change in position is assumed to reflect a constant increase in the delay, and any error due to this assumption is compensated with the process noise. The constant acceleration results in a constant increase in the velocity over a fixed interval, which further results in a constant increase in the Doppler shift. Subsequently, the state model could be designed as an incremental model, i.e.

$$\mathbf{x}_{k+1} = f(\mathbf{x}_k) + \eta_k = \mathbf{x}_k + \Delta\mathbf{x} + \eta_k, \tag{3.76}$$

where $k \in \{1, 2, \cdots, K\}$, $K = ML$ is the total number of discrete samples of returning signal and $\Delta\mathbf{x} = [\frac{T_o}{K}, \frac{1}{T_{\text{PRI}}K}]^T$ is a constant shift in $\mathbf{x}$ between successive samples of returning signal. η_k is additive process noise which models the errors. η_k is assumed to be zero mean Gaussian with covariance Q_k under the Gaussian filtering.

From (3.9), the measurement model can be formulated as

$$\mathbf{r}_{k+1} = h(\mathbf{x}_{k+1}) + \mathbf{v}_{k+1}, \tag{3.77}$$

where v_k represents measurement noise which is added to compensate any error in capturing and/or processing of returning signals and $h(\mathrm{x}_{k+1})$ is given by

$$h(X_{k+1}) = \begin{bmatrix} Re(\zeta \exp(j2\Pi m X_{k+1}(2) T_{PRI}) \exp(-j2\Pi l \Delta f X_{k+1}(1) \exp(j2\Pi X_{k+1}(2) ml(\frac{T_{PRI}\Delta F}{fc}))) \\ Im(\zeta \exp(j2\Pi m X_{k+1}(2) T_{PRI}) \exp(-j2\Pi l \Delta f X_{k+1}(1) \exp(j2\Pi X_{k+1}(2) ml(\frac{T_{PRI}\Delta F}{fc}))) \end{bmatrix} \tag{3.78}$$

Under Gaussian filtering, $\mathbf{v}_k$ is assumed to be zero mean Gaussian with covariance $\mathbf{R}_k$.

3.4.5.2 Bayesian framework for filtering

The Bayesian filtering is performed in two steps:

3.4.5.2.1 Prediction

This step constructs the pdf of states one step forward in time (in reference to the available measurements) using Chapman-Kolmogorov equation [51, 52], i.e.

$$\mathcal{P}(\mathbf{x}_k|\mathbf{r}_{1:k-1}) = \int \mathcal{P}(\mathbf{x}_k|\mathbf{x}_{k-1})\mathcal{P}(\mathbf{x}_{k-1}|\mathbf{r}_{1:k-1})d\mathbf{x}_{k-1}, \tag{3.79}$$

where $\mathcal{P}(\mathbf{x}_k|\mathbf{r}_{1:k-1})$ is commonly known as prior pdf.

3.4.5.2.2 Update

$\mathcal{P}(\mathbf{x}_k|\mathbf{r}_{1:k-1})$

$$P(X_k \mid r_{1:k}) = p(X_k \mid r_{1:k-1}, r_k) = \frac{1}{c_k} p(r_k \mid X_k) p(X_k \mid r_{1:k-1}), \tag{3.80}$$

[d]The ith element of $\mathbf{x}$ is referred to as $\mathbf{x}(i)$.

where $\mathcal{P}(\boldsymbol{r}_k|\mathbf{x}_k)$ is a measurement likelihood which is obtained from (3.75) and c_k is a normalisation constant, i.e.

$$c_k = \mathcal{P}(\mathbf{r}_k|\mathbf{r}_{1:k-1}) = \int \mathcal{P}(\mathbf{r}_k|\mathbf{x}_k)\mathcal{P}(\mathbf{x}_k|\mathbf{r}_{1:k-1})d\mathbf{x}_k. \tag{3.81}$$

The objective of Bayesian filtering is to construct $\mathcal{P}(\mathbf{x}_k|\mathbf{r}_{1:k})$, which is popularly known as posterior pdf.

Hereafter, denoting $\mathcal{P}(\mathbf{x}_k|\mathbf{r}_{1:k-1}) \sim \mathcal{P}(\mathbf{x}_{k|k-1})$ and $\mathcal{P}(\mathbf{x}_k|\mathbf{r}_{1:k}) \sim \mathcal{P}(\mathbf{x}_{k|k})$, which are standard notations used in estimation and filtering literature [51–53].

3.4.5.3 EKF-based estimation of delay and Doppler shift

From the state space model of radar systems ((3.76) and (3.77)), the estimation of τ_o and f_d from returning signal, $r(m, l)$, is simplified as an estimation problem of $\mathbf{x}_k$ from known measurement $\mathbf{r}_k$. A conceptual solution for such a problem is introduced as the Bayesian framework. The EKF is an analytical simplification of Bayesian framework. It assumes the conditional pdfs in the Bayesian framework ((3.79) to (3.81)) as Gaussian, i.e.

$$\mathcal{P}(\mathbf{x}_{k|k-1}) \sim \mathcal{N}_{\mathbb{R}}(\mathbf{x}_{k|k-1}; \hat{\mathbf{x}}_{k|k-1}, \mathbf{P}_{k|k-1}), \tag{3.82}$$

$$\mathcal{P}(\mathbf{x}_{k|k}) \sim \mathcal{N}_{\mathbb{R}}(\mathbf{x}_{k|k}; \hat{\mathbf{x}}_{k|k}, \mathbf{P}_{k|k}), \tag{3.83}$$

where $\hat{\mathbf{x}}_{k|k}$ and $\mathbf{P}_{k|k}$ are mean and covariance of $\mathbf{x}_{k|k}$. Subsequently, the problem is further simplified to determine $\hat{\mathbf{x}}_{k|k-1}$ and $\mathbf{P}_{k|k-1}$ in the prediction step and $\hat{\mathbf{x}}_{k|k}$ and $\mathbf{P}_{k|k}$ in the update step. The computational aspect of the two steps is discussed herewith.

3.4.5.3.1 Prediction

$\hat{\mathbf{x}}_{k|k-1}$and $\mathbf{P}_{k|k-1}$ are obtained from the first and second moments of $\mathbf{x}$, with pdf of $\mathbf{x}$ given in (68). Simplified expressions for $\hat{\mathbf{x}}_{k|k-1}$ and $\mathbf{P}_{k|k-1}$ are given as [51, 52]

$$\hat{\mathbf{x}}_{k|k-1} = f(\hat{\mathbf{x}}_{k-1|k-1}), \tag{3.84}$$

$$\mathbf{P}_{k|k-1} = \mathbf{F}_k\mathbf{P}_{k-1|k-1}\mathbf{F}_k^T + \mathbf{Q}_k, \tag{3.85}$$

where

$$\mathbf{F}_k = \left.\frac{\partial f(\mathbf{x})}{\partial \mathbf{x}}\right|_{\mathbf{x}=\hat{\mathbf{x}}_{k-1|k-1}} \tag{3.86}$$

is Jacobian of $f(\mathbf{x})$ computed at $\hat{\mathbf{x}}_{k-1|k-1}$. The detailed discussion is given in [51–53].

3.4.5.3.2 Update

In the update step, the predicted estimate and covariance, $\hat{\mathbf{x}}_{k|k-1}$ and $\mathbf{P}_{k|k-1}$, are corrected using the information received from new measurement. The correction step requires the statistical information about predicted measurement, which is obtained in the following steps.

- The predicted measurement is obtained as

$$\hat{\mathbf{r}}_{k|k-1} = h(\hat{\mathbf{x}}_{k|k-1}). \tag{3.87}$$

- The error covariance of predicted measurement is obtained as

$$\mathbf{P}^{\mathbf{rr}}_{k|k-1} = \mathbf{H}_k \mathbf{P}_{k-1|k-1} \mathbf{H}_k^T + \mathbf{R}_k, \tag{3.88}$$

where

$$\mathbf{H}_k = \left.\frac{\partial h(\mathbf{x})}{\partial \mathbf{x}}\right|_{\mathbf{x}=\hat{\mathbf{x}}_{k|k-1}}$$

is Jacobian of $h(\mathbf{x})$ computed at $\hat{\mathbf{x}}_{k|k-1}$. Based on the error covariance of the predicted state and measurement, a Kalman gain is computed as

$$\mathbf{K}_k = \mathbf{P}_{k|k-1} \mathbf{H}_k^T (\mathbf{P}^{\mathbf{rr}}_{k|k-1})^{(-1)}. \tag{3.89}$$

On the receipt of a new measurement r_k, the desired estimate and covariance, $\hat{x}_{k|k}$ and $P_{k|k}$, are obtained as

$$\hat{X}_{k|k} = \hat{X}_{k|k-1} + K_k(r_k - \hat{r}_{k|k-1}) \tag{3.90}$$

$$\hat{\mathbf{P}}_{k|k} = \hat{\mathbf{P}}_{k|k-1} - \mathbf{K}_k \mathbf{P}^{\mathbf{rr}}_{k|k-1} \mathbf{K}_k^T. \tag{3.91}$$

For a detailed discussion, please refer [51–53].

The posterior estimate $\hat{\mathbf{x}}_{k|k} = [\hat{\tau}_{o_k} \quad \hat{f}_{d_k}]^T$ provides the desired estimate of delay and Doppler shift. The steps involved in EKF-based estimation of delay and Doppler shift are summarised in Algorithm 6.

3.4.5.4 UKF-based estimation of delay and Doppler shift

The UKF [48, 55, 56] uses a derivative-free implementation for estimating $\mathbf{x}_k$ from known measurement $\mathbf{r}_k$, unlike the EKF. Moreover, it propagates the estimate and covariance through a true system model instead of propagating them through a derivative-based locally approximated model. Note that, the objective of Gaussian filtering is to obtain the estimate and covariance of states for characterising the Gaussian pdfs; $\hat{\mathbf{x}}_{k|k-1}$ and $\mathbf{P}_{k|k-1}$ in the prediction step and, $\hat{\mathbf{x}}_{k|k}$ and $\mathbf{P}_{k|k}$ in the update step. The estimate and covariance are obtained from the first and second moments. For systems with Gaussian assumption on conditional pdfs, the moment computation involves an integral of the form '$\int_{-\infty}^{\infty}$ *arbitrary function*$\times$ *Gaussian pdf*' [57, 58]. The integrals of this form are generally intractable [57, 58], therefore an analytical solution does not exist. The UKF numerically approximates the intractable integrals using unscented transformation-based numerical approximation [48, 55].

Algorithm 6 Estimation of delay and Doppler shift using EKF

Input: $f : \mathbf{x}_k \rightarrow \mathbf{x}_{k-1}$, $h : \mathbf{x}_k \rightarrow \mathbf{r}_k$, $\mathbf{Q}_k$, and $\mathbf{R}_k$
Output: $\hat{\mathbf{x}}_{k|k}$
Initialisation: $\hat{\mathbf{x}}_{0|0}$, $\mathbf{P}_{0|0}$
while $k \leq K$ **do**
  Derivative computation for state dynamics $\mathbf{F}_k = \left.\frac{\partial f(\mathbf{x})}{\partial \mathbf{x}}\right|_{\mathbf{x}=\hat{\mathbf{x}}_{k-1|k-1}}$
  Prediction
  $\hat{\mathbf{x}}_{k|k-1} = f(\hat{\mathbf{x}}_{k-1|k-1})$
  $\mathbf{P}_{k|k-1} = \mathbf{F}_k \mathbf{P}_{k-1|k-1} \mathbf{F}_k^T + \mathbf{Q}_k$
  Derivative computation for measurement model $\mathbf{H}_k = \left.\frac{\partial h(\mathbf{x})}{\partial \mathbf{x}}\right|_{\mathbf{x}=\hat{\mathbf{x}}_{k|k-1}}$
  Update
  $\hat{\mathbf{r}}_{k|k-1} = h(\hat{\mathbf{x}}_{k|k-1})$
  $\mathbf{P}^{\mathbf{rr}}_{k|k-1} = \mathbf{H}_k \mathbf{P}_{k-1|k-1} \mathbf{H}_k^T + \mathbf{R}_k$
  $\mathbf{K}_k = \mathbf{P}_{k|k-1} \mathbf{H}_k^T (\mathbf{P}^{\mathbf{rr}}_{k|k-1})^{(-1)}$
  $\hat{\mathbf{x}}_{k|k} = \hat{\mathbf{x}}_{k|k-1} + \mathbf{K}_k(\mathbf{r}_k - \hat{\mathbf{r}}_{k|k-1})$
  $\hat{\mathbf{P}}_{k|k} = \hat{\mathbf{P}}_{k|k-1} - \mathbf{K}_k \mathbf{P}^{\mathbf{rr}}_{k|k-1} \mathbf{K}_k^T$
end while
Return $\hat{\mathbf{x}}_{k|k}$

The unscented transformation generates a set of $2n + 1$ symmetrically distributed sigma points, with n being the system dimension which is two in the considered problem. It also generates a set of $2n + 1$ weights associated with sigma points. Considering $\hat{\mathbf{x}}$ and $\mathbf{P}$ be the estimate and covariance of random variable $\mathbf{x}$, respectively, the set of sigma points, ξ, can be obtained as [48, 56]

$$\begin{aligned} &\xi_0 = \hat{X}, \\ &\xi_i = \hat{X} + (\sqrt{(n+k)P})_i, \xi_{n+i} = \hat{X} - (\sqrt{(n+k)P})_i \end{aligned} \tag{3.92}$$

where $i = 1, 2, \cdots, n$, κ is a constant (practitioner's choice) and $(\sqrt{(n+k)P})_i$ represents ith column of $(\sqrt{(n+k)P})$. A preferred value of κ is $3 - n$, i.e. $n + k = 3$. The weights are generated as [48, 56]

$$\begin{aligned} &W_0 = k/(n+k) \\ &W_i = W_{n+i} = 1/(2(n+k)) \end{aligned} \tag{3.93}$$

As in (3.88), the sigma points depend on the distribution of random variable (characterised by $\hat{\mathbf{x}}$ and P) which is different for prediction and update steps. Therefore, the UKF generates two different sets of sigma points: $\xi_{k-1|k-1}$ in the prediction step and $\xi_{k|k-1}$ in the update step. $\xi_{k-1|k-1}$ is generated with $\hat{\mathbf{x}} = \hat{\mathbf{x}}_{k-1|k-1}$ and $\mathbf{P} = \mathbf{P}_{k-1|k-1}$, while $\xi_{k|k-1}$ is generated with $\hat{\mathbf{x}} = \hat{\mathbf{x}}_{k|k-1}$ and $\mathbf{P} = \mathbf{P}_{k|k-1}$. The weights are independent of the distribution of a random variable, and hence they are same for both the prediction and update steps. The computational aspect of prediction and update steps for the UKF is discussed herewith.

3.4.5.4.1 Prediction

The predicted estimate and covariance, $\hat{X}_{k|k-1}$and $P_{k|k-1}$, are obtained as

$$\hat{X}k \mid k-1 = \sum_{j=1}^{N_s} W_j \xi^f_{j,k-1|k-1}, \tag{3.94}$$

$$\mathbf{P}_{k|k-1} = \sum_{j=1}^{N_s} \mathbf{W}_j(\xi^f_{j,k-1|k-1} - \hat{\mathbf{x}}_{k|k-1})(\xi^f_{j,k-1|k-1} - \hat{\mathbf{x}}_{k|k-1})^T + \mathbf{Q}_k, \tag{3.95}$$

Algorithm 7 Estimation of delay and Doppler shift using UKF

Input: $f : \mathbf{x}_k \rightarrow \mathbf{x}_{k-1}$, $h : \mathbf{x}_k \rightarrow \mathbf{r}_k$, $\mathbf{Q}_k$, and $\mathbf{R}_k$
Output: $\hat{\mathbf{x}}_{k|k}$
Initialisation: $\hat{\mathbf{x}}_{0|0}$, $\mathbf{P}_{0|0}$
while $k \leq K$ **do**
  Prediction
  $\xi^f_{j,k-1|k-1} = f(\xi_{j,k-1|k-1})$
  $\hat{\mathbf{x}}_{k|k-1} = \sum_{j=1}^{N_s} \mathbf{W}_j \xi^f_{j,k-1|k-1}$

$$\mathbf{P}_{k|k-1} = \sum_{j=1}^{N_s} \mathbf{W}_j(\xi^f_{j,k-1|k-1} - \hat{\mathbf{x}}_{k|k-1})(\xi^f_{j,k-1|k-1} - \hat{\mathbf{x}}_{k|k-1})^T + \mathbf{Q}_k$$

  Update
  $\xi^h_{j,k|k-1} = h(\xi_{j,k|k-1})$
  $\hat{\mathbf{r}}_{k|k-1} = \sum_{j=1}^{N_s} \mathbf{W}_j \xi^h_{j,k|k-1}$

$$\mathbf{P}^{\mathbf{rr}}_{k|k-1} = \sum_{j=1}^{N_s} \mathbf{W}_j(\xi^h_{j,k|k-1} - \hat{\mathbf{r}}_{k|k-1})(\xi^h_{j,k|k-1} - \hat{\mathbf{r}}_{k|k-1})^T + \mathbf{R}_k$$

$$\mathbf{P}^{\mathbf{xr}}_{k|k-1} = \sum_{j=1}^{N_s} W_j(\xi_{j,k|k-1} - \hat{\mathbf{x}}_{k|k-1})(\xi^h_{j,k|k-1} - \hat{\mathbf{r}}_{k|k-1})^T$$

  $\mathbf{K}_k = \mathbf{P}^{\mathbf{xr}}_{k|k-1}(\mathbf{P}^{\mathbf{rr}}_{k|k-1})^{-1}$
  $\hat{\mathbf{x}}_{k|k} = \hat{\mathbf{x}}_{k|k-1} + \mathbf{K}_k(\mathbf{r}_k - \hat{\mathbf{r}}_{k|k-1})$
  $\mathbf{P}_{k|k} = \mathbf{P}_{k|k-1} - \mathbf{K}_k \mathbf{P}^{\mathbf{rr}}_{k|k-1} \mathbf{K}_k^T$
end while
Return $\hat{\mathbf{x}}_{k|k}$

$$\mathbf{P}_{k|k-1} = \sum_{j=1}^{N_s} \mathbf{W}_j(\xi^f_{j,k-1|k-1} - \hat{\mathbf{x}}_{k|k-1})(\xi^f_{j,k-1|k-1} - \hat{\mathbf{x}}_{k|k-1})^T + \mathbf{Q}_k$$

$$\mathbf{P}^{\mathbf{rr}}_{k|k-1} = \sum_{j=1}^{N_s} \mathbf{W}_j(\xi^h_{j,k|k-1} - \hat{\mathbf{r}}_{k|k-1})(\xi^h_{j,k|k-1} - \hat{\mathbf{r}}_{k|k-1})^T + \mathbf{R}_k$$

$$\mathbf{P}^{\mathbf{xr}}_{k|k-1} = \sum_{j=1}^{N_s} W_j(\xi_{j,k|k-1} - \hat{\mathbf{x}}_{k|k-1})(\xi^h_{j,k|k-1} - \hat{\mathbf{r}}_{k|k-1})^T$$

where,

$$\xi^f_{j,k-1|k-1} = f(\xi_{j,k-1|k-1}). \tag{3.99}$$

3.4.5.4.2 Update

The computation of updated estimate and covariance, $\hat{\mathbf{x}}_{k|k}$ and $\mathbf{P}_{k|k}$, is based on statistical information on predicted measurement, which is obtained in the following steps:

- The estimate of predicted measurement is obtained as

$$\hat{\mathbf{r}}_{k|k-1} = \sum_{j=1}^{N_s} \mathbf{W}_j \xi^h_{j,k|k-1}, \tag{3.100}$$

where

$$\xi^h_{j,k|k-1} = h(\xi_{j,k|k-1}). \tag{3.101}$$

- The error covariance of predicted measurement is obtained as

$$\mathbf{P}^{\mathbf{rr}}_{k|k-1} = \sum_{j=1}^{N_s} \mathbf{W}_j(\xi^h_{j,k|k-1} - \hat{\mathbf{r}}_{k|k-1})(\xi^h_{j,k|k-1} - \hat{\mathbf{r}}_{k|k-1})^T + \mathbf{R}_k. \tag{3.102}$$

- The cross-covariance between the predicted state and measurement is computed as

$$\mathbf{P}^{\mathbf{xr}}_{k|k-1} = \sum_j W_j(\xi_{j,k|k-1} - \hat{\mathbf{x}}_{k|k-1})(\xi^h_{j,k|k-1} - \hat{\mathbf{r}}_{k|k-1})^T. \tag{3.103}$$

Based on $\mathbf{P}^{\mathbf{rr}}_{k|k-1}$ and $\mathbf{P}^{\mathbf{xr}}_{k|k-1}$, the Kalman gain is obtained as

$$\mathbf{K}_k = \mathbf{P}^{\mathbf{xr}}_{k|k-1}(\mathbf{P}^{\mathbf{rr}}_{k|k-1})^{-1}. \tag{3.104}$$

The desired parameters, $\hat{\mathbf{x}}_{k|k}$ and $\mathbf{P}_{k|k}$, are obtained by correcting the predicted estimate and covariance on the receipt of new measurement $\mathbf{r}_k$. The correction is based on Kalman gain and obtained as

$$\hat{\mathbf{x}}_{k|k} = \hat{\mathbf{x}}_{k|k-1} + \mathbf{K}_k(\mathbf{r}_k - \hat{\mathbf{r}}_{k|k-1}),$$

$$P_{|} = Px_{|1} - K_k P^{rr}_{|1} K^T_k. \tag{3.106}$$

The estimate $\hat{\mathbf{x}}_{k|k} = [\hat{\tau}_{o_k} \quad \hat{f}_{d_k}]^T$ provides the desired estimate of delay and Doppler shift. The estimation algorithm based on UKF is summarised in Algorithm 7.

3.4.5.5 Comparison between EKF and UKF

EKF is an early development using filtering under the Bayesian framework. As discussed, its implementation involves derivative-based computation, which causes several limitations, like smoothness requirement for system models and poor stability. Though it outperforms the KLMS-modified NC-based estimator and other estimators used in radar systems, it has certain limitations. For instance, the derivative requires a smooth system model; however, it is not guaranteed in the radar systems. Moreover, the propagation of estimate and covariance through locally approximated system models leaves scope for further improvement. Despite all the limitations, it attracts practitioners due to fast computation and implementation simplicity, especially in applications where a small shift in estimation accuracy does not affect the decisiveness [59–61].

UKF offers a derivative-free implementation, which is based on numerical approximation. Due to derivative-free implementation, it shows better stability in comparison to the EKF. It also precludes the approximation due to derivative-based implementation, thus outperforms the EKF in terms of estimation accuracy, especially in complex environments [62–64].

3.4.5.6 Simulation results for EKF- and UKF-based estimators

In this subsection, the performance of the EKF- and UKF-based estimation techniques is validated over LFM radar system and a comparative analysis with existing estimators based on KLMS-modified NC and FT is discussed. The algorithm is implemented over two mono-static LFM radar systems having different parameter values. The parameter values are shown in Table 3.13, where Scenario I [2] and Scenario II [65] refer to the two radar systems. The estimation of delay and Doppler shift is obtained for SNR = 0 dB; however, the comparative analysis is provided for various SNR ranging from −50 dB to 0 dB.

3.4.5.7 Estimation of delay and Doppler shift using estimator-based on EKF and UKF

The EKF- and UKF-based estimators were implemented with simulated data obtained using (3.76) and (3.76) over 5 000 sampling intervals. The true data of states (obtained from (3.76)) are used as reference values for comparison. The NMSE corresponding to the delay and Doppler shift is obtained by implementing $M_c = 100$ Monte Carlo executions, which is given by

$$NMSE_k(i) = \frac{1}{M_c}\sum_{mc=1}^{M_c}\left(\frac{\mathbf{x}_k^{mc}(i)-\hat{\mathbf{x}}_k^{mc}(i)}{\mathbf{x}_k^{mc}(i)}\right)^2, \tag{3.107}$$

where i is the index corresponding to delay or Doppler shift. The NMSEs obtained from different estimators are shown in Figures 3.11 and 3.12 for Scenarios I and II, respectively. The figures show a reduced NMSE as well as a faster convergence for the EKF- and UKF-based estimators compared to the KLMS-modified NC. The reduced NMSE concludes an improved accuracy in estimation of delay and Doppler shift with the estimation techniques. The figures also conclude a relatively better accuracy for the UKF compared to the EKF. Also, as shown in Table 3.14, the relative computational complexity of EKF and UKF is lower than KLMS-modified-NC and FT as $n^3 \ll ML$. Furthermore, the computational time of UKF was observed to be 1.7 times higher as compared to the EKF which reflects an increased computational burden as shown in Table 3.14.

3.4.5.8 Performance analysis with varying SNR

Accuracy of the estimation techniques for various SNRs is evaluated in terms of error variance. The error variance at k^{th} instant is given as

$$\Omega_k^2(i) = \frac{1}{M_c}\sum_{mc=1}^{M_c}\left(\mathbf{x}_k^{mc}(i) - \hat{\mathbf{x}}_k^{mc}(i)\right)^2. \tag{3.108}$$

Table 3.13 LFM radar values of Scenario I [2] and Scenario II [65] used for simulations

Scenario I	Values	Scenario II	Values
Number of pulses (M)	10	Number of pulses (M)	20
Number of frequency intervals (L)	500	Number of frequency intervals (L)	500
Frequency increment (Δf)	10 MHz	Frequency increment (Δf)	10 MHz
Pulse duration (T_0)	5 μs	Pulse duration (T_0)	200 μs
Pulse repetition interval (T_{PRI})	1 ms	Pulse repetition interval (T_{PRI})	0.4 ms
Centre frequency (f_c)	10 GHz	Centre frequency (f_c)	9 GHz

The error variance in the estimation of delay and Doppler shift is evaluated at various SNRs ranging from −50 dB to 0 dB. The variances are compared with the achievable analytical CRLBs for each of the delay and Doppler shift. The variances obtained from the EKF, UKF, KLMS-modified-NC and FT are shown in Figures 3.13 and 3.14. As shown in the figures, the variances obtained with the EKF and UKF are closer to the achievable CRLB in comparison to the KLMS-modified-NC and FT. Moreover, the figures validate a marginally better accuracy for the UKF compared to the EKF, though the computational time is increased.

Summary

In this chapter, first, KLMS-based non-linear estimation algorithms are discussed for LFM and OFDM radar systems. The estimation of the target's unknown parameters is performed via an implicit mapping to RKHS. To facilitate sparse learning without affecting estimator performance, Platt's NC is used to limit the increasing size of training samples. Additionally, a technique is explored for tuning the hyper-parameter σ from observations and is found to be suitable for both LFM and OFDM radar systems in terms of MSE and computational complexity. Moreover, optimisation of the convex cost function over an RKHS makes the KLMS-based estimators viable for generalised radar models perturbed by Gaussian noise. From the simulations, it is observed that the variance of the estimates corresponding to the estimators based on KLMS is lower than the existing non-adaptive estimation techniques like FT and is closer to the achievable CRLB.

Subsequently, in this chapter, two new KMC-based non-linear estimation algorithms are discussed for estimation of target's delay and Doppler shift in the presence of non-Gaussian clutter. Estimation of the target's unknown parameters is performed by maximising a cost function (correntropy) in RKHS, which provides an accurate estimate of delay and Doppler shift in the presence of non-Gaussian clutter. Further, to facilitate sparse learning and for lowering computational complexity without affecting the estimator performance, Platt's NC is used to limit the

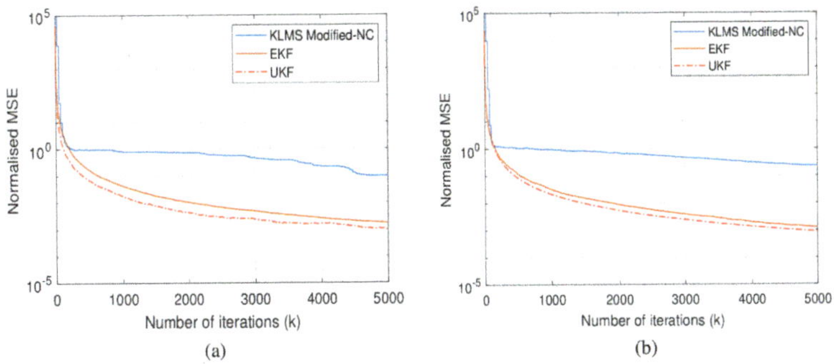

Figure 3.11 NMSE plots of (a) delay and (b) Doppler shift estimation using estimators based on KLMS-modified-NC, UKF and EKF for Scenario I

increasing size of training samples. Additionally, a technique is explored for tuning the hyper-parameter σ from radar returns and an adaptive update equation is derived for its convergence to an appropriate value. Simulations performed over realistic LFM and SF radar systems reveal that the KMC-based estimators in the presence of non-Gaussian clutter, facilitate normalised MSE gain over existing KLMS-based estimators along with lower computational complexity.

Lastly, this chapter introduces two new non-linear estimation techniques, based on EKF and UKF, for both delay and Doppler shift estimation in radar systems. The target localisation is based on the delay and Doppler shift in the returning signal, which can be estimated stochastically. The recent kernel-based approaches provide a comparatively accurate estimate than conventional methods. However, they are based on model parameters specific to target dynamics (which vary in practical problems). Also, they ignore the possibility of uncertainties in the modelling of radar systems. It is shown that EKF- and UKF-based estimators outperform the

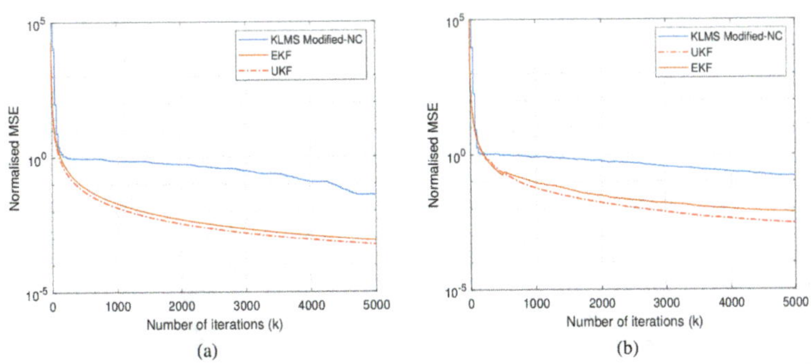

Figure 3.12 NMSE plots of (a) delay and (b) Doppler shift estimation using estimators based on KLMS-modified-NC, UKF and EKF for Scenario II

Table 3.14 Computational complexity of estimators based on FT, KLMS-modified-NC, UKF and EKF

Estimators	Computational complexity
FT	$O(ML \log_2 ML)$
KLMS-modified-NC	$O(ML)$
UKF	$O(n^3)$
EKF	$O(n^2)$

existing estimators in delay and Doppler shift estimation. Additionally, it is also demonstrated that the UKF-based estimator marginally outperforms the EKF, however, at increased computational burden.

The chapter describes three non-linear estimators for the estimation of the target's range and velocity in Gaussian and non-Gaussian environment. All of the described estimators are iterative and capable of providing the online estimates of required parameters. The choice of estimators depends on the type of surveillance environment and application. For instance, if one is seeking for the range and velocity estimation of the target in a Gaussian environment, undoubtedly, KLMS-based estimators are the preferred choice over conventional estimator based on KLMS. On the other hand, if the estimation of target's parameters is required in an environment affected by clutter, estimators based on KMC would be the preferred choice over estimators based on KLMS. In the Gaussian environment, although the estimators based on KLMS provide the efficient estimates of required parameters in comparison to FT, they suffer from a disadvantage of longer convergence time and parameter modelling errors. Therefore, to achieve better estimation accuracy with

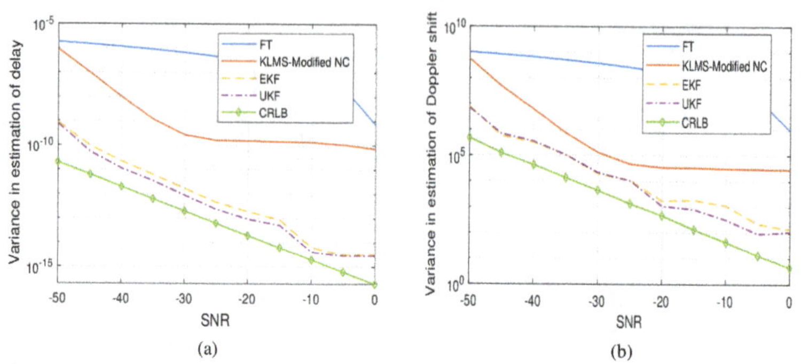

Figure 3.13 Variance in the estimation of (a) delay and (b) Doppler shift using estimation techniques based on UKF, EKF, KLMS-modified-NC and FT for Scenario I

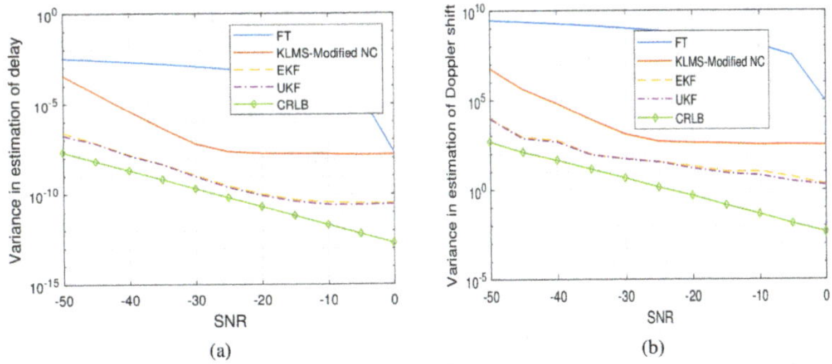

Figure 3.14 *Variance in the estimation of (a) delay and (b) Doppler shift using estimation techniques based on UKF, EKF, KLMS-modified-NC and FT for Scenario II*

low convergence time, EKF and UKF would be the preferred choice over estimators based on KLMS and FT in Gaussian environments.

In future, the other variant of the estimator based on KAF can be explored for the estimation in the multiple-input multiple-output radar system. Subsequently, the other variant of Kalman filter can also be explored for the estimation of parameters in a non-Gaussian environment.

References

[1] Kay S.M. '*Fundamentals of Statistical Signal Processing*'. *Prentice Hall PTR*. 1993.

[2] Abatzoglou T.J., Gheen G.O. 'Range, radial velocity, and acceleration MLE using radar LFM pulse train'. *IEEE Transactions on Aerospace and Electronic Systems*. 1998, vol. 34(4), pp. 1070–83.

[3] Singh U.K., Bhatia V., Mishra A.K. 'Multiple target detection and estimation of range and Doppler for OFDM-radar system'. International Conference on Signal Processing and Integrated Networks (SPIN); 2017. pp. 27–32.

[4] Singh U.K., Mitra R., Bhatia V., Mishra A.K. 'Kernel LMS-based estimation techniques for radar systems'. *IEEE Transactions on Aerospace and Electronic Systems*. 2019, vol. 55(5), pp. 2501–15.

[5] Singh U.K., Mitra R., Bhatia R. 'Target range estimation in OFDM radar system via kernel least mean square technique'. IET Radar Conference; 2017.

[6] Singh U.K., Mitra R., Bhatia V., Mishra A.K. 'Range and velocity estimation using kernel maximum correntropy based nonlinear estimators in non-Gaussian clutter'. *IEEE Transactions on Aerospace and Electronic Systems*. 2020, vol. 56(3), pp. 1992–2004.

[7] Lellouch G., Mishra A.K. 'Multi-carrier based radar signal optimization using genetic algorithm'. *Springer International Conference on Soft Computing for Problem Solving*; 2014. pp. 525–34.

[8] Lellouch G., Mishra A., Inggs M. 'Impact of the Doppler modulation on the range and Doppler processing in OFDM radar'. IEEE Radar Conference; 2014–0803–8.

[9] Richards M.A. '*Fundamentals of radar signal processing*'. *Tata McGraw-Hill Education*. 2005.

[10] Sen S., Nehorai A. 'Target detection in clutter using adaptive OFDM radar'. *IEEE Signal Processing Letters*. 2009, vol. 16(7), pp. 592–5.

[11] Sen S., Nehorai A. 'Adaptive OFDM radar for target detection in multipath scenarios'. *IEEE Transactions on Signal Processing*. 2011, vol. 59(1), pp. 78–90.

[12] Sen S., Nehorai A. 'Adaptive design of OFDM radar signal with improved wideband ambiguity function'. *IEEE Transactions on Signal Processing*. 2010, vol. 58(2), pp. 928–33.

[13] Singh U., Bhatia V., Mishra A. 'Delay and Doppler shift estimation for non-constant envelope modulation in OFDM radar system'. *IET Radar Conference*; 2017.

[14] Rife D., Boorstyn R. 'Single tone parameter estimation from discrete-time observations'. *IEEE Transactions on Information Theory*. 1974, vol. 20(5), pp. 591–8.

[15] Singh U.K., Bhatia V., Mishra A.K. 'Small boat detection using OFDM radar'. *Radio Engineering*. 2019, vol. 28(4), pp. 765–75.

[16] Singh U.K., Mitra R., Bhatia V., Mishra A.K. 'Low-complexity complex KLMS based non-linear estimators for OFDM radar system'. International Conference on Advanced Networks and Telecommunications Systems (ANTS); 2018–1–6.

[17] Singh U.K., Mitra R., Bhatia V., Mishra A.K. 'Vector-valued KLMS based multiple target range and velocity estimation using IEEE 802.11p waveform for autonomous vehicle'. International Conference on Advanced Networks and Telecommunications Systems; 2019–1–6.

[18] Liu W., Pokharel P.P., Principe J.C. 'The kernel least-mean-square algorithm'. *IEEE Transactions on Signal Processing*. 2008, vol. 56(2), pp. 543–54.

[19] Liu W., Principe J.C., Haykin S. *Kernel Adaptive Filtering: A Comprehensive Introduction*. 57. John Wiley & Sons; 2011.

[20] Mitra R., Bhatia V. 'Finite dictionary variants of the diffusion KLMS algorithm'. 2015.

[21] Mitra R., Bhatia V. 'Finite dictionary techniques for MSER equalization in RKHS'. *Signal, Image and Video Processing*. 2017, vol. 11(5), pp. 849–56.

[22] Braun M., Sturm C., Jondral F.K. 'Maximum likelihood speed and distance estimation for OFDM radar'. IEEE Radar Conference; 2010–256–61.

[23] Mitra R., Bhatia V. 'Low complexity post-distorter for visible light communications'. *IEEE Communications Letters*. 2017, vol. 21(9), pp. 1977–80.

[24] Chen B., Zhao S., Zhu P., Príncipe J.C. 'Quantized kernel least mean square algorithm'. *IEEE Transactions on Neural Networks and Learning Systems*. 2012, vol. 23(1), pp. 22–32.

[25] Platt J. 'A resource-allocating network for function'. *Neural computation*. 1991, vol. 3(2), pp. 213–25.

[26] Silverman B.W. '*Density Estimation for Statistics and Data Analysis*'. *CRC press*. 1986, vol. 26.

[27] Chen B., Liang J., Zheng N., Príncipe J.C. 'Kernel least mean square with adaptive kernel size'. *Neurocomputing*. 2016, vol. 191(1), pp. 95–106.

[28] Bouboulis P., Theodoridis S. 'Extension of Wirtinger's calculus to reproducing kernel Hilbert spaces and the complex kernel LMS'. *IEEE Transactions on Signal Processing*. 2011, vol. 59(3), pp. 964–78.

[29] Mitra R., Bhatia V. 'The diffusion-KLMS algorithm'. Information Technology (ICIT), 2014 International Conference; 2014–256–9.

[30] Bhatia V., Mulgrew B., Falconer D.D. 'Non-parametric maximum-likelihood channel estimator and detector for OFDM in presence of interference'. *IET Communications*. 2007, vol. 1(4), pp. 647–54.

[31] Theodoridis S. *Machine Learning: A Bayesian and Optimization Perspective*. Academic Press; 2015.

[32] Mitra R., Bhatia V. 'Adaptive sparse dictionary based kernel minimum symbol error rate post-distortion for nonlinear LEDs in visible light communications'. *IEEE Photonics Journal*. Jun. 2016, vol. 8(4), pp. 1–13.

[33] Zhao S., Chen B., Principe J.C. 'Kernel adaptive filtering with maximum correntropy criterion'. International Joint Conference on Neural Networks; 2011–2012–7.

[34] He Y., Wang F., Yang J., Rong H., Chen B. 'Kernel adaptive filtering under generalized maximum correntropy criterion'. International Joint Conference on Neural Networks; 2016–1738–45.

[35] Singh A., Principe J.C. 'Using correntropy as a cost function in linear adaptive filters'. International Joint Conference on Neural Networks; 2009–2950–5.

[36] Richards M.A., Scheer J., Holm W.A., Melvin W.L. 'Principles of modern radar'. *Citeseer*. 2010.

[37] Rangaswamy M., Weiner D., Ozturk A. 'Computer generation of correlated non-Gaussian radar clutter'. *IEEE Transactions on Aerospace and Electronic Systems*. 1995, vol. 31(1), pp. 106–16.

[38] Jian T., He Y., Su F., Qu C., Ping D. 'Cascaded detector for range-spread target in non-Gaussian clutter'. *IEEE Transactions on Aerospace and Electronic Systems*. 2012, vol. 48(2), pp. 1713–25.

[39] P-l S., Liu M., S-w X. 'Shape-parameter-dependent coherent radar target detection in K-distributed clutter'. *IEEE Transactions on Aerospace and Electronic Systems*. 2016, vol. 52(1), pp. 451–65.

[40] Sahed M., Mezache A., Laroussi T. 'A novel [z log(z)]-based closed form approach to parameter estimation of k-distributed clutter plus noise for radar

detection'. *IEEE Transactions on Aerospace and Electronic Systems*. 2015, vol. 51(1), pp. 492–505.
[41] Shi L., Lin Y. 'Convex combination of adaptive filters under the maximum correntropy criterion in impulsive interference'. *IEEE Signal Processing Letters*. 2014, vol. 21(11), pp. 1385–8.
[42] He R., Zheng W.-S., Hu B.-G. 'Maximum correntropy criterion for robust face recognition'. *IEEE Transactions on Pattern Analysis and Machine Intelligence*. 2011, vol. 33(8), pp. 1561–76.
[43] Chen B., Zhu Y., Hu J., Principe J.C. 'System parameter identification: information criteria and algorithms'. *Newnes*. 2013.
[44] Hou B., He Z., Zhou X., Zhou H., Li D., Wang J. 'Maximum correntropy criterion Kalman filter for α-jerk tracking model with non-Gaussian noise'. *Entropy*. 2017, vol. 19(12), p. 648.
[45] Jin F., Qiu T. 'Adaptive time delay estimation based on the maximum correntropy criterion'. *Digital Signal Processing*. 2019, vol. 88(3)–23–32.
[46] Chen B., Zhao S., Seth S., Principe J.C. 'Online efficient learning with quantized KLMS and L1 regularization'. International Joint Conference on Neural Networks; 2012–1–6.
[47] Kulikov G.Y., Kulikova M.V. 'The accurate continuous-discrete extended Kalman filter for radar tracking'. *IEEE Transactions on Signal Processing*. 2015, vol. 64(4), pp. 948–58.
[48] Julier S.J., Uhlmann J.K. 'New extension of the Kalman filter to nonlinear systems'. *Signal processing, sensor fusion, and target recognition VI*. 3068. International Society for Optics and Photonics; 1997. pp. 182–94.
[49] Cortina E., Otero D., D'Attellis C.E. 'Maneuvering target tracking using extended Kalman filter'. *IEEE Transactions on Aerospace and Electronic Systems*. 1991, vol. 27(1), pp. 155–8.
[50] Farina A., Ristic B., Benvenuti D. 'Tracking a ballistic target: comparison of several nonlinear filters'. *IEEE Transactions on Aerospace and Electronic Systems*. 2002, vol. 38(3), pp. 854–67.
[51] Bar-Shalom Y., Li X.R., Kirubarajan T. *Estimation with Applications to Tracking and Navigation: Theory Algorithms and Software*. John Wiley & Sons; 2004.
[52] Anderson B.D., Moore J.B. 'Optimal filtering'. *Courier Corporation*. 2012.
[53] Brown R.G., Hwang P.Y., *et al. Introduction to Random Signals and Applied Kalman Filtering*; New York: Wiley; 1992.
[54] Arulampalam M.S., Maskell S., Gordon N., Clapp T. 'A tutorial on particle filters for online nonlinear/non-Gaussian Bayesian tracking'. *IEEE Transactions on Signal Processing*. 2002, vol. 50(2), pp. 174–88.
[55] Julier S., Uhlmann J., Durrant-Whyte H.F. 'A new method for the nonlinear transformation of means and covariances in filters and estimators'. *IEEE Transactions on Automatic Control*. 2000, vol. 45(3), pp. 477–82.
[56] Julier S.J., Uhlmann J.K. 'Unscented filtering and nonlinear estimation'. *Proceedings of the IEEE*. 2004, vol. 92(3), pp. 401–22.

[57] Arasaratnam I., Haykin S. 'Cubature Kalman filters'. *IEEE Transactions on Automatic Control*. 2009, vol. 54(6), pp. 1254–69.

[58] Singh A.K., Radhakrishnan R., Bhaumik S., Date P. 'Adaptive sparse-grid Gauss–Hermite filter'. *Journal of Computational and Applied Mathematics*. 2018, vol. 342(6), pp. 305–16.

[59] Athans M., Wishner R., Bertolini A. 'Suboptimal state estimation for continuous-time nonlinear systems from discrete noisy measurements'. *IEEE Transactions on Automatic Control*. 1968, vol. 13(5), pp. 504–14.

[60] Rao S.K. 'Modified gain extended Kalman filter with application to bearings-only passive manoeuvring target tracking'. *IEE Radar, Sonar Navigation*. 2005, vol. 152(4), pp. 239–44.

[61] Taek Song., Speyer J. 'A stochastic analysis of a modified gain extended Kalman filter with applications to estimation with bearings only measurements'. *IEEE Transactions on Automatic Control*. 1985, vol. 30(10), pp. 940–9.

[62] Zhan R., Wan J. 'Iterated unscented Kalman filter for passive target tracking'. *IEEE Transactions on Aerospace and Electronic Systems*. 2007, vol. 43(3), pp. 1155–63.

[63] Jiang Z., Song Q., He Y., Han J. 'A novel adaptive unscented Kalman filter for nonlinear estimation'. 46th Conference Decision Control; 2007–4293–8.

[64] Chang L., Hu B., Li A., Qin F. 'Transformed unscented Kalman filter'. *IEEE Transactions on Automatic Control*. 2013, vol. 58(1), pp. 252–7.

[65] Zhang Y.-X., Hong R.-J., Pan P.-P., Deng Z.-M., Liu Q.-F. 'Frequency-Domain range Sidelobe correction in stretch processing for Wideband LFM Radars'. *IEEE Transactions on Aerospace and Electronic Systems*. 2017, vol. 53(1), pp. 111–21.

Chapter 4

Understanding radar signals in a highly heterogeneous clutter environment

Christian Bräu[1] *and Fernando Rial*[1]

Ground-penetrating radar (GPR) systems operate very often in environments with highly heterogeneous clutter sources and typically low signal-to-clutter ratios (SCRs). These clutter sources are generally divided, according to their origin, as surface and subsurface clutters. Surface clutter is related, particularly, to the reflection produced by the air–soil interface and also by objects located on the surface near the antennas. Subsurface clutter can be originated from all kinds of soil inhomogeneities and by other objects present in the targeted area. Besides dealing with low SCRs, GPR systems have generally limited bandwidth due to soil attenuation. This implies that the obtained resolution is usually not enough to differentiate between targets and clutter sources with similar characteristics. In this chapter, we will show examples as to how clutter affects GPR data and how multistatic acquisitions and polarimetry can be applied in GPR to improve detection and classification attempting to compensate these limitations in resolution and SCR.

4.1 Introduction

In 2017, the science fiction movie *Blade Runner 2049* was released in many countries worldwide. The movie takes place in a future earth. At the beginning of the film, a police officer decides to scan an area under an interesting white tree. The scanning process is carried out by an unmanned aerial vehicle (UAV) presumably equipped with state-of-the-art radar systems. As a result, a 3D data is obtained where the underground root structure of the tree is clearly visible and distinguishable from the soil. In this radar image, another distinctive object can be observed among the roots. This object resembles a box, and in the image provided by the sensor it appears in a different colour (Figure 4.1).

From the point of view of a GPR specialist, the image provided by the sensor has several features that immediately catch your eye in the first view:

[1]Fraunhofer Institute for High Frequency Physics and Radar Techniques (FHR)

Figure 4.1 Image from sci-fi movie Blade Runner 2049 (screenshot from [1] under the fair use doctrine)

- The sensor is able to reconstruct the root system with a very high resolution despite the soil.
- There seems to be no evidence of subsurface or surface clutter.
- The sensor seems to be able to recognize different materials (root system and box).

These three aspects, resolution, clutter distinction and material characterization, represent actual challenges for the successful application of GPR technology in real scenarios.

Most GPR applications deal with low-resolution data when compared with other radar systems that are able to achieve mm and even sub-mm resolution. Soil is usually the main limiting factor. Besides that, information from GPR images is generally qualitative, as quantitative information requires a highly calibrated system and specific knowledge of the environment (soil), which is generally not attainable. The combination of these two aspects, low resolution and lack of knowledge about material properties, constitutes a heavy burden especially when the area under investigation is rich in highly heterogeneous clutter.

Through the rest of this chapter we will deal in more detail with the problem of clutter in GPR systems. As a first step, we will evidence the resolution limitations of these systems and afterwards we will discuss about the characteristics of different clutter sources that can be found in the data. As a last step, we will show examples, based on simulations and real data, of how to deal with clutter in GPR for target detection and classification.

4.2 Understanding resolution limits

4.2.1 Range resolution

Radar systems are essentially time-of-flight sensors. The principle of detection is contrast based. Because the type of sensing signal is an electromagnetic (EM) wave, this contrast is related to the difference in EM properties of the target and its surroundings. Where there is not enough contrast, there will be no detection. But if the contrast is there, a scattering process occurs and some energy is reflected back that

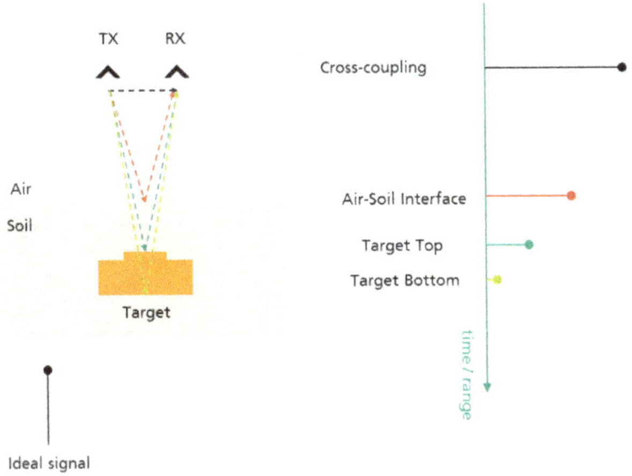

Figure 4.2 Scenario (left). Range profile (A-scan) with ideal signal (right)

can be used for detection. In most cases, this back-scattered energy will be represented by a distorted replica of the transmitted signal.

Under those premises it is clear that one would like to have a very sharp signal in time domain that might allow detecting potential targets that could be very close to each other or to differentiate scattering centres from the same target. Figure 4.2 shows the example of a simple GPR scenario where the signal transmitted is an ideal pulse (infinite bandwidth). The obtained range profile (A-scan) is represented on the right side of the figure, where the scattered signals are clearly differentiated in the time scale. Amplitude of the scattered signal decreases with time as expected in reality. It is important to mention that for the purpose of this discussion additional effects are not considered in the model, for instance, the interaction between target scattering centres (modified Born approximation, see [2]) or creeping waves [3].

Figure 4.3 uses the same model of the previous figure but uses a transmitted signal somewhat close to reality. On the right part of the figure, the scattering centres are added sequentially to make the reader have a first impression of the limitations associated with the range resolution in GPR systems. The reflection from the air–soil interface is generally strong compared with the weak scattering energy coming from a target. It is then easy to understand how the energy coming from the last two scattering centres can be easily mistaken by a residual of the previous scatterer (interface).

The main limitation behind the low-range resolution of GPR signals comes from the soil. The soil acts basically as a low-pass filter where the higher frequencies are usually highly attenuated (see Figure 4.4). The amount of water (soil moisture) plays here an important role. Lower frequencies suffer from less attenuation but, on the other side, the ratio of target to wavelength decreases (Rayleigh region) so the backscattered energy becomes very low for small targets. Besides, lower frequencies generally require bigger antennas, reducing systems' portability [6].

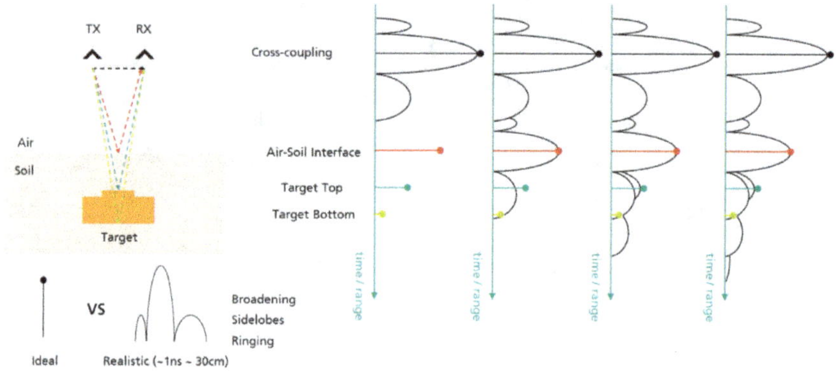

Figure 4.3 *Scenario (left). Range profile (A-scan) with a more realistic signal (right)*

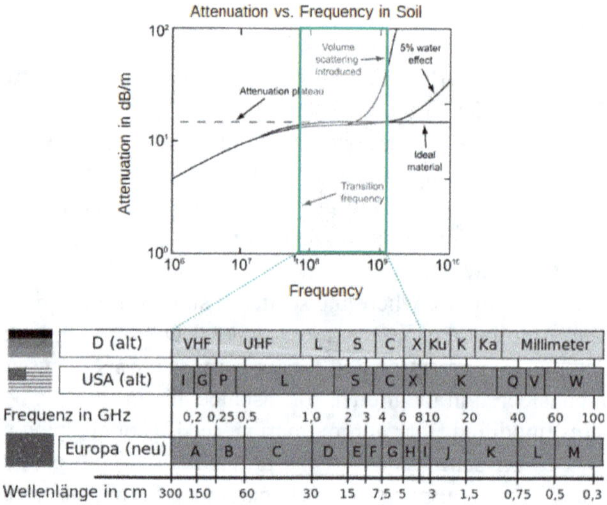

Figure 4.4 *Attenuation versus frequency in GPR (adapted from [4] and [5])*

4.2.2 Spatial resolution

Spatial resolution refers here to the horizontal, lateral or along-track/cross-track resolution.

The vast majority of GPR systems place the antennas in contact with (or very close to) the air–soil interface, maximizing the energy coupled into the soil. In a typical acquisition, GPR antennas interrogate a target from different positions (or target aspects). Due to the soil, the knowledge of the field scattered by this target will be always restricted in target aspect domain and frequency domain. Additionally, some

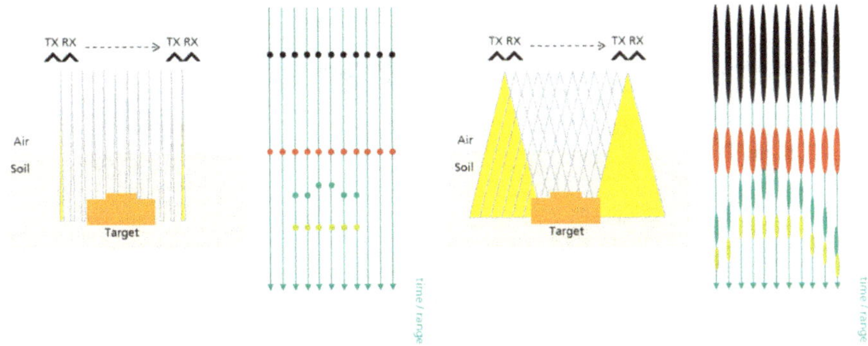

Figure 4.5 *Comparison between B-scans for an ideal (left) and more realistic (right) GPR systems. The limitation in range and spatial resolution are evidenced*

care must be taken during the acquisition in order to get the maximum resolution achievable by the system [7].

Figure 4.5 shows a consecutive acquisition of A-scans in the same scenario as Figure 4.2. The left side of the figure shows the obtained B-scan (A-scan sequence) in an ideal case where the signal is very sharp in time and the antenna has a very narrow beam (laser-like). Then the shape of the target can be easily observed. However, the reality is much closer to the right part of Figure 4.5, where not only the signal has a limited bandwidth but also the antenna has a broader beam. The loss of resolution in the B-scan makes now difficult to appreciate the real shape of the target. Note that in this figure, the information regarding the amplitude of the reflections is not shown.

Another factor that contributes to this loss of spatial resolution is the heterogeneity of the soil. Soil properties are in most cases spatially not homogeneous showing stochastic distributions that can be sometimes mathematically approximated using fractals [8]. The same applies to surface roughness. As shown in Figure 4.6, some kind of fractals can be easily generated by the convolution of Gaussian noise with a simple frequency filter in which exponent p controls the fractal dimension.

This change in soil properties may be significant enough to produce changes in signal velocity and attenuation for different antenna positions around a particular target. Additionally, when the GPR antennas operate very close to the ground in the so-called near-field region, the radiation pattern of the antenna will be also influenced by the soil. Figure 4.7 shows an example of how heterogeneous soil conditions may influence the obtained B-scan in comparison with the homogeneous soil from Figure 4.5.

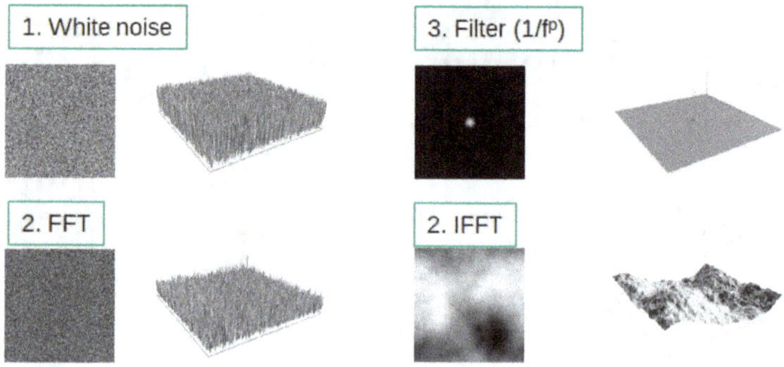

Figure 4.6 *Example of soil roughness-generation using fractals*

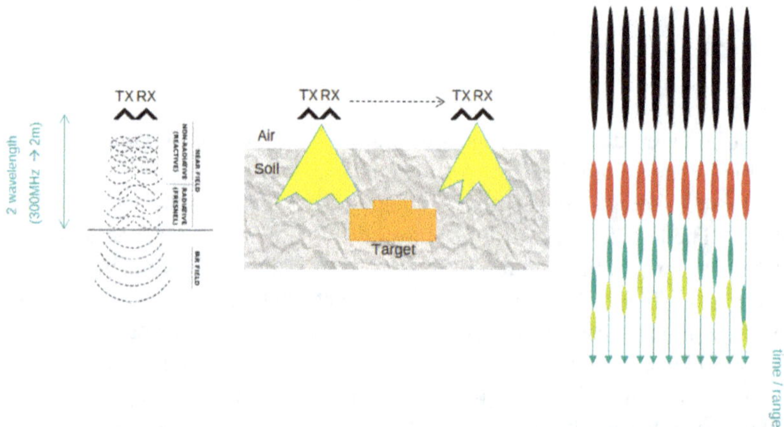

Figure 4.7 *Possible effect of heterogeneous soil in the GPR data compared to Figure 4.5*

4.2.3 Dealing with resolution limits in GPR

In order to improve the spatial resolution, the obtained multiaspect data can be combined using reconstruction algorithms [9, 10] to produce a 3D (or 2D) backscatter image of the interrogated domain including the target. In the radar community, this is generally called synthetic aperture radar (SAR) processing. Basically, the same process is traditionally called migration in seismic imaging. Because the domain of interest of GPR systems is generally the soil, the term migration is commonly used also regarding focusing. Migrated data can be used to derive features sometimes useful in recognition and/or classification tasks. This will be further discussed later, when considering multistatic and full-polarimetric systems.

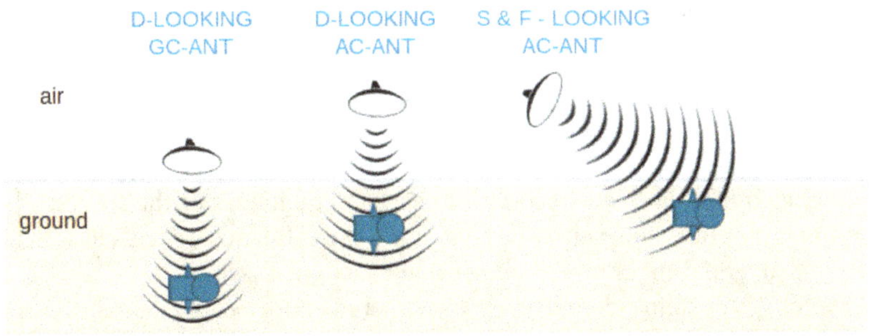

Figure 4.8 *Example of generic GPR systems: down-looking ground-coupled, down-looking air-coupled and forward- and side-looking*

4.3 Understanding clutter sources

As mentioned in the introduction, GPR systems operate very often in environments with highly heterogeneous clutter. These cluttered environments, together with the limitations in resolution discussed in the previous section 4.2, make the problem of detection and classification using GPR data very challenging.

Clutter can be defined as unwanted echoes (signal replicas) that appear in the acquired data. In GPR applications, clutter can be generally divided into subsurface and surface clutters. This definition tries to establish a simple classification of clutter based on its spatial location: either inside the ground (subsurface clutter) or anywhere else (surface clutter). It is important to note that the term surface clutter also integrates here objects that might not be exactly lying on the surface (wall, ceiling, sensor parts, etc.). Other definitions are also common in specialized literature: underground, aboveground, ground clutter, etc.

4.3.1 Surface clutter

GPR systems can be cast into two main groups (Figure 4.8) as down-looking (DL-GPR) and forward/side-looking (FL-GPR, SL-GPR) systems. Most of the DL-GPR systems place their antennas in contact with (or very close to) the surface to improve coupling and maximize power transmitted into the ground. This type of antennas (ground-coupled) is usually encapsulated (shielded) to minimize the power coming from/to other directions different from the ground. In this way, they try to reduce the influence of surface clutter in the data. Some DL-GPR systems predominantly used in traffic infrastructure surveys also use another type of antennas (air-coupled) designed to work at short standoff distances from the road surface. The greatest advantage of air-coupled systems is perhaps their repeatability, since antenna coupling does not change with the changes in soil properties. However, for these antennas, surface clutter may become an important issue.

One common characteristic of surface clutter is that it is not spatially located inside the area of interest (soil) but its echoes arrive at the same time as the ones coming from the targets. Surface clutter includes several sources:

- The signal backscattered at the interface between air and soil (ground-bounce) is included here in this category. The surface roughness of the soil also plays an important role. For very rough surfaces, the influence of this clutter can be significant and difficult to remove during signal processing.
- Antenna (or system) clutter refers to scattered energy originated in the sensor itself and its interaction especially with the soil surface. Because antennas are generally the largest piece of the sensor and they are closer to the ground, their influence in this type of clutter usually dominates.
- Surface clutter refers certainly to objects located in the surface (stones, grass bushes, trees, superficial roots) that may complicate even data acquisition.
- This definition also includes man-made objects in the area nearby (cars, walls, structures, etc.) that are sometimes massive and/or metallic, thus producing strong backscattered signals.

4.3.2 Subsurface clutter

Subsurface clutter refers basically to two types of clutter sources:

- One type corresponds to soil inhomogeneities that produce enough contrast to backscatter the signal. Those inhomogeneities can be organically created or part of a man-made structures such as a road infrastructure and railway bed.
- The other type of subsurface clutter refers to objects located inside the area of interest that are not the ones we are looking for. These objects can be natural (stones, roots, etc.) or man-made (bricks, cans, debris, etc.).

It is important to note that, in a more general context of radar imaging with GPR, subsurface clutter is considered as a target of interest. The reciprocal is also true: a particular target can be considered clutter in a different application. An example of this could be the role of tree roots in the applications: root development (target) and landmine detection (clutter).

4.3.3 Dealing with clutter sources in GPR

In terms of clutter removal/reduction, there are also conceptual differences between surface and subsurface clutters. In most cases with surface clutter, it is generally easy to understand (by visual inspection of the data) that this clutter does not belong to the area of interest (distinctive shape, higher energy). This implies generally that there might be a way to remove this clutter totally or partially using some signal processing (digital filters).

The case of subsurface clutter has a bit different nature. The shape of detected objects may help to classify them into clutter or target when the

Figure 4.9 *Example of clutter sources. A classifier may be needed to separate subsurface clutter from potential targets*

difference is relevant (for instance: road layer vs. pipe), but the lack of resolution and low SCRs (usually close to 1) make the distinction between clutter and target with somewhat close characteristics very challenging. In many cases, a trained classifier will be needed to make a decision about a particular suspicious object in the data (Figure 4.9).

In Sections 4.4 and 4.5, we will focus on the problem of subsurface clutter reduction using multistatic and full-polarimetric approaches.

4.4 Multistatic GPR systems

Radar systems are often categorized as monostatic, bistatic or multistatic (Figure 4.10). In a monostatic configuration, the same antenna transmits a signal and afterwards receives the echoes of this signal. Bistatic systems are the most common configuration in GPR. A bistatic configuration uses a pair of antennas, usually close to each other, one as transmitter and the other as receiver. Due to this proximity between transmitter and receiver, this configuration is also called quasi-monostatic. Both monostatic and bistatic correspond to what is called in radar a SISO (Single Input, Single Output) configuration. When there is only one transmitter but several receivers, the configuration is called SIMO (Single Input, Multiple Output), and when there is only one receiver but several transmitters, it is called MISO (Multiple Input, Single Output). Finally the case with multiple transmitters and receivers is called MIMO (Multiple Input, Multiple Output) configuration. MIMO, SIMO and MISO are examples of multistatic systems, when more than one transmitter or/and receiver are involved in the processing.

Multistatic systems provide a priori with three basic advantages with respect to monostatic systems:

- Improved Signal-to-Noise Ratio (SNR): In ideal conditions (Gaussian noise, dynamic range, target radar cross-section (RCS), etc.), the improvement in SNR is given by $10log_{10}N$, where N is the number of antenna pairs combined. This is

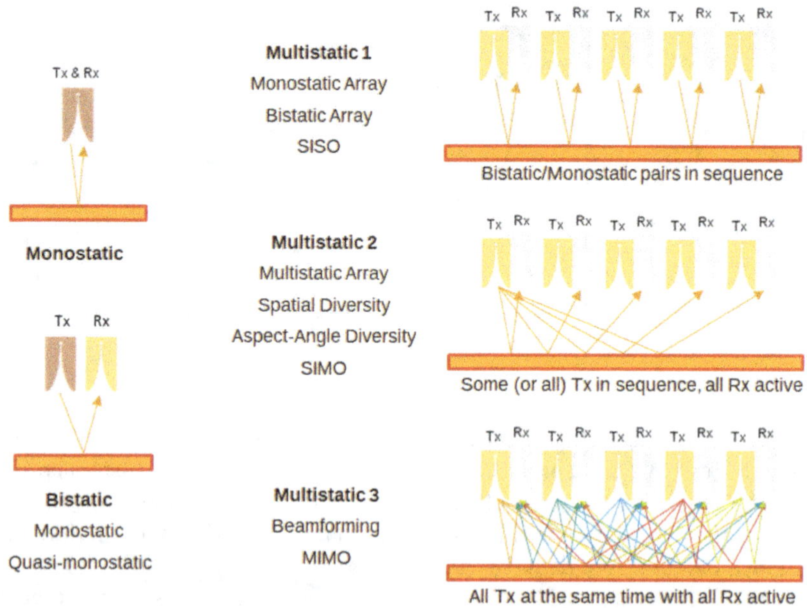

Figure 4.10 Example of antenna configurations in GPR

also considering a perfect calibrated system and spreading losses compensated. Ideal requirements are also often far from on-field measurement systems and real environmental conditions. For those reasons, the improvement given by the formula can be seen as a theoretical maximum which is difficult to reach in practice.

- Improved Aspect Angle Diversity: Intuitively, target imaging will be favoured with wider view-angle diversity. The more view angles from the same object we have, the better our spatial representation of that object will be. This is potentially true especially for complex targets. Depending on the arrangement of the transmitter–receiver pairs, multistatic systems also have the capability in general to provide higher resolution images due to a better spatial orthogonality between antenna pairs.
- Improved Operational Efficiency: Multistatic arrays can be potentially defined to exhibit the same level of the beamforming performances as a synthetic aperture but using fewer elements and, as a consequence, significantly reducing the data acquisition time. This might also have an impact on the cost of the overall system.

It is important to remember that in GPR, the size of the antenna footprint (together with the antenna height, target depth, target shape and soil properties) also establishes a geometrical limit regarding the maximum operational distance between possible transmitter–receiver pairs.

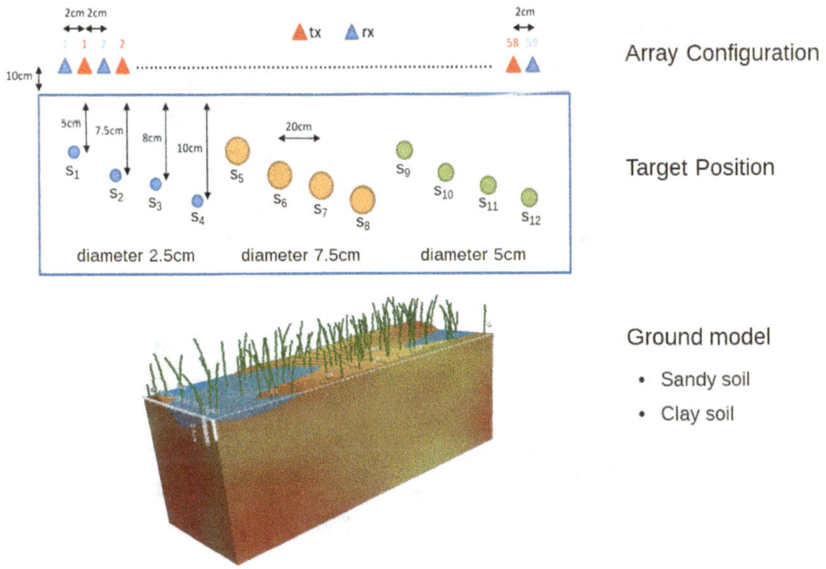

Figure 4.11 *Simulation model showing some of the parameters used: antenna configuration, target position and soil type*

The practical majority of GPR systems using antenna arrays are based in the multistatic configuration 1 (Figure 4.10), where the different antenna pairs work in sequence and are independent of each other. In the multistatic configuration 2, for every transmitted signal, the system obtains information from multiple aspect angles which can help to reduce clutter and improve the overall detection/classification process. Multistatic configuration 3 is represented mainly by systems capable of beamforming in transmission and MIMO approaches, where the level of hardware (and also software) complexity is highly increased.

4.4.1 Example of multistatic systems for clutter reduction

In this section, we will show an example of the advantage of multistatic systems (multi 2 in Figure 4.10) compared with traditional arrays (multi 1 in Figure 4.10). In particular, we will consider the effect of clutter reduction and target definition (energy and sharpness). We will base our analysis on simulated models using the software gprMax [11]. We will deal in this example just with typical subsurface clutter generated by soil inhomogeneities.

Figure 4.11 indicates some parameters of the simulation. The top of the figure shows the multistatic antenna arrangement, where the separation between consecutive transmitters and receivers is 2 cm. In the middle of the figure, the location of the targets of interest is shown. The targets consist of spheres with diameters 2.5, 5 and 7.5 cm, buried at different depths ranging from 5 to 10 cm. Two different soil

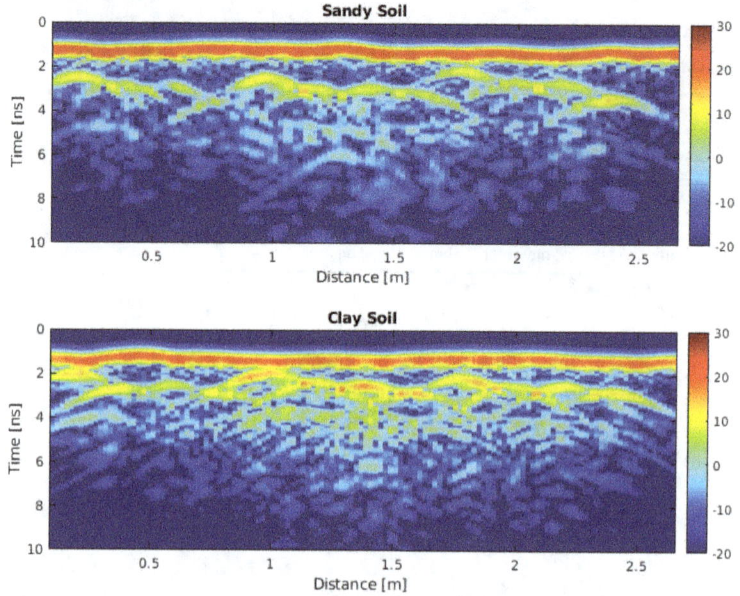

Figure 4.12 B-scans obtained with the multi 1 system for both soils

types were simulated: sandy soil and clay soil, both with similar fractal dimensions and roughness levels.

Figure 4.12 shows the data obtained with the multi 1 configuration for both soils. In the first view, the limitations in the resolution of the system are evident, as we are practically not able to appreciate any difference between targets of different sizes. The clay soil seems to present a slightly higher clutter levels as expected. For comparison, the same dynamic range of 50 dB is adjusted for both images.

Figure 4.13 shows the data after migration. The left part of the figure shows both soils (sandy and clay) after Kirchhoff migration using the data from the

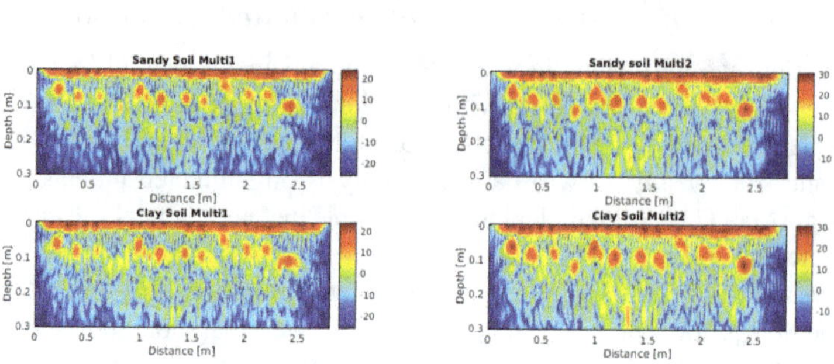

Figure 4.13 Data after migration comparing multi 1 and 2 configurations for both soils

multi 1 configuration. The right side of the figure shows the same data using the multi 2 configuration. A dynamic range of 50 dB below the maximum level is fixed for every image for better comparison. For both configurations in both soils, the targets are mostly visible and better differentiated compared to the raw B-scan data (Figure 4.12). However, we like to point our attention to two of the initial most difficult targets: s_4 and s_{12}. These two targets are particularly challenging because of their small size and location. If we compare these targets between the multi-static configurations (left and right side of Figure 4.13) we can notice some clear differences. Specially in the clay soil for the multi 1, s_4 practically disappears and the shape of s_{12} is highly distorted. In contrast, the multi 2 configuration is able to reconstruct both targets successfully, showing the advantage of this configuration in inhomogeneous media.

4.5 Full-polarimetric GPR systems

Radar polarimetry deals with the full vector nature of polarized EM waves. In this section, we will present some basics for processing polarimetric radar data and show an example of how they can be applied in GPR applications. For a thoughtful introduction on polarimetry applied to radar and further details regarding other applications, we refer the reader to [12].

Our approach is to apply decomposition techniques to full-polarimetric measurements in order to obtain features to classify target signatures. The main goal is to find features for distinguishing clutter objects from the targets of interest. Most decompositions described in [12] were initially developed for airborne or satellite SAR images. Although most of these decompositions can be applied to GPR data, they are not always effective since GPR targets have different features. Dispersive and inhomogeneous soils might also change the polarimetric properties of the signal and hence it is preferable to use decompositions that use a small number of robust components.

In order to use these decomposition techniques, the GPR system must be full polarimetric. A simple way to achieve this is to have a system that is able to transmit and receive two orthogonal polarizations. The most common method is to use linear polarization. In compliance with traditional radar nomenclature, we refer to these two polarizations as horizontal (H) and vertical (V), even though this has no true meaning for GPR data.

Therefore, in a full-polarimetric acquisition we receive four channels which are represented in the so-called scattering matrix or S-matrix

$$S(t,x,y)=\begin{bmatrix} S_{HH}(t,x,y) & S_{HV}(t,x,y) \\ S_{HV}(t,x,y) & S_{VV}(t,x,y) \end{bmatrix}, \tag{4.1}$$

which contains the received data at each spatial location (x, y) (we will omit the arguments in the following). If the antenna characteristics are symmetric for both polarizations, the cross-polar components S_{HV} and S_{VH} should have the same information and the S-matrix becomes symmetric. Because most GPR systems operate in the time domain using short pulses (pulsed radars), the elements of this S-matrix are usually real and define a subspace with dimension 3 being

$$\psi_p = \left\{ \sqrt{2}\begin{bmatrix} 1 & 0 \\ 0 & 1 \end{bmatrix}, \sqrt{2}\begin{bmatrix} 1 & 0 \\ 0 & -1 \end{bmatrix}, \sqrt{2}\begin{bmatrix} 0 & 1 \\ 1 & 0 \end{bmatrix} \right\} \tag{4.2}$$

a possible orthogonal basis. These matrices are called Pauli matrices. This basis is suitable because its elements correspond to physical scattering mechanisms representing elementary targets shown in Table 4.1.

In other words, every scattering can be expressed as a composition of these three elementary scattering mechanisms. A linear object such as a wire located parallel to one polarization would be represented by a scattering matrix that will be a linear combination of planar and dihedral scattering.

4.5.1 Pauli decomposition

The Pauli decompositions is a very simple, coherent decomposition, which directly uses the basis representation (4.2). The vector

$$k_p(t,x,y) = \begin{bmatrix} a \\ b \\ c \end{bmatrix} = \frac{1}{\sqrt{2}} \begin{bmatrix} S_{HH}(t,x,y) + S_{VV}(t,x,y) \\ S_{HH}(t,x,y) - S_{VV}(t,x,y) \\ 2 \cdot S_{HV}(t,x,y) \end{bmatrix} \tag{4.3}$$

formed by the corresponding coefficients is called Pauli vector. These coefficients are a measure of the contribution of each elementary scattering mechanism. They can be used for classification and visual presentation by assigning a different base colour of the RGB (red, green, blue) colour space to each (see Table 4.1).

Table 4.1 Pauli matrices and their corresponding scattering mechanism and colours

Pauli matrix	Scattering mechanism	Elementary target	Colour
$\sqrt{2}\begin{bmatrix} 1 & 0 \\ 0 & 1 \end{bmatrix}$	Planar, single bounce	Sphere, plate	Blue
$\sqrt{2}\begin{bmatrix} 1 & 0 \\ 0 & -1 \end{bmatrix}$	Dihedral, double bounce	Corner reflector	Red
$\sqrt{2}\begin{bmatrix} 0 & 1 \\ 1 & 0 \end{bmatrix}$	Irregular, volume bounce	Irregular object, vegetation	Green

4.5.2 Entropy-α decomposition

The entropy-α decomposition is a non-parametric decomposition which is based on an eigenvalue decomposition. This means that instead of using the basis formed by the Pauli matrices, we search for a basis that decorrelates the data. This decomposition has been already applied to GPR data, for example, in the work of [13]. We start by building the Pauli vector (4.3) for each data point (t, x, y) and consider a window around these data points forming the matrix $\bar{k}_p \in \mathbb{R}^{3\times m}$, where m is the number of data points inside the window. This window should contain nearby points assumed to show the same scattering mechanism. Now, there exists an orthogonal matrix $U \in R^{3\times 3}$ and real eigenvalues $\lambda_1, \lambda_2, \lambda_3$ such that

$$\bar{k}_p\bar{k}_p^T = U^H \mathrm{diag}(\lambda_1, \lambda_2, \lambda_3)U.$$

Hence, the columns of U, i.e. the eigenvectors, are the basis that decorrelates the data and the eigenvalues correspond to their weights. Note that instead of applying an eigenvalue decomposition to the outer product $\bar{k}_p\bar{k}_p^T$, a calculation of the singular value decomposition of $\bar{k}_p$ itself should be preferred since it is numerically more stable and faster. We proceed with calculating the normalized eigenvalues

$$p_i = \frac{\lambda_i}{\sum_{j=1}^{n} \lambda_j}.$$

Furthermore, by writing U as a rotation matrix

$$U_3 = \begin{bmatrix} \cos(\alpha_1) & \cos(\alpha_2) & \cos(\alpha_3) \\ \sin(\alpha_1)\cos(\beta_1)e^{i\delta_1} & \sin(\alpha_2)\cos(\beta_2)e^{i\delta_2} & \sin(\alpha_3)\cos(\beta_3)e^{i\delta_3} \\ \sin(\alpha_1)\cos(\beta_1)e^{i\gamma_1} & \sin(\alpha_2)\cos(\beta_2)e^{i\gamma_2} & \sin(\alpha_3)\cos(\beta_3)e^{i\gamma_3} \end{bmatrix}$$

we obtain the scattering angles α_1, α_2 *and* α_3 and finally

$$H = \sum_{i=1}^{n} -p_i \log_n(p_i) \tag{4.4}$$

$$\alpha = \sum_{i=1}^{n} p_i\alpha_i. \tag{4.5}$$

These values can be interpreted as points in the so-called *H-α* plane, where the elementary scattering mechanisms can be located in specific areas. These are shown in Figure 4.14. We note that because of the reduction from 3 degrees of freedom (Pauli vector) to 2 degrees of freedom *(H, α),* volume and dihedral scattering cannot be distinguished anymore.

- The entropy H can be interpreted as a measure of the disorder or information content. $H = 1$ implies a random scattering mechanism whereas $H = 0$ implies a stable scatterer or Bragg surface. Furthermore, H is increasing with the noise and clutter caused by inhomogeneous soils.

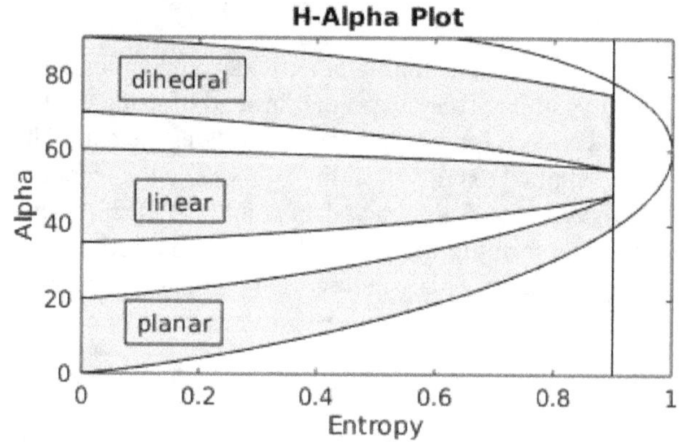

Figure 4.14 *H-α plane and areas corresponding to elementary scattering mechanisms*

- The angle α is a measure of the anisotropy of the scattering. Plane or symmetric scatterers yield a small angle around 0. Linear scatterer yields an angle around 45 and dihedral scatterers like a corner reflector an angle around 90.

4.5.3 Example of full-polarimetric systems for clutter reduction

In this section, we present an example of the advantage of full-polarimetric GPR systems against clutter. The dataset has been obtained with a 2D scanner over sandy soil in a controlled environment. The acquisition followed a 2 cm spacing grid in both dimensions (X and Y) covering a squared area of 1.5 × 1.5 m. Within this area, several objects were buried in the soil with depths ranging from 2 to 15 cm. The objects are shown in Figure 4.15 before they were buried. The variety in shapes, sizes and materials provides a colourful casuistic for the evaluation of classification algorithms.

In this example, we apply the *H-α* decomposition described in Section 4.5.2. Before this step, only data normalization and migration were applied to the acquired data. For each data point, a symmetric window was considered in every dimension (X, Y and Z (depth)) and the decomposition was applied as described. The results of some of the objects are shown in Figure 4.15 as illustration.

From the example, it can be appreciated how the *H-α* plane is able to successfully distinguish objects that have different characteristics. This provides a valuable unsupervised classification method to discard clutter sources using full-polarimetric GPR systems.

4.6 Discussion

GPR systems operate very often in environments with dense and highly heterogeneous clutter. Energy backscattered by surface clutter can be dominant and mask

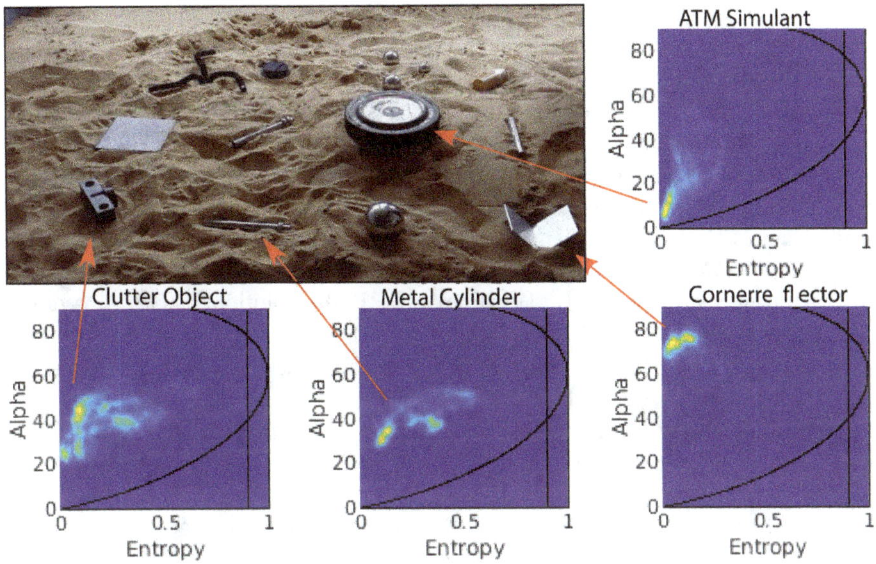

Figure 4.15 *Entropy-α decomposition of measured target signatures*

the weaker energy coming from the targets of interest making detection difficult. However, it is the subsurface clutter that represents usually a bigger challenge especially due to the low resolution of the GPR data. The examples presented in this chapter give a first insight into how multistatic and full-polarimetric configurations provide GPR systems with enhanced capabilities, which can be used to mitigate the clutter problem.

Acknowledgements

This research has been partly funded by the Bundesamt für Ausrüstung, Informationstechnik und Nutzung der Bundeswehr (BAAINBw) under the project number E/K2BJ/HA099/FF042.

References

[1] Screenmusings Movie Screencaps [online]. 2017. Available from https://screenmusings.org/movie/blu-ray/Blade-Runner-2049/ [Accessed March 2020].

[2] Van der Kruk J. 'Three-dimensional imaging of multi-component ground penetrating radar data'. *Ph.D. thesis*. The Netherlands: Delft University of Technology; 2001.

[3] Uschkerat U. 'Comparing UWB GPR measurements and simulation of simple shaped buried targets'. Second International Conference on Detection of Abandoned Land Mines; 1998.

[4] Annan P. *How to select GPR antenna frequency? [online]*. 2016. Available from https://www.sensoft.ca/gpr/antenna-frequency/ [Accessed 2020 March].

[5] Radio spectrum – German version [online]. Available from https://de.wikipedia.org/wiki/Frequenzband [Accessed 2020 March].

[6] Daniels D.J. 'Ground penetrating radar'. *IET Radar, Sonar, Navigation and Avionics Series 15*. 2nd ed. 726; 2007.

[7] Grasmueck M., Weger R., Horstmeyer H. 'Full-resolution 3D GPR imaging'. *Geophysics*. 2005, vol. 70(1), pp. K12–19.

[8] Turcotte D.L. *Fractals and Chaos in Geology and Geophysics*. Cambridge: Cambridge University Press; 1997.

[9] Fortuny-Guasch J. 'A novel 3-D subsurface radar imaging technique'. *IEEE Transactions on Geoscience and Remote Sensing*. 2002, vol. 40(2), pp. 443–52.

[10] Uschkerat U. 'Application of 3D-SAR nearfield imaging algorithms to GPR data'. *2007 IEEE International Geoscience and Remote Sensing Symposium*. 2007, pp. 452–5.

[11] Warren C., Giannopoulos A., Giannakis I. 'gprMax: open source software to simulate electromagnetic wave propagation for ground penetrating radar'. *Computer Physics Communications*. 2016, vol. 209(3), pp. 163–70.

[12] Lee J.S., Pottier E. *Polarimetric Radar Imaging: From Basics to Applications*. USA: CRC Press; 2009.

[13] Yu Y., Chen C.-C., Feng X., Liu C. 'Modified entropy-based fully polarimetric target classification method for ground penetrating radars (GPR)'. *IEEE Journal of Selected Topics in Applied Earth Observations and Remote Sensing*. 2017, vol. 10(10), pp. 4304–12.

Chapter 5

Signature characteristics and discrimination of small airborne targets with a staring radar

Mohammed Jahangir[1], Chris Baker[1], and Gordon Oswald[2]

In recent years, the number of drones operating at low altitudes has increased immensely. Radar provides a means of twenty-four-hour, all-weather surveillance and can act both as a means of countering rogue drones and as facilitating their safe passage through airspace in a manner akin to today's air traffic control. Staring radar provides a unique capability to provide the necessary non-cooperative surveillance of small airborne targets. Staring radar systems[a] can utilize extended dwell times in order to detect small and slowly moving targets such as drones at ranges of several kilometres. However, the sensitivity required to robustly detect such low radar cross-section (RCS)

targets means that many other unsought targets (e.g., birds, surface objects) will also be detected and reported. As the number of false 'alarms' increases, there will be a corresponding reduction in system effectiveness. Consequently, a deeper understanding at the target signature level is central to the ability to develop effective discriminators and hence eliminate false or unwanted targets.

In this chapter, we briefly describe the staring radar concept and show how it can be used to detect and discriminate drones against a severe background of other targets such as birds and ground clutter. In particular, the time-varying responses from drones and competing targets are examined, highlighting inherent differences that can be exploited for discriminating between detected/tracked objects. The relationship between a staring radar mode of operation, target echo attributes and their exploitation for classification is examined. Throughout, examples from real-world measurements will be used to illustrate the form and dynamic nature of

[1]School of Engineering, University of Birmingham, United Kingdom
[2]Gordon Oswald, United Kingdom

[a]An example of staring radar is the Thales Aveillant Gamekeeper L-Band multibeam non-scanning radar [23].

complex echoes as well as demonstrating the target-recognition performance with a supervised-learning-based approach.

The echo signals received by a staring radar include complex spectral components. Over time, these signals, together with range, azimuth and elevation observations, are encoded with unique characteristics, including those relating to target dynamics that differ from those of unwanted targets and ground clutter. These spectral components require specialized analysis to extract the underlying information to subsequently provide the means to classify targets. Here, the critical aspect in reception and processing is the extent of the coherent processing interval (CPI) to capture the salient underlying targets' discriminative features. This interval can be greatly extended as well as varied when using a staring mode of operation. This facilitates efficient exploitation of signal fluctuations through both time and space across a receiving array. These signatures can be used to derive target characteristics and trajectories, which lead to a high recognition performance and hence minimization of false reports. Examples of processing such signatures will be presented.

On the target-recognition aspects of the staring radar, machine learning, data-driven classifiers may be used in conjunction with extended dwell data to remove false or 'confuser' targets such as birds and provide real-time tracks of drones. In this chapter, a simple two-stage supervised-learning method is described that can discriminate drones from other possible confuser targets. This approach is based on a 'decision tree (DT) classifier' and is shown to be effective at filtering out non-drone (ND) targets. Field trials from realistic urban environments show that the two-stage DT classifier provides robust discrimination with minimal false positives (FPs), provided an appropriate learning process.

5.1 Introduction to staring radar

Most radar systems scan a beam through space using either mechanical or electronic means. This is fundamentally different to the concept of a staring radar where no such scanning occurs. However, it is only relatively recently that staring or 'ubiquitous' radar has come into more common usage as part of the overall radar landscape. Indeed, ubiquitous radar is defined according to the IEEE standard radar definitions (2017) as given below.

> *A radar that looks everywhere all the time and which performs multiple functions simultaneously instead of sequentially is a ubiquitous radar. The concept is characterized as having a broad beamwidth (quasi-isotropic) transmitting antenna and a staring array of multiple narrower beamwidth receiving beams covering the same solid angle. The output of each beam has a receiver and processor to provide various radar functions simultaneously and independently in each beam.*

This definition is necessarily broad but captures the essence of staring radar as one that has broad illumination and a set of narrow receive beams matched to the total volume illuminated. It is quite different to, say, a single-target tracking radar, where there are single, coincident, transmit and receive beams that are trained to 'stare' at a moving target as part of providing the best echo data for trajectory tracking. Whilst single-target tracking radar and staring radar are not unrelated, the static nature of staring radar coupled with the use of multiple simultaneous beams makes it quite different from tracking radar. More recently, advanced electronically scanned array radar systems have been developed, which inherently have huge flexibility such that they can mimic staring and many other types of radars. More typically, they can scan either a beam or series of beams and may have a fixed, single transmit and receive beams for special applications such as tracking or weather modes. Electronically scanned arrays can perform multiple functions. However, this means that the radar must make decisions about the nature of the resource and the time at which this should be applied to any given function. This is known as 'resource management' and, along with their very high costs, is a notoriously challenging aspect of this form of radar.

5.1.1 Historical background

It can be argued that the history of 'staring radar' is as old as the history of the radar itself. Christian Huelsmeyer from Germany is usually credited as the inventor of radar. His original 'Telemobiloskop', first demonstrated on 18 May 1904 had fixed illumination and receive beams. An extract from Huelsmeyer's patent is shown in Figure 5.1. Any target entering the illuminated zone triggered a warning alarm such that evasive action to avoid collisions could be taken. Despite his system proving itself in numerous tests and demonstrations, commercial success did not follow. Neither the maritime industry nor the German navy was sufficiently convinced of the role of this 'radar' and Huelsmeyer was forced to apply his considerable talents elsewhere.

It was not until some thirty years later did the concept of radar raise its head again in the form of the 'Daventry' experiment conducted on 26 February 1935. Once the value of the role to be played by radar could be seen, this led quickly to the 'Chain Home' series of systems that provided air defence for British shores. Chain Home was a system that, quite literally, provided floodlight illumination with echoes detected via a series of crossed dipoles connected to a low-noise, high-gain receiver. Figure 5.2 shows part of the Chain Home radar system. Chain Home owed its architecture to the Daventry test and the need to minimize the innovative steps required in transitioning from an exploratory demonstration to a deployed system tasked with a critical role in enabling the air defence of Great Britain. The system proved capable of measuring the direction and range of aircraft and providing the Royal Air Force (RAF) with sufficient warning and positional accuracy to engage the attacking force. However, its architecture was fundamentally different from that underlying almost any subsequent air defence radar, with only a few exceptions.

A combination of developments fuelled by a desire to have more compact systems together with all-round surveillance led to mechanically scanned antennas used for both transmission and reception. The development of the transmit/receive or T/R switch was a key-enabling technology. Subsequently, the simplicity of geometry

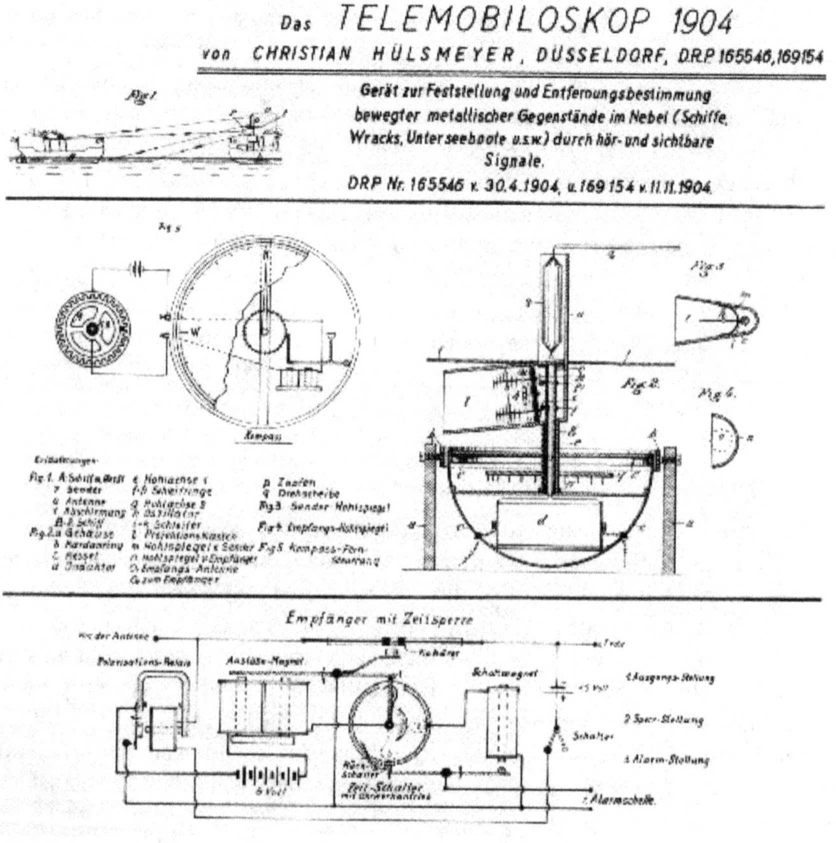

Figure 5.1 Christian Huelsmeyer's original patent for the 'Telemobiloskop'

and use of these radar systems made them the overwhelming design of choice. So much so that, with a few notable exceptions such as imaging radars, mechanically scanned radar systems have become the configuration of choice and are a familiar part of everyday life. An example used for maritime surveillance helping to enable safe navigation for ships is shown in Figure 5.3.

There have also been attempts to explore bistatic radar geometries where the transmitter and receiver are placed in quite separate locations. The more complicated geometries associated with bistatic geometries and the need for synchronization across large distances has inhibited their development. Bistatic radar is also beginning to show signs of coming into more common usage, especially with the increasing popularity of passive radar. In fact, passive radar leads directly to a natural extension of bistatic radar into multistatic or networked systems as multiple transmitters and/or receivers are often exploited. These are more complex still

Figure 5.2 Part of the 'Chain Home' radar system

but are beginning to emerge from research laboratories with increasing take-up by industrial companies. Passive radars are also, usually, examples of staring radars. However, a full description of passive radar lies outside the scope of this text and the interested reader is referred to [1].

As with the bistatic concept, it is only relatively recently that the notion of 'traditional' scanning radar has started to be challenged. Although the basic concept of a staring radar has existed for a long time, it was not until the 1980s that concepts began to emerge that were described in open publications. Three early and significant reports come from Thomson-CSF (now Thales), and ONERA in France refer

Figure 5.3 A marine band mechanically scanned radar transmitting and receiving through a single antenna

Figure 5.4 The RIAS concept as implemented by Thales in 1990

to a concept called 'RIAS' [2–4]. RIAS stands for *Radar a Impulsion et Antenne Synthetique* (or synthetic pulse and antenna radar). RIAS has two concentric arrays: one for transmit and the other one for receive as shown in Figure 5.4. The transmit array simultaneously radiates a set of orthogonal waveforms. As each transmitter uses a separate waveform this can be thought of as an early form of multiple input, multiple output or MIMO radar. The system has total hemispheric instantaneous coverage by virtue of multiple digital beamforming on receive to achieve permanent search in all directions and continuous tracking of the detected target. Experimental results using RIAS were described by Luce *et al.* [5].

RIAS operated in the VHF band, chosen as a counter to stealth technology and, most likely, to reduce the number of required elements and, therefore, system cost and complexity. Indeed, the authors, with great foresight, point out that the price to be paid for such a digital staring array is in the requirement for high computational capacity. This is still true today, but it is a price that has dropped significantly, year on year. In 'Phased Array Radars in France: Present and Future', a review article [6], Colin makes specific reference to RIAS indicating the importance of this program and its contribution to the general development of digital beamforming. More recently, the ideas behind RIAS have been taken up by Jianqi *et al.* from China [7]. They describe a system very similar in concept to RIAS but also emphasize the MIMO nature of the hardware configuration. They call their system synthetic impulse and aperture radar. This work has culminated in a book [8] that describes in some detail the basic concepts along with long-time integration strategies that follow from a staring radar concept and extensions to operation in other frequency bands.

Second, in the mid to late 1980s, Wirth, in Germany, published a series of papers on a concept termed 'Floodlight radar' [9, 10]. Here, the main motivation was to reduce the vulnerability of radar to interception by electronic warfare equipment, especially those associated with anti-radiation missiles. A system was

Figure 5.5 The OLPI receive array

implemented, known as the OLPI (Omni-directional Low Probability of Intercept) radar. The system was CW, operated in S-band and had a transmit power of 10 W. It used sector illumination covering a region that was 120° in azimuth and 20° in elevation via a column of eight vertical dipoles. The broad illumination combined with a CW waveform minimizes the emitted power density and correspondingly minimizes the likelihood of intercept. The receive array (shown in Figure 5.5) consisted of a planar array of 64 columns, each containing eight dipoles making 512 elements in total. Multiple beams were formed in analogue using a 64-port Butler matrix. Each formed beam was then digitized prior to further processing. Overall, the staring nature of the concept coupled with digital beamforming opened up many new processing possibilities such as long term coherent integration for fine Doppler resolution as well as the applications of techniques such as 'sequential processing' for improved target detection. Again, this was part of a broader research effort centred on phased arrays much of which was bought together in a book authored by Wirth published in 2001 [11].

A third and very significant effort was led by Merrill Skolnik of the Naval Research Laboratory (NRL) in the USA. Skolnik's ideas were first introduced in 1998 [12] and later in 1999 [13] as 'a far term concept for a radar that could look everywhere, all the time'. He termed this Ubiquitous radar and a comprehensive account of the concept and its advantages and disadvantages can be found in an NRL report authored by Skolnik [14] and summarized in a 2003 conference paper [15]. In the NRL report, Skolnik describes ubiquitous radar as inherently having fixed low gain or quasi Omni-directional transmission, digital beamforming on receive and a computational capacity to perform multiple signal processing functions in parallel. He also recognizes that a staring radar leads to long coherent integration times, enhanced Doppler filtering and the overall ability to achieve radar capabilities otherwise not possible.

Also, in 2003 Rabideau and Parker explored a variety of ubiquitous radar concepts within a MIMO radar framework as applied to ground surveillance applications [16]. Alter *et al.* [17] implemented a version of ubiquitous radar that operated in L-band and had a peak power of 1 kW. It employed a transmit array of half-wavelength space emitters arranged in 19 rows and 11 columns making 209 elements in all. This gave coverage of a 90° sector in azimuth and 15° in elevation. The receive array consisted of 1 590 elements arranged as 53 rows and 30 columns. This gives a receive beamwidths of 2.7° in elevation and 4.7° in azimuth. Multiple beams could be formed digitally to cover the whole of the illuminated volume. The analysis of results highlights numerous advantages of the ubiquitous radar concept, without necessarily exploring all possibilities.

Subsequently, the ubiquitous radar concept has been extended for use with conformal arrays [18] employing a conical structure. Their concept can be operated with both narrow and broad beams on transmit and can mimic both electronic scanning and staring. In [18] there is a discussion and computation of the requirements for additional gain as the system becomes closer and closer to the ubiquitous case. This extra gain is achieved by long coherent processing times making the range capability of traditional radar and staring radar the same. Harman and Hume [19] have reported a staring radar concept that uses an architecture consisting of several fixed transmit beams with corresponding digital beamforming such that all-round surveillance is possible. This has been embodied in the ALARM [20] and Obsidian radar systems operating in the C and X bands, respectively. The advent of highly capable digitization technology and subsequent signal processing have allowed Skolnik's ideas to be readily implemented and taken to a new level such that all the advantages of staring can be fully exploited.

This chapter goes on to focus on the specifics of the Staring Holographic Radar (SHR) concept that is a particular manifestation of the ubiquitous radar that has been developed by Oswald [21]. The SHR has been commercialized, initially, by Thales-Aveillant, where the radar utilizes long dwells on targets with a multi-receive beam staring radar to achieve a high level of detection sensitivity against weak targets against a background of strong non-stationary clutter. Next, Section 5.2 explains the form and function of the SHR concept. Section 5.3 presents the results of detection and classification performance that have been derived from real-world experimental field trials.

5.2 The physical context for staring radar

SHR irradiates an entire Volume of Regard (VoR) with a single transmitter and forms many receive 'beams' that together cover the totality of the VoR at all chosen ranges, as determined by the power budget. The echo signals received from an object by a SHR, over time, include many complex spectral components. In addition to the range, azimuth and elevation of wanted targets, these encode the characteristics, dynamics and state of operation of both the wanted targets and those of unwanted targets and clutter.

In contrast to traditional monostatic radars, the staring transmitter irradiates a chosen and usually wide azimuth field, using an antenna with relatively low gain in the azimuth dimension. To retain sensitivity, this gain is recovered by the integration of received echo signals over a much-extended CPI. This may include many hundreds or thousands of pulses. The critical benefit of staring is that the spectra of the scattered echo signals are obtained with very high Doppler resolution, and the details of the target and its motion can be extracted, not otherwise available when only intermittent signal data is gathered, as with a scanning radar. The critical aspects in reception and processing are the time duration of the CPI and the effective exploitation of signal fluctuations through that time and across the receiving aperture. The VoR is divided into many resolution cells in range, elevation and azimuth. Appropriate processing allows cells to be 'discovered' whose signal content differs from thermal noise. Subsequently, from these cells, targets are identified by applying sequences of complex amplitudes forming 'signatures' that can be used in solving analytically for target characteristics and trajectories enabling discrimination of wanted and unwanted targets.

In physics, the Uniqueness Theorem (UT), applied to static charges, provides that a potential field that solves Laplace's equation and satisfies the imposed boundary conditions is the unique solution. For a localized, persistent electromagnetic emission within a volume of free space (the VoR) at a known frequency and power density, the Electromagnetic Uniqueness Theorem (EUNIT), applied to Maxwell's equations with retarded potentials, provides that signal measurements over a defined aperture at the boundary of the VoR are uniquely determined by the positions, conductivities and permittivity's of all the objects within the VoR. These form the boundary conditions, and a solution to Maxwell's equations that satisfies these boundary conditions is also unique. These solutions are not dependent on coincidences between objects and events in the VoR with any radar scanning sequence, and in the limit, they lead to analytic solutions for the VoR.

The time extent of the continuing interrogation of a target has two key consequences that need to be balanced. The first is that information available about the target and its trajectory becomes increasingly precise, detailed, significant and robust as the observation extends. In other words, the signal quality and resolution are enhanced. The second is that there is no inherent limit to the volume of computing required to derive approximations to these analytical solutions – that is, it is not limited to the application of specific algorithms to range, frequency and direction estimates, and, at a certain level, every combination of range, azimuth and elevation needs to be evaluated for every pulse transmitted. It is, of course, limited by the parameters defining the radar system. Whilst the computing burden may have an impact on the applicability of staring techniques, unlike the rigid constraints of physical law, Moore's Law has allowed computing devices to grow and its cost to fall with time. Target analysis of the kind envisaged here can be made affordable within the cost-effectiveness bounds of the surveillance business.

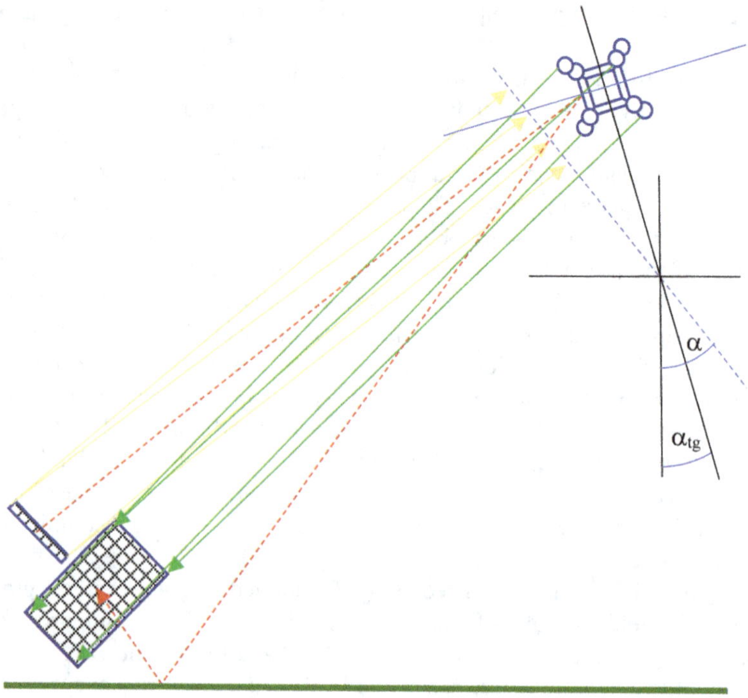

Figure 5.6 *SHR signal propagation geometry*

5.2.1 Geometry of propagation

Figure 5.6 illustrates the network of propagation paths that underlie the formation of radar returns, interacting with transmitter elements, target nodes and receiver elements. The gold arrows indicate transmissions forming a plane wave near the target; green arrows are scattered signals forming plane waves near the receiver. The red trace represents a multipath return. The black axes illustrate the reference axes for radar and target orientations α and α_{tg}.

All these paths will yield superposed return fields with interference effects. The effects vary with α and α_{tg}, with phases for each path and with any time-variation associated with scattering features of the target. Given the complex nature of natural or man-made objects at different geometric scales, these interactions are also complex. However, the complexity can be bounded by appropriate radar characteristics. The first is that for a wavelength λ, features of dimensions smaller than about $\lambda/2\pi$ can be ignored since they cause negligible signal scatters. The second is that the presence of noise in receiver circuits and the limited power of the transmitter, many signals are shrouded by noise unless carefully filtered. The third is that the scale of the receiving antenna limits the geometrical resolution unless time and motion allow aperture synthesis. Processing of data arising from a staring radar is made manageable by controlling these aspects of its design and the VoR.

5.2.2 Fields and information

For small targets such as drones where the object dimensions are of the order of a metre or less, at this scale it means that at the target position the radar signal from the transmitter is effectively a plane wave. The EM fields are described as linearly and vertically polarized, omnidirectional source, transmitting from a location that is effectively a single distant point.

The incident planar wavefront at the target is described in terms of the vertically polarized electromagnetic field $\mathbf{E}_{\mathbf{inc}}(\varepsilon, a, r, t)$:

$$\mathbf{E_{inc}}(\varepsilon, \alpha, \mathrm{r_k}, \mathrm{t}) = \sum\nolimits_{\mathrm{k}} \mathrm{Z_0.I_T.a_k.cos}(\varepsilon)/\mathrm{r}) \,.\, \exp(\mathrm{j}\omega_0(\mathrm{t} - \mathrm{r_k/c}))\,.\,(\hat{\mathbf{i}}.\cos(\varepsilon) - \sin(\varepsilon).(\hat{\mathbf{a}}.\cos(\alpha) + \hat{\mathbf{o}}.\sin(\alpha)), \tag{5.1}$$

where Z_0 is the impedance of free space, I_T is the peak current flowing in the transmitter, $\mathbf{a}_k$ is the vector along which the current flows, the subscript k denotes transmit array elements, ε is the look angle of elevation and α is the look angle in azimuth, ω0 is the angular frequency of transmission, λ is the wavelength, r_k is the range from each transmitter element, $(t - r_k/c)$ describes the retarded potential for each element at range r, and $\hat{\mathbf{i}}$, $\hat{\mathbf{a}}$ and $\hat{\mathbf{o}}$ are the unit vectors in which the VoR and the Poynting vector of transmission are defined.

A drone comprises a rigid body together with many rotating parts such as the rotor blades. Elements of the drone that yield the principal reflections are components containing hard plastics (e.g., carbon fibre) or metals, including batteries, motors, structural beams and rotors whose dimensions exceed λ/2π. For a staring radar operating at L band, this is approximately 4 cm – relevant to a large proportion of civil, commercial or security-relevant drones. A numerical model of a drone can be built as a structure with chosen symmetries, incorporating a number of nodes, each modelled as a superposition of small spheres. The radar cross-section of a sphere is the optical cross-section σ_n. For a conductive sphere, there is a resonance where $\lambda = 2\pi r$, yielding $\sigma_n = 2\pi r^2$, so $\sigma_n = \lambda^2/2\pi$. The model uses effective radar cross-sections for each sphere equal to σ_n, reduced for dielectric materials by a factor between 10 and 100. The backscatter wave from the drone can be modelled using the Huygens' Principle [22] to predict the backscatter echo as a function of the target range, orientation and speed.

The model predictions can be validated using known examples of signatures from drones, with a view to using it to explore the ability of a staring radar to discover and extract relevant target information and to classify different target types. The planar waveform scattered from the target to the receiving array is given in terms of the electric field by:

$$\mathbf{E_{scat}}(\varepsilon, \alpha, \mathrm{r}, \mathrm{t}) = \sum n(-sqrt(\sigma_n).\mathbf{E_{inc}}/r).exp(j\omega_0(t - (2r_n)c)).(\hat{\mathbf{i}}.cos(\varepsilon) - sin(-\varepsilon).(\hat{\mathbf{a}}.cos(\alpha) + \hat{\mathbf{o}}.sin(-\alpha))) \tag{5.2}$$

where n denotes the set of spherical scatterers. Then the signal received at each receiving array element m is described in terms of the scalar E_{scat}:

$$\mathrm{V_{rec}}(\varepsilon, \alpha, \mathrm{r}, \mathrm{t}, \mathrm{m}) = \mathrm{E_{scat}.sqrt}(\mathrm{G_m}(\varepsilon, \alpha))\,.\exp(\mathrm{j}\omega 0(\mathrm{t} - \Delta \mathrm{r_m}(\varepsilon, \alpha)/\mathrm{c})) \tag{5.3}$$

where E_{scat} is the scalar product $\mathbf{E}_{scat}$. (î.cos(ε) – sin(ε).(â.cos(α) + ô.sin(α))

Substituting $\mathbf{E}_{scat}$ from (5.2) and multiplying the unitary outgoing and incoming inverted unitary 'Poynting vectors' and yielding a (negative) unitary scalar. This provides the basis for an analytical solution of the received signal in terms of the transmitted beam from k transmitting elements, the target and its n constituent parts, and the receiving array of m elements. Beams are formed at each range r by applying appropriate phase shifts to each receiver element m and summing. Angular position is estimated by comparing beam amplitudes, yielding the values of ε and α.

A preferred method for this process is to form an initial set of beams by the usual complex Fourier transformation, but then to measure the gradient of the residual phase (–180 < f < 180) across the array (in the case of azimuth measurement) in the peak beam to find the actual direction of reception. This method has the advantage that it is based directly on the complex amplitudes. If the relative element column phases in the Doppler domain are calculated as a function of time, from CPIs offset by chosen fractions of the CPI, then the azimuth angular rate can be calculated directly, rather than relying on amplitude-only measurements that decrease in accuracy as they approach zero. It remains to be seen how these different measures compare in terms of accuracy.

Beyond the ability to measure trajectories analytically, the EUNIT provides that the received signals will contain information leading to analysis of other aspects of the target. Any such information may relate directly to the geometry and the behaviour of the target, potentially providing a direct source for classification. Since many birds have radar cross-sections comparable with those of drones, this can potentially lead to a high FP rate and an ability to differentiate between them will greatly aid classification accuracy.

5.2.3 *Understanding micro-motion characteristics*

The primary mechanism by which information concerning the target is encoded in its radar return is through the Doppler effect. As the target progresses on its trajectory, and as its range to the radar changes due to the radial component (Δr) of its motion, its phase changes according to $\Delta\phi = 2\Delta r \,.\, 2\pi/\lambda$. If the PRI is t_{PRI}, then the rate of change of phase during a sequence of pulses is

$$\Phi/t_{PRI} = 2\pi \,.\, 2f_c \,.\, V_r/c \tag{5.4}$$

which is identically the Doppler shift angular frequency.

The Doppler shift is directly related to the radial speed (V_r) of the target; it can be measured directly and provides coherent integration proportionate to the CPI in the case of constant V_r. Figure 5.7 shows the time domain and Doppler spectrum for a fixed-wing drone at constant radial speed and it can be observed that the target that was buried in noise in the time domain signal is detectable in the Doppler FFT's data.

Figure 5.8 shows the case when the target is accelerating at 10 m/s^2 and where the signal losses coherence resulting in 12 dB reduction. However, zooming in on the Doppler (Figure 5.8 bottom) shows that the effect for an inertial acceleration is phase modulation of the return, spreading it over about 100 Doppler bins. This is not

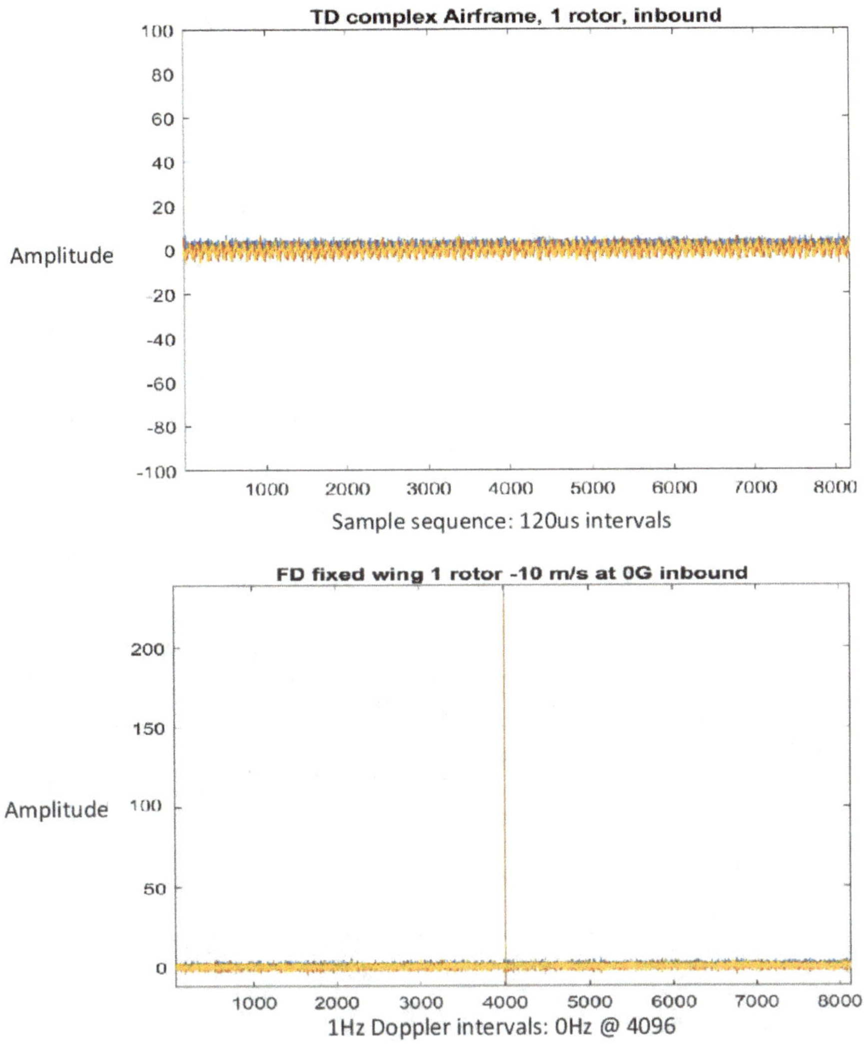

Figure 5.7 Simulated data for a single-rotor fixed-wing drone inbound at 10 m/s. Top: Time domain, horizontal axis is sample sequence and vertical axis is amplitude. Bottom: Frequency domain, horizontal axis is Doppler bin and vertical axis is amplitude.

incoherent and can be recovered by appropriate dispersive filtering. In fact, although the spectral peaks associated with airframe and rotors are suppressed in amplitude, the same signal energy appears in the accelerated version, and inspection of the phases of the components shows that this is a form of phase modulation rather than a loss of coherence. It can be shown that the effects of smooth acceleration forces on an inertial body with reasonable power and control can be detected employing

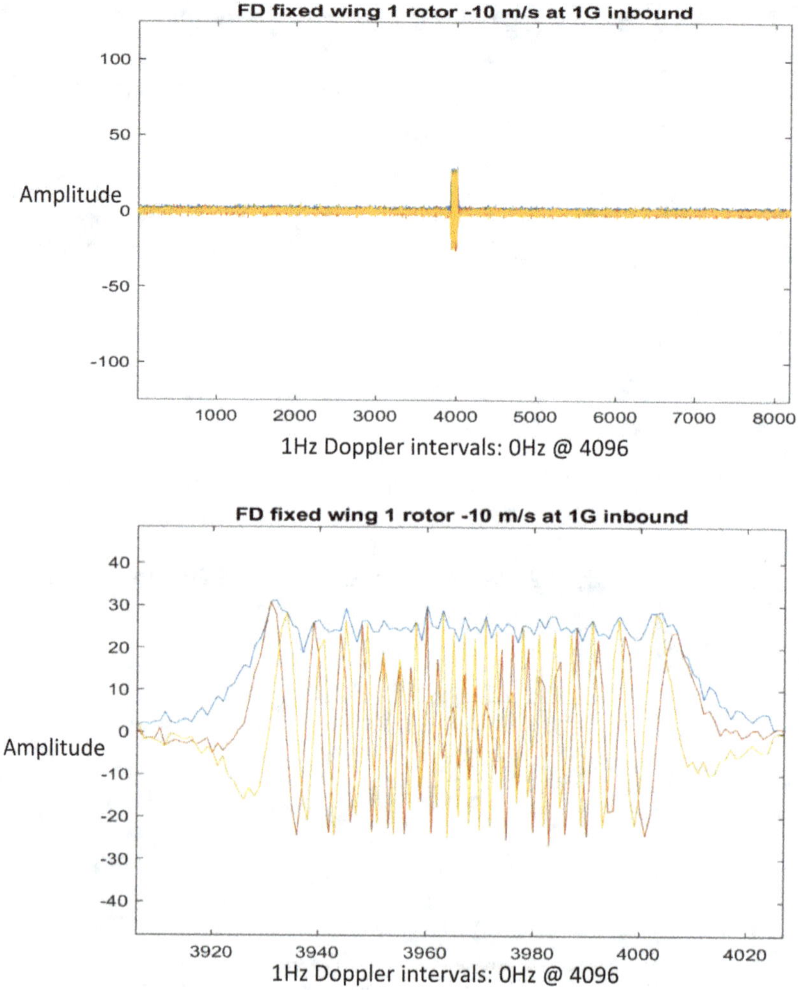

Figure 5.8 Simulated data for a single-rotor fixed-wing drone inbound with acceleration 10 m/s². Top: FFT data. Bottom: Zoomed-in FFT data. Horizontal axis is Doppler bin and vertical axis is amplitude.

simple linear filtering and can be recovered by means of manageable, dispersive digital filtering.

In addition to the rigid body Doppler effect, the returned spectrum represents additional target information. The drones use rotating components to maintain and control flight. Their effect can be described in two ways. Each rotor consists of a number of short, nearly straight mechanical blades, rotating in twos, threes or fours about a centre. Where the blade is comparable with or larger than a half-wavelength at f_c, it scatters and diffracts a beam centred at twice its own angular offset from the radar direction and the reflection, therefore, rotates at twice the rate of the rotor. This effect arises because the reflecting power of the blade is larger when it is aligned with the transverse electric (or the magnetic) field vector of the incident wave (and when, in the model, 'scattering nodes' sum constructively) compared with when it is aligned with the Poynting vector, when scatter components from along the length of the blade (represented by successive scattering nodes in the model and separated by a quarter wavelength or more) sum to a smaller or null total. This swept beam corresponds to an additional component linearly superposed on the drone radar return, with phase determined by its position relative to the scattering centre of the drone. It is typically small compared with the primary scatter; however, it linearly modulates the return amplitude.

The fundamental modulation frequency is at twice the rotation rate of the blade. A 2-bladed, short rotor results in an azimuth-defined beam that is best described by a fundamental, sinusoidal modulation at twice the rotation rate. This will result in upper and lower sidebands in the Doppler spectrum, separated from the airframe Doppler component on each side by twice the rotation rate. A short two-bladed rotor at 6 000 rpm will generate sidebands at 200 Hz above and below the airframe Doppler return.

For longer blades, especially for even number of blades, the rotor beam is expected to be narrower than would be represented by a pure azimuth sinusoid. Where the blade or pair of in-line blades approaches one wavelength, the rotor reflects what are effectively beams rather than sinusoids that sweep the horizon. The effect is that for each rotation of paired blades, the swept beam being narrower than 90,° the amplitude modulation describing the rotor beam rotation includes components at upper harmonics of the fundamental. In that case more than one harmonically related sidebands appear on either side of the airframe Doppler.

The fundamental modulation frequency is related only to the spin rate of the rotor and not to the carrier frequency. It is therefore not strictly an example of the Doppler effect, in which the shift is proportional both to the rotation rate and the carrier frequency. The Doppler effect is seen in the limit in frequency of the sequence of harmonic rotor sidebands that appear with the airframe Doppler and is proportional to both the tip speed of the rotor blades and the carrier frequency. The number of harmonic components is therefore proportional to the ratio between the blade length and the carrier wavelength, which naturally determines the number of diffracted and swept beam components.

These rotor components are small when compared with the drone airframe, batteries, motors, etc. and may not always be detectable based on the radar power budget and threshold. Birds, on the other hand, use wings that beat at much slower and less regular rates than those of drone rotors, and the difference between the arising sidebands may be used to discriminate precisely between drones and birds.

When a drone (airframe and all components) is detected above a given threshold, a search may be made for smaller associated spectral components whose presence would assure the classification of the target as a drone. Provided that the total rate of detections is appropriately limited by the airframe threshold level, the level used to detect these symmetric, harmonic spectrum components can be reduced

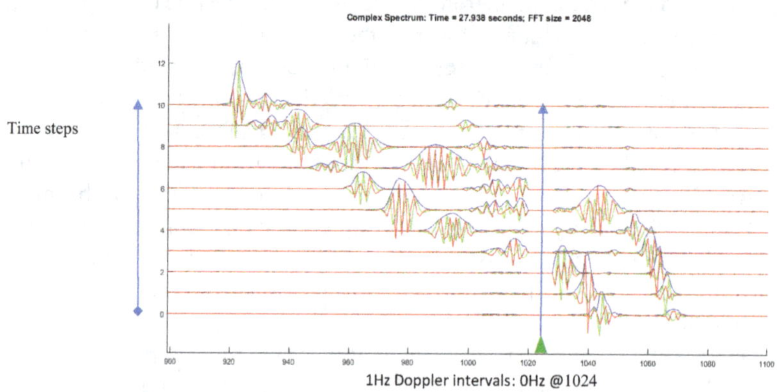

Figure 5.9 *Real spectrogram of sea birds. Horizontal axis is Doppler bin. Vertical axis is time step where step 10 is equal to 5 second lapsed time.*

significantly. To make this a robust part of the process of detecting and classifying drones, the radar power budget must take into account the effective cross-section of the drone, the airframe threshold, the cross-section of the rotor, the minimum level for detection of rotor components or pairs of rotor components and potentially a maximum range at which such classification process is to be applied.

To aid the understanding of the micro-motion Doppler returns two drone types are modelled.

1. a fixed-wing drone with fuselage, wing and tailplane and fin, with a propeller rotating in the vertical plane at its nose
2. a quadcopter with a cruciform airframe, central batteries, motors at each axis and a 2-bladed rotor at each motor.

As a basis for comparison, Figure 5.9 illustrates the Doppler returns of two sea birds. There are two possibly interacting birds within a single range bin of 150 m over 5 seconds. Both targets start with positive Doppler, pass through tangential flight near zero Doppler (where they are suppressed) and appear to converge on to approaching trajectories. These are believed to represent birds, plus some small sea clutter features in addition. These spectral signatures however do not appear to coordinate with each other or with any stable Doppler harmonics. This can be compared with the simulated (Figure 5.10) and real (Figure 5.11) Doppler profile for a single-rotor fixed-wing drone, travelling inbound at 10 m/s and 60° azimuth offset.

The modelled spectrum evolves as the model rotates, varying the radial speed components of the propeller blade tips. Figure 5.12 shows the spectrum at 90° azimuth where the propeller excites the next harmonic of the signature more substantially.

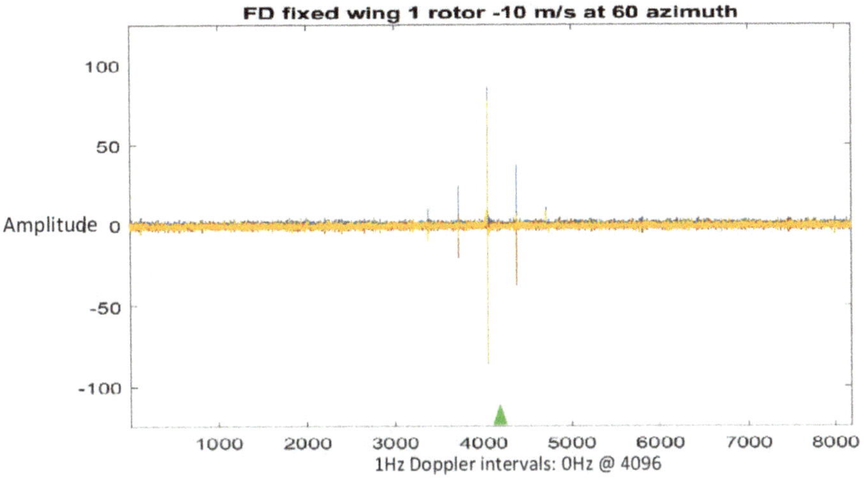

Figure 5.10 *Simulated Doppler FFT data for a single-rotor fixed-wing drone at −10 m/s at 60° offset. Horizontal axis is Doppler bin, vertical axis is amplitude.*

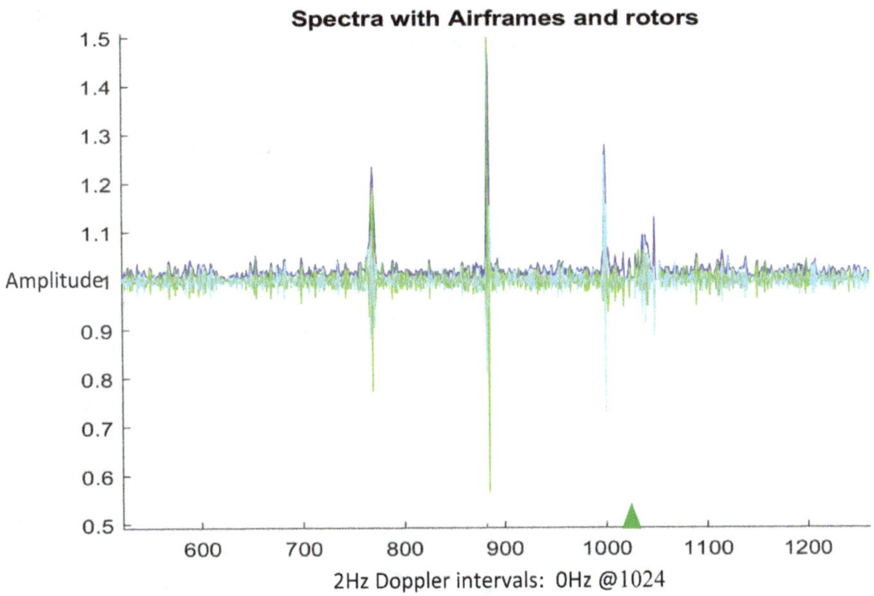

Figure 5.11 *Real Doppler FFT data for a single-rotor fixed-wing drone at −10 m/s at 60° offset. Horizontal axis is Doppler bin, vertical axis is amplitude.*

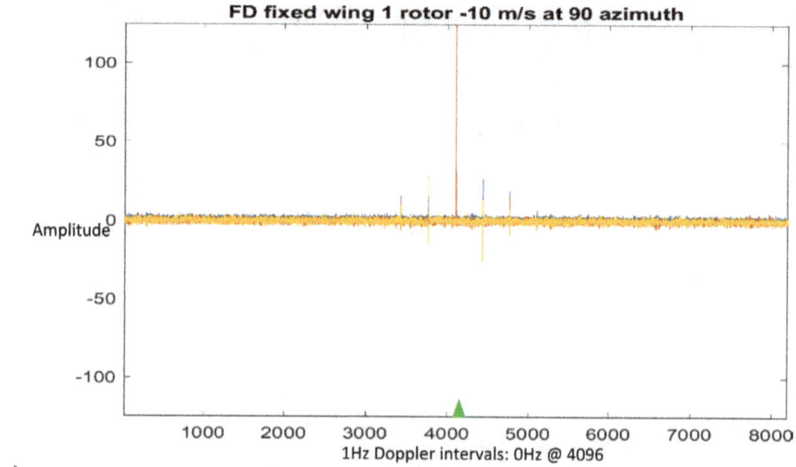

Figure 5.12 *Simulated Doppler for single-rotor fixed-wing drone at –10 m/s and 90° azimuth. Horizontal axis is Doppler bin, vertical axis is amplitude*

The corresponding model data for a multi-rotor drone is shown in Figure 5.13. The frequency-domain data reflect the different Doppler return due to the varying rotor speeds. The example data from a real six-rotor rotary-wing drone is shown in Figure 5.14. Many Doppler sidebands are visible.

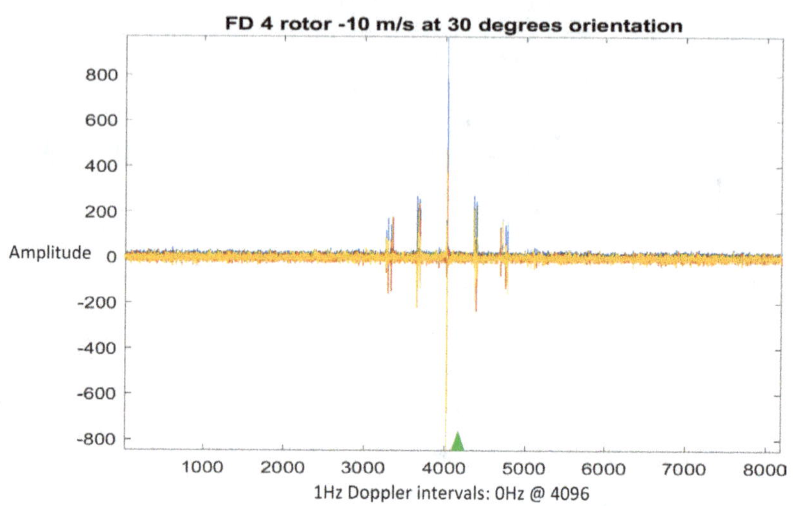

Figure 5.13 *Simulated Doppler for four-rotor Drone. Horizontal axis is Doppler bin, vertical axis is amplitude.*

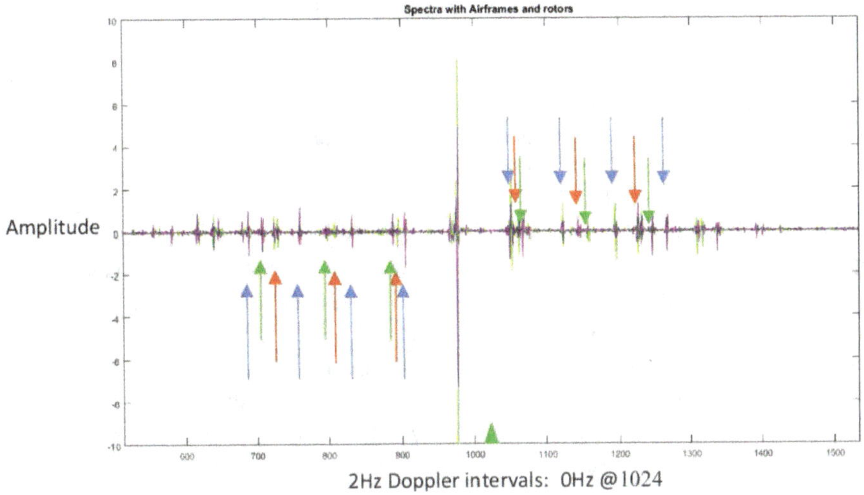

Figure 5.14 *Real Doppler signature of a hexacopter drone. Horizontal axis is Doppler bin, vertical axis is amplitude.*

In Figure 5.14, coloured arrows indicate component pairs and harmonic sequences. In detail, the blue series is separated by 72 bins, the red series by 83 and the green series by 91, indicating rotor rates near 4 320, 4 960 and 5 440 RPM, and suggesting that the six rotors operate in pairs. The red series is less well-coordinated.

Here, several examples of modelling fixed-wing and rotary-wing drones for a staring radar are presented and compared with actual real data for both drones and NDs. Key aspects of the model formation and the signal processing used is that they refer to signal data before the point of detection, and they are linear and do not in principle degrade the signal information content as they progress.

According to the EUNIT they are also unique. That is, a physical process of EM scattering that differs in form from the model – 'an inertially translating and rotatable airframe with spinning rotors' – will generate complex radar returns that differ detectably from returns generated by this model and that are observed in the returns from actual drones. To create an equivalent return from a different airborne body would violate the UT, and we have not identified such an alternate solution. To create an artificial return by other means would not be impossible but would require a sophisticated receiver and signal synthesizer to 'spoof' the radar.

It is possible that measurements of the trajectory of an object beyond those encoded directly within the signals may also be recognizable as one of a number of classes of target – those of interest to the user and those described as 'clutter'. Section 5.3 discusses 'machine learning' approaches that can aid with the discrimination of drones from ND confuser targets.

5.3 Robust drone classification using a staring radar

As mentioned at the beginning of the chapter, the increased prevalence of drones and the huge concern for their safe and secure operation in the presence of manned aviation is an issue of major public interest. The trends in Unmanned Aerial Vehicle (UAV) technologies suggests that in the near feature low altitude airspace could be significantly populated by many kinds of drones. Staring radar is an important design choice for non-cooperative surveillance as it is able to provide high sensitivities and high Doppler resolutions to detect and track drones that have a low RCS and fly at relatively low velocities and altitudes [23–25]. The high sensitivity requirement means that other slow-moving targets with low RCS, such as birds and ground targets, are also detected. These (ND) targets are dubbed *confuser targets* from a drone surveillance system perspective. Consequently, the operator and a supporting secondary system (e.g., optical or RF-based) can be easily overwhelmed if they must analyse or interrogate all objects. Therefore, discriminating drones from other targets is key for successful non-cooperative drone surveillance. In this section, a simple, yet effective, classification approach is presented that can deliver a robust drone classification capability for a staring radar, particularly by minimizing the probability of FPs.

Several studies address the problem of discriminating detected targets from radar echoes that specifically include UAVs [26–33]. A few of the existing approaches, such as in [26–28], use kinematics features of the tracked objects, such as their velocity, acceleration, jolt, curvature. Others rely on micro-Doppler information from rotaries on-board the drone, for instance [30–32]. In this section, both kinematics and the presence of micro-Doppler components are utilized to enhance the classification performance. Hence, unlike [26–28] the approach leverages both types of differentiating features whilst not being limited by the presence of micro-Doppler components as in [30–32]. The latter information might not be available or only intermittently reported due to low signal-to-noise ratio scenario (e.g., at relatively long ranges covered by the staring radar) and/or small or concealed/invisible rotatories (e.g., on fixed-wing drones). Additionally, here real 3-D staring radar data from naturalistic live drone flights at Port of Antwerp undertaken in the SAFIR project [34] are used to validate the introduced classifier, contrary to [26–32]. The SAFIR trials were funded by the EU Single European Sky ATM Research (SESAR) Joint Undertaking [35] programme that is pursuing a range of projects that aim to develop a capability to enable safe and robust complex drone operations with a high degree of automation.

A key component in drone traffic management is to ensure that non-cooperative drones can be tracked reliably and accurately in real-time with minimal false tracks. This requires high levels of classification performance. In this section, results are presented for a simple and easily interpretable classification approach that is implementable in real-time and delivers high performance in practice. Interpretability or transparency is particularly important in the considered application area, especially for debugging and analysing any unexpected classification outcome. Hence, black-box-type solutions based on neural networks, for example, in [28, 30] and [31], are not examined. This is albeit their ability to reliably detect micro-Doppler components and the fact that in theory they can be employed within the staring radar system. A recent overview of various classifiers for UAVs under supervised or unsupervised-training regimes is present in [33].

Here, a simple and easily interpretable two-stage DT classifier is employed. The first-stage tree discards confuser ND targets, that is it incorporates a prior on the characteristics of the sought targets and circumvents overfitting to particularly important features such as heights of the detected targets. The second-stage tree is trained on selected diverse features within a supervised-learning framework. Evaluation using data from the SESAR SAFIR drone trials demonstrates that the introduced approach delivers excellent real-world performance, improving on a one-stage DT as in [25]. For example, it reduces the FPs rate. It is noted that the second-stage tree can be replaced with any other supervised-learning classification technique, for instance, Support Vector Machines (SVM) or random forests using the same identified features. However, such a comparison is outside the scope of this chapter.

5.3.1 Staring radar system overview

5.3.1.1 Platform and capabilities

The Gamekeeper radar (Figure 5.15) has a 4 × 16 receiver array in azimuth and elevation. The transmitter is a vertical stack of eight antennas to narrow the transmit beam in elevation providing more gain at low altitudes. These and other features are incorporated to provide better sensitivity against low altitude, low observable targets. Table 5.1 lists the operating parameters for the Gamekeeper radar.

The radar transmits a pulsed waveform which on receive is digitized at each receiver element level. For each range gate, the samples from all the receiver channels are processed to form multiple receive beams and the pulses are coherently processed to obtain the Doppler samples thus providing a 4-D data matrix in range, azimuth, elevation and Doppler for each frame. CFAR thresholding is applied to each 4-D frame and the detections are localized in position and after some pre-tracker filtering is passed to a tracker that assigns track ID to the detection plots. Feature values are generated for each track for each frame update. The track features are derived from the kinematic features but also contain features relating to the Doppler spectrum. These tracker features are used by a classifier to assign class label and the classification label is updated each frame for all tracks. The radar outputs track with track ID, positional information and classification label per frame update.

5.3.1.2 Target features and discrimination

Features derived from targets tracked by the radar form the basis for discriminating between target categories. The Gamekeeper staring radar is designed to have high sensitivity to detect small drones and consequently also produces a large volume of detections from confuser targets. Figure 5.16 shows an example output tracks produced by the radar over a period of roughly four minutes duration. The radar is positioned at the top right-hand corner of the image that

Figure 5.15 *Gamekeeper L-Band multibeam staring radar showing the radome mounted on top of an International Organization for Standardization (ISO) container that house the processing servers*

Table 5.1 Gamekeeper L-Band multibeam staring radar system parameters

Parameter	Value
Frequency	L band
Bandwidth	~2 MHz
Transmit power	~2 kW
Receiver channels	4 × 16
Azimuth coverage	90°
Elevation coverage	30°
Pulse repetition frequency	~7.5 kHz
Update rate	~0.25 seconds
Polarization	Vertical
Drone detection range	5 km

shows a plan view marking the 90° azimuth field of regard with solid pink lines. The white track is that of a rotary-wing drone. The cluster of tracks over roads are ground vehicles. The rest of the tracks are low altitude airborne targets that are most likely to be birds.

Although the tracks appear quite random, there are certain patterns in the track profiles that can be exploited to differentiate drones from NDs. For example, kinematic features relating to speed, acceleration, height above

Figure 5.16 Gamekeeper radar output overlaid onto map showing a 4 minute history of reported tracks (purple lines). The drone track is shown by the white line.

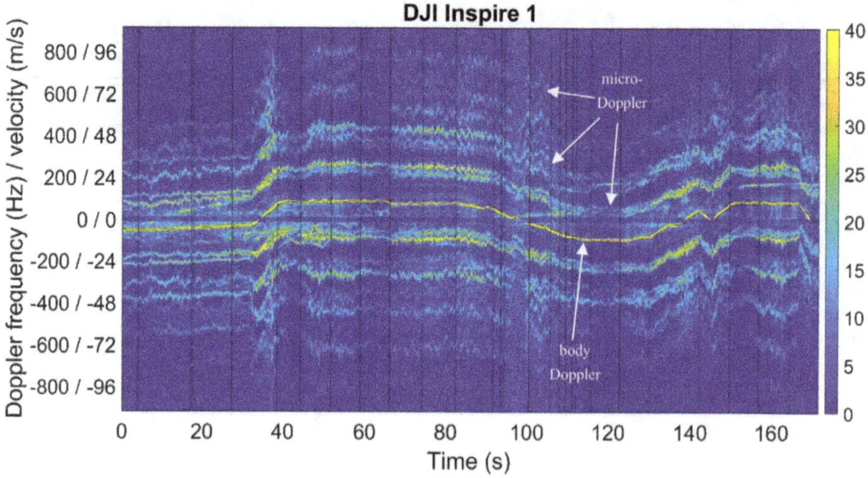

Figure 5.17 Example spectrogram from a rotary-wing drone with the Gamekeeper staring radar

ground and higher-order derivatives can be used to separate between the target categories. Whilst simple rules can be devised to separate ground targets and high-flying fast aircraft from drones, the same is not true for birds versus drones. The reason is that birds and drones occupy airspace in a similar manner, flying at similar altitudes and with similar velocities. Thus, other features where the target characteristics are more discernible are needed to provide good separability between the desired and confuser targets.

The extended dwell achieved with the staring radar enables sufficient Doppler resolution to reveal micro-Doppler characteristics of targets that can assist with targets that otherwise may appear to have similar RCS and flight profile. An example spectrum is shown in Figure 5.17. The vertical axis is the Doppler frequency and velocity in m/s and the horizontal axis is the time in seconds. Thus, there is one spectral line output per frame. Zero Doppler corresponds to static ground clutter and a filter has been applied to remove the zero Doppler return. The target in this case is a rotary-wing drone and its body Doppler along with distinct micro-Doppler signatures resulting from the rotor motion are clearly visible. On the other hand, Figure 5.18 is a corresponding spectrum for a bird that happens to be travelling roughly at the same speed as the example chosen for the drone signature in Figure 5.17. The lack of periodicity in the Doppler can be easily noticed in this figure along with its different overall character compared to that of a drone. Thus, the kinematic features can be augmented by additional features attributed to the Doppler characteristics. In the first instance, this is simply the number of track components that can be attributed to a tracked target. The relationship between the features are still rather complex and machine learning (classification) methods can be utilized to enable the discrimination between drones and confuser targets.

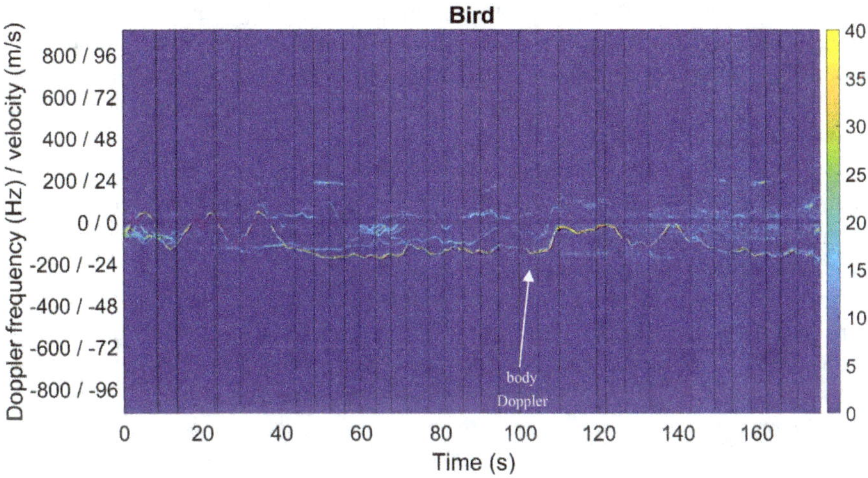

Figure 5.18 *Example spectrogram from a bird with the Gamekeeper staring radar*

5.3.2 Classification: a two-stage tree approach

A simple, easily interpretable, two-stage DT [36] is employed here to discriminate between drones and other confuser targets such as vehicles, pedestrians, birds, airliners and helicopters. An example of such a two-stage tree is depicted in Figure 5.19. It tackles a binary classification problem with each tracked object being identified

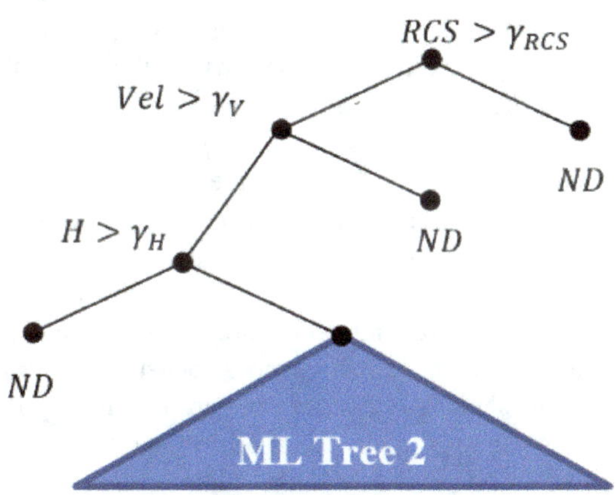

Figure 5.19 *An example two-stage DT showing coarse tree 1 (fulfilling the node condition leads to the child node on the right). RCS, Vel and H are features for target radar cross-section, velocity and height, respectively, and* γ_{RCS} *,* γ_V *and* $N = N_D + N_{ND}$ *are the feature split values that are set a priori in stage-one tree.*

as Drone (D) or ND. This formulation is motivated by the requirements of non-cooperative drone surveillance systems aimed at tracking UAVs as well as the practical difficulties in obtaining ground truth for all present confuser targets, such as birds [37, 38]. Whereas during control trials, drone targets can be identified based on their on-board GPS data.

As in Figure 5.19, the classifier comprises two augmented trees: the first-stage tree is *'hand-crafted'* based on simple rules that use kinematics data derived from the tracker output of the staring radar and the second stage is machine learnt from selected training data containing kinematics and micro-Doppler features. Whilst the radar tracker output includes kinematics information in 3-D (e.g., Cartesian position, velocity and acceleration), the micro-Doppler information, namely the number of detected micro-Doppler components, is provided by a dedicated system module. The rationale behind the two-stage approach and the adopted training regime is discussed next.

5.3.2.1 Stage-one tree (tree 1)

The available training data on selected features, such as the height of the target, which can be key for identifying certain categories of tracked targets (e.g., ground targets), is insufficiently diverse and samples of only a very limited range of possible values are available. For example, a few height values for drone targets are over-represented in conducted trials since a UAV typically flies autonomously from a set of waypoints at a nearly constant height. Thus, using height to train a classifier, for example, DT as in [37, 38] or SVM, can result in overfitting the classifier to the available data and skewing the training process. Additionally, other targets can be easily labelled as ND, for example:

- *Ground target:* if their height above ground is less than a predefined threshold,
- *Large and fast ND targets:* if their RCS is considerably larger than drone targets of interest and they are travelling at unexpectedly high speeds.

Thus, such confuser targets can be eliminated by tree 1, especially that their number can be relatively large as shown in Section 5.3.3. This in effect incorporates a priori known characteristics of the ND targets and salient features (namely those that are not well represented in the available data sets such as height) are excluded from any (machine learning) training procedure to circumvent overfitting to the available training data. This can lead to a more robust classifier as demonstrated when compared with the standard one-stage standard DT in [38]. It is noted that the values of the thresholds (e.g., γ_{RCS} and γ_H) can be set (learnt) from data, for example, using a maximum likelihood procedure. Other simple rules (in addition to height, RCS and velocity) can also be applied at this (first) stage.

5.3.2.2 Machine learnt second-stage tree (tree 2)

At this stage, selected kinematics features (e.g., average velocity and acceleration for a given track) and micro-Doppler information are used to train a DT as in [38].

Supervised learning relies on labelled data where drone on-board GPS data (i.e., ground truth) is utilized to associate a given track with the drone target. Each sample of the training data, for example, for the *n*th radar frame, can be represented by the pair $\left(\mathbf{F}^{(n)}, y^{(n)}\right)$ such that:

1. $\mathbf{F}^{(n)}$ is the features vector of dimension K extracted from the radar current frame (and possibly previous frames) at time t_n. This feature can be derived from a statistical measure, such as average, of time series of a given kinematic (e.g., velocity) or micro-Doppler information (e.g., average number of micro-Doppler components).
2. $y^{(n)}$ is the identity of the object from which $\mathbf{F}^{(n)}$ is extracted, for example $y^{(n)} \in \{0, 1\}$, where 0 = Not Drone and 1 = Drone.

Therefore, the overall training dataset, which consists of N sample data points, pertaining to all tracked targets in the radar field of view, is given by

$$\mathbb{D} = \left\{\left(\mathbf{F}^{(1)}, y^{(1)}\right), \left(\mathbf{F}^{(2)}, y^{(2)}\right), \ldots, \left(\mathbf{F}^{(N)}, y^{(N)}\right)\right\} \tag{5.5}$$

where $N = N_D + N_{ND}$ such that N_D and N_{ND} denote the number of drone and ND data points, respectively.

The classifier is then trained as in [36, 38] with a limited number of nodes/leaves (e.g., optimized based on the Gini index cost function and/or gradient boosting approaches). It also allows for the option of *k-fold* optimization and pruning to prevent over-fitting to the provided training data. We recall that any other standard (supervised learning) classification approach can be used at stage 2, in lieu of a DT. For further details on the training procedure, classifier output and the overall set-up, please refer to [36]. A key motivation of employing a DT classifier is that the learned features are transparent and physically meaningful. One of the previous key limitations of using a single-step DT in [38] was that there were no priors incorporated into the DT and all learning was performed on the training data presented to the classifier (including height). This made it prone to overfitting and not easily generalizable to other unseen (out-of-training) datasets. Introducing a simple rule-based hand-crafted tree leads to a two-stage tree, which is shown to lead to a more robust classifier as illustrated by results from real radar field trials detailed in the next section.

5.3.3 Field trials results

5.3.3.1 Training data

The approach adopted for training and testing is to take data from a reference measurement site and train the classifiers using labelled data and then test the classifier with data from a different site. This is representative of operational systems where the algorithms including the classifier need to be configured prior to the radar deployment at a client site. For this work the reference site used for training is a trial

(a)

(b)

Figure 5.20 Rotary-wing test drone (a) DJI Inspire 1 (b) DJI Inspire 2

site based at an ex-RAF airfield situated in Deenethorpe, Northamptonshire, UK. The test site used in this case is the SAFIR demonstration site in Antwerp, Belgium.

The Deenethorpe site is in a rural location with no large buildings and predominantly comprising open countryside. A DJI Inspire 1 (Figure 5.20 a) was flown based on two flight profiles as depicted in Figure 5.21. In each case the launch point is 0.9 km from the radar. Upon launch, the target climbed to the altitude of 110 m and flew to each waypoint and then descended back to the ground where it returned to the take-off point. In each instance, all track data for several minutes corresponding to the duration of the drone flight are recorded. Each track generated by the radar has a unique track ID and associated track features that are updated every frame. The track ID associated with the GPS truth of the drone is labelled as drone and all other tracks are assigned the label NDs. Altogether the training data was taken from six periods, two instances of each flight profile shown in Figure 5.21 and two where there was no drone target although a light aircraft was contained in the radar field of view. Table 5.2 lists the number of data points for drones and NDs that are contained in the full training set.

The one-stage DT was trained on six features. The same features were used in the two-stage DT with an addition of a further feature of height which is used to separate ground targets from airborne targets. Note, as the two-stage DT is setting the feature split for height a priori, it avoids the pitfall of overtraining on data for the case where all the drone data is from flights at a constant height. Table 5.3 lists the DTs obtained using each method. For the one-stage DT, the training is performed on all the data, but the ND data is reduced from a total of 208 982 labels to maintain a 1 to 3 ratio between the data for the two classes. For the two-stage DT, the first tree excludes a large proportion of the ND samples leaving 95 581 samples for training tree 2. Of the remaining data, a further random selection is used to limit the number of non-drone samples to be no more than three times the drone samples. The next

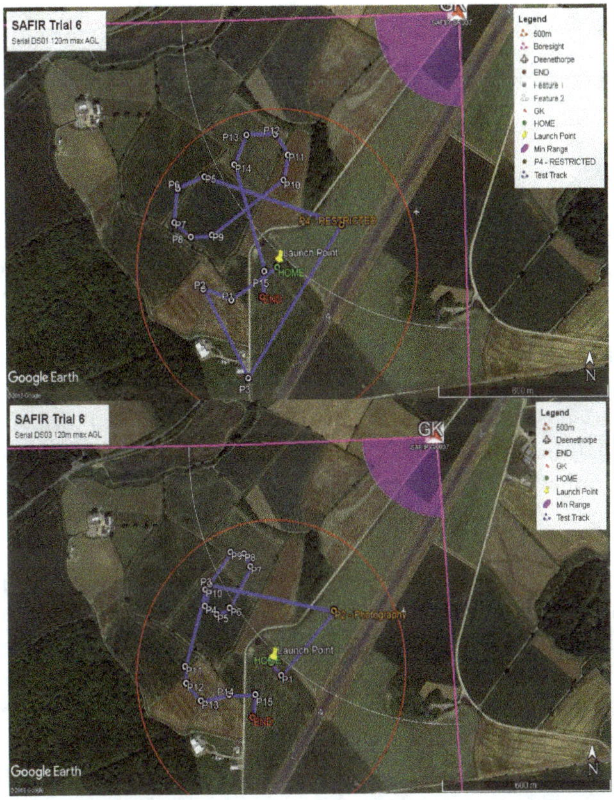

Figure 5.21 Planned flight profiles with the DJI Inspire 1 rotary-wing drone

sub-section reports the classification results obtained with test data taken from a different location and a different flight profile.

5.3.3.2 Test data

The test data is from the SAFIR Demonstration flights where the radar is located to the south of Port of Antwerp and the site comprises a mix of residential and industrial buildings. The radar is shown in the top right-hand corner in Figure 5.22. The blue polygon is the region over which permission was given to fly the drone. The test profile is shown by the waypoints P1 to P7. This represented a scenario of a leisure

Table 5.2 Summary of training data

	Non-drones (N_{ND})	**Drones (N_D)**
Number of data points	208 982	4 626

Table 5.3 Summary of training data

	One-stage DT	Two-stage DT
Features used	6	7
Number of nodes	74	65
N_D training data	4 626	4 622
N_{ND} excluded by stage 1 tree	N/A	113 401
N_{ND} in final training data	9 252 (from 208 982)	9 244 (from 95 581)

drone that is to be monitored for protection against intrusion into restricted airspace for critical infrastructure. The launch point in this case is 1.8 km from the radar and the flight duration is about 3 minutes. Several missions were flown with the same profile with a DJI Inspire 2 rotary-wing drone (Figure 5.20 b). For each mission that was roughly 3-minute duration, test data from three missions was obtained from the radar track output. The results obtained with the two variants of the DT are shown in Table 5.4. It lists the number of labelled data for drones and NDs in the test data sets. For the considered training and testing data, the one-stage DT produces a True Positive (TP) ratio of 81.83% with a corresponding FP ratio of 0.87% for the two-class case. With the two-stage DT using the same training and test data set the results are 85.02% and 0.37% for TP and FP, respectively. Hence, the two-stage DT outperforms the one-stage DT with regard to both correct classification and false reports. The reduction of the latter (more than halved) is particularly important since many ND data points are typically processed. Thereby, a small reduction in the reported FPs can have significant aggregate (positive) impact on the operator (i.e., reducing

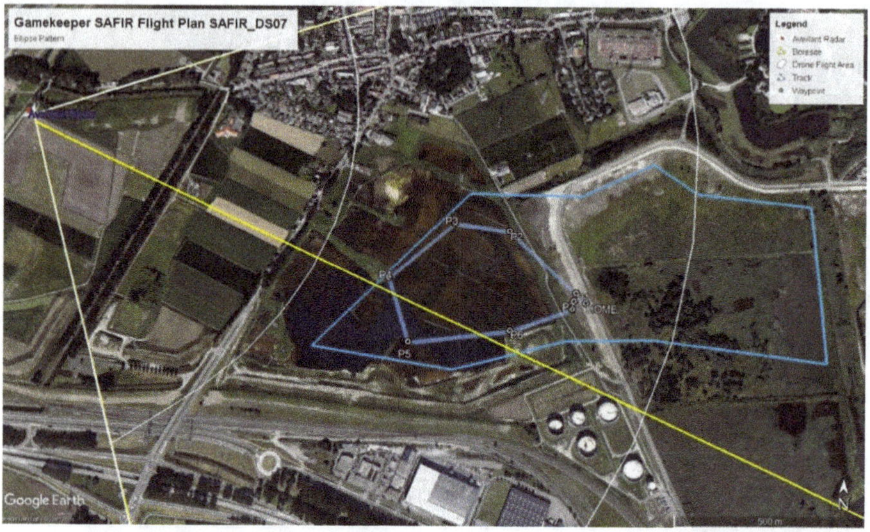

Figure 5.22 Flight profile with the DJI Inspire 2 rotary-wing drone

Table 5.4 Classification results

Test N_{ND}	Test N_D	One-stage DT TP	One-stage DT FP	Two-stage DT TP	Two-stage DT FP
16 150	1 646	81.83%	0.87%	85.02%	0.37%

distractions). This is an improvement on classification approaches that are based on kinematic data only, for example [27] and [29]. Whilst further evaluation might be required, for example, for longer ranges within the radar coverage area and other drone platforms, the achieved enhanced target-recognition performance from the analysed real radar data demonstrates the efficacy of the simple, yet effective, two-stage classification approach.

The classifier was implemented in a real-time processor and used in the live trials at Antwerp. Figure 5.23 shows the radar output from a single drone test flight. It depicts the tracker output with no classifier filtering (Figure 5.23 a) and with ND tracks filtered out and reporting just the tracks classified as drones (Figure 5.23 b). There are no reported false drones, and the true drone is classified correctly. A significant feature aiding the classification is the target micro-Doppler. For this flight, the drone spectrogram is plotted in Figure 5.24. Here, again the micro-Doppler components due to the rotors are clearly visible in the spectrum and are offset from the body Doppler. The distinct micro-Doppler characteristics for the rotary drone are a major factor underpinning the performance of the classifier in this instance.

5.4 Conclusions

In this chapter, we have introduced the concept of staring radar and examined its application to the current challenge of drone surveillance. The fine Doppler resolution inherent in the staring radar concept allows for detailed and continuous information to be collected in the form of spectral properties of target echoes. This spectral information has embedded within it, defining characteristics that facilitate high classification performance. This includes

(a) All tracks (drones and confuser targets) *(b) After filtering out non-drone tracks*

Figure 5.23 Tracker output for the DJI Inspire 2 rotary-wing drone test flight

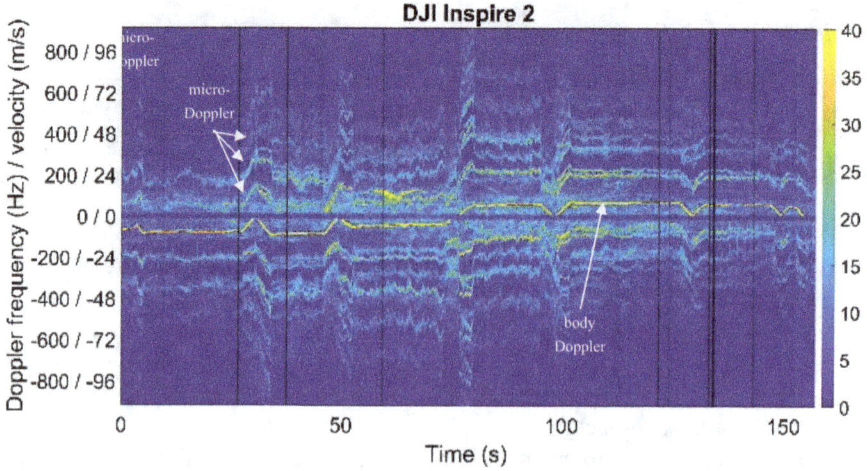

Figure 5.24 *Spectrogram for the drone (DJI Inspire 2) track shown in Figure 5.23*

both short-term characteristics such as those caused by rotating propeller blades and long-term trajectory-based characteristics. Together, these enable drones to be discriminated against the background of other targets such as birds and ground movers and hence providing a robust and powerful surveillance capability. Nevertheless, radar is subject to many of the real-world complications and challenges that are faced by any radar sensor system, staring or scanning. For example, radar is a 'line-of-sight' sensor means that obstacles such as forests and tall buildings will create shadow zones, invisible to the radar system. Thus a drone going into one of these zones would need to be tracked, the zone recognized and the track picked up at an anticipated exit point. This is within the capability of the radar but is a topic for further research. Equally, target to clutter ratios will vary and where they fall detection and location performance will be adversely affected. Again, this is true for any radar system. There are methods such as increased dwell times or Fourier integration that can restore target signal levels if this can be recognized within the signal processing. Such 'cognitive' approaches are, again, the subject of ongoing research.

More generally, the staring radar concept is a much less well-developed technology than its scanning counterpart but is already proving to be a success in several applications and is beginning to be adopted more widely. Most notably, this is occurring in the vehicular radar community where staring radar has been developed in conjunction with a MIMO configuration for fine receive beam resolution. Because of the huge research efforts in vehicular radar, the staring concept is set to make rapid strides in maturation which may well see it adopted for other applications, both existing and emerging.

References

[1] Griffiths H.D., Baker C.J. 'An introduction to passive radar'. *Artech House*. 2017, vol. 13.

[2] Dorey J., Blanchard Y., Christophe F. 'Le project "RIAS": Une approche nouvelle des surveillance aerienne'. *Colloque International sure la Radar, ONERA TP*; Versailles; 1984. pp. 1984–20.

[3] Dorey J., Garnier G. 'The RIAS pulsed synthetic-antenna radar'. *L'Onde electrique*, vol. 69, pp. 36–44.

[4] Dorey J., Garnier G., et Auvray G. RIAS radar a impulsion et antenne synthetique. *Collique International sur le Radar*; April 1989. pp. 556–62.

[5] Luce A.-S., Molina H., Muller D., Thirard V. 'Experimental results on RIAS digital beam-forming radar'. Proceedings of the International Conference on Radar; October 1992. pp. 74–7.

[6] Colin J.-M. 'Phased array radars in France: present and future'. Proceedings of the IEEE International Symposium on Phased Array Systems and Technology; October 1996. pp. 458–62.

[7] Jianqi W., Ruilong H., Kai J. 'Researches of a new kind of advanced metric wave radar'. Proceedings of the IEEE International Radar Conference; 1999. pp. 194–7.

[8] Baixiao C., Jianqi W. 'Synthetic impulse and aperture radar (SIAR)'. *A Novel Multi-Frequency Mimo Radar*. Wiley; 2001.

[9] Wirth W.D. 'Omni-directional low probability of intercept radar'. Proceedings of the International Conference on Radar; Paris; 1989. pp. 25–30.

[10] Wirth W.D. 'Long term coherent integration for a floodlight radar'. Proceedings of IEEE International Radar Conference; 1995. pp. 698–703.

[11] Wirth W.D. 'IET radar, sonar and navigation series'. *Radar Techniques Using Array Antennas*. 2nd edition; 2013.

[12] Skolnik M. 'A different system approach for future air surveillance radars'. The Digest of the Us National Fire Control Symposium; San Diego; 1998.

[13] Skolnik M. 'Improvements for air surveillance radar'. Proceedings of the IEEE Radar Conference; MA; 1999. pp. 18–21.

[14] Skolnik M. Systems aspects of digital beam forming ubiquitous radar. NRL/MR/5007—02-8625; DTIC, ADA403877; 2002.

[15] Skolnik M. 'Attributes of the ubiquitous phased array radar'. Proceedings of the IEEE International Conference on Phased Array Systems and Technology; 2003. pp. 101–6.

[16] Rabideau D.J., Parker P. 'Ubiquitous MIMO multifunction digital array radar'. *proceedings of the 37th Asilomar conference on signals, systems and computers*. 2003, vol. 1, pp. 1057–64.

[17] Alter J.J., White R.M., Kretschmer F.F., Olin I.D., Temes C.L. 'Ubiquitous radar: An implementation concept'. Proceedings of the IEEE Radar Conference; 2004. pp. 65–70.

[18] Carta P., Galati G., Piracci E.G., Madia F., Ronconi R. 'Implementation of the ubiquitous radar concept with a conformal array'. Proceedings of the European Radar Conference; 2015. pp. 441–4.

[19] Harman S.A., Hume A.L. 'Applications of staring radar'. Proceedings of the International Radar Conference; 2015. pp. 0270–3.

[20] Harman S.A. 'A comparison of staring radars with scanning radars for UAV detection: introducing the alarm staring radar'. Proceedings of the European Radar Conference (EuRAD); 2015. pp. 185–8.

[21] Oswald G.K.A. 'Holographic surveillance radar'. *Proceedings of SPIE 7308, Radar Sensor Technology XIII, 73080D*. 29 April 2009.

[22] Huygens' Principle. Available from https://www.mathpages.com/home/kmath242/kmath242.htm.

[23] Jahangir M., Baker C.J. 'Robust detection of micro-uas drones with l-band 3-d holographic radar'. Proceedings of the Sensor Signal Processing for Defense; UK; Sep. 2016.

[24] Jahangir M., Baker C.J., Oswald G.K.A. Doppler characteristics of micro-drones with L-band multibeam staring radar. Proceedings of the IEEE radar conference; May 2017.

[25] Jahangir M., Baker C.J. 'L-band staring radar performance against micro-drones'. Proceedings of the International Radar Symposium. Germany: Bonn; Jun. 2018.

[26] Jochumsen L.W., Østergaard J., Jensen S.H., Clemente C., Ø. Pedersen M. 'A recursive kinematic random forest and alpha beta filter classifier for 2D radar tracks'. *EURASIP Journal on Advances in Signal Processing*. 2016, vol. 2016(1).

[27] Ginoulhac R., Barbaresco F., Schneider J.Y., Pannier J.M., Savary S. 'Target classification based on kinematic data from ais/ads-b, using statistical features extraction and boosting'. Proceedings of the International Radar Symposium; Jun. 2019.

[28] Bækkegaard S., Blixenkrone-Møller J. 'Target classification using kinematic data and a recurrent neural network'. In the Proceedings of the International Radar Symposium. Germany: Bonn; Jun. 2018.

[29] Molchanov P.O., Astola J.T., Egiazarian K.O., Totsky A.V. 'Moving target classification in ground surveillance radar ATR system by using novel bicepstral-based information features'. 2011 8th European Radar Conference; Oct. 2011. pp. 194–7.

[30] Brooks D.A., Schwander O., Barbaresco F., Schneider J.Y., Cord M. 'Complex-valued neural networks for fully-temporal micro-Doppler classification'. Proceedings of the International Radar Symposium. Germany: Ulm; Jun. 2019.

[31] Seyfioglu M.S., Ozbayoglu A.M., Gurbuz S.Z. 'Deep convolutional autoencoder for radar-based classification of similar aided and unaided human activities'. *IEEE Transactions on Aerospace and Electronic Systems*. 2018, vol. 54(4), pp. 1709–23.

[32] Molchanov P., Harmanny R.I.A., de Wit J.J.M., Egiazarian K., Astola J. 'Classification of small UAVs and birds by micro-Doppler signatures'. *International Journal of Microwave and Wireless Technologies*. 2014, vol. 6(3-4), pp. 435–44.

[33] Taha B., Shoufan A. 'Machine Learning-Based Drone detection and classification: state-of-the-art in research'. *IEEE Access*. 2019, vol. 7, pp. 138669–82.

[34] SAFIR - *Safe and flexible integration of initial u-space services in a real environment [online]*. Available from www.sesarju.eu/index.php/projects/safir [Accessed 5/9/2019].

[35] SESAR JU - S*ingle European sky ATM research joint undertaking [online]*. www.sesarju.eu/ [Accessed July 2019].

[36] Jahangir M., Ahmad B. I., Baker C.J. 'Robust drone classification using two-stage decision trees and results from SESAR SAFIR trials'. Proceedings of the IEEE International Radar Conference; 2020. pp. 636–41.

[37] Jahangir M., Baker C.J. 'Characterisation of low observable targets with a multi-beam staring radar'. IET Radar Conference; Belfast; Oct. 2017.

[38] Sim J., Jahangir M., Fioranelli F., Baker C.J., Dale H. 'Effective Ground-truthing of Supervised Machine Learning for Drone Classification'. IEEE Radar Conference; Sep. 2019.

Chapter 6
SAR coherent change detection[a,b]

David Blacknell[1], Daniel Andre[2], and Andrew Bennett[1]

6.1 Introduction

In recent years, many surveillance and reconnaissance applications have exploited the continued development of synthetic aperture radar (SAR) imagery techniques. In particular, the coherent change detection (CCD) technique has enabled the identification of subtle changes in a scene between two imaging passes. This chapter explores the topic of SAR CCD using physics-based derivations to assess various aspects of CCD performance in the default case of monostatic imaging over flat terrain. It then goes on to discuss the impact of the more complex imaging scenario encountered when the terrain is non-flat and introduces advanced processing schemes which can maintain optimum CCD performance in this situation.

Section 6.2 discusses the fundamentals of CCD formation and assesses the sensitivity of the technique to the degree of disturbance in the scene and the CCD processing parameter choices made. This sensitivity analysis is validated through simulation and used to provide a quantification of the detectability of change as a function of the degree of change. Section 6.3 introduces the idea of "frequency support" and analyses, with the help of examples, the impact that non-aligned frequency support between a pair of images can have on the resulting CCD performance. Section 6.4 introduces the advanced concept of CCD formation in the presence of non-flat terrain and develops a spatially variant processing scheme that makes optimal use of the coherent information in this scenario. Section 6.5 then

[1]Dstl, Porton Down, Salisbury SP4 0JQ, UK
[2]Cranfield University, Shrivenham, Swindon SN6 8LA, UK

[a]DSTL/PUB122349

provides experimental validation of the benefits of this advanced processing using a ground-based SAR (GB-SAR) system. A summary and discussion of open questions is provided in Section 6.6.

6.2 Coherent change detection

A CCD image is formed from two SAR images taken of the same scene with approximately the same imaging geometry but at different times. (The first image is called the "reference" image and the second image is called the "mission" image.) It measures the degree of coherence between the complex-valued radar returns associated with the same ground location in the two images. The coherence can take a value from 0 to 1 and it is usual for complete incoherence to be displayed as black and complete coherence to be displayed as white. If the scene is unchanged between the two collects then the coherence will be high; however, if changes have occurred then there will be a loss of coherence. This means that CCD images can detect the occurrence of quite small disturbances between the two image collects.

Figure 6.1 shows an example of a CCD image obtained using the Bright Spark SAR system which operates with a 4 GHz bandwidth at Ka-band [1]. (Bright Spark is a Dstl experimental SAR system designed, built and operated by Thales UK.)

Figure 6.1 Bright Spark CCD showing sheep tracks (Bright Spark is a Dstl experimental SAR system designed, built and operated by Thales UK)

It should be noted that shadows in SAR images will also cause a loss of coherence between two images. In the bottom-right of Figure 6.1, a number of wandering lines of low coherence can be seen. These are the results of sheep walking over the grass in thc time between the two SAR images used to form this CCD image being collected. A zoomed-in segment of this region is shown, with some adjustment of aspect ratio, in which can be seen some of the sheep tracks and also, remarkably, the shadows of two sheep which are certainly responsible for some of the tracks.

The physical principles of CCD have been reported in the open literature and a comprehensive treatment has been given by Preiss and Stacy [2]. The following discussion draws heavily from that reference and full acknowledgement is hereby given. It is seen that disturbances of the order of a few millimetres will produce a significant loss of coherence in an X-band system and are thus potentially detectable.

6.2.1 Single image fundamentals

A SAR image provides a map of the varying radar cross-section (RCS) (scattering amplitude) over the imaged scene. However, importantly, it also provides phase information associated with the return path length from the sensor to a given scattering centre. This phase is determined by the number of cycles of the electro-magnetic wave that fit into this path length or, more specifically, the residual fractional cycle after all complete cycles are discarded (i.e. phase is measured modulo 2π).

The SAR image will be characterised by a resolution length which determines the minimum separation at which two scattering centres can be distinguished as such. Given the two-dimensional (2D) nature of an image, it is convenient to refer to a "resolution cell" as the 2D equivalent of a resolution length. For general terrain types, there will typically be many scattering centres within a resolution cell which cannot be individually distinguished. The effective scene reflectivity, R, associated with the resolution cell will thus be the coherent addition of contributions from a large number of scatterers

$$R = \sum_{k=1}^{N} A_k exp\left(i\phi_k\right) \tag{6.1}$$

where A_k is the amplitude of the kth scatterer and ϕ_k is the phase as discussed above.

The resulting scene reflectivity is a complex number. Both the real and imaginary parts of this complex number arise as the sum of a large number of independent, identically distributed random variables and so each will follow a Gaussian distribution in accordance with the Central Limit Theorem. Under the assumption that the phase is uniformly distributed and that amplitude and phase are independent, it is possible to show [3] that the distribution of the scene reflectivity will be a complex Gaussian distribution, namely

$$p\left(R\right) = \frac{1}{\pi\sigma_R^2} exp\left\{-\left(\frac{R_r^2 + R_i^2}{\sigma_R^2}\right)\right\} \tag{6.2}$$

where σ_R^2 is the expected value of RR^* and is the back-scattered radar power from the resolution cell. This random fluctuation observed in coherent systems is known as "speckle" and is seen in laser applications as well as SAR.

6.2.2 Repeat pass imaging: interferometry and coherent change detection

If the same scene is imaged with a repeat pass to give two SAR images, then the joint distribution of the complex image pair $X = [R, S]^T$ is a random vector with probability density function

$$p(\underline{X}) = \frac{1}{\pi^2 |C|} exp\left(-\underline{X}^H C^{-1} \underline{X}\right) \tag{6.3}$$

where C is the covariance matrix

$$C = \begin{bmatrix} \sigma_R^2 & \sigma_R \sigma_S \gamma exp(i\Phi) \\ \sigma_R \sigma_S \gamma exp(-i\Phi) & \sigma_S^2 \end{bmatrix} \tag{6.4}$$

The complex correlation coefficient between the two images is

$$\gamma exp(i\Phi) = \frac{\langle RS^* \rangle}{\sqrt{\langle |R|^2 \rangle \langle |S|^2 \rangle}} \tag{6.5}$$

where $\langle . \rangle$ is the expectation value, γ is termed the coherence and takes a value in the range [0,1] whilst Φ is termed the interferometric phase and takes a value in the range $[0, 2\pi]$.

The interferometric phase is determined by the baseline offset between the primary and repeat pass collections and the terrain topography plus any bulk displacement of the scattering scene between the two data collections. Change detection based on the measurement of the phase shift associated with the bulk displacement is called differential interferometry and may be used to monitor small-scale surface deformations. This is important in applications such as detecting subsidence.

When there is no bulk displacement but simply a disturbance of the scattering centre positions (i.e. a random zero mean displacement), then change detection can be performed based on the change in the coherence. This is the basis of CCD. It is important to emphasise that CCD is not a measure of bulk displacement. When a truck drives over a gravelled area, it is the disturbance of the stones which causes loss of coherence. Any impression in the ground left by the weight of the truck clearly contributes to the disturbance but is not a necessary condition for a loss of coherence to occur.

6.2.3 CCD image formation

To form the CCD image, it is first necessary to register the mission and reference images so that spatial positions are precisely aligned between the two images. This registration process must be performed to sub-pixel accuracy and will involve warping of the reference image to overlay the mission image thus correcting any spatial location inaccuracies between the two.

The second stage is then to estimate coherence between the two images. This is a statistical estimation procedure which requires the sample coherence over a number of pixel values to be calculated. The standard approach is to use a sliding square window of pixels with the resulting sample coherence being associated with the spatial position corresponding to the centre of the square.

6.2.4 Sensitivity

6.2.4.1 Sensitivity to scattering centre displacement

Zebker [4] has provided an analysis of the sensitivity of the correlation coefficient to disturbances of the positions and relative arrangements of scatterers. He assumed that changes in scatterer positions can be described by independent Gaussian distributions in all three dimensions. The coherence is then given by

$$\gamma = exp\left\{-\frac{1}{2}\left(\frac{4\pi}{\lambda}\right)^2 \left(\sigma_y^2 \cos^2(\theta)\right) + \sigma_z^2 \sin^2(\theta)\right\} \tag{6.6}$$

where σ_y is the standard deviation of displacements in the horizontal plane (i.e. combining displacements in both range and cross-range), σ_z is the standard deviation of displacements in the vertical direction and θ is the angle that the line-of-sight from the radar to the point on the ground makes with the horizontal (the depression angle).

Considering displacements only in the horizontal plane and using a normalised measure of scattering centre displacement (σ_y/λ), the variation of coherence with displacement is shown in Figure 6.2 for depression angles of 15° (typical airborne platform) and 45° (typical space-borne platform). A coherence value of 0.2 can be

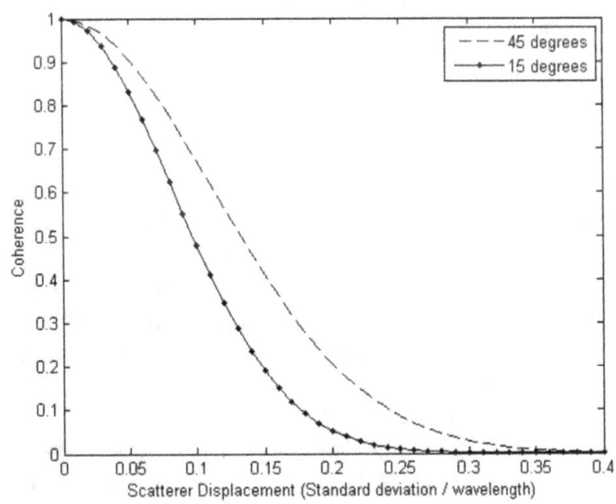

Figure 6.2 Variation of coherence with scattering centre displacement for two imaging geometries. (After Preiss and Stacy [2]).

considered as essentially a loss of coherence. This corresponds to normalised scattering centre displacements of around 0.15 and 0.2 for the two cases considered. Thus, for an X-band system with a wavelength of 3 cm, it can be seen that coherence will be lost in the presence of displacements with standard deviations of around 4.5 mm and 6 mm for the two cases. There will clearly be limitations in the accuracy which can be achieved with models of this kind but this does show that displacements of the order of a few millimetres will significantly reduce coherence. Man-made disturbances such as vehicle tracks which can cause significant scattering centre displacement can thus potentially cause complete loss in coherence without any appreciable change in the mean radar back-scatter. Such changes will clearly show up in a CCD image but will be invisible in the standard radar image.

6.2.4.2 Sensitivity to window size

The coherence is normally estimated by passing a sliding square window containing N pixel over the image extent and calculating the sample coherence

$$\hat{\gamma} = \frac{\left|\sum_{k=1}^{N} R_k S_k^*\right|}{\sqrt{\sum_{k=1}^{N} |R_k|^2 \sum_{k=1}^{N} |S_k|^2}} \tag{6.7}$$

The sample coherence is a random variable whose statistics are determined by the underlying speckle fluctuations. The probability density function of the sample coherence is given by Tough *et al.* [5] as

$$p\left(\hat{\gamma}|\gamma, N\right) = 2\left(N-1\right)\left(1-\gamma^2\right)^N\left(1-\hat{\gamma}^2\right)^{(N-2)} {}_2F_1\left(N, N; 1; \gamma^2\hat{\gamma}^2\right) \tag{6.8}$$

where ${}_2F_1$ is the Gauss hypergeometric function. The general mth statistical moment is then given by Tough *et al.* [4] as

$$\langle\hat{\gamma}^m\rangle = \left(1-\gamma^2\right)^N \Gamma\left(N\right) \frac{\Gamma\left(1+\frac{m}{2}\right)}{\Gamma\left(N+\frac{m}{2}\right)} {}_3F_2\left(1+\tfrac{m}{2}, N; 1, N+\tfrac{m}{2}; \gamma^2\right) \tag{6.9}$$

where ${}_3F_2$ is the generalised hypergeometric function.

The variation of the expected value of the sample coherence as a function of the coherence itself is shown in Figure 6.3 for a range of window sizes. The sample coherence is a biased estimator such that, for low coherence values, the estimated coherence value is significantly larger than the true coherence value. This limits the contrast between low and high coherence values and hence can have a detrimental impact when attempting to identify small regions of low coherence against extended regions of high coherence. It can be seen that this effect is increasingly mitigated as the size of the window is increased.

The variation of the standard deviation of the sample coherence as a function of the coherence itself is shown in Figure 6.4 for a range of window sizes. It is important that this standard deviation should be significantly less than the differences between low and high coherence that it is desired to detect. It can be seen that the standard deviation decreases rapidly as the window size is increased over

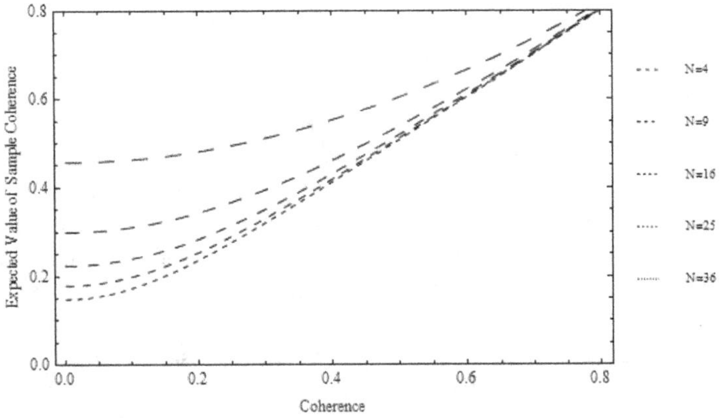

Figure 6.3 *Variation of expected value of sample coherence with true coherence value for a range of window sizes. (After Preiss and Stacy [2]).*

the interval of window sizes shown although gains become less significant as the window size is increased further.

These graphs illustrate that larger window sizes are to be preferred. However, the trade-off is that the averaging required over the extent of the window will result in a loss of resolution in the CCD product. Also, larger window sizes are more likely to include significant scene structures which will corrupt the statistical estimates. The choice of window size will thus be a compromise between these competing effects and will depend on the radar system and fidelity of CCD performance required.

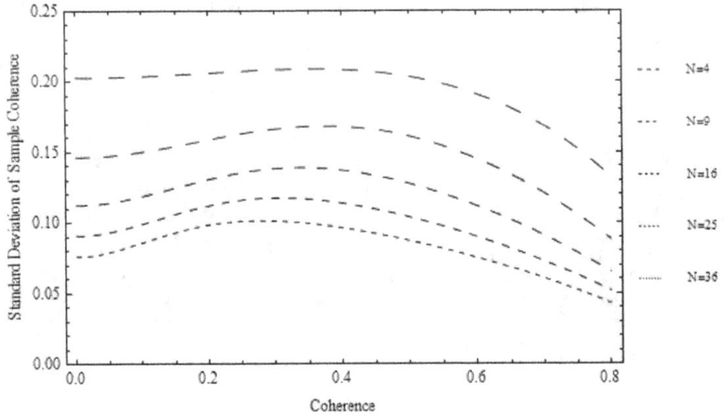

Figure 6.4 *Variation of standard deviation value of sample coherence with true coherence value for a range of window sizes*

6.2.5 *Validation of sensitivity analysis by simulation*

The use of simulations can provide a level of validation of the preceding theoretical sensitivity analysis. In such a simulation approach, a flat scene containing a large number of scatterers is simulated and scatterer displacements are introduced using a Gaussian distribution with a specified standard deviation. The disturbances can be applied over the whole scene extent or just within an area defined by a mask. Pairs of images can then be simulated using these scenes with the option of introducing changes in the imaging parameters used for each image. The CCD image can then be formed from the resulting image pair and used to support performance analysis.

This approach has first been used to validate the variation of coherence with degree of disturbance. It has been seen earlier that it is possible to predict this variation theoretically. To validate this prediction, image pairs have been generated with differing degrees of disturbance over the entire image extent and the corresponding coherence maps generated.

The coherence maps are generated using a sliding square window with the sample coherence being estimated for each window position and output as the coherence value associated with the centre of the window. The mean of the coherence values over the whole scene can then be calculated. There is a complication in that the coherence estimate is biased with the level of bias varying as a function of true coherence and window size. This is shown in Figure 6.3, where it can be seen that the estimated coherence will generally be larger than the true coherence. This needs to be taken into account when comparing with the theoretical predictions.

Images were simulated at 15 cm resolution, using a wavelength of 3 cm, for normalised disturbances (i.e. divided by the wavelength) from 0.0 to 0.4. Identical imaging geometries were used for both images. Sampling was performed at a rate of twice per resolution cell which means that the samples are not independent. For example, for a window size of 6×6, the actual number of independent samples is 3×3, which needs to be borne in mind when using Figure 6.3 in which the window size assumes independent samples. Figure 6.5 shows examples of simulated reference and mission images whilst Figure 6.6 shows coherence maps formed for three different window sizes.

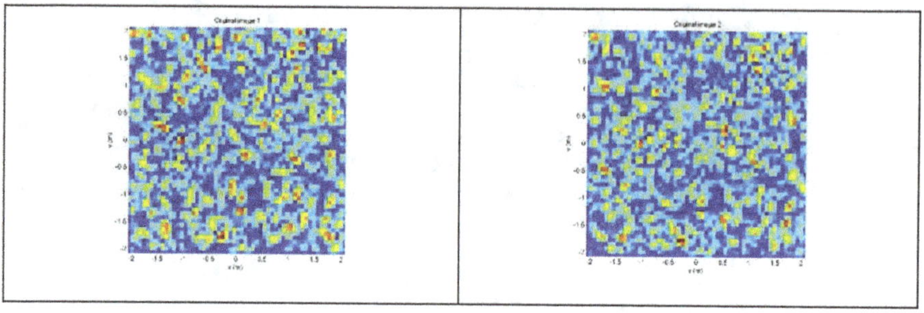

Figure 6.5 Example of reference and mission images for normalised scatterer disturbance of 0.1 and elevation angle of 45°

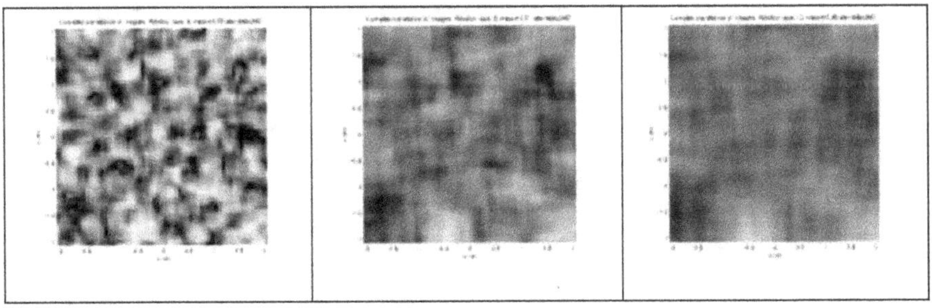

Figure 6.6 Coherence maps using window sizes of 4, 8 and 12 from left to right

Figure 6.7 shows the coherence values obtained by simulation after adjustment for window size bias for the largest window size, i.e. 12 × 12 (circles) overlaid on the theoretical prediction. Results for very low coherence values are very noisy and have not been included. It can be seen that there is excellent correspondence between the coherence values obtained by simulation and the theoretical curves. This provides a degree of validation of the theoretical development and provides confidence that this can be used for exploring a larger parameter space.

6.2.6 Assessment of detectability of change

It is important to understand the degree of coherence which must be present between image collections to allow change to be visually detected. In this section, the simulation approach introduced above is used to attempt to determine thresholds on the coherence value which can be related to the detectability of change. Identical

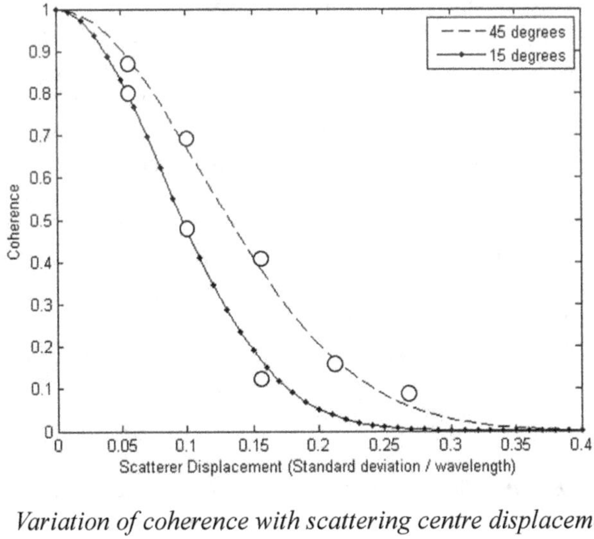

Figure 6.7 Variation of coherence with scattering centre displacement for two imaging geometries

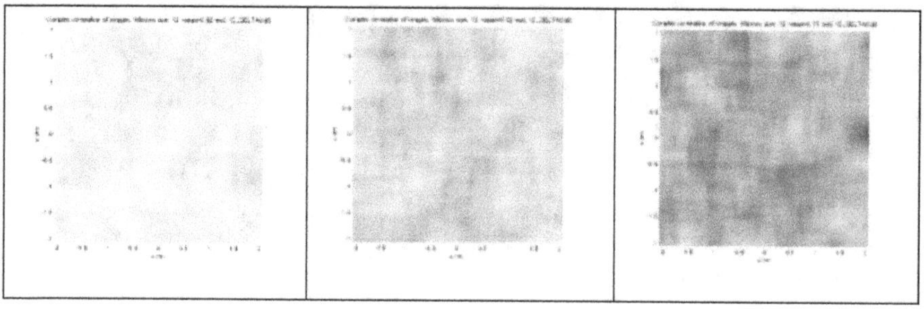

Figure 6.8 Coherence maps for whole scene disturbances of 1.0 mm, 1.5 mm and 2.0 mm using a 12 × 12 window giving mean coherence values of 0.92, 0.82 and 0.71

geometries have been used so that the effect of ground disturbances can be examined in isolation. Also, values of 15° for elevation angle and 10 cm for resolution have been chosen.

An overall loss of coherence between image pairs can occur over time due to the disturbance of the ground resulting from, e.g. environmental conditions. This will be modelled by introducing varying degrees of scatterer disturbance over the whole scene characterised by displacement standard deviations of 1 mm, 1.5 mm and 2 mm. A scene of 5 m × 5 m has been used and Figure 6.8 shows the coherence maps that result from the different levels of disturbance when a window size of 12 × 12 is used (i.e. 6 × 6 independent samples). The bias in the coherence estimate is negligible for this window size when coherence values exceed 0.5 so the estimated values of 0.92, 0.82 and 0.71 should give a good approximation to the true coherence values. Indeed, these values are well matched to those given by the theoretical analysis discussed previously.

To model changes which it may be of interest to detect, the change masks shown in Figure 6.9 have been used. The left-hand mask is intended to represent a pair of tracks with varying separation and thickness of around 10 cm whilst the right-hand mask is intended to represent a localised area of change using a circle with a diameter of around 50 cm. Within these areas, disturbances have been introduced

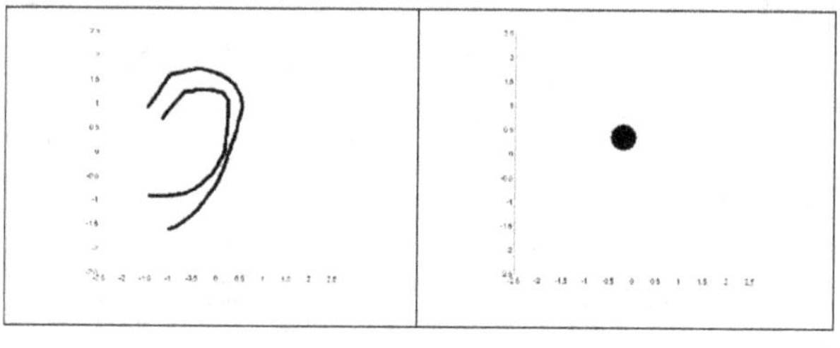

Figure 6.9 Templates for simulating localised disturbances. The tracks (left) have a thickness of about 10 cm whilst the circular area has a diameter of about 50 cm.

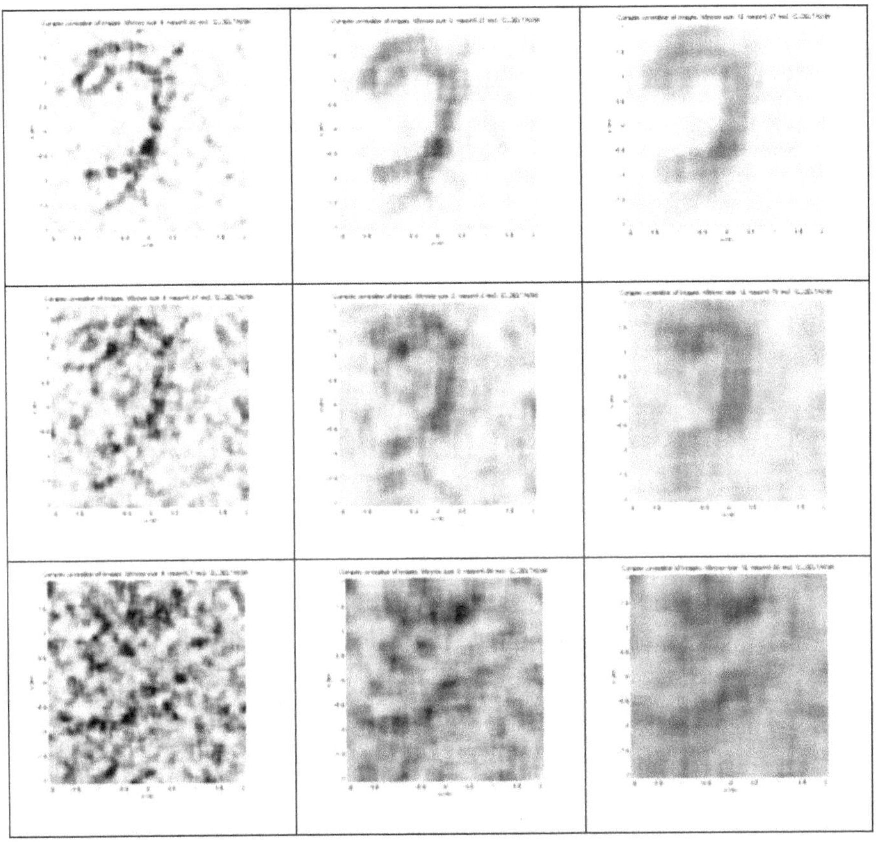

Figure 6.10 *Coherence maps for simulated tracks for scene disturbances of 1.0 mm, 1.5 mm and 2.0 mm (rows) and window sizes of 4 × 4, 8 × 8 and 12 × 12 (columns)*

using a standard deviation of 10 mm which should destroy all coherence based on Figure 6.2. The intention is to determine thresholds for the background coherence against which these changes cannot be detected (red light), can just be detected (amber light) and can clearly be detected (green light).

Figure 6.10 shows coherence maps for the track disturbance for the three levels of background loss of coherence (rows) and window sizes of 4 × 4, 8 × 8 and 12 × 12 (columns). For the highest value of background coherence, i.e. 0.92, the presence of a track-like change is clearly visible. However, it should be noted that the existence of two tracks can clearly be seen for the smallest window size but that this is difficult to determine for the largest window size. For the intermediate value of background coherence, i.e. 0.82, the presence of a track-like change can still be determined but the existence of two tracks is difficult to determine for any window size. For the lowest value of background coherence, it is not possible to reliably determine the presence of a track-like change.

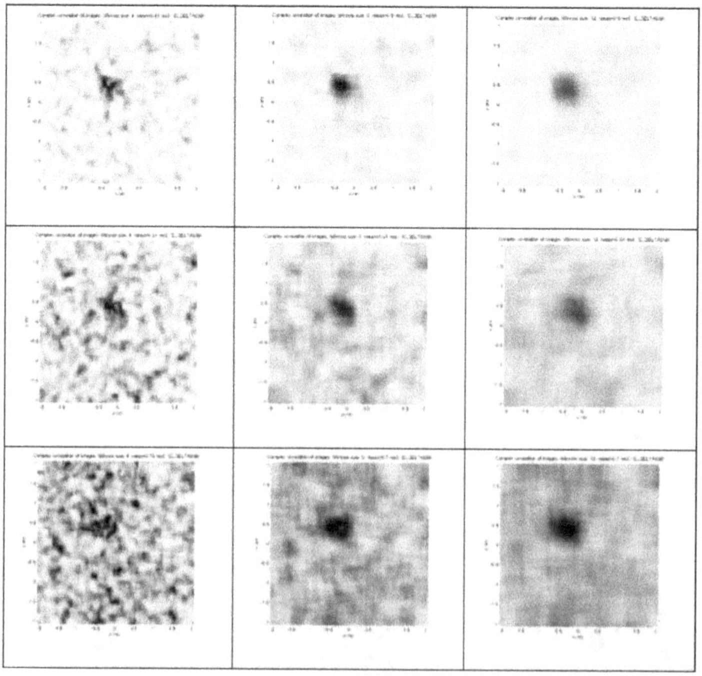

Figure 6.11 Coherence maps for simulated area of disturbance for scene disturbances of 1.0 mm, 1.5 mm and 2.0 mm (rows) and window sizes of 4 × 4, 8 × 8 and 12 × 12 (columns)

Thus, for two tracks of the type simulated, a coarse summary would be that,

- for a background coherence of 0.9, two tracks can be distinguished (green light);
- for a background coherence of 0.8, some form of track can be distinguished (amber light);
- for a background coherence of 0.7, no tracks can be reliably distinguished (red light).

It will be seen later that loss of background coherence can also result when the imaging geometries differ, specifically in terms of grazing angle. It will be shown how it is then possible to relate these coherence thresholds to equivalent grazing angle separations.

Figure 6.11 shows coherence maps for the area disturbance for the three levels of background loss of coherence (rows) and window sizes of 4 × 4, 8 × 8 and 12 × 12 (columns). In this case, visual detection of the change is possible in all cases although it clearly becomes more difficult as the background coherence level reduces. Of particular note is the observation that the change is more easily seen when the larger window size is used which is in contrast to the case for track-like changes. A possible reason for this is as follows. The width of the track-like changes

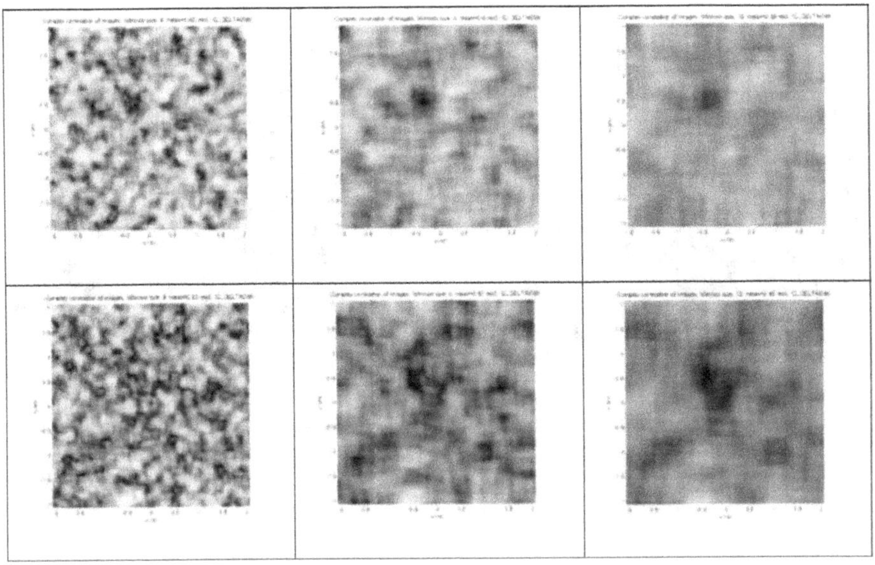

Figure 6.12 *Coherence maps for simulated area of disturbance for scene disturbances of 2.5 mm and 3.0 mm (rows) and window sizes of 4 × 4, 8 × 8 and 12 × 12 (columns)*

is of the same order as the image resolution, thus a high-resolution coherence map is required in order to identify them. On the other hand, the width of the area disturbance is relatively large when compared with the image resolution, thus a larger window size may produce better results. This suggests that there may be benefits in using a window that specifically matches the shape of tracks, e.g. a long, thin window although this approach is not considered here.

In order to determine at what point the area disturbance cannot be detected, higher levels of background disturbance were considered, namely standard deviations of 2.5 mm and 3.0 mm, as in Figure 6.12. From Figure 6.2, these will give background coherence levels of 0.6 and 0.5, respectively. Thus, for the type of area disturbance simulated, a coarse summary would be that,

- for a background coherence of 0.7, the area change can be detected (green light);
- for a background coherence of 0.6, detection of the area change is becoming marginal (amber light);
- for a background coherence of 0.5, detection of the area change is unreliable (red light).

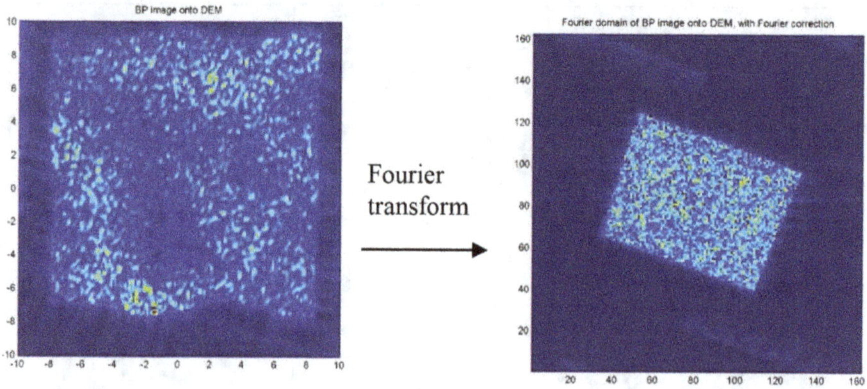

Figure 6.13 *Illustration of constrained frequency support resulting from Fourier transform of SAR image focused onto the flat focal plane, i.e. "standard" processing*

6.3 Frequency support fundamentals

6.3.1 Band-limited images

The CCD process is critically dependent on the correct handling of the "frequency support" of the SAR image. If the Fourier transform of a complex-valued SAR image is formed, then a 2D representation of the frequencies contained in the SAR image will be obtained. This is known as the "frequency domain" or "K-space" representation. For standard SAR image formation, the frequencies will be non-zero only over a finite extent in the range-frequency direction as determined by the bandwidth of the system and will be non-zero only over a finite extent in the cross-range-frequency direction as a result of the finite Doppler bandwidth. Thus, there will be a delimited area of "support" in the frequency domain outside of which there will be no frequency content. This is illustrated in Figure 6.13, where the left-hand image shows a simulated example of a standard SAR image and the right-hand image shows its corresponding frequency support. There are some subtleties arising from the use of a squinted geometry but the fundamental point is that the "support" is a defined region in frequency space embedded in the background of zero values.

The SAR image is thus a 2D band-limited signal. As a result, it can be unambiguously represented by samples taken at or above the Nyquist rate. Increasing the sampling rate in the spatial domain corresponds to extending the bandwidths represented in the Fourier domain. Thus, interpolation can be achieved by padding out the frequency domain with zeros to increase the represented bandwidths. This is equivalent to sinc interpolation in the spatial domain using an infinite length sinc function kernel applied to the periodically repeating image.

Reducing the sampling rate will reduce the bandwidths which will mean that the boundary of zeros surrounding the region of frequency support will become smaller. If the sampling rate becomes too small, the frequency support will become too large

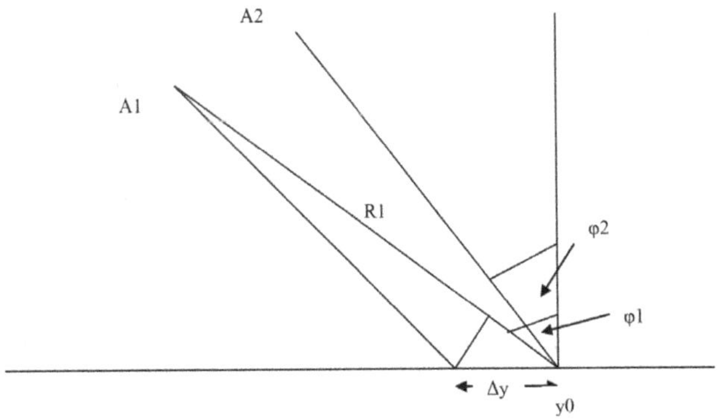

Figure 6.14 Imaging geometry for two platforms at slightly different positions

to be contained within the bandwidth represented in the frequency domain and will begin to wrap around. It is at this point that the Nyquist rate has been violated and the signal is no longer unambiguously represented. If a signal is under-sampled, then padding out the frequency domain with zeros will not produce a valid interpolation and interpolation artefacts may potentially occur.

6.3.2 Coherence theory

As has been discussed earlier, the CCD image is formed by passing a sliding window over the image extent and calculating the cross-correlation of the complex values between the two images. The magnitude of the resulting quantity gives the coherence. The calculation of the coherence is intimately related to the properties of the frequency support and will be discussed in more detail here.

The coherence between interferometric radar signals has been analysed by Zebker and Villasenor [4] and their analysis is followed in this sub-section.

Consider two SAR platforms imaging the same point on the ground at position (x_0, y_0), see Figure 6.14. The complex pixel value associated with this point in the resulting SAR image will be the result of the convolution of the impulse response function with the complex back-scatter, $f(x,y)$ and an appropriate phase terms due to differing path lengths over the neighbourhood of the point of interest (infinite in principal but really determined by the width of the impulse response). This is given by

$$s_1 = \iint f(x_0 - \Delta x, y_0 - \Delta y)\, exp\left(4\pi i \frac{(R_1 - \Delta y \sin(\phi_1))}{\lambda_0}\right) W(\Delta x, \Delta y)\, d(\Delta x)\, d(\Delta y) \tag{6.10}$$

The cross-correlation with the equivalent expression for platform 2 gives

$$s_1 s_2^* = \iint \iint f(x_0 - \Delta x, y_0 - \Delta y) f^*\left(x_0 - \Delta x', y_0 - \Delta y'\right) \tag{6.11}$$

$$\times exp\left(4\pi i\frac{(R_1-R_2+\Delta y\sin(\phi_1)-\Delta y'\sin(\phi_2))}{\lambda_0}\right) \tag{6.12}$$

$$\times W(\Delta x, \Delta y)\,W^*(\Delta x', \Delta y')\,d(\Delta x)\,d(\Delta y)\,d(\Delta x')\,d(\Delta y') \tag{6.13}$$

If it is assumed that the scattering centres are uniformly distributed and uncorrelated then it follows that

$$\langle f(x,y)f^*(x',y')\rangle = \sigma_0\delta\,(x-x',y-y') \tag{6.14}$$

where σ_0 is the average RCS and $\delta\,(.,.)$ is the 2D Dirac delta function. Using this together with letting $\phi_2 = \phi_1 + \delta\phi$ and expanding for small values of $\delta\phi$, the expectation value of the cross-correlation then becomes

$$\langle s_1 s_2^*\rangle = exp\left(4\pi i\frac{(R_1-R_2)}{\lambda_0}\right)\iint exp\left(-4\pi i\frac{\Delta y\cos(\phi_1)\delta\phi}{\lambda_0}\right)|W(\Delta x, \Delta y)|^2\,d(\Delta x)\,d(\Delta y) \tag{6.15}$$

It is the magnitude of the cross-correlation, i.e. the coherence, which is of interest, thus

$$\rho = |\langle s_1 s_2^*\rangle| = \left|\iint exp\left(-4\pi i\frac{\Delta y\cos(\phi_1)\,\delta\phi}{\lambda_0}\right)|W(\Delta x, \Delta y)|^2\,d(\Delta x)\,d(\Delta y)\right| \tag{6.16}$$

The exponential term is linear in Δy and so effectively represents a Fourier transform evaluated at the spatial frequency given by

$$\frac{2\cos(\phi_1)\,\delta\phi}{\lambda_0} \tag{6.17}$$

If the impulse response function is taken to be a sinc function, then the Fourier transform of the resulting sinc squared function will be a triangular function which is the result of convolving a rectangle with itself. For a ground range resolution of r_G this rectangle will extend between $\pm\frac{1}{2r_G}$ with a height of r_G so that the result of convolving it with itself and evaluating at the spatial frequency given above will be

$$\delta\phi = \frac{\lambda_0}{2\cos(\phi_1)\,r_G} \tag{6.18}$$

This is the coherence between the signals received by the two platforms and will fall to zero when

$$\delta\phi = \frac{\lambda_0}{2\cos(\phi_1)r_G} \tag{6.19}$$

An interpretation of this behaviour in the frequency domain will now be discussed.

6.3.3 *Base-banding and analysis of frequency support overlap*

Consider a simple strip-map SAR imaging geometry from an airborne platform flying straight and level. Figure 6.15 shows a representation of the geometry in the range direction with the direction of flight into the page. The points on the ground

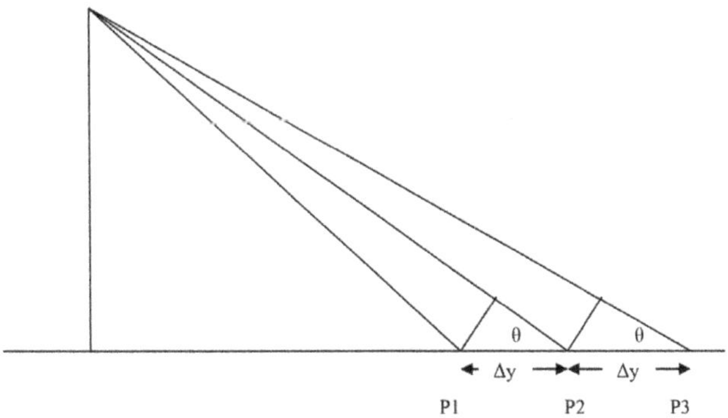

Figure 6.15 *Range imaging geometry*

labelled P1, P2 and P3 represent positions of pixels in the output image. The image is processed with a sampling interval of Δy and the grazing angle is θ. Clearly, the angles shown from radar to different ground positions cannot all be identical. However, given that the extent of the image in range can be assumed small in comparison to the range, this angle will change little and can be assumed constant for the purposes of this discussion.

After range compression, a given complex pixel value will have an amplitude dependent on the RCS of the ground and a phase which will have a deterministic component associated with the path length from platform to ground and back. (There will be random fluctuations in amplitude and phase arising from speckle effects but these are not relevant to the current discussion.) From the range imaging geometry, it can be seen that an extra one-way path length of $\Delta y \cos(\theta)$ is added for each pixel position. Given a two-way path, the phase associated with the nth range sample gives rise to a complex factor

$$exp\left(2\pi i\frac{2n\Delta y \cos(\theta)}{\lambda_0}\right) \tag{6.20}$$

which represents a linear phase ramp as a function of sample number.

The Fourier transform of the slant-plane complex image in the range direction will produce the frequency support which will have an extent in the range-frequency direction equal to the bandwidth of the signal. The phase ramp given by the above phase factor will cause this support to be shifted in the range frequency direction. If the image is multiplied by the complex conjugate of this term, then the phase ramp will be removed and the frequency support will be centred. This process is known as "base-banding".

To form a CCD pair, two images using slightly different grazing angles will be used. It is essential to correct the image phases using the same geometry and so the phase ramp correction applied to the first image must also be applied to the second image. Taking grazing angles θ_1 and θ_2, the correction of the second image using the geometry of the first will leave a residual phase factor given by

$$exp\left(2\pi i\frac{2n\Delta y\left(\cos\left(\theta_2\right)-\cos\left(\theta_1\right)\right)}{\lambda_0}\right). \tag{6.21}$$

Taking $\theta_2 = \theta_1 + \delta\theta$ it can be seen that

$$\cos(\theta_2) - \cos(\theta_1) = \cos(\theta_1 + \delta\theta) - \cos(\theta_1) = \cos(\theta_1)\cos(\delta\theta) - \sin(\theta_1)\sin(\delta) - \cos(\theta_1) \cong -\sin\left(\theta_1\right)\delta\theta \tag{6.22}$$

so that the residual phase factor becomes

$$exp\left(-2\pi i\frac{2n\Delta y\sin\left(\theta_1\right)\delta\theta}{\lambda_0}\right). \tag{6.23}$$

Thus it can be seen that a phase ramp remains which will displace the frequency support of the second image relative to the first image. The size of this displacement is given in spatial frequency terms by

$$\frac{2\sin\left(\theta_1\right)\delta\theta}{\lambda_0}. \tag{6.24}$$

For this displacement to be just sufficient for the two frequency supports to no longer overlap, the shift must equal the (spatial) bandwidth associated with the ground plane resolution, that is

$$\frac{2\sin\left(\theta_1\right)\delta\theta}{\lambda_0} = \frac{B_G}{c} = \frac{1}{r_G} \tag{6.25}$$

which, on re-arranging, gives

$$\delta\theta = \frac{\lambda_0}{2\sin\left(\theta_1\right)r_G} = \frac{\lambda_0}{2\cos\left(\phi_1\right)r_G} \tag{6.26}$$

where $\phi_1 = 90° - \theta_1$ is the incidence angle.

6.3.4 Frequency trimming

Comparing the results from Sections 6.3.2 and 6.3.3, it can be seen that the platform angular separation required for the frequency supports to no longer overlap is the same as that at which the coherence drops to zero.

On reflection, it is clear that the coherence will drop to zero when the frequency supports no longer overlap. The coherence is the magnitude of the complex cross-correlation of one signal with a second signal. In general, the cross-correlation can be written including a potential shift between the signals, namely

$$c\left(x\right) = \int s_1\left(x'\right)s_2^*\left(x'+x\right)dx' \tag{6.27}$$

The Fourier transform of this cross-correlation function will be the multiplication of the Fourier transforms of the two signals, i.e. the multiplication of their frequency supports. The value at zero lag ($x = 0$) is desired which is simply the product of

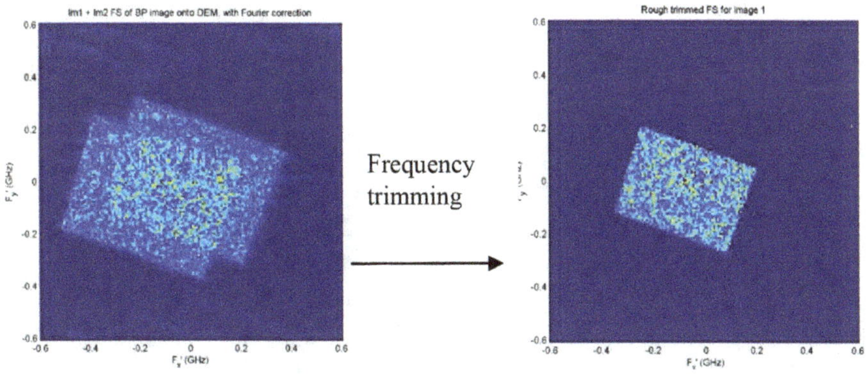

Figure 6.16 Illustration of frequency trimming

the values in the Fourier domain. This product will be non-zero only when the frequency supports overlap.

More generally, it is clear that the coherence will be linearly related to the area of overlap as a proportion of the total area. This assumes that nothing has changed between the two scenes between the images and that there is a significant radar return. In practice there are other sources of loss of coherence and so this relationship represents the maximum coherence that may be observed. Given that the degree of overlap is linearly related to the change in grazing angle between the collects, this means that the maximum coherence will be linearly related to the change in grazing angle.

The above discussion has assumed that the imagery corresponding to the full frequency support is used in the formation of the CCD product. However, it is clear that it is only the region of overlap that is contributing to the coherence and that the non-overlapping regions are contributing noise to the process. An alternative is thus to set the non-overlapping regions to zero. By doing this, the region of overlap is forced to be 100% thus allowing maximum coherence to be achieved if there are no changes between the scenes. This gain in coherence is achieved at the expense of reducing the resolution of the imagery as restriction of the frequency support to only the overlapping region reduces the bandwidth and hence the resolution. This process of setting the non-overlapping regions to zero is known as frequency trimming as illustrated in Figure 6.16 for a simulated data example.

6.3.5 Impact of frequency support considerations on CCD performance

The difference in imaging geometries, Δgeometry, between the mission and reference images will have a significant impact on the potential CCD performance. This geometry change can be decomposed into a change in grazing angle, Δgrazing, and a change in azimuth angle, Δazimuth. It is important to keep the Δgeometry to a minimum in order to maintain a high background coherence against which changes are then more easily detectable. The Δgeometry will impact the relative positions of

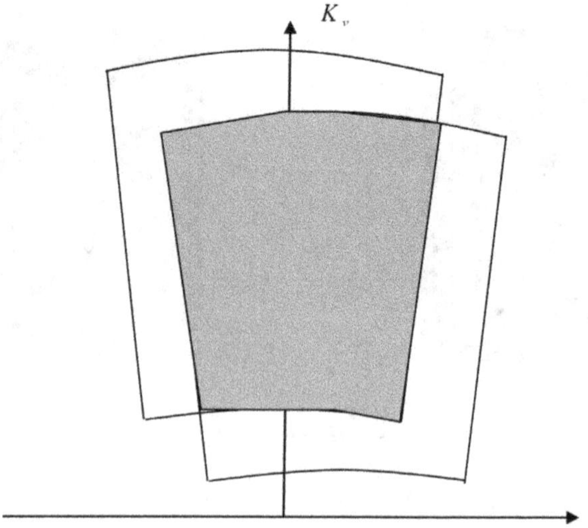

Figure 6.17 Overlapping keystones in K-space from two imaging collects with the region of overlap containing the coherent data shown in grey

the frequency supports from the two collections. In particular, for spotlight SAR, the frequency support takes the shape of a "keystone" and minimising Δgeometry will ensure that the keystones overlap as much as possible, as illustrated in Figure 6.17. As has been previously discussed, it is only the overlap region that contains coherent information (arising from homogeneous, speckled clutter areas in the images) and the background coherence will be reduced if data from outside this region is used. The degradation of background coherence will be linearly dependent on the ratio of overlap area to keystone area, i.e. the fractional overlap [4].

As discussed earlier, frequency trimming can be used in the CCD processing to ensure that only the coherent information is used. This will maximise the coherence in the CCD image although this will be at the expense of resolution which will become proportionately coarser.

In the following approximate analysis, the relationship between coherence and various parameters such as imaging geometry, resolution, bandwidth and frequency is investigated to provide an understanding of the trade-off space for maintaining high coherence.

Figure 6.18 shows the ground-plane frequency support (keystone) for a collection angle of $\Delta\theta$, a grazing angle of ψ and a bandwidth of B. The keystone will be displaced from the origin of co-ordinates by $(4\pi/\lambda)\cos\psi$, [6].

For small fractional bandwidths (such as at X and Ku-bands) and where the collection angle is small, the keystone can be approximated by a rectangle of size ΔX, ΔY, where

$$\rho_{az} \tag{6.28}$$

$$\Delta Y \cong \frac{4\pi B}{c}\cos\psi = \frac{2\pi}{\rho_r}\cos\psi \tag{6.29}$$

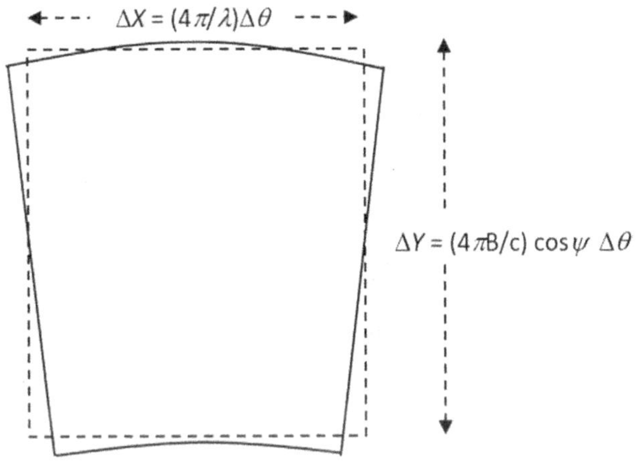

Figure 6.18 Keystone and rectangular approximation which is valid for small fractional bandwidths

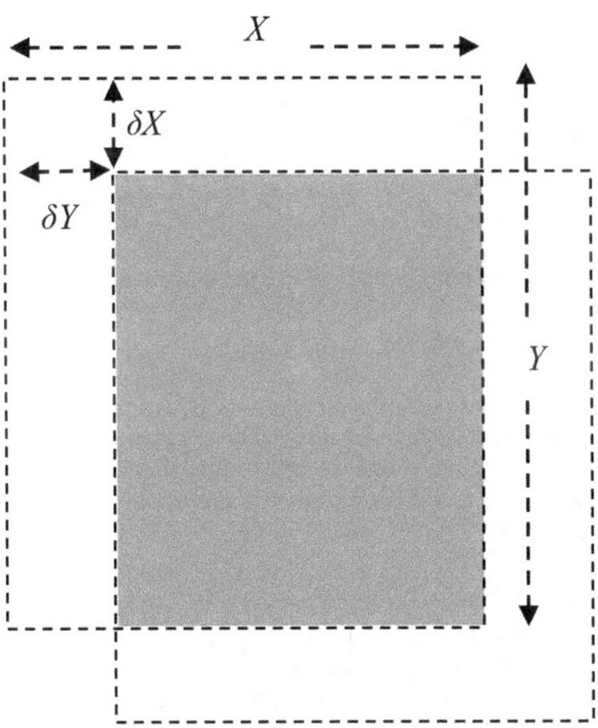

Figure 6.19 Overlapping keystones using rectangular approximation

and ρ_{az} and ρ_r are the image resolutions in azimuth and range, respectively [4].

Figure 6.19 provides an illustration of the overlap of two rectangular approximations to the frequency support where the change in geometry between the two collections has resulted in shifts of δX and δY between the two rectangular supports. As has been discussed previously, it is only the overlap region which contains coherent information and the fractional overlap can be used as a measure of loss of coherence. This is given by

$$F = \frac{(\Delta X - \delta X)(\Delta Y - \delta Y)}{\Delta X \Delta Y} = 1 - \frac{\delta Y}{\Delta Y} - \frac{\delta X}{\Delta X} + \frac{\delta X \delta Y}{\Delta X \Delta Y}. \tag{6.30}$$

For ρ_r and $\delta Y \ll \Delta Y$, to first order the final term can be neglected giving

$$F \cong 1 - \frac{\delta Y}{\Delta Y} - \frac{\delta X}{\Delta X}. \tag{6.31}$$

If grazing angles of ψ_r and ψ_m are used for the reference and mission images and given that the keystone is displaced from the co-ordinate origin by $(4\pi/\lambda)\cos\psi$ then

$$\delta Y = \frac{4\pi}{\lambda}\left(\cos\psi_m - \cos\psi_r\right) \tag{6.32}$$

Defining $\delta\psi = \psi_m - \psi_r$ and again using a small angle approximation then this becomes

$$\delta Y \cong \frac{4\pi}{\lambda}\sin\psi_r \delta\psi \tag{6.33}$$

where the reference image has nominally been chosen as a baseline. Similarly as $\Delta X \ll (4\pi/\lambda)$ then

$$\delta X \cong \frac{4\pi}{\lambda}\cos\psi_r \delta\theta \tag{6.34}$$

where $\delta\theta = \theta_m - \theta_r$. It then follows that the fractional overlap is

$$F = 1 - \frac{2}{\lambda}\left(\rho_r \tan\psi_r \delta\psi + \rho_{az}\cos\psi_r \delta\theta\right). \tag{6.35}$$

This equation will only be valid when there is not great mismatch between mission and reference images and for high-frequency radars because of the approximations used. Nonetheless, it captures the essential elements that describe how the loss of coherence can be related to various parameters. From the examination of the equation, it can be seen that

- F decreases as the angular collection errors, $\delta\theta$ and $\delta\psi$, increase (i.e. the reference and mission collections must be constrained as tightly as possible to achieve high coherence) (Figure 6.20).
- F increases as the image resolutions ρ_{az} and ρ_r are made smaller i.e. for finer resolution systems the fractional overlap is larger than for corresponding coarser resolution systems (for given angular deltas, $\delta\theta$ and $\delta\psi$) (Figure 6.21, left-hand

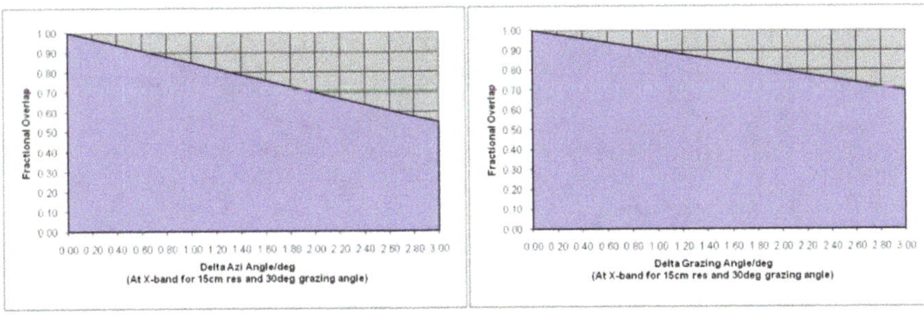

Figure 6.20 *Fractional overlap as a function of* Δ*azimuth (left) and* Δ*grazing (right) for X-band, 15 cm res. SAR at 30° grazing.*

side). Clearly there will also be a reduction in interpretability and sensitivity for coarser resolutions but this has not been taken into account here.

- *F* increases with wavelength i.e. an X-band system will be able to achieve a larger fractional overlap than a Ku-band system (for a given resolution and angular deltas, $\delta\theta$ and $\delta\psi$) (Figure 6.21, right-hand side). It should be noted that the reduction in performance is not as severe as for Δgeometry and resolution. There will be a sensitivity improvement of Ku-band systems against X-band systems [6] but this has not been taken into account here.

The relationship between *F* and grazing angle sees fractional overlap decreasing for larger grazing angles (Figure 6.22). In practice, the situation is likely to be more complex as SNR, which will increase for large grazing angles, is not taken into account in the analysis.

6.3.6 Performance trade-off example

Previously, it was seen that changes due to exemplar tracks can be confidently detected (green light) given a background coherence of 0.9 whilst changes due to

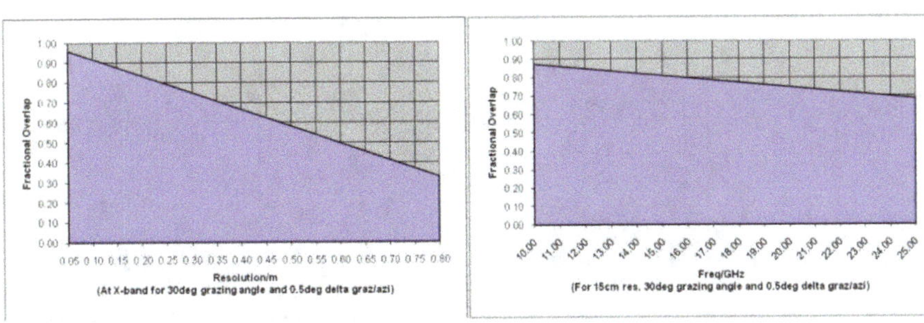

Figure 6.21 *Fractional overlap as a function of resolution (left) and frequency (right) for X-band, 15 cm res. SAR at 30° grazing with* Δ*azimuth and* Δ*grazing = 0.5°*

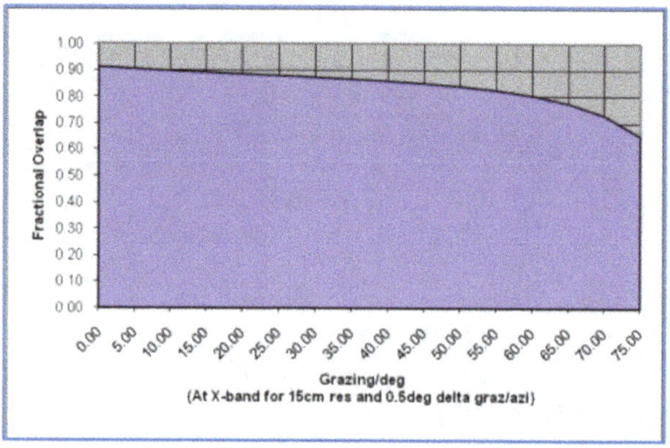

Figure 6.22 *Fractional overlap as a function of grazing angle for X-band, 15 cm res. SAR with* Δ*azimuth and* Δ*grazing = 0.5°*

an exemplar localised area of disturbance can be confidently detected given a background coherence of 0.7. The trade-space of radar parameters to achieve a given coherence can be explored using these figures based on the analytical model for overlap given imaging parameters and the theoretical curve relating disturbance to coherence. For example, the trade-off between ground disturbance and delta graze (with all other parameters kept equal) to achieve the green light coherence values given above are shown in Figure 6.23. The trade-off for a coherence of 0.9 is shown on the left-hand side where it can be seen that a difference in grazing angles of 3° can be tolerated if there is no disturbance of the background between the collections. However, this needs to be reduced to 0.5° if the background average ground disturbance is of the order of 1 mm. In the case when the green light coherence value is 0.7, the geometry constraints are less severe and average ground disturbances of the order of 2 mm can be accommodated given a difference in grazing angles of 0.5°. In this case, if there is no background disturbance, the change in grazing angle can be

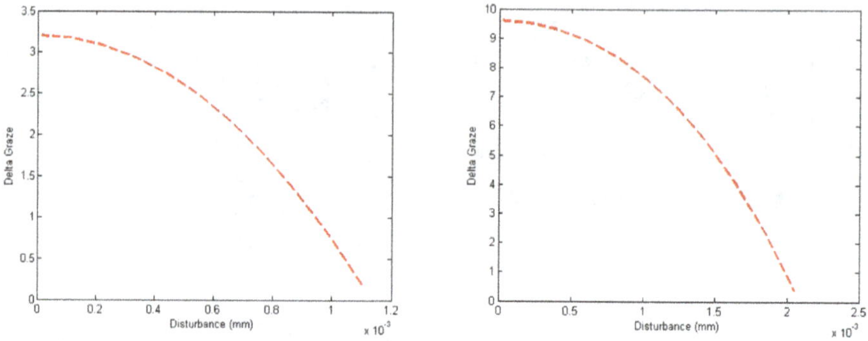

Figure 6.23 *Trade-off of delta graze and ground disturbance to maintain background coherence of 0.9 (left) and 0.7 (right)*

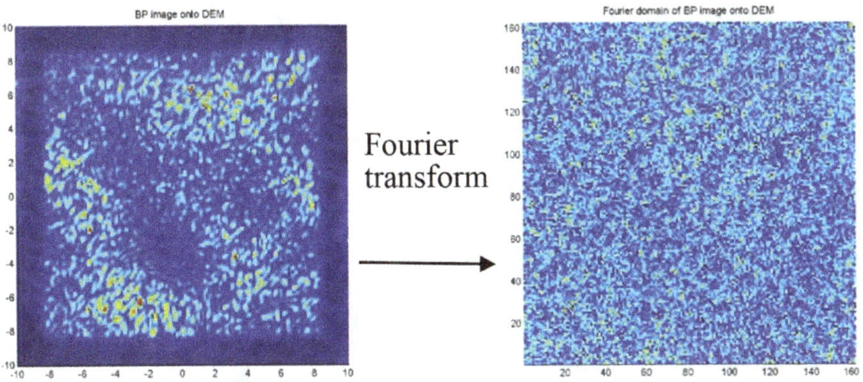

Figure 6.24 *Image formed onto the representation of terrain heights together with corresponding frequency support*

close to 10° although the validity of the model may not hold for such a large change in geometry.

6.4 Advanced frequency support considerations

6.4.1 Topographic base-banding

Having introduced some frequency domain concepts, the relevance of these to image formation onto a non-flat focal surface will now be discussed. SAR image formation onto a non-flat focal surface (e.g. one defined by a digital elevation model) can be achieved using the back-projection approach [6] in which the radar energy for each pulse is coherently projected back to all possible points of origin on the focal surface. For a given point, the energy over all pulses originating from that point will sum coherently thus providing focused imagery.

If image formation is performed using a flat focal surface, then the frequency support will be localised in frequency space as was illustrated in Figure 6.13. However, if back-projection is used to focus the image onto a representation of the varying terrain height, then the resulting frequency support will occupy more of frequency space than that associated with the nominal fast time and slow time bandwidths. Depending on the extent to which the terrain is non-flat and the sampling rate used, the frequency support can occupy the entire frequency space as is illustrated in Figure 6.24.

To understand this phenomenon, consider a simple spotlight imaging geometry with a straight line trajectory imaging a ground surface with varying terrain height $h(x)$, where x is cross-range ground position.

The difference in range $R(x,x')$ from the platform position, x', to the ground position, x, in comparison to the nominal range, R_0, is given below.

$$\begin{aligned}\Delta R(x,x') &= R(x,x') - R_0 \\ &= \left[(R_0 - h(x))^2 + (x-x')^2\right]^{1/2} - R_0 \\ &\cong (R_0 - h(x))\left[1 + \frac{(x-x')^2}{2(R_0 - h(x))^2}\right] - R_0 \\ &= -h(x) + \frac{(x-x')^2}{2(R_0 - h(x))} \\ &\cong -h(x) + \frac{(x-x')^2}{2R_0}\left(1 + \frac{h(x)}{R_0}\right) \\ &\cong -h(x) + \frac{(x-x')^2}{2R_0}\end{aligned} \tag{6.36}$$

For a point scatterer at $x = x_p$, the received signal as a function of platform position is

$$s(x') = \exp\left(-2\pi i\frac{2h(x_p)}{\lambda}\right)\exp\left(2\pi i\frac{(x_p-x')^2}{\lambda R_0}\right) \tag{6.37}$$

To form an image, this is convolved with the pulse replica for a general ground position

$$r(x') = \exp\left(-2\pi i\frac{2h(x)}{\lambda}\right)\exp\left(2\pi i\frac{(x-x')^2}{\lambda R_0}\right) \tag{6.38}$$

This convolution (with a normalisation for the aperture length) gives

$$\begin{aligned}c(x) &= \exp\left(2\pi i\frac{2(h(x)-h(x_p))}{\lambda}\right)\exp\left(2\pi i\frac{(x_p^2-x^2)}{\lambda R_0}\right)\frac{1}{L}\int_{-}^{L/2}\exp\left(-4\pi i\frac{(x_p-x)x'}{\lambda R_0}\right)dx' \\ &= \exp\left(2\pi i\frac{2(h(x)-h(x_p))}{\lambda}\right)\exp\left(2\pi i\frac{(x_p^2-x^2)}{\lambda R_0}\right)\operatorname{sinc}\left(\frac{2\pi L(x_p-x)}{\lambda R_0}\right)\end{aligned} \tag{6.39}$$

The resulting complex image consists of a sinc function, a quadratic phase term and the first term which has a phase that is dependent on the height. It is this term that is of interest, i.e.

$$T(x) = \exp\left(2\pi i\frac{2(h(x) - h(x_p))}{\lambda}\right) \tag{6.40}$$

It can be seen that a linear height variation $h(x) = h(x_p) + xh'(x_p)$ (which would be the first-order approximation at a given point, i.e. the tangent to the slope) will give rise to a linear phase ramp, i.e. constant frequency offset, which would cause a shift in the frequency domain by this frequency offset. This shift will be a function of position and hence will vary across the extent of the image.

To mitigate this effect, the phase term due to the height variation needs to be removed from the complex image. This can be achieved through multiplication by the complex conjugate of the phase factor, i.e.

$$\exp\left(4\pi i\frac{h(x_p)}{\lambda}\right) \tag{6.41}$$

However, it should be noted that the imaging geometry used has considered everything as projected into the slant plane. For a three-dimensional (3D) broadside geometry, the required phase correction term will be given by the height, z, projected onto the direction of the radar line-of-sight to the scene centre using the elevation angle ϕ, i.e.

$$\exp\left(4\pi i\frac{z\cos(\phi)}{\lambda}\right) \tag{6.42}$$

At this point, it is worth recalling that the base-banding correction discussed earlier for the phase ramp resulting from the inclination of the flat focal plane to the radar line-of-sight is given by

$$exp\left(2\pi i\frac{2n\Delta y\cos(\theta)}{\lambda}\right) = \exp\left(4\pi i\frac{y\cos(\theta)}{\lambda}\right) \tag{6.43}$$

where θ is the grazing angle and $\theta + \phi = 90^{\mathrm{o}}$. This can be recognised as a phase factor which arises from the projection of the range of the point of interest onto the radar line-of-sight.

For a non-broadside geometry, it is natural to infer that a phase correction dependent on the cross-range position will also be required which would be formed in a similar manner. Combining all these terms into a single correction then gives the required phase correction for a position (x, y, z) on the focal surface to be

$$\exp\left(4\pi i\frac{\alpha x}{\lambda}\right)\exp\left(4\pi i\frac{\beta y}{\lambda}\right)\exp\left(4\pi i\frac{\gamma z}{\lambda}\right) \tag{6.44}$$

where (α, β, γ) are the direction cosines of the line-of-sight vector from the scene centre to the platform position at the mid-point of the aperture. This term combines the base-banding term and the term required to correct for height variations, i.e. topography, and so the process has been termed "topographic base-banding".

The effect of topographic base-banding is illustrated in Figure 6.25. It can be seen that all the supports associated with the different spatial positions which had been shifted in different directions (giving rise to Figure 6.24) have all been shifted to the frequency origin to provide a compact frequency support. This is crucial because it is now possible to perform valid re-sampling of the complex image to implement any warping required to register the images to sub-pixel accuracy.

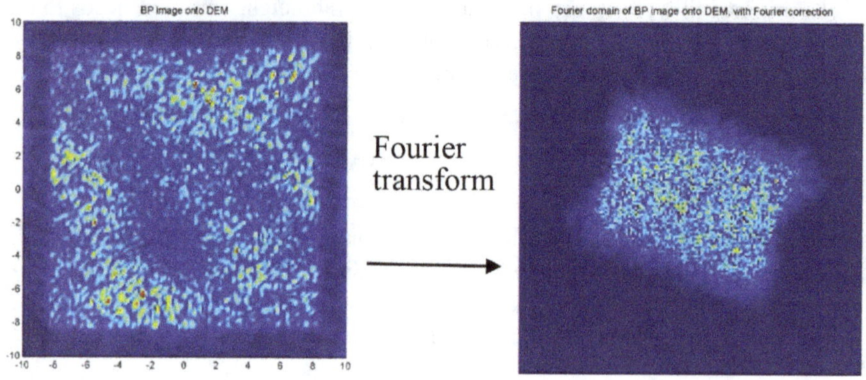

Figure 6.25 *Image formed onto the representation of terrain heights together with corresponding frequency support after topographic base-banding has been applied*

6.4.2 Spatially-variant frequency trimming

6.4.2.1 Spatially-variant frequency support overlap

In Section 6.4.1, the impact of using a non-flat focal surface for image formation was discussed with respect to maintaining good frequency support behaviour. In particular, the technique of topographic base-banding was introduced to ensure that the frequency support remains within a constrained area of frequency space, i.e. it is band-limited, so that the Nyquist criterion is satisfied and hence interpolation can proceed with adequately sampled data.

A second frequency support consideration arising from the use of a non-flat focal surface will be discussed in this section. This concerns the frequency trimming process, specifically how to make best use of the coherent information between the CCD collections whilst rejecting the incoherent information. This results in a spatially variant frequency trimming process. The concept is developed here and validated against experimental data in Section 6.5.

The complexities arising in the frequency domain when using a non-flat focal plane are illustrated in Figure 6.26. The frequency support can essentially be viewed as a planar object (or, more generally, a 2D manifold) in 3D space (see inset 3D cube in Figure 6.26) which is projected down onto two dimensions [6]. The direction in which this projection is performed will depend on the local tangent plane to the focal surface. For a flat focal plane, the local tangent plane will always be in the same direction and so the projections will all be in the same direction and will hence all lie on top of each other. However, for an undulating surface, the direction of projection will be spatially varying and so the projections will be in many different directions. This means that the amalgamated frequency support will not be compact in the frequency domain. In Figure 6.26, the red and blue outlines represent the local frequency supports for two image collections. It can be envisaged that the amalgamation of the projected red supports (or the amalgamation of the projected blue

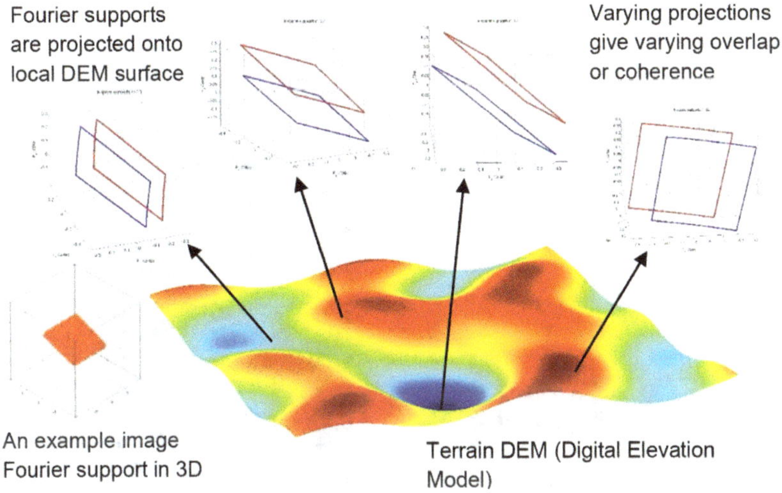

Figure 6.26 *Illustration of frequency support shifting and varying overlap resulting from terrain undulations*

supports) will not be compact. This gives rise to the spreading of frequency supports discussed earlier which is mitigated by topographic base-banding.

A second effect is that the degree of overlap between the red and blue frequency supports is spatially varying. This means that it is not a trivial process to perform the frequency trimming as it essentially needs to be done on a pixel-by-pixel basis. Indeed, it should be emphasised that standard frequency trimming will be sub-optimal for any non-flat surface. It should also be noted that the maximum possible coherence that can be obtained will vary spatially. For some positions, there may be no overlap between the frequency supports in which case there will be no coherence even if there has been no disturbance in the scenes.

A simulated example of an image pair to be used for CCD formation obtained from image formation onto a non-flat focal plane is shown in Figure 6.27. A random distribution of scatterers has been placed on an undulating surface with RCS modulated by the angle of the surface relative to the radar line-of-sight such that the RCS is brighter when the incidence is closer to normal. This is not meant to provide an accurate representation of terrain imagery but rather to provide some structure within the simulated scene.

The top-left image shows the image draped over the surface which has been used as the focal surface. An important point to note at this stage is that the radar energy has been focused at its correct spatial position. This is in contrast to the use of a flat focal plane when the energy from elevated positions will be focused at the incorrect spatial position resulting in the "layover" phenomenon. Thus a key advantage of image formation onto a non-flat focal plane is that it naturally produces correctly ortho-rectified imagery with no layover.

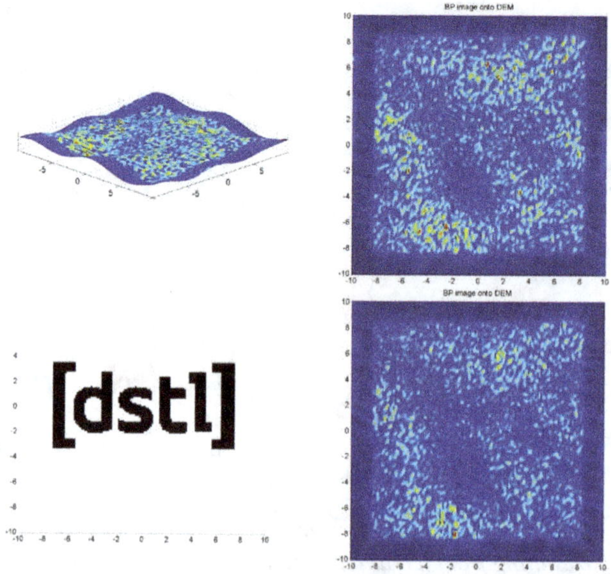

Figure 6.27 *Simulated image formed by back-projection onto a DEM draped over undulating focal surface (top-left) or without height information as 2D image (top-right). Scatterer disturbances were introduced in the dark region shown (bottom-left) and a second simulated image was obtained (bottom-right).*

In practice, it is more convenient to remove the height co-ordinate when displaying the imagery to give a 2D image as shown at the top-right. A second simulation was performed for which the positions of certain of the scatterers have been displaced to mimic the type of changes which can be observed in CCD images. The area of disturbance in this case is in the shape of the [dstl] logo as illustrated at the bottom-left. The image resulting from this is shown at the bottom-right. It can be seen that there is effectively no evidence of the disturbances in the intensity image that is displayed.

To obtain the best CCD image from this pair of SAR images, it is necessary to deal with the problem of a spatially varying overlap region. This situation is illustrated in Figure 6.28, which shows frequency supports from two images taken at different imaging geometries after topographic base-banding (top-left), the superposition of spatially varying templates defining the non-overlapping regions containing incoherent energy (top-right) and the superposition of spatially varying templates defining the overlapping regions containing coherent energy (bottom-left). In Section 6.4.2.2, it is discussed how to perform frequency trimming in this situation to ensure that only coherent information is used.

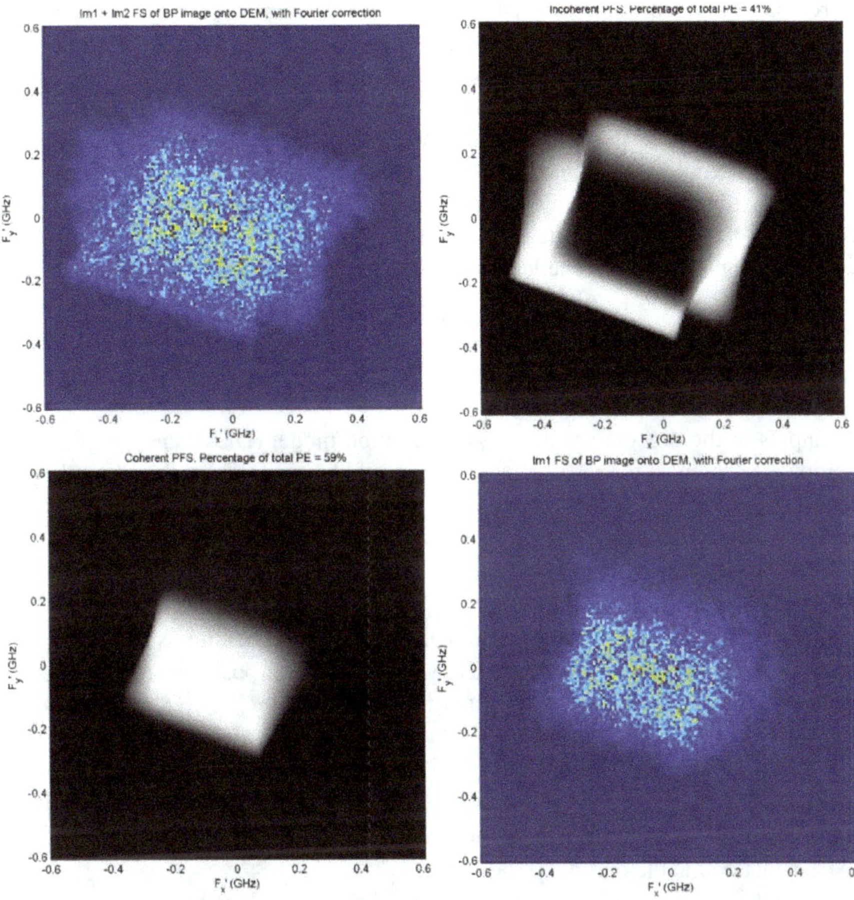

Figure 6.28 *Illustration of spatially variant trimming showing frequency supports from two images taken at different imaging geometries after topographic base-banding (top-left), the superposition of spatially varying templates defining the non-overlapping regions containing incoherent energy (top-right), the superposition of spatially varying templates defining the overlapping regions containing coherent energy (bottom-left) and the frequency support obtained after spatially variant trimming (bottom-right)*

6.4.2.2 Spatially variant ("per-pixel") trimming

One approach to frequency trimming for the non-flat focal surface case would be to use the union of coherent energy templates. This can be implemented post-image formation but is not optimal since some incoherent energy will still be present. The optimum approach is to perform the frequency trimming during image formation so that the raw radar data used to compute a particular pixel value is restricted to only the region of overlap in the frequency domain for that particular pixel. This means that the image formation must be formulated specifically for a particular CCD pair of images and also that the resolution will vary spatially as a result of the varying size of the overlap region. However, it does mean that no incoherent energy is subsequently included in the CCD calculation. This spatially variant ("per-pixel") trimming procedure is explained in detail below but, in anticipation, the frequency support resulting from spatially variant trimming can be seen on the bottom-right of Figure 6.28. The benefit in terms of CCD performance will be illustrated after a detailed explanation of the practicalities of implementing spatially variant trimming.

Procedure for spatially variant frequency trimming

Inputs

The inputs to the process are the digital elevation model (DEM) and the platform position at each pulse for each collection. It should be ensured that these are all provided in a consistent co-ordinate system. Also required are the centre frequency, the frequency bandwidth and the frequency step size.

Step 1: Normal vector to DEM

The unit normal vector to the DEM for each output pixel position is required. One approach would be to interpolate the DEM onto the pixel positions and then use a standard technique for calculating the normal directions.

Step 2: Define the shape of support in 3D K-space for each collect

The frequency supports for each collect will form surfaces in 3D K-space as illustrated in Figure 6.29. These shapes need to be captured in terms of positions in K-space that define the outline of the shape.

Consider the upper support and the edge of the shape labelled A. This edge is associated with the first pulse of the collection. The unit vector defining the line on which this edge lies is given by the vector from the scene centre to the position of the platform when the first pulse is transmitted. (This is a consequence of the Fourier slice theorem that angles are preserved through the Fourier transform operation.) If the unit vector for the first pulse is $\underline{u}_1$ then the first position of edge A is given by $f_{min}\underline{u}_1$ and the last is given by $f_{max}\underline{u}_1$ using the minimum and maximum frequencies. Intermediate points are defined by the frequency samples within the pulse in a natural way.

The edge labelled by B is associated with the maximum frequency. It begins with $f_{max}\underline{u}_1$ and ends with $f_{max}\underline{u}_P$, where P is the total number of pulses and hence $\underline{u}_P$ is the last collected pulse. Intermediate points are defined by $f_{max}\underline{u}_n$ for the nth pulse in a natural way.

The definitions of edges C and D follow naturally. The whole shape can then be described as a sequence of 3D co-ordinates defining the edges A, B, C and D. It is convenient for these to follow around the boundary in sequence which is why the arrow directions define a complete circuit.

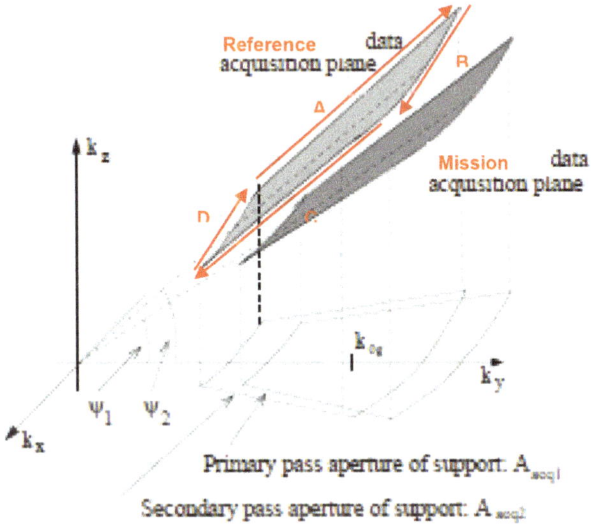

Figure 6.29 *Illustration of frequency supports for the two collections in 3D K-space showing how the shapes are defined in terms of vectors (red arrows)*

Step 3: Base-banding

The support shapes as defined are now base-banded, i.e. the centre of the support needs to be translated to the origin of frequency co-ordinates. Importantly, this is performed for one of the collects (this should be the image which will subsequently be warped onto the other image so that the support is centred and surrounded by zeros so that interpolation can be done correctly) and then the identical translation applied to the other image. It should be noted that this step (and the reverse operation at Step 6) is not strictly necessary. However, it is useful to move the support shapes to the origin to aid in visualisation.

The position of the centre of the support can be defined by taking the mean unit vector over all pulses and multiplying by the centre frequency, i.e.

$$f_0 \frac{1}{P} \sum_{n=1}^{P} \underline{u}_n \tag{6.45}$$

This position vector (calculated for the reference image) should be subtracted off every position vector in the sequence defining the outline of the shape for the reference image. The identical position vector should be used to translate the shape for the mission image. This is illustrated in Figure 6.30.

It should be noted that this operation embodies the "topographic base-banding" operation which is simply represented in this 3D K-space domain.

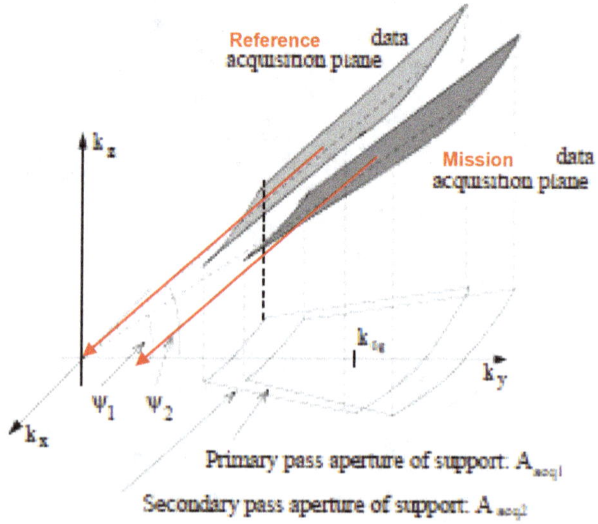

Figure 6.30 *Illustration of frequency supports for the two collections in 3D K-space showing the base-banding operation (red arrows)*

Step 4: Rotation of co-ordinate system to align with surface normal for a given pixel

The two support shapes are now defined in a co-ordinate system in which the unit vectors k_x and k_y lie in a horizontal plane and k_z gives vertical height. Simple deletion of the k_z component provides the projection of the shape onto the default horizontal plane. For a given output pixel, the projection onto the tangent plane to the DEM at this pixel position is required. If the co-ordinate frame is rotated so that the new z-direction k_z' is aligned with the surface normal (previously calculated) for the pixel of interest, then simple deletion of this new k_z' component will achieve the required projection.

Note that in Figure 6.31, which illustrates the situation, the supports are shown prior to base-banding for convenience.

The axis of rotation is given by the cross-product of k_z and k_z' (i.e. the surface normal) and the angle through which rotation is required is given by $\cos^{-1}\left(k_z.k_z'\right)$.

Step 5: Determine region of overlap

Once the two support shapes have been projected onto the appropriate tangent plane for a given pixel, then the region of overlap as illustrated in Figure 6.32 is of interest.

For a given output pixel and each pulse, the frequency samples are projected onto the tangent plane following essentially the same procedure as defined above. It is then determined whether the projected sample lies within both the red support and the green support. If this is the case then that sample is used to contribute to the output pixel value.

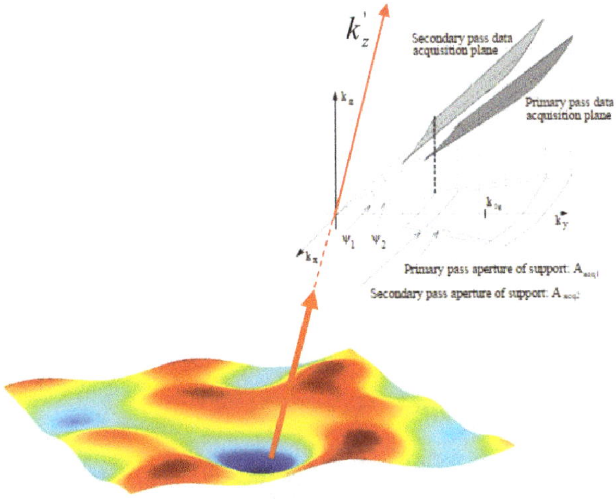

Figure 6.31 Illustration of rotation of co-ordinate system to align with the surface normal for a given pixel (red arrow)

An alternative approach is to determine the sequence of 2D co-ordinates that define the boundary of the region of overlap and reverse the process of projection, rotation and base-banding for each collect to map this region into the pulse/frequency sample space for each collect.

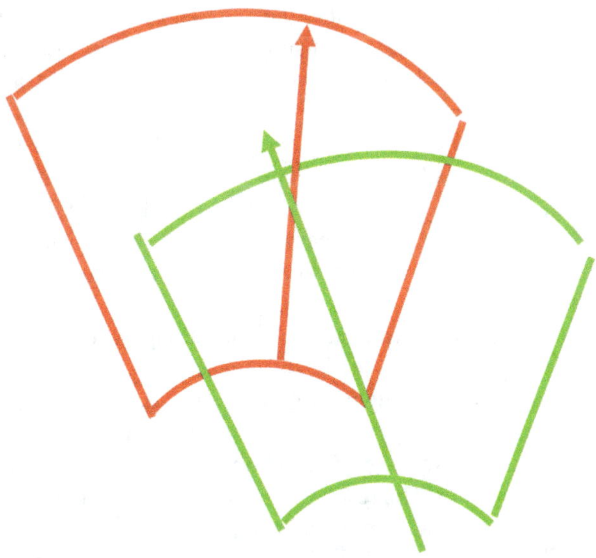

Figure 6.32 Illustration of the region of overlap between two frequency supports projected onto the appropriate tangent plane

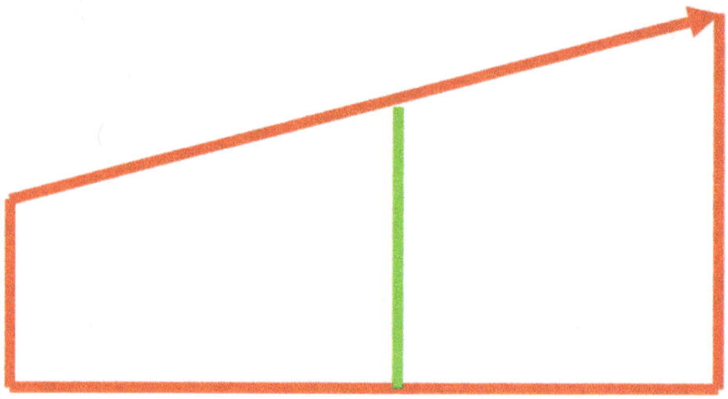

Figure 6.33 Illustration of a cut through red K-space along the direction corresponding to a single pulse

The sequence of overlap co-ordinates can be found by running through the green co-ordinates and keeping those that lie within the red shape, then running through the red co-ordinates and keeping those that lie within the green shape.

It is worth noting that if the focal surface were flat, this process would result in the standard Fourier-based frequency trimming discussed in Section 6.3.

Step 6: Reverse process to define overlap shape in pulse/frequency domain

The 2D overlap co-ordinates arising from the green support can be moved back into the 3D K-space corresponding to the green collect by simply replacing the z co-ordinate that was removed to perform the projection. This applies similarly for moving the red support samples back into the 3D K-space corresponding to the red support. However, to move the 2D green support co-ordinates into the 3D K-space corresponding to the red collect (and vice versa) is not as straightforward since those z co-ordinates were never available.

A solution to this is to consider cuts through red K-space aligned with pulse directions. The z co-ordinates (heights) will be known at the start and end of this cut corresponding to f_{min} and f_{max}. The height will vary linearly along this cut and so any intermediate height can be found by linear interpolation. In particular, the height in 3D K-space for the red collect of a position of the 2D green support outline can be determined in this way.

Figure 6.33 shows the cut through red K-space along the direction corresponding to a single pulse. The height of the intersecting green support can be calculated by linear interpolation from the heights at the f_{min} and f_{max} end positions.

Once the overlap shape has been moved (un-projected) back into 3D K-space, it is then straightforward to remove the rotation to return to the default geometry and remove the base-banding. The result of this process is that the overlap region is defined over the pulse/frequency sample domain.

It is now known which samples contribute to each pixel in the image. The image formation process is thus implemented using a modified version of back-projection which uses only the required samples. More specifically, a pixel-dependent pulse

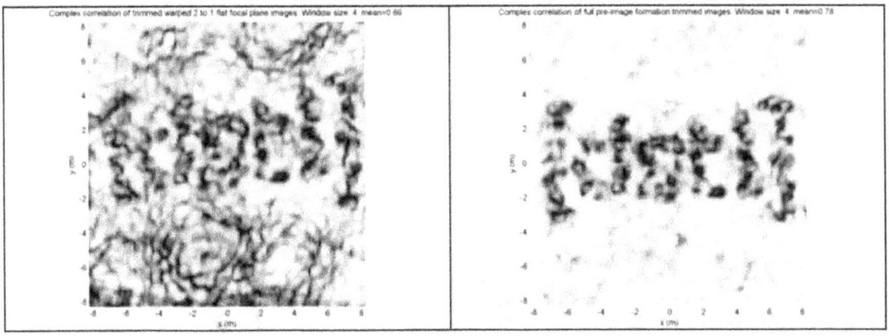

Figure 6.34 *CCD images resulting from standard flat focal surface processing (left) and back-projection to a DEM image formation with optimal frequency trimming (right)*

compression is used in which the compression is restricted to only those samples which overlap for the pixel under consideration. (There is, of course, a computational cost in doing this but the aim here is to achieve the best CCD performance in the presence of non-flat terrain.) Topographic base-banding is then applied to the resulting images which can then be used to form the CCD image.

The frequency support resulting from this spatially variant trimming can be seen on the bottom-right of Figure 6.28. This contains the base-banded amalgamation of the frequency overlaps for each pixel.

Figure 6.34 shows CCD results using the simulated data for undulating terrain that has been discussed. The CCD image on the left is the result of applying standard flat focal plane processing to the radar data obtained from the undulating terrain whilst the CCD image on the right is the result of applying back-projection to a DEM image formation with optimal frequency trimming applied during image formation. It can be seen that, when using this approach, it is much more evident that the coherence is high over those regions where no change has taken place. As a consequence, the incoherence due to small disturbances in the scatterer positions is much more easily detected. It should be noted that this process is readily extended to the bistatic case [7].

6.5 Experimental results

6.5.1 Radar measurement facility and experimental set-up

Cranfield University in the UK operates an experimental radar measurement facility based on a stepped frequency, continuous-wave imaging system. This GB-SAR can be used for experimental data gathering both indoors and outdoors. For the experiments described here, an indoor set-up was used as shown in Figure 6.35 with the gravel area being the scene to be imaged. A horizontal rail, 2.75 m above the floor, can be seen along which an antenna horn is incrementally stepped, using a

Figure 6.35 The GB-SAR set up indoors with a gravel scene to be imaged

predefined step-size, to form a 3.5 m imaging aperture. The bandwidth used was 8.25–12 GHz and the slant range to the centre of the scene was 7 m. This means that the 3 m square gravel scene is essentially in the SAR near field.

In order to gather reference and mission images to support CCD image formation, the SAR rail was used in two positions 50 cm apart horizontally and both at a height of 2.75 m. The reference image was obtained from the front position and the mission image was obtained from the rear position. Aperture sizes of 1.5 m were used in both cases with the starting position differing by 1 m between the collections in order to provide a variation in imaging geometry whilst still maintaining some coherence.

Figure 6.36(a) shows a ground-truth representation of the set-up, with Figure 6.36(b) providing the plan view, in which the extent gravel scene to be imaged is shown. The collection transceiver positions for the reference and mission collections are indicated in the images by the adjacent horizontal lines. It should be noted that angular apertures and differences between the two collections vary according to scene position since the scene is in the near field. The grazing angle is approximately 33° although there is a difference of 3° between reference and mission geometries. Both azimuth apertures are nominally 20° wide but the azimuth angles differ by 11°.

A photograph of the gravel scene is shown in Figure 6.37(a) in which it can be seen that the terrain is flat apart from in one quadrant in which there is a significant height variation. A DEM is required for the processing that has been described earlier and this was obtained using interferometric processing. Two monostatic images were obtained, one from each rail position, and interferometry was used to obtain

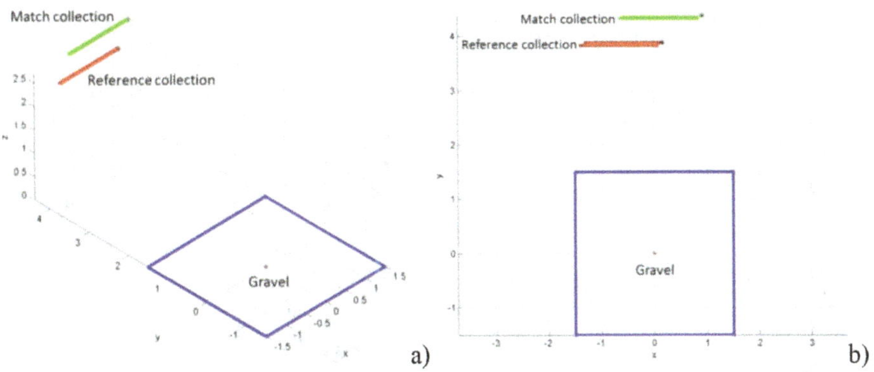

Figure 6.36 *The ground truth diagrams (a) and (b) show the imaged 3 m square scene denoted "gravel". The reference monostatic SAR collection and the mission (match) SAR collection transceiver positions are indicated in the images and represented by the adjacent horizontal lines*

the DEM shown in Figure 6.37(b). This DEM covers a 4 m square although is most accurate over the central 3 m square. It can be seen to consist of a large flat area at ground level with two plateaus which reach a height of approximately 35 cm.

6.5.2 Standard CCD processing

Figure 6.38(a) shows the reference image obtained using the set-up described earlier. When standard (flat focal plane) CCD processing is used to combine this with the mission image and when no frequency trimming is applied, the CCD image shown in Figure 6.38(b) results. It can be seen that there is a level of coherence

Figure 6.37 *The non-flat 3 m square gravel scene was constructed (a) with a large area at ground level and with two plateaus. A monostatic SAR interferogram-derived DEM (in m) was formed (b), which is most accurate in the central 3 m square. The DEM is extrapolated towards the edges*

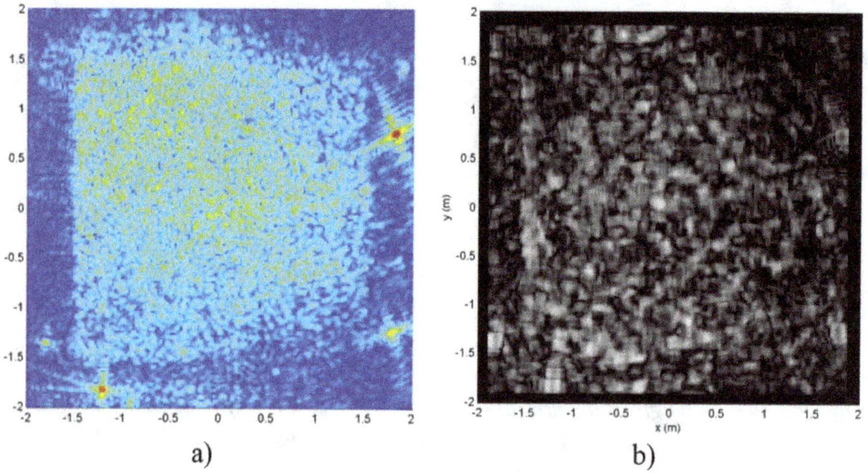

Figure 6.38 *Standard image formation involves the use of a flat focal surface, giving rise to the SAR image in (a). When no trimming is applied, the resulting CCD (b) shows a low coherence of mean value 0.36.*

(whiteness) in Figure 6.38(b) but changes would be difficult to detect against this level of background coherence.

As was discussed in Section 6.3.4, there will be incoherent energy used in the formation of the CCD unless it is removed using frequency trimming. Frequency trimming can be applied in the Fourier (K-space) domain by using a mask corresponding to the area of overlap of the two frequency supports. This is optimum for a flat focal plane in the far field and can be implemented using a single global mask and so will be referred to as global frequency trimming (as opposed to spatially variant trimming). An equivalent process is to restrict image formation to use only those radar pulses within fixed frequency bands known to correspond to the overlap region.

A complication arises for a near-field situation since the process of "global" frequency trimming requires determination of the radar pulses and frequency bands that overlap on a pixel-by-pixel basis over the whole scene. Once this has been done, the extreme values for pulse number and frequency are selected and used for image formation. The CCD can then be formed and is the near-field equivalent of the globally trimmed CCD for the far-field case.

Using this process, the reference image shown in Figure 6.39(a) is obtained which is of lower resolution than the reference image in Figure 6.38(a) since the extent of the frequency support has been reduced. However, it can be seen from the resulting globally trimmed CCD in Figure 6.38(b) that the coherence has been improved in what is the near field equivalent of the standard CCD product.

In order to quantify overall improvements in coherence, the mean coherence over a central square of edge size 2.6 m is calculated. For the non-trimmed CCD, the mean coherence is 0.36 whilst for the globally trimmed image it is 0.44, which is a

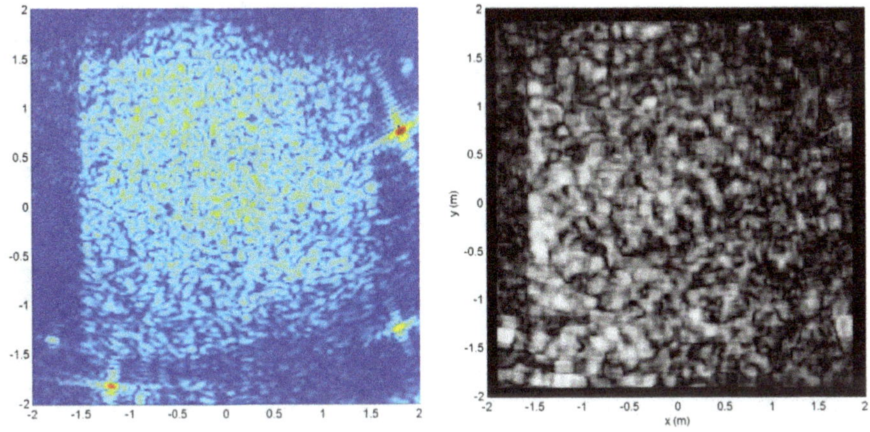

a) b)

Figure 6.39 *Standard CCD processing involves the application of global trimming to the radar collections (limiting the aperture and the frequency band) and here it is applied before image formation. The resulting reference SAR image (a) is of a lower resolution than previously (Figure 6.38(a)), but there is an improvement in coherence (b) when compared to the untrimmed result (Figure 6.38(b)), achieving a mean coherence of 0.44. Substantial regions of incoherence can be seen on the terrain slopes, however, making detection of changes difficult here.*

small improvement. It should be noted that the improvement is not uniform over the scene, but is concentrated on the left side of the image where the global incoherence trimming is found more suitable. In particular, substantial regions of incoherence can be seen on the terrain slopes for both CCD images, making detection of changes difficult here.

6.5.3 CCD processing using spatially variant trimming

To implement spatially variant trimming, the first stage is to perform spatially variant back-projection (i.e. back-projection that accumulates only radar data associated with overlapped portions of locally projected frequency supports) onto a DEM model of the scene thus placing the radar energy at the correct height at each pixel position. Figure 6.40(a) shows the resulting reference image in which it can be seen that, in comparison to Figure 6.38(a), the bands of higher reflectivity resulting from the terrain slopes have been moved down the image. This reflects the removal of the layover effect since reflections from higher positions are now being placed in their correct spatial position. The corresponding untrimmed CCD image is shown in Figure 6.40(b) but it is found that simply focusing onto a DEM without addressing

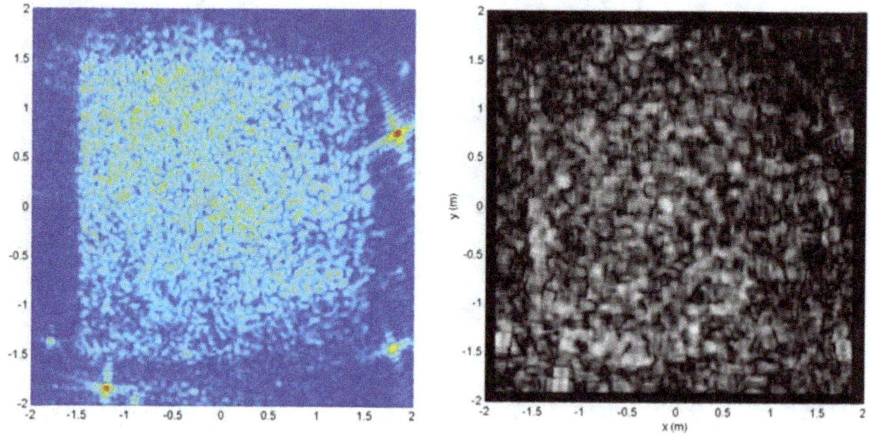

a) b)

Figure 6.40 *A more accurate approach to image formation involves focusing the radar energy to a focal surface coincident with the DEM (seen in Figure 6.37(b)) giving rise to the reference image (a). The corresponding CCD (b), however, has a mean coherence of 0.37, only very slightly greater than for the untrimmed standard approach (Figure 6.38(b)).*

trimming requirements has barely changed the coherence level, the mean level being 0.37 in comparison to 0.36 for Figure 6.38(b).

As with standard image formation for the near field scenario described in Section 6.5.2, the spatially variant trimming process involves finding the spatial frequency image support overlap for each pixel position. However, as has been discussed earlier, for the situation with a non-flat focal surface, projection of supports onto the DEM local tangent plane for the pixel position must be first performed. The extreme values of support overlap in radar pulse number are recorded for each pixel position, and the extreme frequency values of support overlap are recorded for each radar pulse and pixel position.

Figure 6.41(a) illustrates the process, where the image supports for the top-right pixel are shown, and the final pulse of SAR collect 1 (the left-hand outline) is interrogated to find which frequency points are to be kept. This procedure is also carried out for SAR collect 2. Once all this data is calculated over the whole scene, the coherent portions of the radar energy are known and can be employed for image formation or to provide an expected coherence overview, such as that seen in Figure 6.41(b). The coherence overview shows the degree of frequency support overlap due to radar geometry and topography across a 2.6 m square in the centre of the scene and is measured in units of GHZ^2, which is a measure of spatial frequency area. This in turn provides a measure of expected coherent image resolution. It can be seen that two curved bands of incoherence are present, and these can be seen to

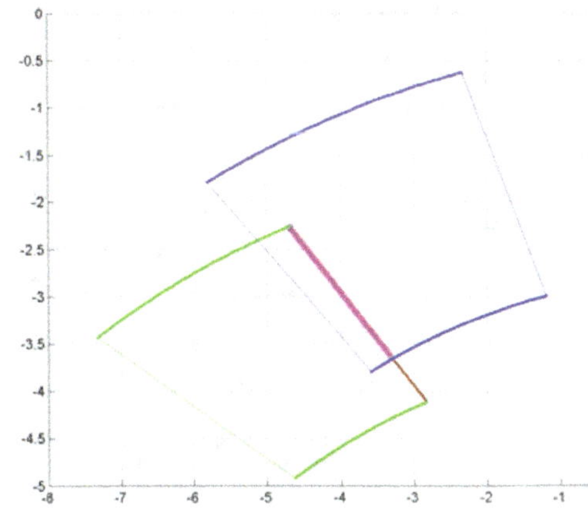

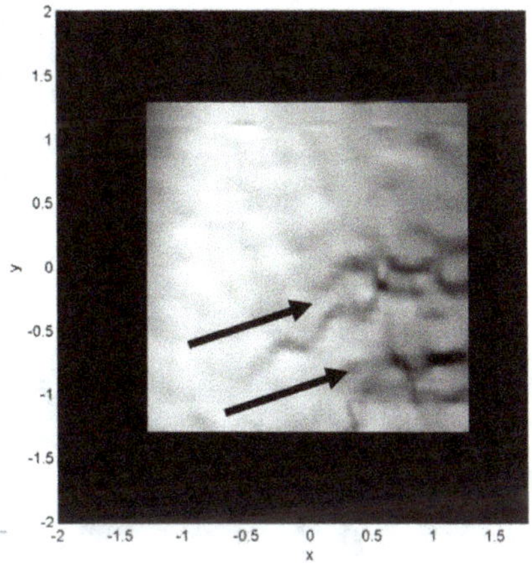

Figure 6.41 *The spatially variant trimming process involves finding the overlapping portions of the image spatial frequency supports when projected onto the DEM local tangent surface and carrying out this procedure across the whole scene. An outline of the projected supports for the two SAR collections is shown (a) to illustrate the procedure. This specific case is for the pixel position on the top-right of the imaged scene. In the illustration, the final radar pulse for collect 1 (left-hand outline) is being interrogated by the algorithm. The pulse is indeed one of those in the overlapping region; however, some of the lowest frequencies in the pulse are not in the overlap region and will be excluded. When this information is collected over the whole scene, a coherence overview can be formed (b) showing the amount of overlap available for each pixel in the scene (over a central 2.6 m square), in units of GHz*^{2} *(black implies no overlap). The overview shows that there are two bands of low overlap, and therefore incoherence, in the scene, coincident with the terrain slopes. These regions are indicated by the arrows.*

coincide both with regions of higher RCS in the SAR image in Figure 6.40(a) and with the slopes seen in Figure 6.37(b). The expectation therefore is that coherence due to topography and radar geometry will be lowest on the slopes. It is further noted that the expected bands of incoherence were also evident in the standard CCD product, seen in Figure 6.39(b).

The result of inputting only the expected coherent energy to the back-projection algorithm and processing to a DEM can be seen in Figure 6.42. Inspection of the SAR image in Figure 6.42(a) shows that the speckle features are indeed of lower resolution on the terrain slopes. Moreover, the corresponding CCD in Figure 6.42(b) has lower coherence on the slopes, but otherwise coherence levels have improved substantially. The new mean coherence level is 0.72, whereas the standard approach

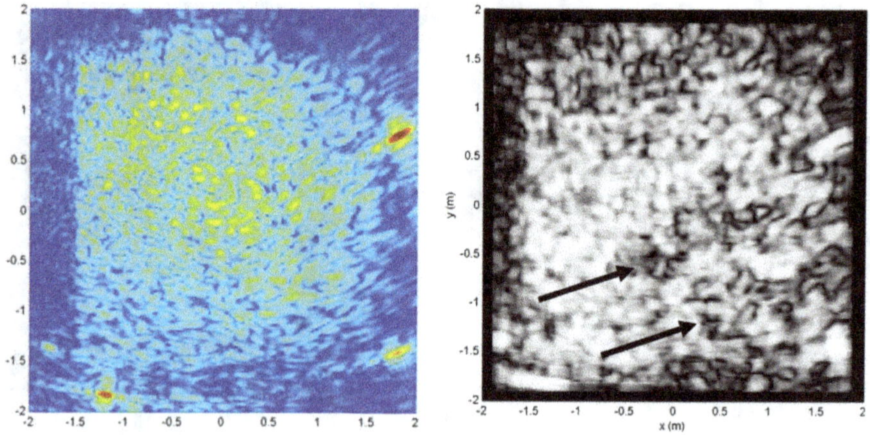

a) b)

Figure 6.42 *When the outputs of the spatially variant trimming process are input to the back-projection imaging algorithm, with a DEM focal surface in such a way that only the coherent portions of the SAR collections are input to the imaging process, the reference image (a) is obtained. The corresponding CCD (b) is of the highest coherence yet, achieving a mean coherence of 0.72. Some regions of incoherence remain, indicated by arrows; however, these are mostly in the regions predicted to have low coherence due to the local topography and radar geometry, as can be seen by comparison with Figure 6.41(b).*

gave rise to a mean coherence level of 0.44. It should be noted that the overall coherence levels in the image region with non-flat terrain has improved substantially, and where coherence has not improved this can be understood from the inspection of the output of the spatially variant trimming calculation in Figure 6.41(b) as being mostly due to the local topography and radar geometry.

The improvement in coherence level from 0.44 to 0.72 is particularly significant, as it was seen in Section 6.2.6 that localised regions of change can be clearly detected when the background coherence is 0.7 but detection is unreliable for background coherence levels of 0.5 (and lower). Thus, spatially variant trimming has raised the background coherence level such that detection of localised areas of change is now possible. The detection of such change will be highlighted in Section 6.5.4.

6.5.4 Detection of change using spatially variant trimming

Direct implementation and observation of changes on the ground help to validate the spatially variant trimming algorithm and to illustrate the underlying processes. The full SAR rail extent 40° aperture spatially variant trimming SAR CCDs shown

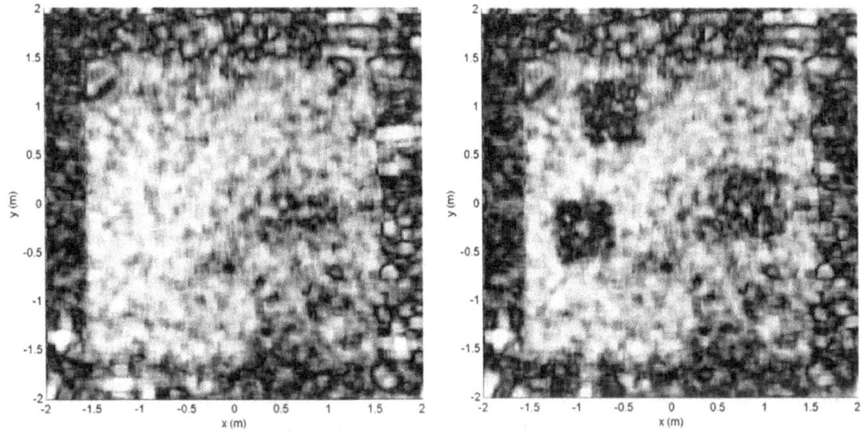

a) b)

Figure 6.43 *Full 40° aperture SAR image spatially variant trimming CCDs of the scene, showing the case where no man-made changes have been implemented on the gravel (a), and where they have (b). The changes were implemented on 30 cm square regions through minimal disturbances of the gravel and the right-hand change was implemented on a slope on the terrain.*

in Figure 6.43 clearly show the effect of implementing three 30 cm square regions of change on the ground. It is estimated that the 40° aperture provides a maximum resolution of approximately 3.5 cm. The changes were implemented through minimal disturbances of the gravel. Figure 6.43(a) is a spatially variant trimming CCD free from man-made changes, and regions of incoherence correspond to slopes on the terrain as predicted from the coherence or overlap map in Figure 6.41(b). Figure 6.43(b) shows the corresponding spatially variant trimming CCD where the changes have been implemented. The two leftmost squares are clearly visible and are located on flat ground. Although the rightmost square is also visible, its outlines are blurred by incoherence due to the local topography (a slope) and radar geometry as predicted.

6.6 Summary

CCD is a powerful tool for detecting the change in SAR image pairs taken of the same region, at similar geometries but separated in time. This chapter has provided an overview of the fundamentals of CCD formation including the need for frequency trimming to ensure that any incoherent information is excluded so as to maximise coherence. An analysis of the detectability of change has been provided which quantifies the level of background coherence that needs to be maintained between collections in order to enable the detection of particular types of change. A further analysis has been provided of how the radar and imaging geometry parameters will impact the background coherence which then allows the trade-off space over such parameters to be investigated.

Following the discussion on fundamentals, the chapter has then considered the challenge of maximising CCD performance over non-flat terrain. The use of back-projection processing onto a DEM has been discussed and the need to extend frequency trimming to be a spatially variant process has been established. A detailed explanation of how spatially variant frequency trimming can be implemented is provided. The benefits of maximising background coherence in this way have been demonstrated using both simulation and real SAR data obtained using Cranfield University's GB-SAR system in a controlled laboratory environment.

Many challenges remain for CCD processing as SAR systems become more sophisticated. In particular, swarms of SAR sensors acting coherently will produce complicated K-space coverage which needs to be taken into account when attempting to maximise CCD performance. How this should be done is currently an open challenge.

References

[1] Stevens M., Jones O., Moyse P., Tu S., Wilshire A. 'Bright spark: Ka-band SAR technology demonstrator'. International Conference on Radar Systems (Radar 2017); Belfast; 2017. pp. 1–4.

[2] Preiss M., Stacy N.J.S. *Coherent Change Detection: Theoretical Description and Experimental Results, Technical Report*. DSTO-TR-1851. Australia: Defence Science and Technology Office; 2006.

[3] Goodman J.W. *Statistical Optics*. Wiley; 2000.

[4] Zebker H.A., Villasenor J. 'Decorrelation in interferometric radar echoes'. *IEEE Transactions on Geoscience and Remote Sensing*. 1992, vol. 30(5), pp. 950–9.

[5] Tough R.J.A., Blacknell D., Quegan S. 'A statistical description of polarimetric and interferometric synthetic aperture radar data'. *Proceedings of the Royal Society A*. 1995, vol. 449(1937), pp. 567–89.

[6] Jakowatz C.V Jr., Wahl D.E., Eichel P.H., Ghiglia D.C., Thompson P.A. *Spotlight-Mode Synthetic Aperture Radar: A Signal Processing Approach*. Kluwer Academic Publishers; 1999.

[7] André D.B., Blacknell D., Morrison K. 'Spatially variant incoherence trimming for improved bistatic SAR CCD'. IEEE Radar Conference, 1st-5th April 2013; 2013.

Chapter 7

Modern GPR target recognition methods

Fabio Giovanneschi[1], Kumar Vijay Mishra[2], and Maria Antonia Gonzalez-Huici[1]

In some humanitarian, commercial, and military applications, information on imaging, detection, and localization of shallow-buried targets is highly desirable [1]. Often there is a necessity to sense and retrieve this information reliably and safely in, for instance, landmine recognition, archaeological excavations, planetary expeditions, and construction engineering. While a wide variety of technologies are available for subsurface exploration, namely radiometric, seismic, and electromagnetic (EM), only a ground penetration radar (GPR) provides non-invasive, safe, efficient, and high-resolution sensing [1]. This advantage has led to significant advances in research on GPR acquisition and information retrieval for the past two decades.

In an *impulse GPR* system [2], the transmit antenna emits an ultra-wideband (UWB) EM pulse at the surface. As the signal propagates through the subsurface layers, their magnetic and electrical properties such as permittivity, permeability, and conductivity induce changes in the phase and amplitude of the scattered wave. While the amplitude of the reflected wave is directly proportional to the complex reflectivity of the buried target, the phase provides information on the target's location relative to the radar mounted on the surface. The GPR receiver collects and processes the reflected echoes to extract the reflectivities and ranges of several targets.

In GPR applications, a target is located within the close range of the radar. As a result, the received signal is composed of multiple reflections from different parts of the same object. This *extended* target model implies retrieval of multiple ranges and amplitudes of portions corresponding to the same larger object. By exploiting the spatial correlation among the point values of the backscattered signal, GPR generates a *target signature* that is usually specific to the object being explored and, hence, useful to identify it [3]. In this chapter, we focus on this *target recognition* aspect of GPR processing.

[1]Fraunhofer FHR, Germany
[2]United States Army Research Laboratory, USA

Often terms such as target recognition, classification, and identification are interchangeably used for techniques to classify targets from measured radar data. However, for the purposes of this chapter, *target classification* implies isolating the target to a general class, while *target identification* is used to distinguish the target more precisely. For example, labeling a target as a landmine or soil clutter is a classification step. Identifying a landmine target as PMN/PMA (i.e. two particular types of Standard Test Targets, see Section 7.3.2) is its identification. The target recognition encompasses both classification and identification [4].

In practice, the recorded GPR signals generally suffer from several unwanted contributions arising from system effects such as antenna coupling, multiple surface reflections, system instability, timing jitter, limited spatial resolution, and amplitude variations [5]. Further, the rough ground surface in which the targets of interest are buried generates the undesired *clutter* signals that are comparable in strength to the target signatures. As a result, GPR target recognition is a difficult task.

Traditional GPR target recognition methods (see [3] for a review) include preprocessing the data by removal of noisy signatures, dewowing (high-pass filtering to remove low-frequency noise), filtering, deconvolution, and migration (correction of the effect of survey geometry) and can rely on the simulation of GPR responses. These techniques usually suffer from the loss of information, inability to adapt from prior results, and inefficient performance in the presence of strong clutter and noise.

To address these challenges, several advanced processing methods have been developed over the past decade to enhance GPR target recognition. In this chapter, we provide an overview of these modern GPR processing techniques. In particular, we focus on the following methods:

- *adaptive receive processing* of range profiles depending on the target environment [6]
- adoption of *learning-based methods* so that the radar utilizes the results from prior measurements [6, 7]
- application of methods that exploit the fact that the target scene is *sparse* in some domain or dictionary [8, 9]
- application of advanced *classification* techniques [10]
- *convolutional coding* that provides succinct and representatives features of the targets [11].

We describe each of these techniques or their combinations through a representative application of landmine detection.

The rest of the chapter is organized as follows. In the next section, we provide a brief overview of GPR technology and the complex scattering behavior of EM waves in the soil. We describe the GPR signal model and list both classical and modern processing methods in Section 7.2. We explain the dataset for our illustrative application of landmine detection in Section 7.3. Section 7.4 focuses on various sparse representation (SR) techniques that are enabled by dictionary learning (DL) algorithms detailed in Section 7.5. We outline the adaptive statistical evaluation carried out by the radar processor in order to selectively apply one of the algorithms in Section 7.6. Finally, we present the results of all of these techniques for target recognition in Section 7.7 before concluding in Section 7.8.

Throughout this chapter, we reserve boldface lowercase and uppercase letters for vectors and matrices, respectively. The *i*th element of vector y is $\mathbf{y}_i$ while the (*i,j*) th entry of the matrix Y is $\mathbf{Y}_{i,j}$. We denote the transpose by $(\cdot)^T$. We represent the set of real and complex numbers by $\mathbb{R}$ and $\mathbb{C}$, respectively. The notation $\|\cdot\|_p$ stands for the *p*-norm of its argument; $\|\cdot\|_F$ is the Frobenius norm; $\langle\cdot,\cdot\rangle$ indicates the dot product between the two arguments; and $|\cdot|$ is the cardinality of the set argument. A subscript in the parenthesis such as $(\cdot)_{(t)}$ is the value of the argument in the *t*th iteration. The convolution product is denoted by *. The function diag(·) outputs a diagonal matrix with the input vector along its main diagonal. We use Pr{·} to denote probability, E{·} is the statistical expectation, and $|\cdot|$ denotes the absolute value. The functions max(·) and sup(·) output the maximum and supremum value of their arguments, respectively and $\mathrm{i} = \sqrt{-1}$ is the imaginary unit; $\vec{(\cdot)}$ is a vector field; $\nabla\cdot$ and $\nabla\times$ denote the divergence and curl vector operators, respectively.

7.1 GPR for buried target recognition

The GPR is a sophisticated technology with decades of development history. We refer the reader to excellent expositions in [1, 3]. In this section, we summarize the key principles of GPR transmission and acquisition that are relevant to this chapter.

7.1.1 Subsurface EM propagation

A GPR probes the underground by repeatedly transmitting EM pulses into the subsurface at a wavelength λ of near-constant power P_t. The wave travels through the medium and reflects back to the radar when there is a change in the dielectric properties caused by variation in the medium [1, 12].

Define the vector $\vec{E}$ (in V/m) as the electric field intensity, $\vec{D}$ (C/m^2) as the electric flux density, $\vec{B}$ (*T*) as the magnetic flux density, $\vec{J}$ (A/m^2) as the electric current density, and ρ_c (C/m^3) as the electric charge density. The ε_0 (F/m) and μ_0 (H/m) are the electric and magnetic field constants, respectively; ε_r and μ_r are the relative dielectric permittivity and the relative magnetic permeability, respectively; σ_r (S/m) is the electrical conductivity and the parameter ω (rad/s) is the angular frequency. Provided some boundary conditions are defined and assuming the general case of anisotropic (dispersive) media, the electric and magnetic fields vectors of the transmit EM wave at a given point in space characterized by the position vector $\mathbf{r}$ at time instant *t* is described by the Maxwell equations [12, 13]

$$\nabla \times \vec{E}(\mathbf{r},\omega) = -\mathrm{i}\omega\vec{B}(\mathbf{r},\omega) \tag{7.1}$$

$$\nabla \times \vec{H}(\mathbf{r},\omega) = \left(\sigma(\mathbf{r},\omega) + \mathrm{i}\omega\varepsilon_0\varepsilon_r(\mathbf{r},\omega)\right)\vec{E}(\mathbf{r},\omega) \tag{7.2}$$

$$\nabla \cdot \left(\varepsilon_0\varepsilon_r(\mathbf{r},\omega)\vec{E}(\mathbf{r},\omega)\right) \tag{7.3}$$

$$\nabla \cdot \vec{B}(\mathbf{r},\omega) = 0, \tag{7.4}$$

and the related *constitutive relations*

$$\vec{D}(\mathbf{r},\omega) = \varepsilon_0\varepsilon_r(\mathbf{r},\omega)\vec{E}(\mathbf{r},\omega) \tag{7.5}$$

$$\vec{B}(\mathbf{r},\omega) = \mu_0\mu_r(\vec{r},\omega)\vec{H}(\mathbf{r},\omega) \tag{7.6}$$

$$\vec{J}(\mathbf{r},\omega) = \sigma(\mathbf{r},\omega)\vec{E}(\mathbf{r},\omega). \tag{7.7}$$

We choose to provide the frequency-domain representations of these equations because, unless the medium of propagation is non-dispersive (e.g. vacuum), the constitutive equations (7.5)–(7.7) do not generally hold for time-dependent fields.

The quantities ε_r, μ_r and σ are generally complex and frequency-dependent. Assuming the media to be homogeneous, they are spatially independent. However, for a typical GPR scenario with operating frequency from 10 MHz to a few GHz, the magnetic permeability is negligible ($\mu_r = 1$), the imaginary part of the electric conductivity is ignored and its real part is considered frequency-independent and equal to the direct current (DC) conductivity [14]. The real part of ε_r (with $\varepsilon_r = \varepsilon_r' + i\varepsilon_r''$) represents the electric permittivity of the soil and the imaginary part incorporates the losses for the conductivity and frequency.

The speed of the EM wave inside the soil is affected by its dielectric properties. In general, the higher the dielectric permittivity, the greater is the reduction in the speed [1]. The *phase velocity* (in m/s) of the EM wave in the subsurface is $v_p = \frac{1}{\sqrt{\varepsilon\mu}} \approx \frac{c_0}{\sqrt{\varepsilon_r}}$, where $c_0 = \frac{1}{\sqrt{\varepsilon_0\mu_0}}$ is the phase velocity in free space with $c_0 = 3 \times 10^8$ m/s. For most GPR applications, the imaginary part of ε_r is ignored and only the real part affects the attenuation and phase constant of the transmitted EM wave. Hence, v_p is inversely proportional to the square root of the real part of the dielectric permittivity $v_p \approx c_0/\sqrt{\varepsilon_r'}$.

In practice, GPR media are not homogeneous and often an EM wave encounters interfaces of different dielectric media. The classical optical geometry equations are not sufficient to describe the EM waves' behavior at the interface when the shallow objects are smaller than the GPR wavelength or interfaces in the close proximity of the illuminating source are present. The electrical properties of the soil, which affect the GPR signal propagation, depend on various factors such as the volumetric water content, texture of the soil particles, bulk density, and temperature. Among these, the water content is the most predominant factor influencing the electrical permittivity of the soil. The spatial variability expressed by the correlation length of the soil, i.e. the measure of the range over which fluctuations in one region of space are correlated with those in another region, also characterizes the EM propagation in GPR. Finally, the roughness of the air-ground interface influences the magnitude of the backscattered energy and depends on the surface characteristics and radar's wavelength.

The operation of GPR is similar to conventional UWB radar systems except that the GPR signal propagates into the subsurface. GPR systems can be employed for frequencies above 1 MHZ, where the EM behavior is not inductive and can be described by EM waves. At these frequencies, other factors such as the moisture content of the subsurface or, more generally, the dielectric properties of the soil hamper the penetration depth. In applications such as landmine detection, higher bands (> 1 GHz) are employed because they provide sufficiently wide bandwidth to achieve the necessary range resolution for discriminating very small targets [1] at the cost of a lower penetration depth. The attenuation coefficient (α) of an EM wave traveling in the subsurface is [1]

$$\alpha = \sqrt{\frac{\mu}{\varepsilon}}\,\frac{\sigma}{2} \approx Z_0\,\frac{\sigma}{2\,\sqrt{\varepsilon_r}}, \tag{7.8}$$

where $\varepsilon = \varepsilon_r\varepsilon_0$ is the permittivity of the soil, $\mu = \mu_0\mu_r$ is its magnetic permeability, and $Z_0 = \sqrt{\frac{\mu_0}{\varepsilon_0}}$ is the free space impedance (in Ω). Note that we employed the approximation $\mu_r \approx 1$ for the operating frequencies here.

The radar signal is reflected when there is a change in the EM properties of the soil, especially ε. The reflected signal is the superposition of the contributions of a specific target at a certain depth and several unwanted effects as listed below.

Ground reflection

The antennas of GPR systems are usually mounted just above the surface leading to significantly strong reflections from the air-ground interface.

Antenna cross-coupling

The cross-coupling between the receive and the transmit antennas causes the transmit signal to interfere with the received echoes. This is mitigated only partially by using absorber materials in the radar.

Antenna ringing

This happens when transmit antenna continues radiating after the exciting source has expired. In the popular GPR bow-tie antenna, internal reflections of the charges that occur at the ends of the bow-tie wings cause ringing. At the excitation points, this corrupts the transmit pulse waveform and reduces range resolution.

Surface roughness

As per the Rayleigh criterion [15], the roughness of the surface generates undesired scattering patterns that are not associated with subsurface features. This is especially true for shallow surface exploration (e.g. landmine detection) where the frequencies of operations are ~1–2 GHz.

Moisture content

The signal attenuation caused by the water content depends on the frequency of operation and is strictly related to the dielectric permittivity of the soil [1].

Low permittivity contrast

The intensity of reflections depends on the permittivity contrast between the soil and the features of interest [12]. Small non-metallic targets (such as landmines) have very low permittivity contrast thereby making it difficult to identify them.

7.1.2 GPR systems

Broadly, GPR systems are classified as time- and frequency-domain systems. In the former, GPR transmits a short pulse and the receiver applies the time-domain processing to the backscattered reflection. In the latter, the radar emits a series of

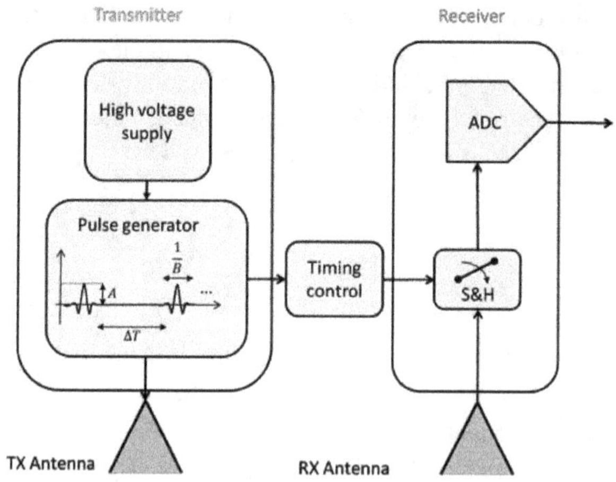

Figure 7.1 *Simplified block diagram of a pulsed UWB GPR. The pulse generator employs a technique of rapid discharge of the stored energy (from the high-voltage supply) into a short transmission line for generating the pulse. The sample-and-hold (S&H) receiver quantizes the signal to obtain digital samples using a flash analog-to-digital converter (ADC). The pulsed GPR antennas are usually dipoles and bow-ties because they are wideband, easy to design, non-dispersive, and linearly polarized [16]. They are sealed in a shielding box filled with the absorbing material to prevent coupling.*

individual frequencies and receives the signal via a frequency conversion receiver. In this chapter, we focus on time-domain pulsed GPR systems.

A pulsed GPR generates a series of short pulses (with pulse widths ranging from 200 ps to 50 ns) with a repetition interval of the order of hundreds of μs to 1 ms. A UWB GPR pulse (Figure 7.1) has a central frequency (f_c) (few MHz to 1 GHz) with bandwidth equal to Δf. The depth resolution Δ_z (in m) of a GPR along the depth axis (z in Cartesian coordinate system) is $\Delta_z = \frac{v_p}{B}$. Since the resolution depends on the phase velocity, it is affected by the dielectric permittivity of the material in which the target is embedded. Hence, a radar with receiver bandwidth of 2 GHz resolves targets spaced 1–10 cm apart from each other depending on the value of ε_r.

Figure 7.2 illustrates the geometry of GPR operation. The radar moves along a survey line. For each horizontal position (indicated with crosses), the radar sends down a pulse (or a series of pulses) and records the backscatter (or integrates a series of received echoes). The GPR could be moved by a single operator or pulled by a vehicle. The acquisition is usually triggered as a function of the traveled distance (for instance, by an odometer connected to the wheel of the moving platform). For each horizontal position, we obtain a single range profile known as one-dimensional (1-D) or A-scan (Figure 7.2) usually

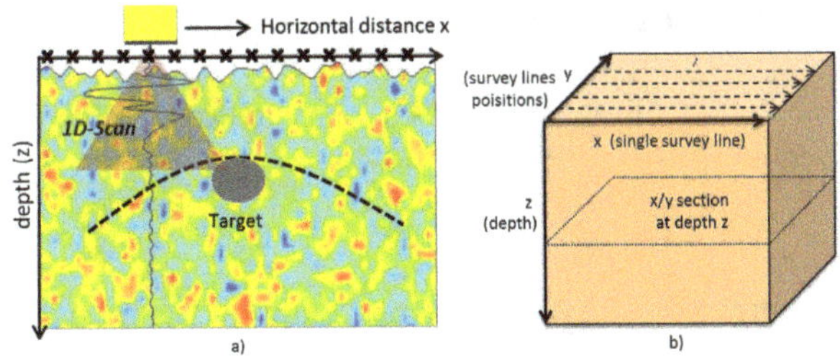

Figure 7.2 *(a) Illustration of a down-looking GPR system operation indicating a single range profile. (b) Directions of a single survey line (x-axis), survey lines positions (y-axis), and depth (z-axis) in a Cartesian coordinate system. The 2-D section (C-scan) is indicated at depth z.*

generated from the integration of many received pulses in order to increase the signal-to-noise ratio (SNR). If we stack many range profiles together along the survey line, we obtain a 2-D visualization or a B-scan (Figure 7.3a). Due to the wide antenna beamwidth, the GPR receives reflections from a target before it is right above it. This results in the well-known hyperbola-like response in the B-scan where the edge of the hyperbola is associated with the real depth of the target. If the acquisition has been repeated for many survey lines, we could stack all the data together to form a *data-cube*. Figure 7.3b shows an X/Y section of the aforementioned data-cube for a certain depth. This visualization is called C-scan.

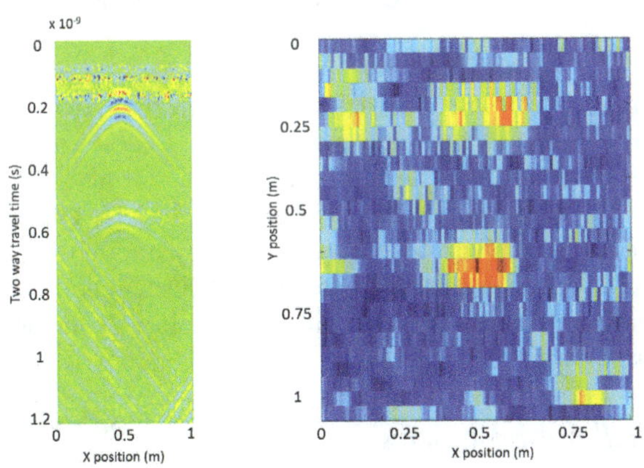

Figure 7.3 *(a) B-scan of a simulated spherical target buried in sand material. (b) C-scan from real data at 15 cm depth of two landmine simulants buried in highly non-homogeneous soil.*

7.2 GPR signal processing

One of the most employed transmit pulses of time-domain GPR systems is the *monocycle*. Given a Gaussian waveform

$$s_G(t) = Ae^{-2\pi^2 f_c{}^2(t-\tau)^2},\ t \in [0, +\infty], \tag{7.9}$$

where f_c is the central frequency, A is the peak amplitude, and $\tau = 1/f_c$; the monocycle waveform is its first derivative

$$s_T(t) = -4\pi^2 f_c{}^2 A(t-\tau)e^{-2\pi^2 f_c{}^2(t-\tau)^2},\ t \in [0, +\infty]. \tag{7.10}$$

In these UWB systems, both the central frequency and the bandwidth are approximately the reciprocal of the pulse length. The scattering of UWB radar signals from complex targets that are composed of a finite number of scattering centers can be described in terms of the channel impulse response (CIR). Here, the CIR is considered as a linear, time-invariant, causal system, which is a function of the target shape, size, constituent materials, and scan angle. Without taking into account the influence of the soil, the CIR $h(t)$ of a GPR target, with M scatterers, can be expressed as a series of time-delayed and weighted Gaussian pulses

$$h(t) = \sum_{m=1}^{M} \alpha_m e^{-4\pi\left(\frac{t-t_m}{\Delta T_m}\right)^2} \tag{7.11}$$

where each scatterer located at range r_m from the radar is characterized by the reflectivity α_m, duration ΔT_m, relative time shift $t_m = 2r_m/v_s$, where $v_s = c/\sqrt{\varepsilon_r}$ is the speed of the EM wave in the soil, $c = 3 \times 10^8$ m/s is the speed of light, and ε_r is the dielectric constant, which depends on the soil composition and moisture.

The response of the target to the Gaussian monocycle is then the received signal

$$y(t) = s_T(t) * h(t) \tag{7.12}$$

also regarded as the *range profile*. For each X/Y position, the system receives a radar echo (range profile) from the transmitted pulse. In order to deal with the exponential signal attenuation during the propagation through the soil medium, the dynamic range of the signal is enhanced via stroboscopic sampling [1, 17]. This technique comprises integrating N receiver samples (generated by transmitting a sequence of N pulses) at the ADC receiver sampling rate but with a small time offset δ for each of them. To achieve the desired stroboscopic sampling rate T_s, the time offset must be selected accordingly, i.e. $\delta = T_s/N$.

7.2.1 Classical pre-processing

In order to extract useful information from the raw data contaminated with noise and system instability, classical pre-processing in time-domain pulsed GPR is grouped into the following four categories: *basic handling* (editing/gating, dewow, background removal, and resampling) [3], *filtering* (spatial filtering and predictive deconvolution) [18], *imaging* (velocity analysis, elevation correction, backprojection, and Stolt migration) [5], and *modeling* (2-D forward modeling and simulation of GPR responses) [19]. We provide details of some of these techniques below.

Time gating

In practice, the location of ground reflections and/or starting depths of targets may be known from prior measurements. Then, time gating is used to edit the range/time profiles to limit the prospection depth to a specified interval. However, this may also end up removing shallow target (i.e. landmines) responses. Time gating is also used in post-processing.

Dewow

This process comprises a running average filter to remove the initial DC and low-frequency component. A common way to perform dewow is to calculate the mean of a few GPR profiles (A-scans) along a section and then subtract it from every single profile in the section. The length of the section depends on GPR wavelength.

Time gain

Some GPR architectures incorporate a time-varying gain to compensate for the attenuation of the signal during its propagation through the ground. The gain curve is customized in order to account for a more smooth or edge transition along the depth. This could also be applied in post-processing to improve data visualization.

Background subtraction

Background subtraction is commonly implemented by subtracting the mean profile (over an entire survey line) from each profile in the line, usually through Principal Component Analysis (PCA) [20]. The PCA performs singular value decomposition (SVD) to extract the first principal features of the data associated with the background and subtract them from profiles under consideration. Note that background removal is different from dewow because it takes an entire survey line instead of a "moving window".

Migration

Due to the beamwidth of transmit and receive antennas and the differences in round-trip travel time of the pulse caused by the movement of the antenna along the measurement line, the reflections from scatterers will appear as hyperbolic curves in the recorded data. These hyperbolic structures can be migrated (focused) into the real position of the corresponding scatterer via different migration techniques [5] to increase SNR and better localize/detect the targets. To apply these algorithms successfully a correct estimation of the velocity structure of the propagation medium, i.e. the dielectric permittivity of the soil, needs to be done. The phase-shift migration is a Fourier transform-based technique, which is also referred to as frequency-wavenumber migration (F-K migration). It exploits the wave equation to back-propagate the received signal into the soil back to the scattering source, and obtain an image of the subsurface reflectors. In practice, the 2-D case and a monostatic setting are generally assumed (a valid approximation when the transmitter and receiver are close to each other) but the extension to 3-D is straightforward. Consider a 2-D spatio-temporal Fourier transform of the range profile $y(z_0 = 0, x, t)$ at time t along the x-axis

$$D\left(z_0 = 0, k_x, \omega\right) = \iint d\left(z_0 = 0, x, t\right) e^{jk_x x} e^{-j\omega t} \mathrm{d}x\mathrm{d}t \tag{7.13}$$

where ω is the angular frequency, z_0 is the antenna vertical position, and k_x is the wavenumber along the x-axis. Here, it is assumed that y satisfies the wave equation. Then, to determine the field at a range of depths, a phase shift is applied, which depends on the propagation constant. This phase shift operation is an extrapolation along the x-axis, such that at$z = z_1 = z_0 + \Delta z$,

$$D\left(z_1, k_x, \omega\right) = D\left(z_0 = 0, k_x, \omega\right) e^{j\sqrt{\frac{4\omega^2}{v^2 p} - k_x^2}\Delta z} \tag{7.14}$$

By recursively extrapolating the field in steps Δz and using the result of each step as input for the next iteration, the F-K distribution of the field is reconstructed. Finally, the migrated data $\tilde{y}$ are obtained via the inverse fast Fourier transform (IFFT) of the wavenumber data over k_z and ω for the imaging condition

$$\tilde{y}(z = z_1, x, t = 0) = \frac{1}{4\Pi^2} \iint D\left(z = z_1, k_x, \omega\right) e^{jk_x x} \mathrm{d}k_x \mathrm{d}\omega \tag{7.15}$$

where z_1 is the depth of the migrated scene. In order for the IFFT to solve the above integral, the data matrix D previously needs to be evenly mapped into the k-space via interpolation. When v is assumed constant along the depth, it is called *Stolt migration* or *Stolt mapping* [5].

7.2.2 Modern GPR processing techniques

The classical processing techniques involve several approximations and heuristics that lead to loss of information during editing and filtering operations. Often a constant intervention of the operator is required to ensure correct interpretation of the data. When the sampled data is limited, it is very difficult to classify targets. Modeling based on EM equations leads to inaccuracies in the presence of coupling and ringing effects. To address these drawbacks, we leverage recent advances in signal processing to enhance target recognition. We consider the following techniques in subsequent sections.

Role of sparsity

In many radar applications, the received signal is sparse in a certain domain or dictionary [8]. This property is not only useful for applying compressive sensing-based algorithms in signal reconstruction obtained at reduced rates but also helpful in classifying the signal with fewer samples [6]. In Section 7.4, we introduce the basics of SR to construct the dictionary in which the GPR signal occupies fewer basis points. The selection of the dictionary matrix is crucial for obtaining a sparse(r) representation of a signal, i.e. a representation with a minimum number of non-zero coefficients. Depending on the application, one can build a dictionary with an arbitrary basis (such as Fourier, Wavelets) or by collecting empirical or synthetic target signatures. The SR has shown improvement in augmented resolution, clutter reduction, and target classification. Note that SR and CS share the same framework of techniques except that the former tries to exploit the inherent sparsity of data by representing it on an appropriate basis without necessarily reducing the number of measurements.

Learning-based methods

In recent years, learning methods have been widely adopted in radar processing. Since GPR signals are not naturally sparse in the common domains of time, space, and frequency, it is critical to learn the arbitrary sparse basis of such signals. In Section 7.3, we present DL algorithms aimed at target classification using GPR range profiles (A-scans). Here, we are particularly interested in *online* DL (ODL) algorithms because of their fast computation times.

Another state-of-the-art learning technique for classification is the use of deep neural networks. These methods have the ability to extrapolate new features from a limited set of features contained in a training set and, thus, are valuable tools in the face of non-availability of critical parameters. Architectures such as Convolutional Neural Network (CNN) are very effective in object classification. However, defining these networks involves many design decisions. Further, huge datasets are required to train the network, which may be a major difficulty in downlooking GPR applications [21–23]. Possible solutions may be to extend the training set using synthetically generated images [24] or to augment the training dataset with suitable transformations or augmentation strategies [25]. Later in the chapter, we compare both CNN- and DL-based classifications when the samples of the original range profiles are randomly reduced. Comparisons of the classification performance using CNN reveal that sparse decomposition-based techniques with DL generally perform better than CNN alone when the input signals are randomly sub-sampled.

Adaptive and cognitive processing

The cognitive processing in radar entails a framework of techniques to realize the sense-learn-adapt cycle [26] that operates at different layers of the radar architecture to achieve pre-determined tasks tuned to the changes in the target environment. The lower level of skill-based performances includes all techniques that permit the adaptation of both optimal waveforms and other design parameters. At the level of rule-based behavior, extraction of adaptive features from the received data along with applications such as target identification and classification are included. This often involves the use of learning techniques and artificial intelligence. The final knowledge-based processing layer [27] includes the modes of operation, such as specific scans, beam scheduling, and waveform selection, employed to enable cognition [28, 29].

The aforementioned adaptive architectures have also been investigated and developed for GPR. The localizing GPR system described in [30] employs online data processing to keep autonomous ground vehicles in a lane by feeding them with a map of subsurface features. This map-based vehicle localization using GPR works in conjunction with the existing sensors such as a Global Positioning System, lidar, and camera mounted at the bottom of the vehicle. When these devices fail to provide accurate maps because of low visibility in the presence of snow, dust, gravel, or dirt on the road, the GPR cognitively functions to provide complementary data to the vehicle. In [31], supervised machine learning with pre-trained models was used in a GPR system to probe glacier surfaces and automatically identify crevasses. It employs hidden Markov models to adaptively prescreen the data and mark locations in the collected data. Then, only marked files are used for crevasse detection.

Fully cognitive GPRs employ a feedback mechanism to dynamically tune radar operational parameters and continuously improve sensing performance. In this context, real-time GPR data processing requires significant computing and storage capability, often limiting its applicability. To tackle this problem, edge computing is used to reduce the computing latency by pushing the computation, communication, and storage resources from a remote data center to the edge of the network [32]. Recently, [33] describes the development of an edge computing and reinforcement learning framework that enables autonomous cognitive GPR.

The GPR target environment is riddled with various contamination sources. There is, therefore, interest in making the data processing adaptive based on the received data so as to maximize the probability of detection [26]. In this context, time-domain GPRs are preferred over the frequency-domain systems. The latter, such as the ones employing stepped-frequency continuous-wave (SFCW) transmitter, are more efficient than impulse GPRs but offer a poorer resolution. The frequency-modulated continuous-wave GPR has the advantage of a low-rate ADC at the receiver because of the low bandwidth of the beat signal [34]. However, these systems are expensive, inflexible for adaptive processing, and, in case of multiple transmitters, suffer from high level of self-interference.

The impulse radar is more suitable for enabling cognition and scenario-awareness in modern GPR systems, which use digital transmit and receive modules built using field-programmable gate arrays to impart high reconfigurability. The receiver directly samples backscattered echoes at radio frequency. At present, GPRs generate only 1-D-scans and contiguous linear scans must be stacked together after acquiring more measurements. The manual errors that creep in while choosing the start/ending point generate drifts in the acquired data. To scan different depths, transmit-receive pairs from antenna array elements in a multiplexed manner [30] or separate antennas [35] are employed.

In [36], an illumination system that uses pseudorandom codes with ideal autocorrelation properties is proposed for GPR. This enables receiver processing based on pulse-compression that improves the range resolution and maximum depth [37]. To reduce scan drifts and improve target localization, this system deploys multistatic antenna with eight transmitters and eight receivers.

In the context of DL, we describe an adaptive technique in Section 7.6, where we evaluate various DL algorithms based on statistical metrics to select the optimal input parameters for DL. For the specific application of landmine detection, we demonstrate the adaptive assessment using statistical distances [38]. These metrics allow greater fine-tuning of parameters with respect to the conventional bulk statistics such as root mean square error (RMSE).

Advanced target recognition and classification

GPR classification approaches aim to discriminate target/anomalies from the ground clutter, possibly indicating their precise position in the analyzed surveys. Unlike GPR imaging approaches (such as backprojection and migration) that aim at improving the visualization of the targets in the final GPR image, the output of

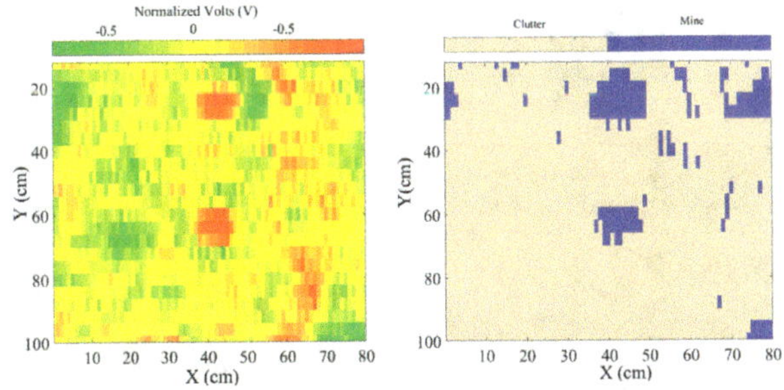

Figure 7.4 *Left: Raw data at 15 cm depth from an L-band GPR contains two landmine simulants. Right: Corresponding classification map of the survey area.*

classification methods can be seen as a map of declared classes along the survey area (see, e.g. Figure 7.4, which shows such a map from real data).

GPR classification approaches work on raw or pre-processed data, in time or frequency domain, and evaluate single A-scans or a collection of received GPR data (such as entire B- or C-scans). A variety of signal processing algorithms have been proposed for the detection of low metal-content targets in realistic scenarios; approaches based on feature extraction and classification are found to be the most successful (see e.g. [39]), yet false-alarm rates remain very high. Many classification approaches rely on a database of GPR signatures. This can be synthetically generated (e.g. by a modeling software like GprMax [40] or Comsol) taking into account all the necessary parameters (soil, antenna, transmitted signals, etc.) of the scenario under test. However, obtaining a complete and general database of GPR signatures is usually very challenging; in many cases, it is more convenient to extract salient features from a *representative* database of the target of interest and use them as an input for the chosen classifier. A representative database is smaller with respect to a general database because it contains only an accurate selection of the classes of interest for the desired application. Some GPR signal representation that were promising for buried targets/structures discrimination purposes include wavelets [7], target resonances and top/bottom reflection ratios [10], edge histogram descriptors [41], and SR coefficients [42].

In this chapter, we show the use of low-dimensional features for classification instead of entire B-scans or range profiles. Specifically, we employ SR coefficients extracted from the GPR range profiles. These coefficients strongly depend on the selected basis for the representation, which is a key for the success of the classification approach. Figure 7.4 schematically shows the proposed strategy for landmine classification. The final detection could be based on hypotheses testing [43] or data-dependent heuristics [6]. Figure 7.5 shows the processing flow for GPR target classification.

Convolutional methods

It is possible to further improve the performance of SR, which conventionally assumes that different input data (A-scans) are independent of each another. In

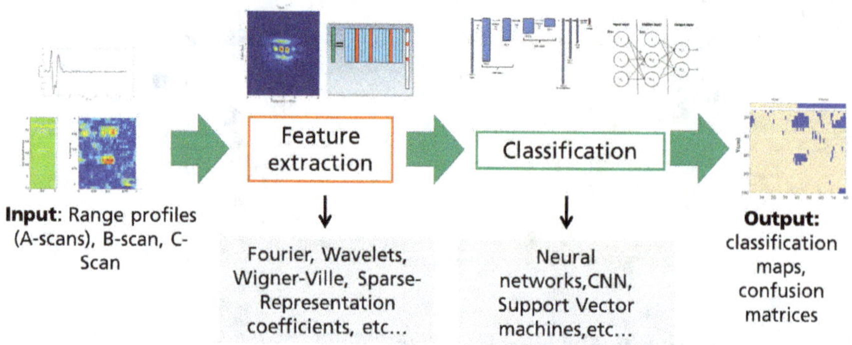

Figure 7.5 Simplified flow of modern GPR target classification. An appropriate data representation is necessary to extract features from the range profiles, which are then fed to an advanced classifier to separate and identify targets from clutter.

practice, successive GPR scans are spatially correlated and, therefore, processing larger datasets (e.g. using B-scans) may be more appropriate. Application of SR and DL to massive datasets is computationally expensive because of tight memory constraints. Although patch-based SR (DL) [44] reduces the computations by processing smaller data blocks, they ignore the inter-patch correlation and, consequently, the overall reconstructed signal could be erroneous.

Recently, convolutional sparse coding (CSC) [45] has attracted interest because it allows regularization to be performed directly on the global signal by imposing a banded convolutional structure to the dictionary. The dictionary is arranged as a sequence of convolution filters, each of which represents a single convolution with the given data. The decomposition generates more succinct data features for target recognition and/or clutter suppression.

To learn dictionaries with the aforementioned structure, convolutional dictionary learning (CDL) strategies have been proposed [46]. In the most common formulation for the sparse coding via CSC or the dictionary update via CDL, a - minimization is solved through different variations of the Alternate Direction Method of Multipliers (ADMM) algorithm [47]. While CSC has been applied to computer vision and pattern recognition, its application to GPR is more recent. In [11], separation of target returns from clutter is formulated as a constrained optimization problem to estimate the low-rank and sparse components of the data. Whereas the low-rank portion contains a strong ground surface reflection and is extracted with a dictionary based on discrete cosine transform, the sparse part is associated with the targets and is estimated by CSC with a dictionary comprising convolutional Ricker wavelet filters.

We now illustrate the aforementioned techniques for GPR-aided landmine detection. For this purpose, in Section 7.3, we briefly describe the deployed system and field campaign to obtain data.

7.3 GPR measurements for landmine recognition

The detection of anti-personnel landmines (APMs) is appropriate not only from our representative application point of view but also its general utility for humanitarian purposes. APMs are a serious threat to the civilian population and one of the worst kinds of global pollution nowadays. Broadly, APMs include blast mines, fragmentation mines, or even improvised explosive devices; the latter are not constructed by conventional military designs and are usually associated with insurgent guerrillas and commando forces. The UWB GPR is a promising non-invasive technology to tackle APM detection. Mine detection GPR usually operates in L-band (1–2 GHz) with UWB transmit signals that allow resolving small targets (5–10 cm diameter) at shallow depths (~15–30 cm) [10, 48]. The constituting material of many models of landmines is largely plastic and has a very weak response to radar signals because of its low dielectric contrast with respect to soil. In the following, we describe the specific GPR system and its field measurement campaign to obtain mine data for our processing application.

7.3.1 Deployed system

We employ a commercially available Surface Penetrating Radar unit called *SPRScan* manufactured by ERA Technology. The SPRScan is an L-band, impulse waveform, UWB radar that is mounted on a movable trolley platform (see Figure 7.6). Pulsed GPRs are more effective in terms of offering penetration depth and wide bandwidth with respect to the standard SFCW systems. The former is also more robust to electronic interference and does not suffer from unequal balancing of antenna signals.

Table 7.1 lists the most important operational parameters of the system. The radar uses a 8×8 cm dual bow-tie dipole antenna for both transmit (Tx) and receive (Rx) sealed in a metallic shielding filled with an internal absorber. The central frequency of the system (f_c) and its bandwidth (Δf) are 2 GHz.

The PRF and the sampling of the receiver ADC is 1 MHz. The scanning system has a resolution of 1 cm towards the perpendicular broadside (or X-direction) and 4 cm towards the cross-beam (Y-direction). In our field campaigns, the SPRScan system moves along the survey area over a rail system that allows accurate positioning of the sensor head in order to obtain the aforementioned resolution in X and Y (see also Section 7.1).

Our GPR system employs stroboscopic sampling to reach a pseudo sampling frequency of $f_s = 1/T_s = 40$ GHz (much above the Nyquist rate) to yield the discrete-time signal $y[n] = y(nT_s)$. The receiver has the ability to acquire a maximum of 195 profiles per second, each one consisting of 512 time samples. Prior to the A/D conversion, the signal is averaged to improve the SNR. A time-varying gain correction can be applied to compensate for the soil attenuation and increase the overall dynamic range of the system. The receiver averages 100 range profiles for each antenna position.

7.3.2 Test field

We use the measurement data from a 2013 field campaign at Leibniz Institute for Applied Geophysics (LIAG) in Hannover (Germany) [49]. Figure 7.7 shows the test

Table 7.1 Technical characteristics of impulse GPR

Parameter	**Value**
Operating frequency	2 GHz
Pulse repetition frequency (PRF)	1 MHz
Pulse length	0.5 ns
Sampling time	25 ps
Spatial sampling along the beam	1 cm
Cross-beam resolution	4 cm
Antenna height	5–9 cm
Antenna configuration	Perpendicular broadside
Samples/A-scan	512

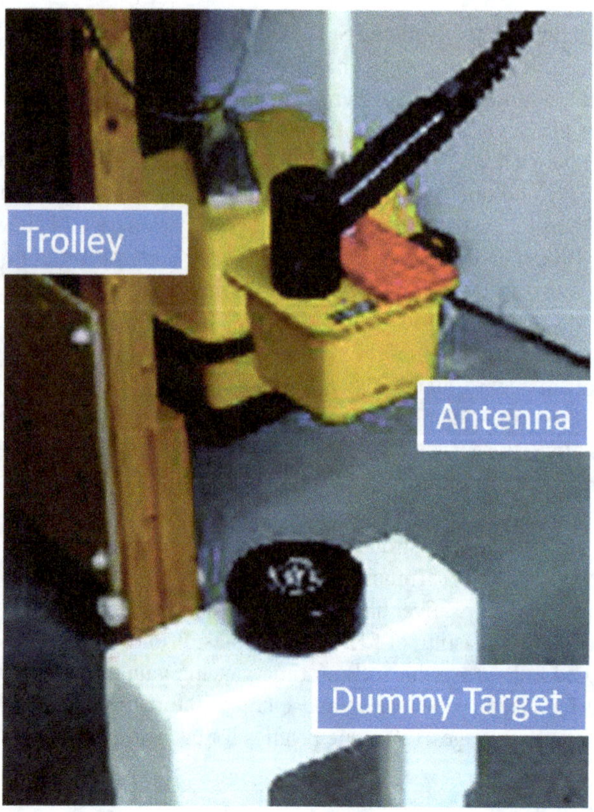

Figure 7.6 The L-band GPR system is attached to a movable trolley platform. It is mounted along a rail system and scans the target from above.

Figure 7.7 *The LIAG test field in Hannover (right) along with its layout (left). The scan directions X and Y of the radar are indicated in the photograph and layout. The radar coverage region is indicated by solid red lines with a red circle showing the origin of the scan. The white arrows in the photograph indicate specific lanes scanned in the X-direction that are separated in the Y-direction by 4 cm. In the layout, each gray dot represents the location of a buried test target. An individual survey area unit of 1 m × 1 m that contains 2 targets is also indicated in the layout (solid black lines) and the photograph (dotted black lines). The solid black arrow over the middle rail in the photograph is where the SPRScan was mounted.*

field for detailed ground truth information. The soil texture was sandy and highly inhomogeneous (due to the presence of material such as organic matter and stones), thereby leading to high variability in the electrical parameters. The dielectric constant at three different locations of the testbed was measured with a Time Domain Reflectometer to obtain an estimate of its mean value and variability. The average value oscillated between 4.6 and 10.1 with 15% standard deviation and *correlation length* [49] of 20 cm. These big variations in soil dielectric properties pose difficulties in mine detection.

During the field tests, the SPRScan system moved on two plastic rails with the scan resolution in the X and Y directions being 1 cm and 4 cm, respectively. The entire survey lane was divided into 1 × 1 m sections (see Figure 7.7), each

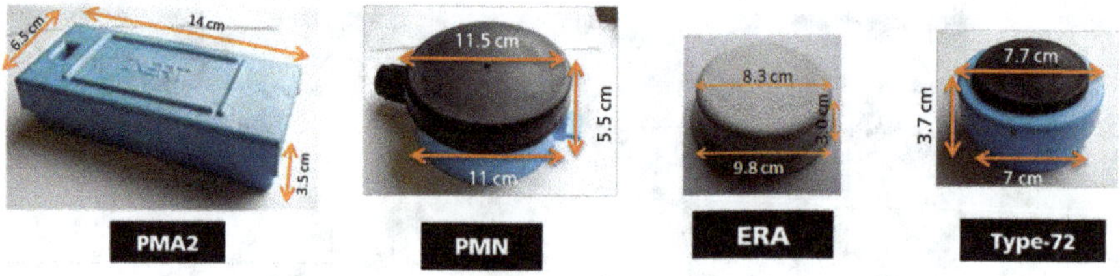

Figure 7.8 Details of the SIMs and the STT buried in the test field

containing two targets in the center. The targets on the left and right sides of the lane were buried at approximately 10 and 15 cm depths, respectively.

Our testbed contains standard test targets (STTs) and simulant landmines (SIMs) of different sizes and shapes. An STT is a surrogate target used for testing landmine detection equipment. It is intended to interact with the equipment in an identical manner as a real landmine does. A SIM has the representative characteristics of a specific landmine class although it is not a replica of any specific model. Here, we study three STTs (PMA2, PMN, and Type-72) and one SIM (ERA). All of these test objects are buried at a depth of 10–15 cm in the test field [12]. For classification purposes, we group PMN and PMA2 together as the largest targets while T72 mines are the smallest (Figure 7.8).

7.4 Sparse representation techniques

We now consider the SR techniques that are aimed at the solution of an under-determined linear system. The regularization process comprises minimizing a cost function that enforces the sparsity of the vector $\mathbf{x}$ (i.e. the number of its non-zero elements):

$$\hat{\mathbf{x}} = \underset{\mathbf{x}}{\operatorname{argmin}} \ \|\mathbf{x}\|_0 \ \text{subject to} \ \mathbf{y} = \mathbf{D}\mathbf{x} \tag{7.16}$$

where $\mathbf{y} \in \mathbb{C}^{M\times 1}$ represents the measurements of the observed scene, arranged in an M-dimensional column vector, $\mathbf{D} \in \mathbb{C}^{M\times K}$ is a $M \times K$ matrix called *dictionary*, and $\mathbf{x} \in \mathbb{C}^{1\times K}$ is a coefficient vector that represents the signal information according to the model specified in $\mathbf{D}$. The dictionary $\mathbf{D}$ contains a collection of realizations of the signal of interest in a certain domain. For image processing and classification purposes, every realization (atom) could represent, e.g. a particular patch of an image or a certain signal class associated with a particular target. The vector $\mathbf{x}$ represents the information to be recovered and $\|\mathbf{x}\|_0$ is its ℓ_0-norm.

If the observed scene is sparse, then it has been shown that (7.16) has a unique solution, i.e. one can find a matrix $\mathbf{D}$ that gives a perfect reconstruction of the measurements with probability close to one [50]. The number of non-zero elements of the vector $\mathbf{x}$ is called the *support* of the signal $\mathbf{y}$; if $\mathbf{y}$ is mapped on the vector $\mathbf{x}$ using S non-zero elements then it is said to be *S-sparse*. The problem in (7.16) is NP-hard. However, techniques based on greedy algorithms and convex optimization are shown to be efficient in relaxing this problem.

Assume that matrix $\mathbf{D} \in \mathbb{C}^{M\times K}$ has $M < K$ and that $\mathbf{y}$, the measured signal, is *sparse* or *compressible*. A signal is sparse in a certain domain (e.g. space, time, and frequency) if there exists a basis (or frame) in which this signal is represented using a very limited number of elements of the aforementioned basis. A signal is compressible when it can be well-approximated by sparse signals; this is relevant for noisy real-world signals. We define the compressibility of order p of a signal $\mathbf{y}$ as $\sigma_S(\mathbf{y})_p$ by calculating the error in approximation, i.e.

$$\sigma_S(\mathbf{y})_p = \min_{\hat{\mathbf{y}}\in\Sigma_S} \|\mathbf{y} - \hat{\mathbf{y}}\|_p \tag{7.17}$$

where Σ_S is a set of S-sparse signals, i.e. with $\|x\|_0 \leq S$.

Popular greedy approaches to solve (7.16) such as matching pursuit, orthogonal matching pursuit (OMP), and block orthogonal matching pursuit [51]) approximate the ℓ_0-norm solution in (7.16) using iterative strategies. These methods are computationally fast but may not lead to a global optimum solution. State-of-the-art algorithms for SR such as basis pursuit (BP), basis pursuit denoising (BPDN) [52], least absolute shrinkage and selection operator (LASSO) [53], and least-angle regression (LARS) [54] aim to approximate (7.16) using less restrictive constraints (ℓ_1 and ℓ_2) that also assure a sparse solution for $\mathbf{x}$ [55]. These methods based on convex optimization are more robust to noise and, therefore, may lead to a better reconstruction of $\mathbf{y}$ even although they are computationally more demanding.

In Sections 7.4.1–7.4.4, we discuss the following SR techniques: orthogonal matching pursuit (OMP) along with its batch version (batch-OMP) [56]. We also include convex optimization methods (such as BPDN, LASSO, and LARS) because they are used for the dictionary update step of some of the DL algorithms mentioned later in Section 7.5.

7.4.1 OMP

Greedy strategies for SR are iterative approaches that search for the best local optimal solution for each iteration (i.e. the *greediest* one) hoping that it coincides with the global best one. They can approximate the solution of the problem in (7.16). These algorithms choose only the most appropriate elements in $\mathbf{x}$ according to the input constraint (which is the expected sparsity or the reconstruction error) to reconstruct the measurement vector $\mathbf{y}$. The OMP problem has a dual formulation:

$$\hat{\mathbf{x}} = \operatorname*{argmin}_{\mathbf{x}} \|\mathbf{y} - \mathbf{D}\mathbf{x}\|_2 \text{ subject to } \|\mathbf{x}\|_0 \leq S. \tag{7.18}$$

Alternatively,

$$\hat{\mathbf{x}} = \operatorname*{argmin}_{\mathbf{x}} \|\mathbf{x}\|_0 \text{ subject to } \|\mathbf{y} - \mathbf{D}\mathbf{x}\|_2 \leq \delta. \tag{7.19}$$

The constraint in (7.18) is based on the sparsity S while the one in (7.19) employs the ℓ_2-norm of the residual by the parameter δ. The OMP is an iterative process; at each iteration, it searches for an atom $\mathbf{d}_i$ that gives the best projection onto the residual signal $\mathbf{y} - \mathbf{Dx}$ and obtains the value of the corresponding coefficient $\mathbf{x}_\mathrm{i}$ by least squares. Either S or δ is used as a stopping criterion.

The least square problem in OMP is computationally demanding and usually computed using a Cholesky factorization or a QR decomposition [56]. To speed up computations we propose a faster variant of OMP (see Algorithm 1). The difference in this implementation is that the newly found atom is orthogonalized against all the previously found one $\mathbf{d}_\mathrm{j}$ with $j = 1, \cdots, i-1$ via Gram-Schmidt process (see step 10) [57] before the coefficient update. Having an orthonormal basis among the newly found atoms is beneficial for the coefficient update where, instead of solving a least square problem, we can directly obtain the updated coefficients $\mathbf{x}$ by projecting them over the residual $\mathbf{r}_\mathrm{i}$ (see step 13). After the stopping criterion is met, we adjust the coefficient vector obtained at the last iteration ($\mathbf{x}_\mathrm{i}$) solving just one least square problem without orthogonalizing before; this makes possible that the coefficients $\mathbf{x}$ have the right coefficient values with respect to $\mathbf{y}$. We define the set of new atoms to be updated during the iterations as $\mathcal{D}$.

Algorithm 1: Fast OMP

Input: Measurement vector ($\mathbf{y} \in \mathbb{C}^{mx1}$), Dictionary ($\mathbf{D} \in \mathbb{C}^{mxn}$), Sparsity number ($S$) or Residual error threshold (δ)

Output: Sparse coefficients vector ($\mathbf{x} \in \mathbb{C}^{1xn}$)

1 Initialize iteration count $i = 1$
2 Initialize the residual as the measurement vector: $\mathbf{r}_\mathrm{i} = \mathbf{y}$
3 Initialize the vector of sparse coefficients: $\mathbf{x} = \underline{\mathbf{0}} \in \mathbb{C}^{1xn}$
4 Initialize the matrix of new found atoms: $\mathcal{D}_0 = \varnothing$
5 **Loop until** $i = S$ **or** $||\mathbf{r}_\mathrm{i} - \mathbf{Dx}||_2 \geq \delta$
6   Find the atom $\mathbf{d}_\mathrm{i}$ which gives the maximum dot product with $\mathbf{r}$: $\mathbf{di} = \underset{\mathbf{d}_\mathrm{i}}{\mathrm{argmax}} \langle \mathbf{r}_\mathrm{i}, \mathbf{d}_\mathrm{i} \rangle$
7   Normalize the new found atom by the norm-2 : $\mathbf{d}_\mathrm{i} = \frac{\mathbf{d}_\mathrm{i}}{||\mathbf{d}_\mathrm{i}||_2}$
8   Include $\mathbf{d}_\mathrm{i}$ in the set of new found atoms: $\mathcal{D}_i = \mathcal{D}_{i-1} \cup \mathbf{d}_\mathrm{i}$
9   **for** $j = 1, \cdots, i-1$
10   Orthogonalize $\mathbf{d}_\mathrm{i}$ against all previously found $\mathbf{d}_\mathrm{j}$: $\mathbf{d}_\mathrm{i} = \mathbf{d}_\mathrm{i} - \langle \mathbf{d}_\mathrm{j}, \mathbf{d}_\mathrm{i} \rangle \mathbf{d}_\mathrm{j}$
11   **end for**
12   **for** $j = 1, \cdots, i$
13   Coefficients update $\mathbf{x}_\mathrm{i} = \langle \mathbf{x}_\mathrm{j}, \mathbf{r}_\mathrm{j} \rangle$
14   **end for**
15 **EndLoop**
16 UPDATE x_i by solving a least square problem: $\mathbf{x}_\mathrm{i} = \underset{\mathbf{x}}{\mathrm{argmin}} ||\mathbf{y} - \mathcal{D}_i \mathbf{x}||_2$
17 Update residual: $\mathbf{r}_\mathrm{i} = \mathbf{r}_{\mathrm{i}-1} - \mathcal{D}_i \mathbf{x}_\mathrm{i}$

7.4.2 Batch-OMP

Batch-OMP is a variant of the classical OMP and it is used when a large number of signals must be represented over the same dictionary. The intuition behind this method is that it is not required to compute the residual $\mathbf{r}$ at each iteration for the atom selection step, only $\mathbf{D}^T\mathbf{r}$ is required. The atom selection step can be rewritten in order to exploit its dependence only on the dictionary at the iteration i (called $\mathbf{D}_\mathrm{i}$) and the measurement vector $\mathbf{y}$ without calculating $\mathbf{D}^T\mathbf{r}$.

Calling $\alpha = \mathbf{D}^T\mathbf{r}$, $\alpha^0 = \mathbf{D}^T\mathbf{y}$ and $\mathbf{G} = \mathbf{D}^T\mathbf{D}$, one can write

$$\alpha = \mathbf{D}^T(\mathbf{y} - \mathbf{D}_I(\mathbf{D}_I)^+\mathbf{y}) \tag{7.20}$$
$$= \alpha^0 - \mathbf{G}_I(\mathbf{D}_I)^+\mathbf{y} \tag{7.21}$$
$$= \alpha^0 - \mathbf{G}_I(\mathbf{D}_I^T\mathbf{D}_I)^{-1}\mathbf{D}_I^T\mathbf{y} \tag{7.22}$$
$$= \alpha^0 - \mathbf{G}_I(\mathbf{G}_{I,I})^{-1}\alpha_I^0. \tag{7.23}$$

Given the pre-computed α^0 and $\mathbf{G}$, one can compute α without explicitly computing $\mathbf{r}$. However, if the residual is not computed, it is impossible to set a stopping criterion based on the error. In [56] an extention of batch-OMP where the residual ℓ_2 norm $\|\mathbf{r}\|_2^2$ at the iteration i is calculated by an incremental formula; this permits the use of the residual as a stopping criterion without explicitly computing it; see [56] for the implementation details.

7.4.3 Basis pursuit denoising, LASSO, and penalized least square

A convex optimization problem deals with minimizing convex functions over a convex set; the convexity makes the minimization easier since the local minimum of the function is also the global one. Convex optimization methods for SR, contrary to greedy approaches, aim to minimize the ℓ_1-norm of the coefficient vector x.

$$\hat{\mathbf{x}} = \underset{x}{\operatorname{argmin}}\ \|\mathbf{x}\|_1 \text{ subject to } \mathbf{Dx} = \mathbf{y}. \tag{7.24}$$

The objective function $\|.\|_1$ is in fact a convex and tractable, whereas $\|.\|_0$ is non-convex and generally very difficult to solve. The problem in (7.24) is known as BP. It can be solved efficiently using linear programming techniques and it is demonstrated [50] that, under certain conditions, it still leads to a sparse approximation of the underdetermined system solution in (7.16).

The BPDN accounts for the noise in the data through an additional approximation:

$$\hat{\mathbf{x}} = \underset{x}{\operatorname{argmin}}\ \|\mathbf{x}\|_1 \text{ subject to } \|\mathbf{y} - \mathbf{Dx}\|_2 \le \delta. \tag{7.25}$$

This does not exactly solve (7.24) but makes an approximation based on a positive parameter δ whose amplitude (between 0 and 1 if the data is normalized) corresponds to the noise level. This parameter plays the same role as δ in OMP (see Algorithm 1).

Through a different derivation, LASSO [53] solves (7.25) as

$$\hat{\mathbf{x}} = \underset{x}{\operatorname{argmin}}\ \|\mathbf{y} - \mathbf{Dx}\|_2 \text{ subject to } \|\mathbf{x}\|_1 \le \tau. \tag{7.26}$$

Another technique is penalized least squares, which models the problem as

$$\hat{\mathbf{x}} = \underset{x}{\operatorname{argmin}}\ \|\mathbf{y} - \mathbf{Dx}\|_2 + \lambda\|\mathbf{x}\|_1, \tag{7.27}$$

where λ is a regularization parameter. For an appropriate choice of δ, τ, and λ the solutions of these three approaches coincide.

One of the most popular algorithms for solving the problems (7.25) and (7.26) was proposed by Van Der Berg and Friedlander and it is based on the relation between LASSO and BPDN, defined by the Pareto curve. The Pareto curve indicates the optimal trade-off between l_2-norm of the residual ($\|\mathbf{y} - \mathbf{Dx}\|_2$) and the one-norm of the solution **x**. The solution of a BPDN problem using this approach consists of solving a sequence of LASSO problems using spectral projected gradient and uses the Newton method applied to the Pareto curve to observe how much the solution of (7.26) gets close to (7.25); the process stops when a satisfactory threshold is met [52].

7.4.4 LARS

Homotopy approaches such as LARS [54] solve the BPDN problem by repeatedly solving (7.27) for all possible values of λ. LARS is applied for both variable and coefficient selection and has been proposed for ℓ_1 minimization. It is convenient for solving the penalized least square problem in (7.27) because of its speed (comparable to forward selection algorithms) and producing a full piece-wise linear solution path (i.e. for every λ). Moreover, LASSO can be derived by LARS. In fact, LASSO is viewed as a variation of a ridge-regression problem with the constraint of some coefficients set to zero. Algorithm 2 summarizes this LARS–LASSO technique [54].

7.5 Dictionary learning framework and algorithms

We now describe the theory and algorithms of some important DL methods that are useful for GPR target recognition [6, 58]. We consider both batch- [56] and online-learning [59, 60] approaches. We then analyze in detail a novel ODL approach proposed in our prior work [6]. This is an improvement over the state-of-the-art ODL approaches.

The dictionary **D** may be learned from the data it is going to represent. The DL techniques aim to create adapted dictionaries that provide the sparsest reconstruction for given training sets, i.e. a representation with a minimum number of constituting atoms. DL methods are critical building blocks in many applications such as deep learning, image denoising, and super-resolution (see [61]).

The first step in DL consists of building a training database $\mathrm{Y}=\begin{bmatrix}\mathbf{y}_1 & \cdots & \mathbf{y}_L\end{bmatrix} \in \mathbb{R}^{MxL}$ of L vectors, each with M elements. We assume that every vector $\mathbf{y}_i$ is generated by a linear combination of the K atoms of a certain dictionary matrix $\mathbf{D} \in \mathbb{R}^{MxK}$ and an associated sparse vector of coefficients $\mathbf{x}_i$, having $\mathrm{X}=\begin{bmatrix}\mathbf{x}_1 & \cdots & \mathbf{x}_L\end{bmatrix} \in \mathbb{R}^{KxL}$. The core problem of DL is to find the dictionary **D** that gives the set of sparsest solution **X**; we have then:

$$\mathbf{Y}_{(MxL)} = \mathbf{D}_{(MxK)}\ \mathbf{X}_{(KxL)} \tag{7.28}$$

where the dimension L is usually larger than the number of the dictionary elements ($L > K$). The DL problem has a dual formulation, whether it constraints sparsity (7.29)

$$\begin{aligned} \hat{\mathbf{D}}, \hat{\mathbf{X}} &= \underset{\mathbf{D},\mathbf{X}}{\operatorname{argmin}}\ \|\mathbf{Y} - \mathbf{D}\mathbf{X}\|_F \\ \text{subject to}\ &\|\mathbf{x}_i\|_0 \leq S,\ 1 \leq i \leq L. \end{aligned} \tag{7.29}$$

or the model deviation (7.30)

$$\hat{\mathbf{D}}, \hat{\mathbf{X}} = \underset{\mathbf{D},\mathbf{X}}{\operatorname{argmin}} \quad \|\mathbf{Y} - \mathbf{DX}\|_F \qquad \text{subject to} \quad \|\mathbf{y}_i - \mathbf{D}\mathbf{x}_i\|_2 \le \varepsilon, > 1 \le i \le L. \tag{7.30}$$

where S is the sparsity number, i.e. the desired number of the non-zero elements in the decomposed vector, and ε is the residual error.

Since both **D** and **X** are unknown, a common approach is to use alternating minimization in which we start with an initial guess of **D** and then obtain the solution iteratively by alternating between two stages: *SR* and *dictionary update* [62] as follows:

(1) *SR*: Obtain $\mathbf{X}_{(t)}$ for each $\mathbf{y}_i$ as:

$$\mathbf{X}_{(t)} = \underset{\mathbf{X}}{\operatorname{argmin}} \quad \left\|\mathbf{Y} - \mathbf{D}_{(t-1)}\mathbf{X}\right\|_F \qquad \text{subject to} \left\|\mathbf{x}_{i(t-1)}\right\|_p \le S, 1 \le i \le L, \tag{7.31}$$

where $\mathbf{X}_{(t)}$ is the SR in tth iteration. This can be solved using greedy algorithms such as OMP ($p = 0$) or convex relaxation methods like BPDN ($p = 1$).

(2) *Dictionary update*: Given $\mathbf{X}_{(t)}$, update $\mathbf{D}_{(t)}$ such that

$$\mathbf{D}_{(t)} = \underset{\mathbf{D}\in\mathcal{D}}{\operatorname{argmin}} \quad \left\|\mathbf{Y} - \mathbf{D}\mathbf{X}_{(t)}\right\|_F \tag{7.32}$$

where $\mathcal{D}$ is a set of all dictionaries with unit column-norms, $\left\|\mathbf{d}_j\right\|_2 = 1$ for $1 \le j \le K$. This subproblem is solved by methods such as SVD or gradient descent [59, 63].

Conventional DL techniques deal with the entire training set in each iteration, making the learning process slow for high-dimensional datasets; this approach is called batch DL. ODL techniques deal with the training set by considering one element at a time or in mini-batches, making the processing of learning faster than batch-DL and adaptive to variations. For batch-DL, we focus on the popular K-times singular value decomposition (K-SVD) [56] and the more recent low-rank shared DL (LRSDL) [64]. The latter is capable of generating a class-discriminative dictionary. Among the ODL techniques, we analyze the state-of-the-art ODL [59], recently proposed correlation-based weighted least square update (CBWLSU) [59, 60], and our novel ODL strategy named drop-off mini-batch online dictionary learning (DOMINODL) [6, 65].

7.5.1 K-SVD

K-SVD is a batch-DL algorithm that iteratively alternates a sparse decomposition and a dictionary update step solving a minimization over the number of non-zero elements in the set of representation vectors **X**, and one over **D** for updating the dictionary. For the sparse coding step at the iteration t, K-SVD solves OMP for each element $\mathbf{y}_i$ in the training set **Y**:

$$\mathbf{x}_i = \underset{\mathbf{x}_i}{\operatorname{argmin}} \quad \|\mathbf{x}_i\|_0 \qquad \text{subject to} \quad \left\|\mathbf{y}_i - \mathbf{D}_{(t-1)}\mathbf{x}_i\right\|_2^2 \le \delta, \ \forall 1 \le i \le L, \tag{7.33}$$

where i is the index that represents the training set elements, $\mathbf{D}_{(t-1)}$ is the dictionary computed at the previous iteration $(t-1)$, and δ is the maximum residual error.

K-SVD shares the same strategy for the sparse decomposition step with another batch-DL method called method of optimal directions (MOD) [66] but their respective dictionary update rules are different. Let K be an input parameter of K-SVD that indicates the number of columns of the learned dictionary. For the dictionary update step, K-SVD solves the global minimization problem in (7.32) via K sequential minimization problems, wherein every column $\mathbf{d}_k$ of $\mathbf{D}$ and its corresponding row of coefficients $\mathbf{X}_{\text{row},k}$ of $\mathbf{X}$ are updated. Let us assume that we are at the iteration t and we already performed the sparse decomposition step using the dictionary that has been obtained in the iteration $t-1$. For each kth dictionary atom, we calculate the error term $\mathbf{Y}_r = \mathbf{Y} - \sum_{l \neq k} \mathbf{d}_{l_{t-1}} \mathbf{X}_{\text{row},l_{(t-1)}}$ and extract a subset $\mathcal{Y}_k$ of it, which comprises the elements of $\mathbf{Y}_r$ that use the selected atom (k). Then we use SVD to find the closest rank-1 approximation of $\mathcal{Y}_k$ to obtain $\mathbf{d}_k$ subjected to the constraint $\|\mathbf{d}_{k_{(t)}}\|_2 = 1$. K-SVD global optimization will terminate after a series of sparse decomposition plus dictionary update steps depending on the changes on $\|\mathbf{Y} - \mathbf{D}\mathbf{X}_{(t)}\|_F^2$.

In this chapter, a particular K-SVD implementation was used [56]. This version employs a faster approximation for the SVD step in the dictionary update and uses batch-OMP for the sparse decomposition step, making it more feasible to deal with large sets of signals. The performance of K-SVD can be improved if the learning process enforces constraints such as hierarchical tree sparsity [67], structured group sparsity (StructDL) [68], Fisher discrimination (FDDL) [69], and low-rank-and-Fisher ($D^2L^2R^2$) [70].

7.5.2 LRSDL

LRSDL [64] is one of the latest evolution of (batch) discriminative DL algorithms; we tested it on our classification approach due to its promising capabilities for class recognition. Here, we provide the basic idea of discriminative DL and briefly describe the theoretical background on which LRSDL is based on.

Discriminative DL algorithms, as D-KSVD [71] and LC-KSVD (Label Consistent-KSVD) [72], employ a learning strategy that promotes the generation of a dictionary $\mathbf{D}$ that is separated into blocks of atoms associated to different classes as $\mathbf{D} = [\mathbf{D}_1, \cdots, \mathbf{D}_C] \in \mathbb{R}^{M \times K}$ where C is the number of classes present in the training set $\mathbf{Y}$. The resultant coefficient matrix $\mathbf{X}$ is close to be sparse with all non-zeros being one while satisfying a block diagonal structure.

However, objects belonging to different classes often have common features, therefore the assumption of non-overlapping subspaces done by such algorithms is often unrealistic in practice. Techniques such as DL with structured incoherence and shared features [73], separating the commonality and the particularity [74], and convolutional sparse DL [75] exploit common patterns among different classes even though different objects possess distinct class-specific features. These methods produce an additional constituent $\mathbf{D}_0$, which is shared among all classes so that

$\mathbf{D} = [\mathbf{D}_1, \cdots, \mathbf{D}_C, \mathbf{D}_0] \in \mathbb{R}^{M \times K}$. The drawback of these strategies is that the shared dictionary may also contain class-discriminative features.

To avoid this problem, LRSDL [64] requires that the shared dictionary must have a low-rank structure and that its sparse coefficients have to be almost similar. LRSDL learns both **D** and **X** by solving a minimization problem with a cost function that is closely related to another DL algorithm called fisher discriminative dictionary learning (FDDL) [76]. The LRSDL dictionary update step employs ADMM [47] and fast iterative shrinkage-thresholding algorithm (FISTA) [77] for the sparse decomposition step. Once the data is sparsely represented with such dictionaries, a sparse-representation-based classification (SRC) [42] is used to predict the class of new data.

7.5.3 ODL

The ODL is an interesting alternative for inferring a dictionary from large training sets or ones that change over time [59]; like K-SVD this algorithm also updates the entire dictionary sequentially but draws one element of training data at a time for the dictionary update. ODL assumes that the training set **Y** is composed of i.i.d. samples of a distribution $\mathbf{p}(\mathbf{x})$. In many practical applications, this condition is usually not fulfilled and **Y** is obtained by a collection of measurements that has been randomly permuted and drawn consecutively for each iteration (we follow a similar procedure in Section 7.6).

The input parameters of ODL are the initial dictionary $\mathbf{D}_0 \in \mathbb{C}^{M \times K}$, the regularization parameter λ, the dimension of the learned dictionary K, and the number of iterations T (which will also correspond to the number of training set elements that will be used for learning **D**). The first step at the iteration t, with $t = 1 \ldots T$, is to draw an example of the training set $\mathbf{y}_t$ from **Y**. The sparse decomposition step is done using the dictionary obtained at the previous operation $\mathbf{D}_{(t-1)}$ via a Cholesky-based implementation of the LARS–LASSO algorithm, which solves a ℓ_1-regularized least-squares problem as indicated in 7.34. In the dictionary update we consider all the training set elements analyzed so far: $\mathbf{y}_i$ with $i = 1 \ldots t$.

$$\hat{\mathbf{x}}_{(t)} = \underset{\mathbf{x} \in \mathcal{C}^n}{\operatorname{argmin}} \ \frac{1}{2} \|\mathbf{y}_t - \mathbf{D}_{(t-1)} \mathbf{x}\|_2^2 + \lambda \|\mathbf{x}\|_1 \tag{7.34}$$

The next step is the dictionary update, and this step requires the input dictionary:

$$\mathbf{D} = [\mathbf{d}_1, \mathbf{d}_2, \cdots, \mathbf{d}_k] \in \mathbb{R}^{M \times K} \tag{7.35}$$

and two matrices:

$$\mathbf{A} = [\mathbf{a}_1, \mathbf{a}_2, \cdots, \mathbf{a}_k] = \sum_{i=1}^{t} \mathbf{x}_i \mathbf{x}_i^t \in \mathbb{R}^{K \times K} \tag{7.36}$$

$$\mathbf{B} = [\mathbf{b}_1, \mathbf{b}_2, \cdots, \mathbf{b}_k] = \sum_{i=1}^{t} \mathbf{y}_i \mathbf{x}_i^t \in \mathbb{R}^{M \times K} \tag{7.37}$$

This algorithm updates each column of $\mathbf{D}$ sequentially using block coordinate descent with a "warm restart" (which consists in the dictionary calculated at the previous step $\mathbf{D}_{t-1}$). The following equations are used for updating the jth column of $\mathbf{D}$ while keeping the other ones fixed:

$$\mathbf{u}_j \leftarrow \frac{1}{\mathbf{A}_{jj}}(\mathbf{b}_j - \mathbf{D}\mathbf{a}_j) + \mathbf{d}_j \tag{7.38}$$

$$\mathbf{d}_j \leftarrow \frac{1}{\max\left(\|\mathbf{u}_j\|_2\right)}\mathbf{u}_j \tag{7.39}$$

This approach gives the solution to the dictionary update problem:

$$\mathbf{D}_t = \underset{\mathbf{D}\in C}{\operatorname{argmin}}\ \frac{1}{t}\left(\sum_{i=1}^{t}\|\mathbf{y}_i - \mathbf{D}\mathbf{x}_i\|_2^2 + \lambda\|\mathbf{x}_i\|_1\right) \tag{7.40}$$

$$= \underset{\mathbf{D}\in C}{\operatorname{argmin}}\ \frac{1}{t}\left(\frac{1}{2}\mathrm{Tr}\left(\mathbf{D}^T\mathbf{D}\mathbf{A}_t\right) - \mathrm{Tr}\left(\mathbf{D}^{\mathbf{T}}\mathbf{B}_t\right)\right) \tag{7.41}$$

where $\mathbf{C}$ is a convex set, i.e. we have to impose that the column of the dictionary matrix have an ℓ_2-norm less than or equal to one, then

$$C \triangleq \left\{\mathbf{D} \in \mathbb{R}^{M\times K} \text{ subject to } \forall\ j = 1, \ldots, k\ \mathbf{d}_j^T\mathbf{d}_t \leq 1\right\} \tag{7.42}$$

The ODL algorithm is faster than the batch K-SVD and, since it uses $\mathbf{D}_{(t-1)}$ as warm restart for computing $\mathbf{D}_t$, a few iterations could be enough for the correct reconstruction.

7.5.4 CBWLSU

A recent study [60] notes that even though online processing reduces computational complexity compared to batch methods, ODL performance can be further improved if only useful information from previous data is used for updating the atoms. In this study, a new ODL called CBWLSU was proposed. CBWLSU is an online method that introduces an interesting alternative for the dictionary update step. Like ODL, CBWLSU evaluates one new training data at a time, $\mathbf{y}_t$. However, to update the dictionary, it searches among all previous training data and uses only the ones that share the same atoms with $\mathbf{y}_t$. Let $Y\_{\mathcal{Q}_t}$=$\left[\mathbf{y}_{l_1} \ \cdots \ \mathbf{y}_{l_{|\mathcal{Q}_t|}}\right]$, where $l_1, \cdots, l_{|\mathcal{Q}_t|} \in \mathcal{Q}_t$, be the set of previous training elements at iteration t. Define $\mathcal{N}_t = \left\{l : 1, \left\langle \mathbf{x}_l^T, \mathbf{x}_t \right\rangle \neq 0\right\} \subset \mathcal{Q}_t$ as the set of indices of all previous training elements that are correlated with the new element such that $|\mathcal{N}_t| = N_{pt}$. The new index set is $\mathcal{N}_t = \mathcal{N}_t \cup \{t\}$ so that the training set becomes $Y\_{\mathcal{N}_t}$=$\left[\mathbf{y}_{l_1} \ \cdots \ \mathbf{y}_{l_{|\mathcal{N}_t|}}\right]$, where $l_1, \cdots, l_{|\mathcal{N}_t|} \in \mathcal{N}_t$. CBWLSU then employs a weighting matrix $\mathbf{W}(\mathbf{y}_t)$ to evaluate the influence of the selected previous elements for the dictionary update step and solves the optimization problem therein via weighted least squares (WLS). The sparse coding in CBWLSU is achieved via batch OMP.

7.5.5 DOMINODL

We now introduce our DOMINODL approach for ODL, which not only leads to a dictionary (**D**) that is tuned to sparsely represent the training set (**Y**) but is also faster than other online algorithms. The key idea of DOMINODL is as follows: When sequentially analyzing the training set, it is pertinent to leverage the memory of previous data in the dictionary update step. However, algorithms such as CBWLSU consider *all* previous elements. Using all previous training set samples is computationally expensive and may also slow down convergence. The samples that have already contributed to the dictionary update do not need to be considered again. Moreover, in some real-time applications (such as highly correlated range profiles of GPR), their contribution may not be relevant anymore for updating the dictionary.

In DOMINODL, we save computations by considering only a small batch of previous elements that are correlated with the new elements. The two sets are defined as correlated if, in their sparse decomposition, they have at least one common non-zero element. The time gained from considering fewer previous training elements is used to consider a mini-batch of new training data (instead of a single element as in ODL and CBWLSU).

The sparse coding step of DOMINODL employs batch OMP, selecting the maximal residual error δ in (7.33) using a data-driven entropy-based strategy as described later in this section. At the end of each iteration, DOMINODL also drops off those previous training set elements that have not been picked up after a certain number of iterations, N_u. The mini-batch drawing combined with dropping off training elements and entropy-based criterion to control sparsity results in an extremely fast ODL algorithm that is beneficial for real-time radar operations.

We initialize the dictionary **D** using a collection of K training set samples that can be randomly chosen from **Y** (alternatively one can also use random vectors with a given distribution) and then perform a sparse decomposition of **Y** with the dictionary **D**; the algorithm then scans the entire training set sequentially. Define the mini-batch of $N_b + 1$ new training elements as $Y\_{\mathcal{B}_t}$=$\left[\mathbf{y}_{l_1} \quad \cdots \quad \mathbf{y}_{l_{N_b+1}}\right]$ with $l_1, \cdots, l_{N_b+1} \in \mathcal{B}_t$ such that the index set $\mathcal{B}_t = \{l : t \leq l + N_b\}$. When $t - N_b$, we simply take the remaining new elements to constitute this mini-batch.[a] We store the set of dictionary atoms participating in the SR of the signals in $\mathbf{Y}_{\mathcal{B}_t}$ as $\mathbf{D}_{\mathcal{B}_t}$. Let the coefficient vectors associated with the SR of $\mathbf{Y}_{\mathcal{B}_t}$ are indicated by $\mathbf{X}_{\mathcal{B}_t}$ and

$$\mathbb{I}_{\mathbf{X}_{\mathcal{B}_t}} = \sum_{p \in \mathcal{B}_t} \mathbf{x}_p \tag{7.43}$$

being an indicator vector whose non-zero elements indicate the atoms of **D** being used by $\mathbf{Y}_{\mathcal{B}_t}$.

[a]In numerical experiments, we observed that the condition rarely occurs because DOMINODL updates the dictionary and converges in very few iterations. The algorithm also ensures that the number of previous samples before the dictionary update. If this condition is not fulfilled, then it considers all previous training samples.

Define $Y\_{\mathcal{Q}_t} = [\mathbf{y}_{l_1} \ \cdots \ \mathbf{y}_{l_t}]$ with $l_1, \cdots, l_t \in \mathcal{Q}_t$ as the collection of previous training elements with the index set $\mathcal{Q}_t = \{l : 1 \leq l - 1\}$. Consider $Y\_{\mathcal{M}_t} = [\mathbf{y}_{l_1} \ \cdots \ \mathbf{y}_{l_{N_r}}]$ with $l_1, \cdots, l_{N_r} \in \mathcal{M}_t \subset \mathcal{Q}_t$ as a randomly selected mini-batch of N_r previous elements. The coefficient vectors associated with the SR of $\mathbf{Y}_{\mathcal{M}_t}$ are indicated with $\mathbf{X}_{\mathcal{M}_t}$ and

$$\mathbb{I}_{\mathbf{X}_{\mathcal{M}_t}} = \sum_{l \in \mathcal{M}_t} \mathbf{x}_l \tag{7.44}$$

is an indicator vector whose non-zero elements indicate the atoms of $\mathbf{D}$ being used by $\mathbf{Y}_{\mathcal{M}_t}$.

Define $Y\_{\mathcal{A}_t} = [\mathbf{y}_{l_1} \ \cdots \ \mathbf{y}_{l_{|\mathcal{A}_t|}}]$ with $l_1, \cdots, l_{|\mathcal{A}_t|} \in \mathcal{A}_t$, where

$$\mathcal{A}_t = \left\{ l : l \in \mathcal{M}_t, \left\langle \mathbb{I}_{\mathbf{X}_{\mathcal{B}_t}}{}^T, \mathbb{I}_{\mathbf{X}_{\mathcal{M}_t}} \right\rangle \neq 0 \right\} \subset \mathcal{M}_t \tag{7.45}$$

is a subset of previous training elements that are correlated with the mini-batch of new elements. In order to avoid multiple occurrences of the same element in consecutive mini-batches, DOMINODL ensures that $\mathcal{M}_t \cap \mathcal{M}_{t-1} = \varnothing$ providing that a sufficient number of previous training set elements is available. Our new training set is $\mathbf{Y}_{\mathcal{C}_t} = \mathbf{Y}_{\mathcal{A}_t} \cup \mathbf{Y}_{\mathcal{B}_t}$. Both mini-batches of new and previous elements are selected such that the entire training set size $(N_b + N_r)$ is still smaller than that of CBWLSU, where it is $N_{p_t} + 1$ (see above on CBWLSU explanation).

The dictionary update subproblem then reduces to considering only the sets $\mathbf{Y}_{\mathcal{C}_t}$, $\mathbf{D}_{\mathcal{C}_t}$, and $\mathbf{X}_{\mathcal{C}_t}$:

$$\hat{\mathbf{D}}_{\mathcal{C}_t} = \underset{\mathbf{D}_{\mathcal{C}_t} \in \mathcal{D}}{\operatorname{argmin}} \ \|\mathbf{Y}_{\mathcal{C}_t} - \mathbf{D}_{\mathcal{C}_t}\mathbf{X}_{\mathcal{C}_t}\|_F^2. \tag{7.46}$$

Assume that the sparse coding for each example is known and define the errors as

$$\mathbf{E}_{\mathcal{C}_t} = \mathbf{Y}_{\mathcal{C}_t} - \mathbf{D}_{\mathcal{C}_t}\mathbf{X}_{\mathcal{C}_t} = [\mathbf{e}_1, \cdots, \mathbf{e}_{N_r}]. \tag{7.47}$$

We can update $\mathbf{D}_{\mathcal{C}_t}$, such that the above error is minimized, with the assumption of fixed $\mathbf{X}_{\mathcal{C}_t}$. A similar problem is considered in MOD where error minimization is achieved through least squares. Here, we employ WLS inspired by the fact that it has shown improvement in convergence over standard least squares [60]. We compute the weighting matrix $\mathbf{W}_{\mathcal{C}_t}$ using the SR error $\mathbf{E}_{\mathcal{C}_t}$

$$\mathbf{W}_{\mathcal{C}_t} = \operatorname{diag}\left(\frac{1}{\|\mathbf{e}_1\|_2^2}, \ldots, \frac{1}{\|\mathbf{e}_{N_r}\|_2^2} \right) \tag{7.48}$$

and then solve the following optimization problem

$$\hat{\mathbf{D}}_{\mathcal{C}_t} = \underset{\mathbf{D}_{\mathcal{C}_t} \in \mathcal{D}}{\operatorname{argmin}} \ \|(\mathbf{Y}_{\mathcal{C}_t} - \mathbf{D}_{\mathcal{C}_t}\mathbf{X}_{\mathcal{C}_t})\mathbf{W}_{\mathcal{C}_t}^{\frac{1}{2}}\|_F^2. \tag{7.49}$$

This leads to the WLS solution

$$\hat{\mathbf{D}}_{\mathcal{C}_t} = \mathbf{Y}_{\mathcal{C}_t}\mathbf{W}_{\mathcal{C}_t}\mathbf{Y}_{\mathcal{C}_t}^T(\mathbf{Y}_{\mathcal{C}_t}\mathbf{W}_{\mathcal{C}_t}\mathbf{Y}^{\mathbf{T}}{}_{\mathcal{C}_t})^{-1}. \tag{7.50}$$

The dictionary **D** is then updated with the atoms $\hat{\mathbf{D}}_{C_t}$ and its columns are normalized by their ℓ_2-norms. The **D** is then used for updating the sparse coding of $\mathbf{Y}_{C_t}$ using batch OMP. Algorithm 3 summarizes all major steps of DOMINODL.

7.5.6 Comparison of DL algorithms

Table 7.2 summarizes the important differences between DOMINODL and other related algorithms. Like MOD and CBWLSU, DOMINODL uses a WLS solution in the dictionary update. The proof of convergence for the alternating minimization method in MOD was provided in [78], where it is shown that alternating minimization converges linearly as long as the following assumptions hold true: sparse coefficients have bounded values, sparsity level is on the order of $\mathcal{O}(M^{1/6})$ and the dictionary satisfies the RIP property. In [60], these assumptions have been applied for CBWLSU convergence. Compared to CBWLSU, the improvements in DOMINODL include mini-batch-based data selection and data reduction via drop-off strategy but the update algorithms remain the same. Numerical experiments in Section 7.6 suggest that DOMINODL usually converges faster than CBWLSU.

Computational complexity of DOMINODL has a very low order compared to other online approaches. Let us indicate the order of complexity as OC_{DL}, with the subscript DL indicating the employed DL algorithm. As mentioned earlier, there are N atoms in the dictionary. Assume that every signal is represented by a linear combination of K_s atoms, $K_s \ll N$. Empirically, among all possible combinations of K_s atoms from N, the probability to have a common atom in the SR is K_s/N. Given L training elements, the number of training data that has a specific atom in their representation is proportional to LK_s/N. Suppose our mini-batch has elements that reduce the number of training data by a factor $\beta < 1$ (depending on the value of N_b and N_r). Further, assume that the dropping off step reduces the training set elements by a factor $\rho < 1$. The number of used training data L_t in the tth iteration

Table 7.2 Comparison of DL steps

DL step	K-SVD	LRSDL	ODL	CBWLSU	DOMINODL
Training method	Batch	Batch	Online	Online	Online
Sparse coding method	OMP	FISTA	LARS	Batch OMP	Batch OMP
Dictionary update	Entire **D**	Entire **D**	Entire **D**	Entire **D**	Partial **D** adaptively
Samples per iteration	Entire **Y**	Entire **Y**	$\mathbf{Y}_t$	$\mathbf{Y}_{Nt}$	$\mathbf{Ye}_t$
Optimization method	SVD	ADMM	Gradient descent	WLS	WLS
Dictionary pruning	Yes	No	Yes	No	No
Training-set drop-off	No	No	No	No	Yes

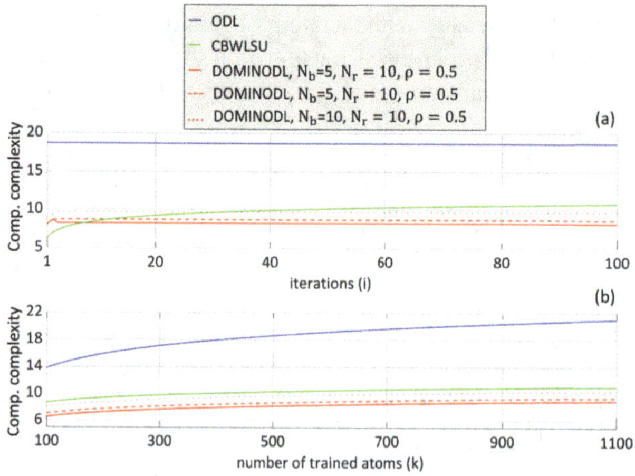

Figure 7.9 *Computational complexity of ODL strategies for increasing number of (a) iterations and (b) trained atoms*

is proportional to $\beta\rho t K_s/N$. Then, the worst estimate of DOMINODL's computational complexity is due to the sparse coding batch OMP, which is of the order $OC_{DOMINODL} = \mathcal{O}(L_t N^2) = \mathcal{O}(\beta\rho t K_s N) \approx \mathcal{O}(\beta\rho t N)$. This is much smaller than the complexity of ODL ($OC_{ODL} = \mathcal{O}(N^3)$) or CBWLSU ($OC_{CBWLSU} = \mathcal{O}(tN)$) [59, 60].

Figure 7.9 illustrates the computational complexity of ODL approaches compared to DOMINODL. Here, we generally indicate with $OC_{DL}(t)$ and $OC_{DL}(k)$ the order of complexity as a function of the number of iterations and atoms of the dictionary. Figure 7.9(a) shows that, for a fixed number of iterations ($t = 60$), the general trend of complexity with respect to the increase in the number of atoms (K) is similar for all algorithms. However, the complexity of ODL is higher than CBWLSU and DOMINODL, the latter being the least complex. When the number of iterations is increased, the complexity of ODL and CBWLSU have a similar increasing trend (see Figure 7.9(b)). In case of DOMINODL, its complexity is similar to the increasing trend of CBWLSU and determined largely by N_b. When DOMINODL iterations begin accounting for N_r previous elements, its complexity stays constant. The value of β changes for every iteration, while ρ depends on the data itself. In general, after a few dozen of iterations, DOMINODL's complexity always stays lower than CBWLSU.

The proposed DL methods are heuristic, i.e. they do yield over-complete dictionaries but with no theoretical guarantee of obtaining coherent dictionaries. Several state-of-the-art results that outline DL algorithms with concrete performance guarantees require stronger assumptions on the observed data.

An efficient SR that accurately represents the scattering behaviors related to soil type and targets serves a twofold purpose. First, it allows application of CS in reducing sensing data and associated scan time. Second, the literature [79] indicates that SR is effective in extracting the mid- or high-level features in image classification.

For example, in the context of landmines classification with GPR, our prior works [42, 58] have shown that frameworks based on SR improve the classifier performance. SR has been frequently applied to data from synthetic aperture radar (SAR) [80]), inverse SAR [81]), and interferometric SAR [82] because these are not naturally sparse in the range-time domain. In the particular case of GPR, [83–86] have proposed CS-based imaging for various GPR waveforms.

Techniques to reduce the acquired radar samples using CS have been previously used to address the bottleneck of long scan times in conventional radar systems. The estimation accuracy of target parameters is greatly affected by radar's dwell time [87], i.e. the time duration a directional radar beam spends hitting a particular target. But, at the same time, this negatively affects the ability of the radar to look at targets in other directions thereby prolonging the total scan time. For example, [88] employs matrix completion to reduce scan time in a weather radar. Similar techniques have also been shown to be useful in radar imaging applications [89].

7.6 Adaptive statistical DL evaluation

Adaptive processing is crucial to enhance the radar performance. In the specific context of DL-based classification, we optimize the learned dictionary to sparsely represent GPR measurements. This optimization is based on computing certain statistical properties of the radar data. For example, in our mine detection application, this procedure is applied to the experimental measurements from a GPR test field with buried landmine-simulants (see Section 7.3.2). The sparse vectors obtained by DL are fed to a support vector machine (SVM) classifier [90], which then discriminates between different types of mines and clutter (shown later in Section 7.7).

The selection of different input parameters for the aforementioned DL algorithms determines their success in sparsifying the data. Hence, it is not useful to directly apply DL with arbitrary parameter values. Prior works set these parameters through hit-and-trial or resorting to metrics that are unable to discriminate the influence of different parameters [7]. Here, we propose a method to investigate the effect of the various input parameters on learning performance and then preset the parameters to *optimal* values that yield the dictionary **D** (for each DL method) optimized to sparsely represent our GPR data, therefore improving the quality of the features for classification (i.e. the sparse coefficients).

Table 7.3 *DL parameters*

DL algorithm	**Input parameters**
K-SVD	N_t, K
ODL	N_t, K
CBWLSU	K
DOMINODL	K, N_b, N_r, N_u

Table 7.3 lists these input parameters. The K-SVD and ODL admit the number of iterations N_t and the number of trained atoms K as input parameters while CBWLSU uses only K. The DOMINODL parameters are the dimensions of the mini-batch of new (N_b) and previous (N_r) training elements for each iteration and the drop-off value N_u, which indicates after how many iterations the algorithm should discard each unused training set element. We applied K-SVD, ODL, CBWLSU, and DOMINODL separately on the training set for different combinations of parameter values.

In the next subsection, we provide details of the training set obtained from the LIAG survey and follow it by various criteria used for our adaptive DL evaluation such as similarity measure and statistical metrics.

7.6.1 Training set generation

The entire LIAG dataset described in Section 7.3 consists of 27 survey sections (or simply, "surveys", see Section 7.3.2 for a description of the test field) of size 1 × 1 m. Every survey consists of 2 500 range profiles. We divided the data into two sets: *training set* ($\mathbf{Y}$) to be used for both DL and classification and the *test set* ($\mathbf{Y}_{\text{TEST}}$) to evaluate the performance of the classification.

The training set $\mathbf{Y} \in \mathbb{R}^{M \times L}$ is a matrix whose L columns $\{\mathbf{y}_i\}_{i=1}^{L}$ consist of sampled range profiles $\mathbf{y}_i = [y[0], \cdots, y[M-1]]^T$ of M range samples each. The profiles are selected from different surveys and contain almost exclusively either a particular class of landmine or clutter. In total, we have 463, 168, 167, and 128 range profiles for clutter, PMA2/PMN, ERA, and Type-72, respectively, see Table 7.4. An accurate separation of these classes was very challenging because of the contributions from the non-homogeneous soil clutter that often masked the target responses completely. A poor selection would lead the DL to learn a dictionary that is appropriate for sparsely representing clutter, instead of landmines. The test set $\mathbf{Y}_{\text{TEST}} \in \mathbb{R}^{M \times J}$ is a matrix with $J = 15\,000$ columns $\{\mathbf{y}_{\text{TEST},i}\}_{i=1}^{J}$ that correspond to sampled range profiles from six surveys, two for each target class. The test and training sets contain data from separate surveys to enable fair assessment of the classification performance. We denote by the matrices $\mathbf{X} \in \mathbb{R}^{N \times L}$ and $\mathbf{X}_{\text{TEST}} \in \mathbb{R}^{N \times J}$ as the SRs of $\mathbf{Y}$ and $\mathbf{Y}_{\text{TEST}}$, respectively, and by K the number of atoms of the learned dictionary $\mathbf{D} \in \mathbb{R}^{M \times K}$.

Table 7.4 Training set classes

Target class	**Number of elements**
Clutter	463
PMN	168
ERA	167
T72	128

7.6.2 Similarity measure

In order to compare the dictionaries obtained from various DL algorithms, we use a *similarity measure* that quantifies the closeness of the original training set $\mathbf{Y}$ with the reconstructed set $\hat{\mathbf{Y}}$ obtained using the sparse coefficients of the learned dictionary $\mathbf{D}$. From these similarity values, empirical probability density functions (EPDFs) for any combination of parameter values are obtained; we evaluate these EPDFs using statistical metrics described in Section 7.6.3. These metrics efficiently characterize the similarity between $\mathbf{Y}$ and $\hat{\mathbf{Y}}$ and lead us to an optimal selection of various DL input parameters for our experimental GPR dataset. Consider the cross-correlation between a training set vector $\mathbf{y}_i$ and its reconstruction using a learned dictionary $\hat{\mathbf{y}}_i$: The cross-correlation between $\mathbf{y}_i$ and $\hat{\mathbf{y}}_i$ can be defined as:

$$\mathbf{r}_{\mathbf{y}_i,\hat{\mathbf{y}}_i}[l] = \sum_{n=-\infty}^{+\infty} \mathbf{y}_i[n]\hat{\mathbf{y}}_i[n+l] \tag{7.51}$$

whereas its normalized version (normalized cross-correlation) is given by:

$$\overline{\mathbf{r}_{\mathbf{y}_i,\hat{\mathbf{y}}_i}}[m] = \frac{\mathbf{r}_{\mathbf{y}_i,\hat{\mathbf{y}}_i}[m]}{\sqrt{\mathbf{r}_{\mathbf{y}_i,\mathbf{y}_i}[0]\mathbf{r}_{\hat{\mathbf{y}}_i,\hat{\mathbf{y}}_i}[0]}}. \tag{7.52}$$

For the vector $\mathbf{y}_i$, we finally define the similarity measure s_i as

$$s_i = \max_m \left|\overline{\mathbf{r}_{\mathbf{y}_i,\hat{\mathbf{y}}_i}(m)}\right| \tag{7.53}$$

where a value of s_i closer to unity demonstrates greater similarity of the reconstructed data with the original training set element (i).

We compute $\{s_i\}_{i=1}^L$ for all vectors $\{\mathbf{y}_i\}_{i=1}^L$ and then obtain the normalized histogram or EPDF of all similarity measures. In statistics, an EPDF is defined as the distribution function associated with the empirical measure of a set of data (in our case a set of similarity measures). We indicate the EPDF of a particular DL algorithm as $p_{s_{DL}}$. Here, the subscript DL represents the algorithm used for learning $\mathbf{D}$, e.g. “K-SVD”, “LRSDL”, “ODL”, “CBWLSU,” and “DOMINODL”.

Various parameter combinations for a specific DL method result in a collection of EPDFs. For a given DL method, our goal is to compare the EPDFs by varying these parameters and arrive at the thresholds of parameter values after which the changes in $p_{s_{DL}}$ are only incremental. For instance, Figure 7.10 shows the EPDFs of $\{s_i\}_{i=1}^L$ obtained from the GPR mines data where optimal parameters for different DL methods were determined using the statistical methods described in the following subsection. In the figure we also indicate which parameters were affecting each particular distribution (in parenthesis). We note that the ODL approaches ($p_{s_{ODL}}$, $p_{s_{CBWLSU}}$, and $p_{s_{DOMINODL}}$) yield distributions that are more skewed towards unity than K-SVD ($p_{s_{K-SVD}}$). The distribution associated with LRSDL is concentrated on a single peak with high values of similarity, nevertheless, it does not exhibit the extremely close to unity values of ODL distributions.

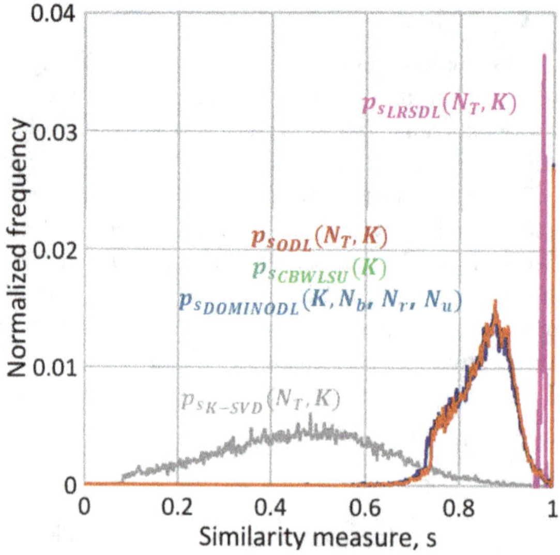

Figure 7.10 *Normalized histograms of similarity measure using the following optimal parameters for the DL algorithm: $N_t = 100$, $K = 640$, $N_b = 30$, $N_r = 10$, and $N_u = 10$. See Section 7.6.4 for the process to select these optimal values.*

7.6.3 Statistical metrics

Our goal is to find parameter values for which $p_{s_{DL}}$ is skewed towards unity and has small variance. The individual comparisons of mean (μ) and standard deviation (σ) of EPDFs, as used in previous GPR DL studies [7], are not sufficient to quantify the observed dispersion in the EPDFs obtained by varying any of the parameter values. Some DL studies [7, 91] rely on bulk statistics such as the normalized root mean square error but these quantities are less sensitive to changes in parameter values and, therefore, not so helpful in fine-tuning the algorithms. For this evaluation, we will use three different metrics: the coefficient of variation, the two-sample Kolmogorov–Smirnov (K–S) distance, and the Dvoretzky–Kiefer–Wolfowitz (DKW) inequality.

Coefficient of variation

We choose to simultaneously compare both mean (μ) and variance (σ) of a single EPDF by using the *coefficient of variation*:

$$CV = \sigma/\mu \tag{7.54}$$

In our analysis, it represents the extent of variability in relation to the mean of the similarity values.

Two-sample Kolmogorov–Smirnov distance

In the context of our application, it is more convenient to work with the cumulative distribution functions (CDFs) rather than with EPDFs because the well-developed statistical inference theory allows for convenient comparison of CDFs.

Therefore, our second metric to compare similarity measurements obtained by successive changes in parameter values is the *two-sample K–S distance*, which is the maximum distance between the two given empirical cumulative distribution functions (ECDFs). Larger values of this metric indicate that samples are drawn from different underlying distributions. Given two distributions s_1 and s_2 taken at L discrete points, suppose $\hat{F}_{s_1}$ and $\hat{G}_{s_2}$ are their ECDFs of the same length and correspond to their EPDFs $\hat{f}_{s_1}$ and $\hat{g}_{s_2}$, respectively. Denote Ω as the set of L observations used to evaluate both distributions. Then, the discrete two-sample K–S distance is

$$d_{ks}(\hat{F}_{s_1}, \hat{G}_{s_2}) = \sup_{i \in \Omega} |\hat{F}_{s_1}(i) - \hat{G}_{s_2}(i)| \tag{7.55}$$

where sup(·) denotes the supremum over all distances. In our case, L is the number of range profiles in the training set. We first compute a reference ECDF ($\hat{G}_{s_{\text{ref}}}$) for each DL algorithm with fixed parameter values. For our purposes, this reference ECDF will be obtained by a particular combination of input parameters of the selected DL algorithm. Then, we vary parameter values from this reference and obtain the corresponding ECDF $\hat{F}_{s_{\text{test}}}$ of similarity measure. Finally, we calculate the K–S distance d_{ks} of each $\hat{F}_{s_{\text{test}}}$ with respect to $\hat{G}_{s_{\text{ref}}}$ as

$$d_{ks} = d_{ks}(\hat{F}_{s_{\text{test}}}, \hat{G}_{s_{\text{ref}}}) = \sup_{1 \le i \le L} |\hat{F}_{s_{\text{test}}}(i) - \hat{G}_{s_{\text{ref}}}(i)|. \tag{7.56}$$

For our evaluation, d_{ks} states how much the selection of certain input parameters of DL changes the ECDFs of similarity values (i.e. how different is the result of DL) w.r.t. the reference one. In other words, using this metric in combination with CV gives us information regarding the variation and the quality of the obtained dictionary according to the selected input parameters.

Dvoretzky–Kiefer–Wolfowitz inequality

As a third metric, we exploit the *DKW inequality* [38], which precisely characterizes the rate of convergence of an ECDF to a corresponding exact CDF (from which the empirical samples are drawn) for any finite number of samples.

Let $d_{ks}(\hat{G}_s, F_s)$ be the K–S distance between ECDF $\hat{G}_s$ and the continuous CDF F_s for a random variable s and L samples. Since $\hat{G}_s$ changes with the change in the L random samples, $d_{ks}(\hat{G}_s, F_s)$ is also a random variable. We are interested in the conditions that provide desired confidence in verifying if F and G are the same distributions for a given finite L. If the two distributions are indeed identical, then the DKW inequality bounds the probability that d_{ks} is greater than any number ε, with $0 < \varepsilon < 1$ as follows:

$$\Pr\left\{d_{ks}\left(\hat{G}_s, F\right) > \varepsilon\right\} \le 2e^{-2L\varepsilon^2}. \tag{7.57}$$

The corresponding asymptotic result that as $L \to \infty$, $d_{ks} \to 0$ with probability 1 is due to the Glivenko–Cantelli theorem [92].

Consider a binary hypothesis testing framework where we use (7.57) to test the null hypothesis $\mathcal{H}_0 : F = \hat{G}$ for a given ε. The DKW inequality bounds the probability of rejecting the null hypothesis when it is true, i.e. the Type I statistical error.

The probability of rejecting the null hypothesis when it is true is called the *p*-value of the test and is bounded by the DKW inequality. Assuming the *p*-value is smaller than a certain confidence level α, the following inequality must hold with probability at least $1 - \alpha$ [93]:

$$d_{ks}\left(\hat{G}_s, F\right) \leq \sqrt{-\frac{1}{2L}\ln\left(\frac{\alpha}{2}\right)}. \tag{7.58}$$

Our goal is to use the DKW inequality to compare two ECDFs $\hat{F}_{s_{\text{test}}}$ and $\hat{G}_{s_{\text{ref}}}$ as in (7.56), to verify if they are drawn from the same underlying CDF. By the triangle inequality, the *K–S* distance

$$d_{ks}(\hat{F}_{s_{\text{test}}}, \hat{G}_{s_{\text{ref}}}) = d_{ks}(\hat{F}_{s_{\text{test}}}, F) + d_{ks}(\hat{G}_{s_{\text{ref}}}, F) \tag{7.59}$$

where *G* and *F* are the underlying CDFs corresponding to . We now bound the right side using DKW

$$d_{DKW}(\hat{F}_{s_{\text{test}}}, \hat{G}_{s_{\text{ref}}}) \leq \sqrt{-\frac{1}{2L}\ln\left(\frac{\alpha}{2}\right)} + \sqrt{-\frac{1}{2L}\ln\left(\frac{\alpha}{2}\right)} = \sqrt{-\frac{2}{L}\ln\left(\frac{\alpha}{2}\right)} \tag{7.60}$$

which is the maximum distance for which and are identical with probability . The *DKW metric* is the difference

$$d_{DKW} = \sqrt{-\frac{2}{L}\ln\left(\frac{\alpha}{2}\right)} - d_L(\hat{F}_{s_{\text{test}}}, \hat{G}_{s_{\text{ref}}}). \tag{7.61}$$

Larger values of this metric imply greater similarity between the two ECDFs; a negative value implies that the null hypothesis is not true. For our purposes, d_{DKW} is an alternative way to state the variation of ECDFs with respect to d_{KS}; however, d_{DKW} also tells us if two distributions are coming from the same underlying CDFs or not (whether the metric is positive or negative, i.e. fulfilling or not the null hypothesis), thus giving a stronger indication on how the input parameters affect the final result of DL.

7.6.4 Evaluation results

We evaluated the performance of all DL algorithms using the metrics explained in Section 7.6.3 for the reconstruction of the training set **Y**. We refer to Table 7.3 to indicate what are the parameters that affect the presented DL approaches. In particular, the number of iterations N_t is not relevant to CBWLSU and DOMINODL while the latter requires additional parameters for the mini-batch dimensions and the iterations required to drop off unused training set elements. The role of these parameters in providing an optimized dictionary is summarized in Table 7.5.

Note that we compute the K–S distance and the DKW metric for all methods with respect to a reference distribution p_{ref}, as explained in Section 7.6.3. This reference is obtained using the following parameters, as applicable: $N_t = 1$, $K = 300$, $N_b = 30$, $N_r = 10$, and $N_u = 10$.

Influence of the number of iterations

Figure 7.11 (a)–(c) shows the effect of N_t on the *CV*, K–S test distance (d_{ks}) and the DKW metric (d_{dkw}) for K-SVD and ODL and LRSDL. We have skipped

Table 7.5 *Outlook on the influence of the different input parameters for the proposed DL approaches*

	N_t	K	N_b	N_r	N_u
K-SVD	N.I.	I.	N.U.	N.U.	N.U.
ODL	S.I.	I.	N.U.	N.U.	N.U.
CBWLSU	N.U.	I.	N.U.	N.U.	N.U.
DOMINODL	N.U.	I.	S.I.	S.I.	I.

I, important; N.U., not used; S.I., slightly important.

CBWLSU and DOMINODL from this analysis because they do not accept N_t as an input. For ODL, the CV remains relatively unchanged with an increase in N_t. However, the K-SVD CV exhibits an oscillating behavior and generally high values. In case of the K–S distance, ODL shows slight increase in d_{ks} while K-SVD oscillates around a mean value that is higher than ODL. The DKW metric provides better insight: even though the ODL distributions differ from p_{ref} with increase in the iterations, the null hypothesis always holds because d_{DKW} remains positive. The d_{DKW} for K-SVD is also positive but much smaller than ODL. It also does not exhibit any specific trend with an increase in iterations. We also observed a similar behavior with the mean of similarity values. The influence of the number of iterations in LRSDL had the same oscillating behavior as in K-SVD but with larger variation. We conclude that the number of iterations N_t does not significantly influence the metrics for both algorithms and choose $N_t = 100$.

Influence of the number of trained atoms

Figure 7.12 (a)–(c) compares all three metrics with a change in the number of trained atoms K, a parameter that is common to all DL methods. We observe that CV generally decreases with an increase in K. This indicates an improvement in the similarity between the reconstructed and the original training set. K-SVD shows an

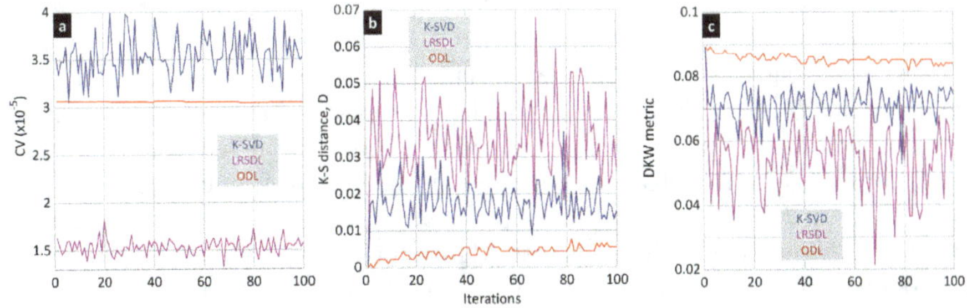

Figure 7.11 *(a) CV, (b) K–S distance, and (c) DKW metric for K-SVD, LRSDL, and ODL parameter analyses as a function of the number of iterations N_t*

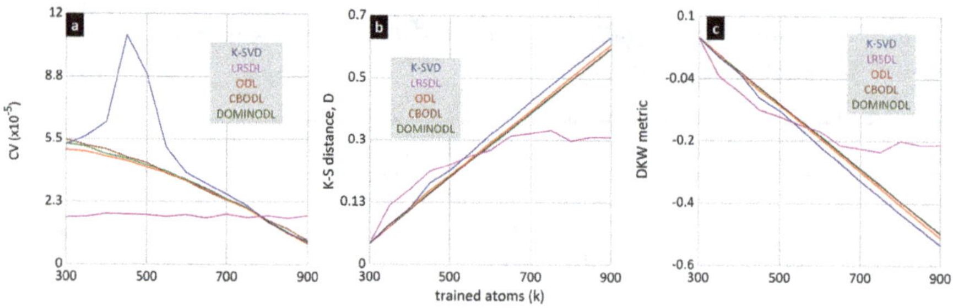

Figure 7.12 (a) CV, (b) K–S distance, and (c) DKW metric for various DL algorithms as a function of the number of trained atoms K

anomalous pattern for lower values of K but later converges to a trend that is identical to other DL approaches. The K–S distance exhibits a linear change in the distributions with respect to the reference. Since d_{ks} quantifies the difference between the distributions rather than stating which one is better, combining its behavior with CV makes it evident that an increase in K leads to better distributions of similarity values. The DKW metric d_{DKW}, calculated with the same reference, expectedly also shows a linear change. It is clear that even a slight change in K leads to more negative values of d_{DKW} implying that the null hypothesis does not hold true. This shows the significant influence of the parameter K on the distributions. It was interesting to see a slight improvement for the coefficient of variation when using LRSDL with respect to the other strategies. However, K–S distance and DKW metric indicated that the distributions of similarity values for LRSDL were sensitive to the number of trained atoms only up to a certain value. The value of K is finally chosen such that the dictionary is consistently overcomplete, e.g. the number of atoms is three times greater than the number of samples ($K = 640$ vs $M = 211$).

DOMINODL input parameters selection

It is difficult to evaluate DOMINODL EPDFs by varying all four parameters together. Instead, we fix the parameter that is common to all algorithms, i.e. the number of trained atoms K, and then determine optimal values of N_b, N_r, and N_u.

Figure 7.13 shows the coefficient of variation CV of the distribution of similarity values as a function of DOMINODL parameters. The drop-off value N_u appears to have a greater influence with respect to the mini-batch dimensions N_b and N_r. To select these parameters we made considerations based on the computational time and the way the algorithm is initialized. The computational time of DOMINODL is essentially independent of N_r and N_u but slightly increases with N_b. This was verified and expected because, with N_b we are also increasing the number of SR steps (see Algorithm 3) at every iteration, which is the source of bulk of computations in DL algorithms [60]. Further, in order to ensure that the correlation and the drop-off steps kick off from the very first iteration, DOMINODL should admit several new samples for each iteration thereby increasing N_b as well as the number of previous elements accordingly. Taking into account these observations, we choose $N_b = 30$ and $N_u = 10$.

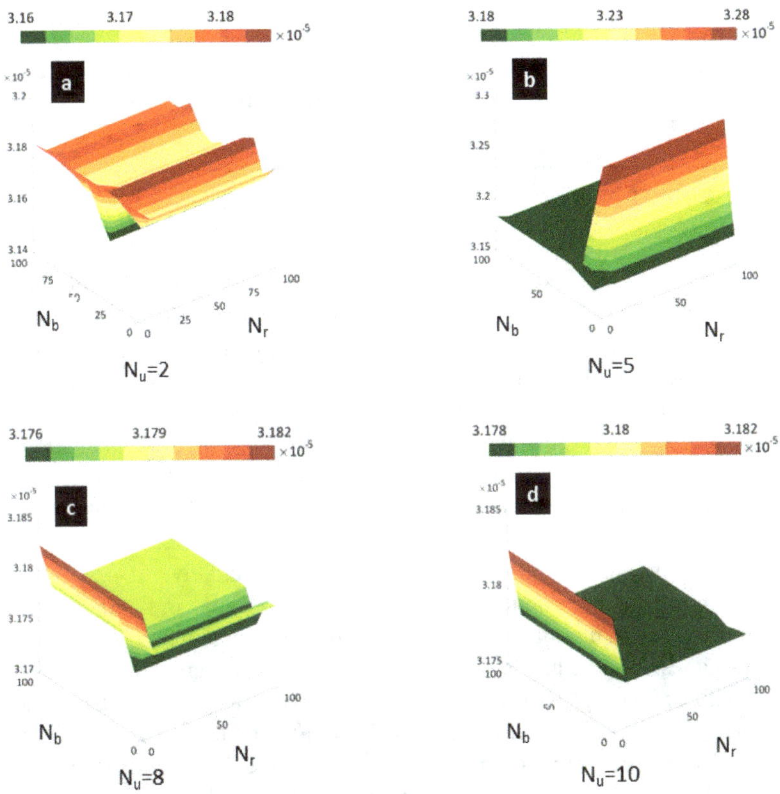

Figure 7.13 *CV as a function of DOMINODL input parameters for* $k = 640$ *and* N_u *as (a) 2, (b) 5, (c) 8, and (d) 10*

According to the results of the parametric evaluation, we choose the following combination of "optimal" parameters for testing our DL strategies: $N_t = 100$, $K = 640$, $N_b = 30$, $N_r = 10$, and $N_u = 10$.

Considerations on computational efficiency of DL algorithms

We used a MATLAB® platform on an 8-Core CPU Windows 7 desktop PC to clock the times for DL algorithms. The ODL algorithm from [59] is implemented as mex executable and, therefore, already fine-tuned for speed. For K-SVD, we employed the efficient implementation from [56] to improve computational speed. Table 7.6 lists the execution times of the four DL approaches when using optimal input parameters. The LRSDL is the slowest of all while ODL is more than four times faster than K-SVD. The CBWLSU provided better classification results but is three times slower than ODL. This could be because the dictionary update step always considers the entire previous training set elements that correlate with only one new element (i.e. there is no mini-batch strategy). This makes the convergence in CBWLSU more challenging.

Table 7.6 Computational times for DL algorithms

	DOMINODL	CBWLSU	ODL	K-SVD
Time (s)	1.75	16.49	5.75	25.8

Blue denotes the best performance among all DL algorithms.

The DOMINODL is the fastest DL method clocking 3× speed than ODL and 15× than K-SVD. This is because the DOMINODL updates the dictionary by evaluating only a mini-batch of previous elements (instead of all of them as in CBWLSU) that correlate with a mini-batch of several new elements (CBWLSU uses just one new element). Further, DOMINODL drops out the unused elements leading to faster convergence. We note that, unlike ODL and K-SVD implementations, we did not use mex executables of DOMINODL, which can further shorten current execution times. From Table 7.6, the reduction in DOMINODL computational time over K-SVD is $((25.8 - 1.75) \times 100)/25.8 \approx 93\%$. The reduction for ODL and CBWLSU are computed similarly as 8% and 36%, respectively.

The computational bottleneck of mines classification lies in the training times. In comparison, the common steps of sparse decomposition and SVM-based classification during testing take just 0.4 s and 1 s, respectively, for an entire survey (1 m × 1 m area with 2 500 range profiles). Thus, time taken per range profile in *ca.* 0.59 ms. The average scan rate of our GPR system is 0.19 m/s (or 1 cm/52.1 ms). This can go as high as 2.7 m/s (or 1 cm/3.61 ms) in other GPRs used for landmines application. Therefore, the test times do not impose much computational cost.

7.7 Advanced GPR target recognition

We now show the combination of SR, DL, and adaptive processing on GPR target recognition for the case of landmine detection. The general procedure for this target recognition has been illustrated earlier in Section 7.2.2. Here, we also compare DL-based classification with CNN and evaluate the effect of reduced samples. This is reflective of modern adaptive GPR processing that adopts a *swiss-knife* approach based on blending many techniques.

After selecting the input parameters of the proposed DL strategies (see Table 7.5), we use the resulting dictionaries for the sparse decomposition of both training and test sets. The resulting sets of sparse coefficients are the input to the SVM classifier. Figure 7.14 summarizes the flowchart of methodology. Note that the labeled training set $\mathbf{Y}$ is used both for DL and classification.

The threshold C and the kernel function parameter γ for SVM have been selected through cross-validation. Our key objective is to assess whether ODL algorithms (and in particular DOMINODL) lead to an improvement in the classification accuracy over batch learning strategies or not. As a comparison with a popular state-of-the-art classification method, we also show the classification results with a

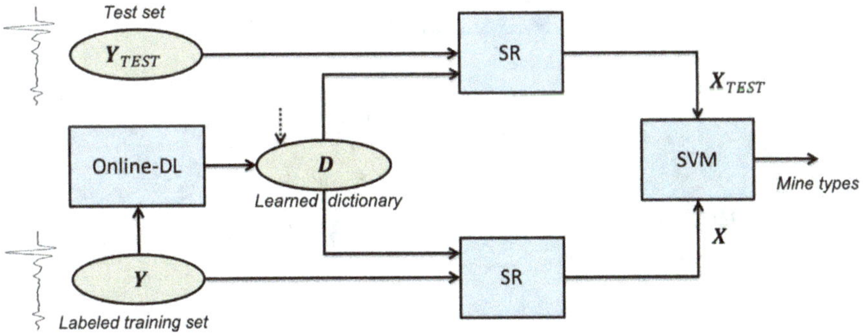

Figure 7.14 Flowchart for adaptive classification strategy

deep-learning approach based on CNN. Finally, we will show classification performances when the original samples of the range profiles are randomly reduced.

Classification with optimal parameters

For a comprehensive analysis of the classification performance, we provide both classification maps and confusion matrices for the test set $\mathbf{Y}_{\text{TEST}}$ using the optimal DL input parameters that we selected following our parametric evaluation in Section 7.6. The classification maps depict the predicted class of each range profile of the survey under test. The pixel dimension of these maps is dictated by the sampling of the GPR in X and Y directions (see Table 7.1). We stacked together three of the six surveys from the test set $\mathbf{Y}_{\text{TEST}}$ where each survey had two buried landmines of a specific class (PMN/PMA2, ERA, and Type-72).

The support of the measurements on our learned dictionaries varies from one to four. Due to coherency, these few non-zero elements may appear in the same position for different vector classes but with slightly different values. This proved to be enough for the classifier to correctly discriminate between different classes.

Figure 7.15 shows the classification maps for different DL methods along with the raw data at depth 15 cm. The raw data in Figure 7.15(a) shows that only four of the six mines exhibit a strong reflectivity while the other two mines have echoes so weak that they are not clearly visible in the raw data. Figure 7.15(b)–(d) shows the results of the SRC approaches using DL. All methods clearly detect and correctly classify the large PMN/PMA2 mines. In case of the medium-size ERA, the echoes are certainly detected as non-clutter but some of its constituent pixels are incorrectly classified as another mine. It is remarkable that the left ERA mine is recognized by our method even though it cannot be discerned visually in the raw data. Most of the false alarms in the map belong to the smallest Type-72 mines. This is expected because their small sizes produce echoes very similar to the ground clutter. On the other hand, when T-72 is the ground truth, it is correctly identified.

Using accurate ground truth information, we defined *target halos* as the boundaries of the buried landmines. The dimension of the target halos varied depending on the mine size. Let the number of pixels and the declared mine pixels inside the target halo (for a certain class) be n_t and n_m, respectively. Similarly, we denote the number

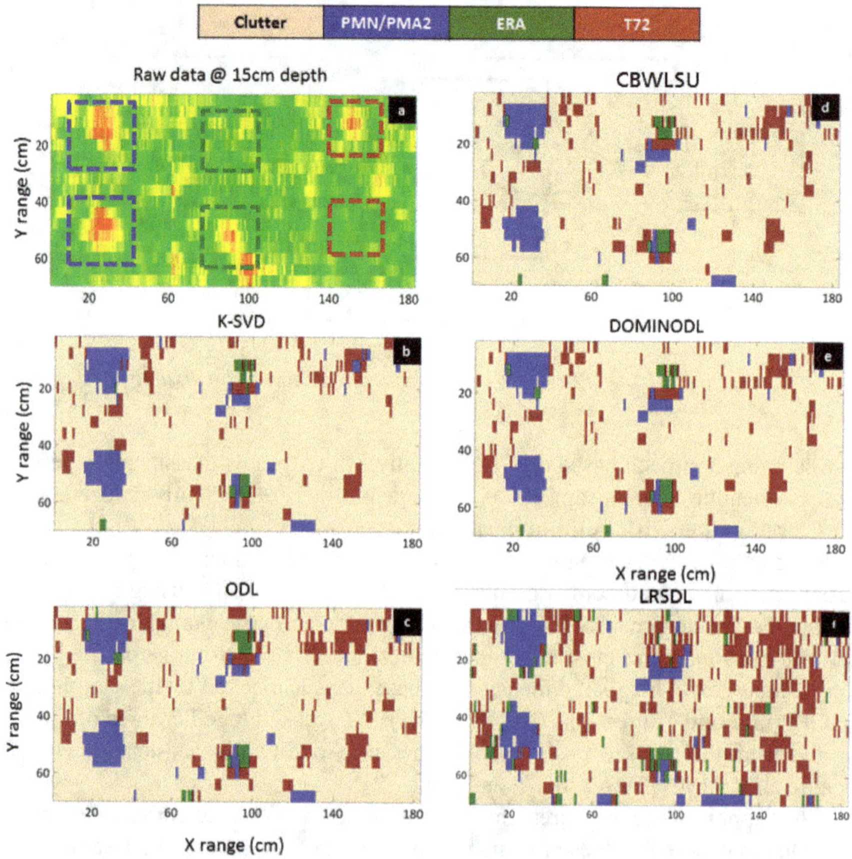

Figure 7.15 *(a) Raw data at 15 cm depth. The classification maps of the same area containing six buried landmines (enclosed by rectangles with dotted boundaries) using an SR-based approach with dictionary learned using (b) K-SVD, (c) ODL, (d) CBWLSU, (e) DOMINODL, and (f) LRSDL algorithms with optimally selected input parameters.*

of true and declared clutter pixels outside the target halo by n_c and n_d, respectively. Then, the probabilities of correct classification (P_{CC}) for each target class and clutter are, respectively,

$$P_{CC_{\text{mines}}} = \frac{n_m}{n_t} \text{ and } P_{CC_{\text{clutter}}} = \frac{n_d}{n_c}. \tag{7.62}$$

The P_{CC} being the output of a classifier should not be mistaken as the radar's probability of detection P_d, which is the result of a detector. A detector declares the presence of a mine when only a few pixels inside the halo have been declared as mine. For statistical detectors, we refer the reader to [43]. Here, we describe a more

Table 7.7 Probability of detection and false alarm for various DL methods when optimal parameters are used

	K-SVD	ODL	CDW LSU	DOMINODL	LRSDL
P_d	0.75	0.75	0.833	0.833	0.75
P_{fa}	0.124	0.128	0.129	0.108	0.565

Blue denotes the best performance among all DL algorithms.

practical approach. P_{CC} provides a fairer and more accurate evaluation of the classification result. This per-pixel information can be easily used to improve the final detection result. For instance, the operator could set a threshold for the minimum number of pixels to be detected in a cluster so that a circle with center at the cluster centroid could be used as the detected mine. However, such a circle may exclude some of the mine pixels leading to a potential field danger. The per-pixel classification is then employed to determine the guard area around the mine circle.

As an additional test, we computed the final detection result using two-third of pixels in the entire target halo to declare a successful detection. These values are based on only six mines in the field. It will be, obviously, more useful to repeat this procedure for a broader field with higher number of mines for conclusive P_d/P_{fa} results. Table 7.7 lists P_d and probability of false alarm P_{fa} for the case when optimal parameters were used for DL algorithms.

A *confusion matrix* is a quantitative representation of the classifier performance. The matrix lists the probability of classifying the ground truth as a particular class. The classes listed column-wise in the confusion matrix are the ground truths while the row-wise classes are their predicted labels. Therefore, the diagonal of the matrix is the P_{CC} while off-diagonal elements are probabilities of misclassification.

For the classification map of Figure 7.15, Table 7.8 shows the corresponding confusion matrices for each DL-based classification approach. In general, we observe an excellent classification of PMN/PMA2 landmines ($\sim$98%), implying that almost every range profile in the test set that belongs to this class is correctly labeled. The P_{cc} for the clutter is also quite high ($\sim$90%). This can also be concluded from the classification maps where the false alarms within the actual clutter regions are very sparse (i.e. they do not form a cluster) and, therefore, unlikely to be interpreted as an extended target. As noted previously, most of the clutter misclassification is associated with the Type-72 class. The ERA test targets show some difficulties with correct classification. However, most of the pixels within its target halo are declared at least as some type of mine (which is quite useful in terms of issuing safety warnings in the specific field area). This result can be explained by the fact that ERA test targets (being SIMs) do not represent a specific mine but have general characteristics common to most landmines. The Type-72 mines exhibit a P_{cc}, which is slightly higher with respect to ERA targets. This is a remarkable result

Table 7.8 Confusion matrix with optimal DL input parameter selection

		Clutter	**PMN/PMA2**	**ERA**	**Type-72**
K-SVD	Clutter	0.892	0.044	0.25	0.37
	PMN/PMA2	0.022	0.938	0.166	0.074
	ERA	0.021	0.017	0.472	0.018
	Type-72	0.064	0	0.111	0.537
LRSDL (SRC)	Clutter	0.435	0.061	0.111	0.351
	PMN/PMA2	0.155	0.289	0.319	0.259
	ERA	0.172	0.372	0.361	0.278
	Type-72	0.237	0.272	0.208	0.111
ODL	Clutter	0.871	0	0.194	0.333
	PMN/PMA2	0.022	0.973	0.139	0
	ERA	0.018	0.026	0.583	0.018
	Type-72	0.088	0	0.083	0.648
CBWLSU	Clutter	0.872	0.017	0.181	0.314
	PMN/PMA2	0.023	0.973	0.153	0
	ERA	0.025	0.008	0.528	0
	Type-72	0.08	0	0.138	0.685
DOMINODL	Clutter	0.876	0.017	0.167	0.315
	PMN/PMA2	0.023	0.974	0.138	0
	ERA	0.027	0.008	0.58	0
	Type-72	0.077	0	0.11	0.685

Gray denotes the value for a specified class and DL algorithm.

because Type-72 targets were expected to be the most challenging to classify due to their small size.

Conventionally, as mentioned in [64], LRSDL is associated with a SRC technique. However, applying this approach to our problem resulted in very low accuracy (an average of *ca.* 20% across all classes as evident from Table 7.8) and semi-random classification maps (Figure 7.15). This can be explained by the extreme similarity between the training set examples of different classes; mines and clutter are only slightly dissimilar in their responses and mine responses are generally hidden in the ground reflections. Each learned "block" D_c differed only slightly from the other and, therefore, poor classification results are achieved with this dataset. LRSDL will not be used for further evaluations.

All DL algorithms used for our sparse classification approach show very similar results for the clutter and PMN/PMA2 classes. However, ODL methods show higher P_{CC} for the ERA and Type-72 targets with respect to K-SVD. From Table 7.8, the detection enhancement using the best of the ODL algorithms for PMN/PMA2 over K-SVD is $((0.974 - 0.938) \times 100)/0.938 \approx 4\%$. The improvements for ERA and T-72 are computed similarly as 23% and 28%, respectively.

We also evaluated the performance of the conventional classification approach, i.e. SVM without DL-SR. In this case, the confusion matrix Table 7.9 shows that the clutter recognition does not deteriorate. However, the smaller mines (ERA and

Table 7.9 *Confusion matrix for SVM-based classification*

	Clutter	PMN/PMA2	ERA	Type-72
Clutter	0.960	0.184	0.625	0.685
PMN/PMA2	0.024	0.807	0.305	0.203
ERA	0.005	0.008	0.069	0.111
Type-73	0	0	0	0

T-72) are incorrectly classified and the accuracy for PMN/PMA2 is poorer than DL-based approaches. It follows that it is critical to extract features prior to classification. This step is included in our methods, which combine SR with DL to generate discriminative features for the particular case of APM classification.

Classification with non-optimal parameters

In order to demonstrate how the quality of the learned dictionary affects the final classification, we now show the confusion matrices for a non-optimal selection of input parameters in different DL algorithms. Our goal is to emphasize the importance of learning a good dictionary by selecting the optimal parameters rather than specifying how each parameter affects the final classification result. We arbitrarily selected the number of trained atoms K to be only 300 for all DL approaches, reduce the number of iterations to 25 for ODL and KSVD, and, for DOMINODL, we use $N_r = 30$, $N_b = 5$, and $N_u = 2$. Table 7.10 shows the resulting confusion matrix. While the clutter classification accuracy is almost the same as in Table 7.8, the P_{CC} for PMN/PMA2 landmines decreased by ~10% for most of the algorithms except ODL where it remains unchanged. The classification accuracy for ERA and Type-72

Table 7.10 *Confusion matrix with non-optimal DL input parameter selection*

		Clutter	PMN/PMA2	ERA	Type-72
K-SVD	Clutter	0.853	0.07	0.305	0.222
	PMN/PMA2	0.037	0.851	0.222	0.111
	ERA	0.032	0	0.194	0.241
	Type-72	0.077	0.078	0.277	0.426
ODL	Clutter	0.86	0.017	0.181	0.444
	PMN/PMA2	0.016	0.973	0.097	0
	ERA	0.022	0.008	0.638	0
	Type-72	0.1	0	0.083	0.555
CBWLSU	Clutter	0.887	0.078	0.319	0.352
	PMN/PMA2	0.019	0.877	0.097	0
	ERA	0.018	0.043	0.541	0
	Type-72	0.074	0	0.042	0.648
DOMINODL	Clutter	0.888	0.078	0.319	0.352
	PMN/PMA2	0.019	0.877	0.097	0
	ERA	0.018	0.043	0.54	0
	Type-72	0.074	0	0.042	0.648

mines is only slightly worse for ODL approaches. However, in the case of K-SVD, the P_{CC} reduces by ~30% and ~10% for ERA and Type-72, respectively. Clearly, the reconstruction and correct classification of range profiles using batch algorithms such as K-SVD is strongly affected by a non-optimal choice of DL input parameters. As discussed earlier in Section 7.6.4, this degradation is likely due to the influence of K rather than N_t.

Comparison with deep-learning classification

The core idea of SRC is largely based on the assumption that signals are linear combinations of a few atoms. In practice, this is often not the case. This has led to a few recent works that suggest employing deep learning for radar target classification. However, these techniques require significantly large datasets for training.

We compared the classification results of our methods with a deep-learning approach. In particular, we constructed a CNN because these networks are known to efficiently exploit structural or locational information in the data and yield comparable learning potential with far fewer parameters. We modeled our proposed CNN framework as a classification problem wherein each class denotes the type of mine or clutter. The training dataset for our CNN structure is the matrix $\mathbf{Y}$ (see Section 7.6.1).

Building up a synthetic database is usually an option for creating (or extending) a training set for deep-learning applications. However, accurately modeling a GPR scenario is still an ongoing challenge in the GPR community because of the difficulties in accurately reproducing the soil inhomogeneities (and variabilities), the surface and underground clutter, the antenna coupling and ringing effects, etc. Even though some applications have been promising [94], this remains a cumbersome task. The input layer of our CNN took 1-D range profiles with 211 samples. It was followed by two convolutional layers with 20 and 5 filters of size 20 and 10, respectively. The output layer consisted of four units wherein the network classifies the given input data as clutter or one of the three mines. There were rectified linear units (ReLUs) after each convolutional layer; the ReLU function is given by $\mathrm{ReLU}(x) = \max(x, 0)$.

The architecture of the CNN was selected through an arduous process of testing many combinations of layers/filters and hyperparameters, which would lead to better accuracy during training. A deeper network slightly increased the accuracy in the training phase but led to poorer performance when classifying new data (i.e. the test set $\mathbf{Y}_{test}$). Since our data are limited, adding more layers (i.e. more weights) only led to overfitting and made the network incapable to generalize on new datasets. A multi-dimensional CNN formed by clustering 2-D and 3-D data would have further reduced the training set. Augmenting the data was also envisioned but commonly used transformations such as scaling/rotations are not useful in our case because the mines were always in the same inclination and their dimension defines the class itself. We also attempted adding different levels of noise but this did not lead to better results considering the available data are already very noisy.

We trained the network with the labeled training set $\mathbf{Y}$, selecting ~20% of the training data for validation. Specifically, the validation set employed 100, 25, 25,

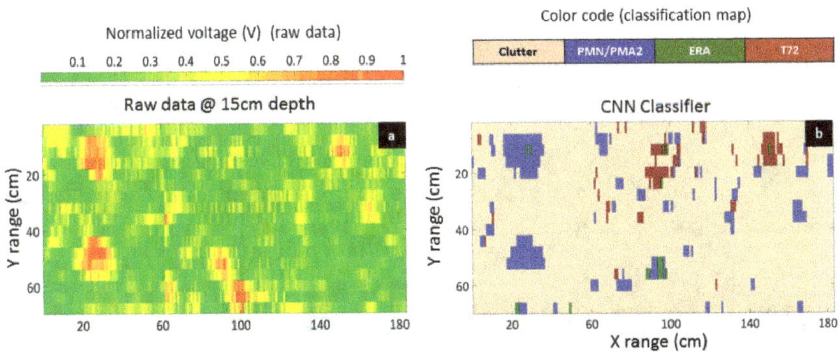

Figure 7.16 *(a) Raw data at 15 cm depth. (b) Classification maps of the same area containing six buried landmines using CNN-based classification.*

and 25 range profiles for clutter, PMN/PMA2, ERA, and Type-72, respectively. We used a stochastic gradient descent algorithm for updating the network parameters with the learning rate of 0.001 and mini-batch size of 20 samples for 2 000 epochs. We realized the proposed network in TensorFlow on a Windows 7 PC with 8-core CPU. The network training took 3.88 minutes. Figure 7.16 shows the classification map obtained using CNN. The corresponding confusion matrix is listed in Table 7.11. We note that the CNN classifier shows worse P_{CC} than our SR-based techniques, particularly for ERA and Type-72 target classes.

Classification with reduced range samples

We now analyze the robustness of our DL-based adaptive classification method to the reduction of the number of samples in the raw data. Assuming the collected data $\mathbf{Y}_{\text{TEST}}$ is sparse in dictionary $\mathbf{D}$, we undersampled the original raw data $\mathbf{Y}_{\text{TEST}}$ in range to obtain its row-undersampled version $\tilde{\mathbf{Y}}_{\text{TEST}}$ by randomly reducing the samples. We then applied the same random sampling pattern to the dictionary $\mathbf{D}$ for obtaining the sparse coefficients. We also analyzed the CNN classifier when the signals are randomly reduced in the same way. Figure 7.17 illustrates the classification map for all DL approaches when the sampling is reduced by 50%. Table 7.12 shows the corresponding confusion matrix.

Table 7.11 *Confusion matrix for CNN-based classification*

	Clutter	PMN/PMA2	ERA	Type-72
Clutter	0.909	0.14	0.38	0.574
PMN/PMA2	0.036	0.807	0.181	0
ERA	0.022	0.053	0.319	0.056
Type-72	0.033	0	0.111	0.370

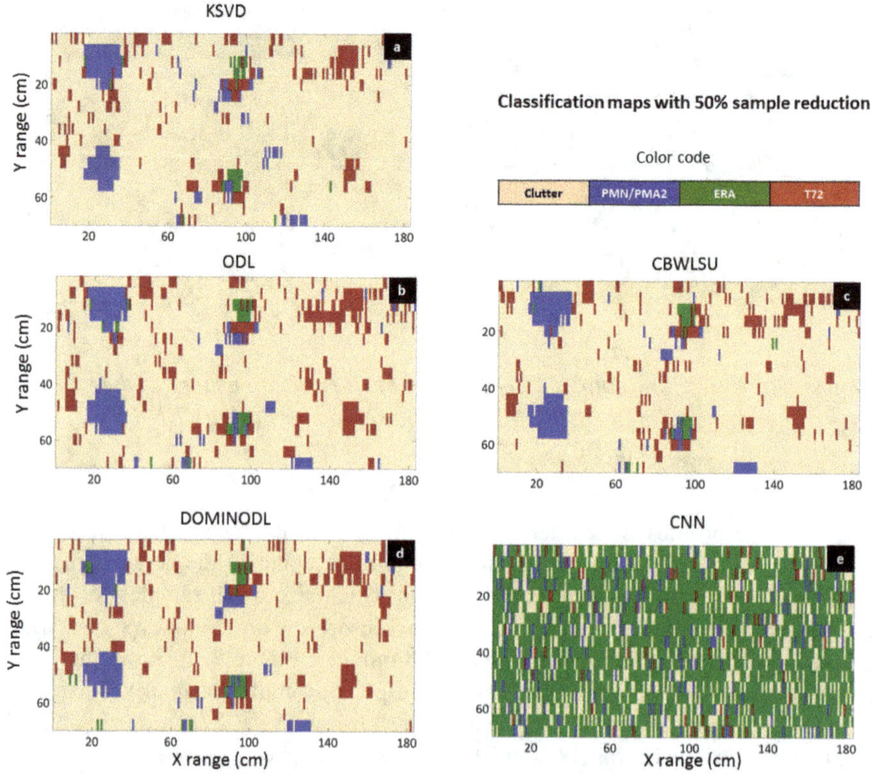

Figure 7.17 The classification maps of the same area containing six buried landmines using an SR-based approach with dictionary learned using (a) K-SVD, (b) ODL, (c) CBWLSU, and (d) DOMINODL algorithms. The input parameters were optimally selected and the number of samples was reduced by 50%. (e) The corresponding result with reduced samples for CNN-based classification.

In comparison to the results in Table 7.8, which used all samples of the raw data, the DL approaches maintain similar classifier performance even when we reduce the samples by 50% (i.e. just 105 samples in total). In contrast, the CNN classifier fails completely for 50% sampling rate.

Reducing the number of signal samples when using a dictionary that minimizes the number of non-zero entries in the SR still assures an exact reconstruction of the signal itself and, consequently, its correct classification. The features for classifying the traces are thus robust to the reduction of the original samples. Deep-learning strategies use the signal samples directly as classification features. They also require enormous amount of data for training. Therefore, the degradation in their performance is expected.

Table 7.12 *Confusion matrices for different DL algorithms and CNN with 50% samples reduction*

		Clutter	PMN/PMA2	ERA	Type-72
K-SVD	Clutter	0.882	0.026	0.291	0.37
	PMN/PMA2	0.018	0.947	0.153	0.037
	ERA	0.021	0.026	0.5	0
	Type-72	0.078	0	0.055	0.592
ODL	Clutter	0.868	0	0.208	0.333
	PMN/PMA2	0.021	0.965	0.18	0.018
	ERA	0.018	0.035	0.5	0
	Type-72	0.09	0	0.111	0.648
CBWLSU	Clutter	0.872	0.017	0.25	0.40
	PMN/PMA2	0.023	0.973	0.111	0
	ERA	0.02	0.008	0.541	0
	Type-72	0.083	0	0.097	0.592
DOMINODL	Clutter	0.868	0.035	0.194	0.296
	PMN/PMA2	0.023	0.929	0.138	0
	ERA	0.024	0.035	0.527	0.018
	Type-72	0.083	0	0.138	0.685
CNN	Clutter	0.265	0.166	0.181	0.148
	PMN/PMA2	0.062	0.096	0.069	0.018
	ERA	0.645	0.728	0.736	0.75
	Type-72	0.027	0.088	0.014	0.074

7.8 Summary

In this chapter, we reviewed modern methods for GPR target recognition that includes SR, data learning methods, adaptive processing, and advanced classifiers. Most applications combine these techniques for specific applications. In particular, we presented effective ODL strategies for sparse decomposition of GPR traces. The online methods outperform K-SVD thereby making them a good candidate for SRC. Our algorithm DOMINODL is always the fastest providing near real-time performance and high clutter rejection while also maintaining a classifier performance that is comparable to other ODL algorithms. DOMINODL and CBWLSU generally classify smaller targets better than ODL and K-SVD. Unlike previous works that rely on RMSE, we used metrics based on statistical inference to tune the DL parameters for enhanced operation.

Most importantly, we illustrated application of each technique using real data from GPR for landmine detection. Our analyses show that adoption of SR-based DL classification is not only feasible in real applications but also fast enough to be implemented in real-time. We believe that such practical implementations pave the way towards the next step of cognition in modern GPR operation, wherein the system uses previous measurements to optimize the processing performance and is capable of sequential sampling adaptation based on the learned dictionary.

References

[1] Daniels D.J. *Ground Penetrating Radar*. IET; 2004.

[2] Persico R. *Introduction to Ground Penetrating Radar: Inverse Scattering and Data Processing*. John Wiley & Sons; 2014.

[3] Jol H.M. *Ground penetrating radar theory and applications. Elsevier Science*; 2009.

[4] Tait P. *Introduction to Radar Target Recognition. Vol. 18 of Radar, Sonar and Navigation*. IET; 2005.

[5] González-Huici M.A., Catapano I., Soldovieri F. 'A comparative study of GPR reconstruction approaches for landmine detection'. *IEEE Journal of Selected Topics in Applied Earth Observations and Remote Sensing*. 2014, vol. 7(12), pp. 4869–78.

[6] Giovanneschi F., Mishra K.V., Gonzalez-Huici M.A., Eldar Y.C., Ender J.H.G. 'Dictionary learning for adaptive GPR landmine classification'. *IEEE Transactions on Geoscience and Remote Sensing*. 2019, vol. 57(12), pp. 10036–55.

[7] Shao W., Bouzerdoum A., Phung S.L. 'Sparse representation of GPR traces with application to signal classification'. *IEEE Transactions on Geoscience and Remote Sensing*. 2013, vol. 51(7), pp. 3922–30.

[8] Mishra K.V., Eldar Y.C., Maio Yc A.D., Maio A.D., *et al*. 'Sub-Nyquist radar: principles and prototypes' in Maio A.D., Eldar Y.C., Haimovich Y.C. (eds.). *Compressed sensing in radar signal processing*. Cambridge University Press; 2019. pp. 1–48.

[9] Krueger K.R., McClellan J.H., Scott Jr W.R. 'Compressive sensing for radar imaging of underground targets.' in Amin M. (ed.). *Compressive Sensing for Urban Radar*. 87. CRC Press; 2014.

[10] Giovanneschi F., González-Huici M.A., Uschkerat U. 'A parametric analysis of time and frequency domain GPR scattering signatures from buried landmine-like targets'. *SPIE Defense, Security, and Sensing*; 2013.

[11] Tivive F.H.C., Bouzerdoum A., Abeynayake C. 'GPR signal classification with low-rank and convolutional sparse coding representation'. IEEE Radar Conference; 2017. pp. 1352–6.

[12] González-Huici M.A. Accurate ground penetrating radar numerical modeling for automatic detection and recognition of antipersonnel landmines [PhD dissertation]. Universitäts-und Landesbibliothek Bonn; 2013.

[13] Balanis C.A. *Antenna Theory Analysis and Design*. John Wiley & Sons; 2005.

[14] Knight R.J., Endres A.L. 'An introduction to rock physics principles for near-surface geophysics' in Dk B. (ed.). *Near-Surface Geophysics*; 2005. pp. 31–70.

[15] Lambot S., Antoine M., Vanclooster M., Slob E.C. 'Effect of soil roughness on the inversion of off-ground monostatic GPR signal for noninvasive quantification of soil properties'. *Water Resources Research*. 2006, vol. 42(3).

[16] Lestari A.A., Yarovoy A.G., Ligthart L.P. 'Adaptive wire bow-tie antenna for GPR applications'. *IEEE Transactions on Antennas and Propagation*. 2005, vol. 53(5), pp. 1745–54.

[17] Pasculli D., Manacorda G. 'Real-Time, pseudo real-time and stroboscopic sampling in time-domain GPRs'. IEEE International workshop on advanced ground penetrating radar; 2015. pp. 1–4.

[18] Kim J.H., Cho S.J., MJ Y. 'Removal of ringing noise in GPR data by signal processing'. *Geosciences Journal*. 2007, vol. 11(1), pp. 75–81.

[19] Bitri A., Grandjean G. 'Frequency-wavenumber modelling and migration of 2D GPR data in moderately heterogeneous dispersive media'. *Geophysical Prospecting*. 2008, vol. 46(3), pp. 287–301.

[20] Wall M.E., Rechtsteiner A., Rocha L.M., Berrar D.P. 'Singular value decomposition and principal component analysis' in Berrar D.P., Dubitzky W., Granzow M. (eds.). *A Practical Approach to Microarray Data Analysis*. Springer; 2003. pp. 91–109.

[21] Sakaguchi R.T., Morton Jr K.D., Collins L.M. 'Recognizing subsurface target responses in ground penetrating radar data using convolutional neural networks'. *SPIE Detection and Sensing of Mines, Explosive Objects, and Obscured Targets XX*. 9454; 2015. p. 94541A.

[22] Besaw L.E., Stimac P.J. 'Deep convolutional neural networks for classifying GPR B-scans'. *SPIE Detection and Sensing of Mines, Explosive Objects, and Obscured Targets XX*. 2015, vol. 9454, p. 945413.

[23] Bralich J., Reichman D., Collins L.M. 'Improving convolutional neural networks for buried target detection in ground penetrating radar using transfer learning via pretraining'. *SPIE Detection and Sensing of Mines, Explosive Objects, and Obscured Targets XXII*. 10182; 2017. p. 101820X.

[24] Lameri S., Lombardi F., Bestagini P. 'Landmine detection from GPR data using convolutional neural networks'. IEEE European Signal Processing Conference; 2017. pp. 508–12.

[25] Reichman D., Collins L.M., Malof J.M. 'Some good practices for applying convolutional neural networks to buried threat detection in ground penetrating radar'. IEEE International Workshop on Advanced Ground Penetrating Radar; 2017. pp. 1–5.

[26] Mishra K.V., Shankar M.B., Ottersten B. 'Toward metacognitive Radars: concept and applications'. IEEE International Radar Conference; 2020. pp. 77–82.

[27] Gurbuz S.Z., Griffiths H.D., Charlish A., Rangaswamy M., Greco M.S., Bell K. 'An overview of cognitive radar: past, present, and future'. *IEEE Aerospace and Electronic Systems Magazine*. 2019, vol. 34(12), pp. 6–18.

[28] Ender J., Brüggenwirth S. 'Cognitive radar-enabling techniques for next generation radar systems'. IEEE International Radar Symposium; 2015. pp. 3–12.

[29] Bruggenwirth S. 'Design and implementation of a three-layer cognitive radar architecture'. Asilomar Conference on Signals, Systems and Computers; 2016. pp. 929–33.

[30] Cornick M., Koechling J., Stanley B., Zhang B. 'Localizing ground penetrating radar: a step toward robust autonomous ground vehicle localization'. *Journal of Field Robotics*. 2016, vol. 33(1), pp. 82–102.

[31] Williams R.M., Ray L.E., Lever J.H., Burzynski A.M. 'Crevasse detection in ice sheets using ground penetrating radar and machine learning'. *IEEE Journal of Selected Topics in Applied Earth Observations and Remote Sensing*. 2014, vol. 7(12), pp. 4836–48.

[32] Mishra S.K., Puthal D., Rodrigues J.J.P.C., Sahoo B., Dutkiewicz E. 'Sustainable service allocation using a metaheuristic technique in a FOG server for industrial applications'. *IEEE Transactions on Industrial Informatics*. 2018, vol. 14(10), pp. 4497–506.

[33] Omwenga M.M., Wu D., Liang Y. 'Autonomous cognitive GPR based on edge computing and reinforcement learning'. IEEE International Conference on Industrial Internet; 2019. pp. 348–54.

[34] Mishra K.V., Bhavani Shankar M.R., Koivunen V., Ottersten B., Vorobyov S.A. 'Toward millimeter-wave joint radar communications: a signal processing perspective'. *IEEE Signal Processing Magazine*. 2019, vol. 36(5), pp. 100–14.

[35] GSSI Road-Scan30 Brochure. Geophysical Survey Systems, Inc. 2020. Available from https://www.geophysical.com/wp-content/uploads/2018/01/GSSI-RoadScan30Brochure.pdf.

[36] Srivastav A., Nguyen P., McConnell M., Loparo K.A., Mandal S. 'A highly digital multiantenna ground-penetrating radar (GPR) system'. *IEEE Transactions on Instrumentation and Measurement*. 2020, vol. 69(10), pp. 7422–36.

[37] Mishra K.V., Mulleti S., Eldar Y.C. 'RaSSteR: random sparse step-frequency radar. arXiv preprint arXiv'. 2020, vol. 200405720.

[38] Massart P. 'The tight constant in the Dvoretzky-Kiefer-Wolfowitz inequality'. *The Annals of Probability*. 1990, vol. 18(3), pp. 1269–83.

[39] Giannakis I., Giannopoulos A., Yarovoy A. 'Model-based evaluation of signal-to-clutter ratio for landmine detection using ground-penetrating radar'. *IEEE Transactions on Geoscience and Remote Sensing*. 2016, vol. 54(6), pp. 3564–73.

[40] Giannopoulos A. 'Modelling ground penetrating radar by GprMax'. *Construction and Building Materials*. 2005, vol. 19(10), pp. 755–62.

[41] Frigui H., Gader P. 'Detection and discrimination of land mines in ground-penetrating radar based on edge histogram descriptors and a possibilistic *k*-nearest neighbor classifier'. *IEEE Transactions on Fuzzy Systems*. 2009, vol. 17(1), pp. 185–99.

[42] Giovanneschi F., González-Huici M.A. 'A preliminary analysis of a sparse reconstruction based classification method applied to GPR data'. IEEE International Workshop on Advanced Ground Penetrating Radar; 2015. pp. 1–4.

[43] Pambudi A.D., Fauß M., Ahmad F. Forward looking GPR-Based Landmine detection using a robust likelihood ratio test. In: European Signal Processing Conference; 2019. pp. 1–5.

[44] Papyan V., Elad M. 'Multi-scale patch-based image restoration'. *IEEE Transactions on Image Processing*. 2015, vol. 25(1), pp. 249–61.

[45] Rey-Otero I., Sulam J., Elad M. 'Variations on the convolutional sparse coding model'. *IEEE Transactions on Signal Processing*. 2020, vol. 68, pp. 519–28.

[46] Garcia-Cardona C., Wohlberg B. 'Convolutional dictionary learning: a comparative review and new algorithms'. *IEEE Transactions on Computational Imaging*. 2018, vol. 4(3), pp. 366–81.

[47] Boyd S., Parikh N., Chu E. 'Distributed optimization and statistical learning via the alternating direction method of multipliers'. *Foundations and Trends® in Machine learning*. 2011, vol. 3(1), pp. 1–122.

[48] Yarovoy A.G., van Genderen P., Ligthart L.P. 'Ultra-wideband ground penetrating impulse radar' in Smith P.D., Cloude S.R. (eds.). *Ultra-Wideband, Short-Pulse Electromagnetics 5*. Springer; 2002. pp. 183–9.

[49] González-Huici M.A., Giovanneschi F. 'A combined strategy for landmine detection and identification using synthetic GPR responses'. *Journal of Applied Geophysics*. 2013, vol. 99, pp. 154–65.

[50] Donoho D.L., Compressed sensing. 'IEEE transactions on information theory'. 2006, vol. 52(4), pp. 1289–306.

[51] Tropp J.A., Gilbert A.C. 'Signal recovery from random measurements via orthogonal matching pursuit'. *IEEE Transactions on Information Theory*. 2007, vol. 53(12), pp. 4655–66.

[52] Van Den Berg E., Friedlander M.P. 'Probing the Pareto frontier for basis pursuit solutions'. *SIAM Journal on Scientific Computing*. 2008, vol. 31(2), pp. 890–912.

[53] Tibshirani R. 'Regression shrinkage and selection via the LASSO'. *Journal of the Royal Statistical Society: Series B*. 1996, vol. 58(1), pp. 267–88.

[54] Efron B., Hastie T., Johnstone I., Tibshirani R. 'Least angle regression'. *The Annals of Statistics*. 2004, vol. 32(2), pp. 407–99.

[55] Candès E.J., Romberg J.K., Tao T. 'Stable signal recovery from incomplete and inaccurate measurements'. *Communications on Pure and Applied Mathematics*. 2006, vol. 59(8), pp. 1207–23.

[56] Rubinstein R., Zibulevsky M., Elad M. 'Efficient implementation of the K-SVD algorithm using batch orthogonal matching pursuit'. *Technion - Israel Institute of Technology*. 2008.

[57] Cheney W., Kincaid D. *Linear Algebra: Theory and Applications*. Jones & Bartlett Learning; 2009.

[58] Giovanneschi F., Mishra K.V., Gonzalez-Huici M.A. 'Online Dictionary Learning Aided Target Recognition In Cognitive GPR'. IEEE International Geoscience and Remote Sensing Symposium; 2017–4813–16.

[59] Mairal J., Bach F., Ponce J. 'Online dictionary learning for sparse coding'. International Conference on Machine Learning; 2009. pp. 689–96.

[60] Naderahmadian Y., Beheshti S., Tinati M.A. 'Correlation based online dictionary learning algorithm'. *IEEE Transactions on Signal Processing*. 2016, vol. 64(3), pp. 592–602.

[61] Sparse E.M., Redundant Representations. *From Theory to Applications in Signal and Image Processing*. Springer; 2010.
[62] Elad M., Aharon M. 'Image denoising via sparse and redundant representations over learned dictionaries'. *IEEE Transactions on Image processing.* 2006, vol. 15(12), pp. 3736–45.
[63] Aharon M., Elad M., Bruckstein A. 'K-SVD: an algorithm for designing overcomplete dictionaries for sparse representation'. *IEEE Transactions on Signal Processing*. 2006, vol. 54(11), pp. 4311–22.
[64] TH V., Monga V. 'Fast low-rank shared dictionary learning for image classification'. *IEEE Transactions on Image Processing*. 2017, vol. 26(11), pp. 5160–75.
[65] Giovanneschi F. 'Online dictionary learning for classification of antipersonnel landmines using ground penetrating radar'. *[phd dissertation] Universitäts Siegen*. 2020.
[66] Engan K., Aase S.O., Husoy J.H. 'Method of optimal directions for frame design'. IEEE International Conference on Acoustics, Speech and Signal Processing, 5; 1999. pp. 2443–6.
[67] Varshney K.R., Çetin M., Fisher J.W., *et al.* 'Sparse representation in structured dictionaries with application to synthetic aperture radar'. *IEEE Transactions on Signal Processing*. 2008, vol. 56(8), pp. 3548–61.
[68] Suo Y., Dao M., Srinivas U., Monga V., Tran T.D. 'Structured dictionary learning for classification'. *arXiv preprint*. 2014.
[69] Yang M., Zhang L., Feng X., Zhang D. 'Sparse representation based Fisher discrimination dictionary learning for image classification'. *International Journal of Computer Vision*. 2014, vol. 109(3), pp. 209–32.
[70] Li L., Li S., Fu Y. 'Learning low-rank and discriminative dictionary for image classification'. *Image and Vision Computing*. 2014, vol. 32(10), pp. 814–23.
[71] Zhang Q., Li B. 'Discriminative K-SVD for dictionary learning in face recognition'. IEEE Conference on Computer Vision and Pattern Recognition; 2010. pp. 2691–8.
[72] Jiang Z., Lin Z., Davis L.S. 'Label consistent K-SVD: learning a discriminative dictionary for recognition'. *IEEE Transactions on Pattern Analysis and Machine Intelligence*. 2013, vol. 35(11), pp. 2651–64.
[73] Ramirez I., Sprechmann P., Sapiro G. 'Classification and clustering via dictionary learning with structured incoherence and shared features'. IEEE Conference on Computer Vision and Pattern Recognition; 2010. pp. 3501–8.
[74] Kong S., Wang D. 'A dictionary learning approach for classification: separating the particularity and the commonality'. European Conference on Computer Vision; 2012. pp. 186–99.
[75] Gao S., Tsang I.W.H., Ma Y. 'Learning category-specific dictionary and shared dictionary for fine-grained image categorization'. *IEEE Transactions on Image Processing*. 2014, vol. 23(2), pp. 623–34.
[76] Yang M., Zhang L., Feng X. 'Fisher discrimination dictionary learning for sparse representation'. International Conference on Computer Vision; 2011. pp. 543–50.

[77] Beck A., Teboulle M. 'A fast iterative shrinkage-thresholding algorithm for linear inverse problems'. *SIAM Journal on Imaging Sciences*. 2009, vol. 2(1), pp. 183–202.

[78] Agarwal A., Anandkumar A., Jain P. 'Learning sparsely used overcomplete dictionaries'. Conference on Learning Theory; 2014. pp. 123–37.

[79] Wright J., Ma Y., Mairal J., Sapiro G., Huang T.S., Yan S. 'Sparse representation for computer vision and pattern recognition'. *Proceedings of the IEEE*. 2010, vol. 98(6), pp. 1031–44.

[80] Cetin M., Stojanovic I., Onhon O., *et al.* 'Sparsity-driven synthetic aperture radar imaging: reconstruction, autofocusing, moving targets, and compressed sensing'. *IEEE Signal Processing Magazine*. 2014, vol. 31(4), pp. 27–40.

[81] Ender J.H. 'On compressive sensing applied to radar'. *Signal Processing*. 2010, vol. 90(5), pp. 1402–14.

[82] Hongxing H., Bioucas-Dias J.M., Katkovnik V. 'Interferometric phase image estimation via sparse coding in the complex domain'. *IEEE Transactions on Geoscience and Remote Sensing*. 2015, vol. 53(5), pp. 2587–602.

[83] Gurbuz A.C., McClellan J.H., Scott W.R. 'Compressive sensing of underground structures using GPR'. *Digital Signal Processing*. 2012, vol. 22(1), pp. 66–73.

[84] Gurbuz A.C., Teke O., Arikan O. 'Sparse ground-penetrating radar imaging method for off-the-grid target problem'. *Journal of electronic imaging*. 2013, vol. 22(2), p. 021007.

[85] Suksmono A.B., Bharata E., Lestari A.A., Yarovoy A.G., Ligthart L.P. 'Compressive Stepped-Frequency continuous-wave ground-penetrating radar'. *IEEE Geoscience and Remote Sensing Letters*. 2010, vol. 7(4), pp. 665–9.

[86] Soldovieri F., Solimene R., Lo Monte L., Bavusi M., Loperte A. 'Sparse reconstruction from GPR data with applications to rebar detection'. *IEEE Transactions on Instrumentation and Measurement*. 2010, vol. 60(3), pp. 1070–9.

[87] Skolnik M.I. *Radar Handbook. 3rd.* McGraw-Hill; 2008. p. 021007.

[88] Mishra K.V., Kruger A., Krajewski W.F. 'Compressed sensing applied to weather radar'. IEEE International Geoscience and Remote Sensing Symposium; 2014. pp. 1832–5.

[89] Akhtar J., Torvik B., Olsen K.E. 'Compressed sensing with interleaving slow-time pulses and hybrid sparse image reconstruction'. IEEE Radar Conference; 2017. pp. 0006–0010.

[90] Chang C.C., Lin C.J. 'LIBSVM: a library for support vector machines'. *ACM Transactions on Intelligent Systems and Technology*. 2011, vol. 2(3), p. 27.

[91] Chen J., Jiao L., Ma W., Liu H. 'Unsupervised high-level feature extraction of SAR imagery with structured sparsity priors and incremental dictionary learning'. *IEEE Geoscience and Remote Sensing Letters*. 2016, vol. 13(10), pp. 1467–71.

[92] Glivenko V. 'Sulla determinazione empirica DELLA Legge di probabilit Giornale Dell'Istituto Italiano Degli Attuari'. *In Italian. Anno IV*. 1993, pp. 92–9.

[93] Lillacci G., Khammash M. 'Model selection in stochastic chemical reaction networks using flow cytometry data'. IEEE Conference on Decision and Control and European Control Conference; 2011. pp. 1680–5.

[94] Giannakis I., Giannopoulos A., Warren C. 'A realistic FDTD numerical modeling framework of ground penetrating radar for landmine detection'. *IEEE Journal of Selected Topics in Applied Earth Observations and Remote Sensing*. 2016, vol. 9(1), pp. 37–51.

Chapter 8

Jet engine recognition via sparse decomposition of ISAR images

Simon Wagner[1] and Joachim Ender[1]

Recognition and identification of targets is an important part of radar systems. Due to the high resolution, imaging radars are a well suitable choice for this task. This chapter presents a framework, which is based on sparse decomposition of radar images, to identify different kinds of scattering mechanisms. The aim of this framework is the identification of specific parts of certain targets. The basic procedure of the framework is shown exemplarily by a simulation using different canonical shapes, i.e. points, squares, circles, crosses and lines, which will be separated by the framework. As a practical application, the recognition of jet engines in radar images is presented. For this application, a scattering model based on waveguides is introduced and an example using data from the Tracking and Imaging Radar (TIRA) of Fraunhofer FHR is shown.

8.1 Introduction

The use of surveillance systems, like air traffic control or driver assistance to avoid collisions on the street, has become a part of everyday life. In all of these systems, it is necessary to know the number and types of other traffic participants to decide how to act next. In tracking systems, classifiers can be used to support the tracker in ambiguous situations [1] and by that enhance the safety of transport and navigation systems [2].

In modern air traffic, the use of transponders has become standard, and in many airspaces mandatory, as an identification method. These transponders are used to communicate with the so-called secondary radar systems and are therefore communication rather than radar systems. First, the airplane is detected by the primary radar, the actual radar system, and after that, the secondary radar identifies the target. These transponders are a sort of *Identification Friend or Foe* (IFF) system, which are used in civil and military airplanes [3]. Unfortunately, there

[1]Fraunhofer Institute for High Frequency Physics and Radar Techniques, Germany

(a) Target shape

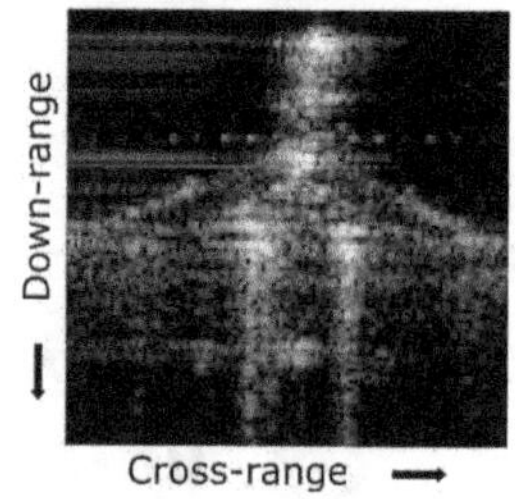

(b) ISAR image

Figure 8.1 Example of a real ISAR image

are some examples in history where the IFF system was not working properly and caused disastrous accidents. One example is the shoot-down of Iran Air 655 in 1988 [4], where according to the official statement, the Airbus was mistaken as an F-14 fighter aircraft due to a wrong identification [2]. A more recent example of an accidental downing is the Ukraine International Airlines flight PS752 from Tehran, Iran, to Ukraine in January 2020, which has been mistaken as a cruise missile [5].

These examples show that the IFF system is not completely secure and that further methods, which do not rely on communication between the radar and the target, are necessary. These methods are so-called *noncooperative methods* and the method presented in this chapter belongs to this class of methods. The algorithm itself is not aiming at the identification of specific targets but at the identification of specific scattering mechanisms, i.e. the air intake of jet engines. The algorithm exploits an effect that appears in inverse synthetic aperture radar (ISAR) images. An example is depicted in Figure 8.1.

Figure 8.1(b) shows a real ISAR image of an airplane and the corresponding contour is shown in Figure 8.1(a). Given the image and the contour, it becomes visible that behind the engine a trail of scattering centers appears that is not a real part of the target. These scattering centers are called *range extended returns* and appear behind cavities in general. The aim of this chapter is to find a model for these kinds of scatterers and an algorithm that is able to recognize this scattering mechanism in an ISAR image. The methods and results of this chapter are based on preliminary results, which have been presented in [6] and [7].

The results of this kind of algorithm can be used to gather information about the structure of targets and can thus support the final target identification. A short overview of available methods for scattering identification is given in the next section. The waveguide model and the sparse decomposition framework that are used in this chapter are introduced in Sections 8.3 and 8.4, respectively. Results with simulated and real data are shown in Section 8.5 and finally Section 8.6 concludes the chapter.

8.2 State of the art in scattering identification

This chapter focuses on scattering identification in high-resolution radar images, therefore, a review on this topic is given here. Yet the use of scattering properties is not limited to high-resolution imaging, also resonance effects that appear in the lower frequency bands can be used for identification. An example is given in [8], which itself includes a short review on this field. The general idea is to use a radar with a wavelength in the region of the size of the air-intake of an aircraft, since this will cause resonances that can be detected in the echo signal. These resonances can be sufficient to identify targets from a list of potential targets using a matching process between the measured data and a database.

In the field of imaging radar, the oldest and thus most mature technique to identify scattering mechanisms is the use of polarimetric data with an appropriate decomposition [9]. To use these decomposition theorems, fully polarimetric measurements are necessary, i.e. two coherent transmitters and two receivers with vertical and horizontal polarization of the antenna. With this constellation, a 2×2 scattering matrix consisting of the echos in co-polarized and cross-polarized combinations can be defined for each pixel in the radar image. For this scattering matrix, several coherent decomposition theorems exist that allow the identification of single, double and multiple bounce scattering centers. An example with two different decompositions is given in [10], which presents the decomposition of measurements of a bridge during construction and after completion. In the given example, the difference between the cables, the deck and the supporting structures is clearly visible and shows the potential of fully polarimetric measurements.

Nevertheless, the use of fully polarimetric systems is still not very widespread due to the high costs and the demanding calibration needed for these systems. Therefore, most research related to scattering identification over the last years was focused on single-channel radar systems.

One of the first works pointing in this direction is [11] that introduced a parametric signal model for *attributed scattering centers*. These scattering centers are namely corners, edges, flat plates and curved surfaces and the response of them is derived by the geometric theory of diffraction. Additionally to these models also a feature extraction method was introduced and some examples with SAR images are shown to validate the derived models. The application of these parametric scattering center models for target recognition can be found, e.g. in [12], which uses a SAR dataset consisting of ten targets to perform classification. For the presented algorithm, a target is modeled as a collection of attributed scattering centers, whose position and type are extracted from the measurements and afterward matched with the predicted features based on a target hypothesis. The application of attributed scattering centers can also be found in more recent publications like [13] or [14]. A three-dimensional extension of these models, also for bistatic observations, was published in [15], which also gives a maximum likelihood estimator to identify these scattering centers in 3-D radar images.

An application that is using sparse decomposition of radar measurements to identify different scattering targets is presented in [16]. In this publication, a through-the-wall radar is used to perform stand-off measurements of a house and a sparse decomposition algorithm is used to identify corners and planar areas of the walls inside the house. The model of these scattering centers does not only rely on the magnitude, rather the phase is exploited as well. A planar wall should produce a linear phase, while a corner is modeled with a quadratic phase term.

Another example that uses sparse representation of radar images is [17], which presents a framework that is able to handle several different scattering models in the complex measurement domain. This work is, although it focuses on the improvement of the imaging result and is limited to wavelets and point scattering centers, together with [16] one of the closest to the one presented here.

8.3 Waveguide scattering model

In this section, the signal model that is used for jet engine detection is derived. Since this model is physically motivated, the fundamentals of the propagation of electromagnetic waves in waveguides are shortly repeated in Section 8.3.1 and visualized by an example in Section 8.3.2. The actual model that will be used later is shown in Section 8.3.3 and an example for a scattered LFM waveform is given in Section 8.3.4.

8.3.1 Propagation of electromagnetic waves in waveguides

The propagation of electromagnetic waves in waveguides is an application of the Maxwell equations with certain boundary conditions. Since jet engines are filled with air, the simplified *Maxwell equations* for the source-free case

$$\nabla \times \mathbf{E} = -\frac{\partial \mathbf{B}}{\partial t} \tag{8.1}$$

$$\nabla \times \mathbf{H} = \frac{\partial \mathbf{D}}{\partial t} \tag{8.2}$$

$$\nabla \cdot \mathbf{D} = 0 \tag{8.3}$$

$$\nabla \cdot \mathbf{B} = 0 \tag{8.4}$$

can be used for further derivations. The quantities **E** and **H** are the electric and magnetic *fields*, while **D** and **B** describe the electric and magnetic *flux densities*.

The electric and magnetic flux densities **D** and **B** are related to the electric and magnetic field by the *permittivity* ϵ and *permeability* μ of the medium via

$$\mathbf{D} = \epsilon \mathbf{E} \tag{8.5}$$

$$\mathbf{B} = \mu \mathbf{H}. \tag{8.6}$$

Since only propagation in the air is considered in this chapter, the medium parameters can be set to the free space parameters, i.e. $\mu = \mu_0$ and $\epsilon = \epsilon_0$. In this case, the speed of propagation in unbounded free space is given as the speed of light

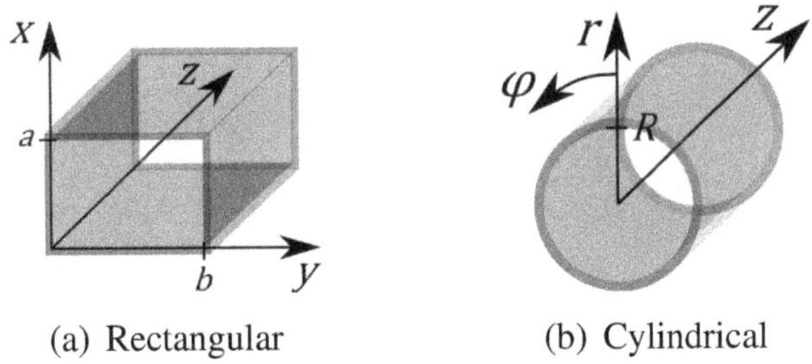

Figure 8.2 *Waveguide shapes and the corresponding coordinate systems*

$$c = \frac{1}{\sqrt{\epsilon_0 \mu_0}} \tag{8.7}$$

and the general forms of the free space propagating wave for **E** and **H** in z-direction are given as

$$\mathbf{E}(z,t) = \mathbf{E}(0,0)e^{j(\omega t - k_z z)} \qquad \mathbf{H}(z,t) = \mathbf{H}(0,0)e^{j(\omega t - k_z z)}. \tag{8.8}$$

The expressions in (8.8) show that the values of **E** and **H** depend in free space only on the direction of propagation, which is z throughout this section on waveguides, i.e. $\frac{\partial}{\partial x}, \frac{\partial}{\partial y}\mathbf{E} = 0$ and $\frac{\partial}{\partial x}, \frac{\partial}{\partial y}\mathbf{H} = 0$. The variable ω denotes the angular frequency and k_z the z-component of the wavenumber, which is defined by

$$k_z = \frac{2\pi}{\lambda} = \frac{\omega}{c} \tag{8.9}$$

using λ as *wavelength.*

With these formulations of the free space conditions, a derivation for the propagation inside waveguides is possible. This derivation is only roughly sketched here and the results are given. Details can be found in textbooks on electromagnetic theory like [18]. Results will be shown for rectangular and circular waveguides, which are depicted in Figure 8.2 with the corresponding coordinate system.

The main assumptions of the waveguides are:

1. The waveguide is filled with air, i.e. $\mu = \mu_0$ and $\epsilon = \epsilon_0$.
2. Confining walls are perfect electric conductor (PEC) materials, i.e. electric field lines are perpendicular and magnetic field lines are parallel to the conducting surface.
3. **E**and **H** do not only depend on the direction of propagation, which remains the z-axis, i.e. $\frac{\partial}{\partial x}, \frac{\partial}{\partial y}\mathbf{E} \neq 0$ and $\frac{\partial}{\partial x}, \frac{\partial}{\partial y}\mathbf{H} \neq 0$. Further, none of the field components is assumed to be zero, i.e. $E_x, E_y, E_z \neq 0$ and $H_x, H_y, H_z \neq 0$ in general.
4. A lossless wave propagation along the z-direction is possible and represented by the factor $e^{-jk_z z}$.

The main result is that electromagnetic waves can only propagate under certain conditions on the electric and magnetic fields, which define modes of these fields.

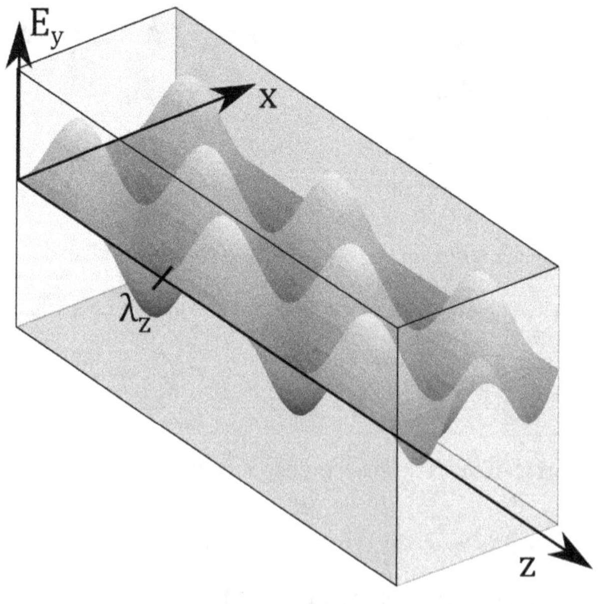

Figure 8.3 E_y*component inside a waveguide at a fixed value of* y

Each mode describes a different shape of the fields, which are in the case of a rectangular waveguide given by

$$E_z^{(m,n)}(x,y,z) = \hat{E}\sin(k_x x)\sin(k_y y)e^{-jk_z z} \tag{8.10}$$

$$H_z^{(m,n)}(x,y,z) = \hat{H}\cos(k_x x)\cos(k_y y)e^{-jk_z z}. \tag{8.11}$$

The superscripts m and n are the mode numbers and in combination with the dimensions a and b they determine k_x and k_y as

$$k_x = m\frac{\pi}{a} \quad m = 0, 1, 2, \ldots \tag{8.12}$$

$$k_y = n\frac{\pi}{b} \quad n = 0, 1, 2, \ldots. \tag{8.13}$$

The case $m = 0$ or $n = 0$ is only possible for the magnetic field. In case of the electric field the minimal mode is given for $m = n = 1$. The case $m = n = 0$ is not possible, since the field would be a static magnetic field with $E_z = 0$ and $H_z = \hat{H}$.

To visualize these abstract formulas, an example of the y component of an $\mathbf{E}^{(3,n)}$ mode for a certain y-value is depicted in Figure 8.3. The wavenumber x component can be obtained directly from the image, since $\frac{k_x}{2\pi} = \frac{1.5}{a}$ and following (8.12) gives the order $m = 3$. The y component instead cannot be obtained from the picture as it would require plotting the electric field x component, but the principle is the same. Further, the order n is not unique for the image in Figure 8.3, in fact all n up to the later defined cutoff wavenumber are possible to produce this qualitative image.

The visualization in Figure 8.3 was used to explain the transversal field perpendicular to the direction of propagation z. For the model used later, the most important part is the wavenumber k_z inside the waveguide and implications of the boundary conditions on the group velocity of a signal. Therefore, the so-called *cutoff wavenumber* κ_c is defined by

$$\kappa_c^2 = k_x^2 + k_y^2. \tag{8.14}$$

Using the common notation k_0 for the free space wavenumber

$$k_0 = \frac{\omega}{c} \tag{8.15}$$

the wavenumber k_z becomes

$$k_z = \sqrt{k_0^2 - \kappa_c^2} = \sqrt{\frac{\omega^2}{c^2} - k_x^2 - k_y^2}. \tag{8.16}$$

The expression in (8.16) gives the propagation factor for a certain frequency ω and has different values for the different values of k_x and k_y. The term k_z is also called *waveguide-wavenumber*. It is also visible that k_z can become imaginary if k_0 is smaller than κ_c, which means that the propagation term $e^{-jk_z z}$ becomes real and will result in exponential damping of the wave along the z-direction. Due to this connection, the value of κ_c gives a lower limit for the frequencies that can propagate without losses inside a waveguide and is thus called cutoff wavenumber. Modes with a cutoff wavenumber higher than k_0 are called *evanescent modes*, which means that for every k_0 a limited number of *propagating modes* with κ_c below k_0 exist, but always an infinite number of evanescent modes. Analogous to the cutoff wavenumber, a cutoff wavelength λ_c can be defined as

$$\lambda_c = \frac{2\pi}{\kappa_c}. \tag{8.17}$$

To simplify notation in the upcoming part, a dimensionless factor Λ is defined here as

$$\Lambda := \sqrt{1 - \left(\frac{\kappa_c}{k_0}\right)^2} = \sqrt{1 - \left(\frac{\lambda_0}{\lambda_c}\right)^2}, \tag{8.18}$$

where λ_0 is the free space wavelength that belongs to k_0. For $k_0 > \kappa_c$, the factor Λ is real-valued with a magnitude smaller than one and below the cutoff, i.e. $k_0 < \kappa_c$, it becomes imaginary. With this expression, formulas for k_z an λ_z can also be written as

$$k_z = \sqrt{\frac{\omega^2}{c^2} - \kappa_c^2} = k_0 \Lambda \tag{8.19}$$

$$\lambda_z = \frac{2\pi}{k_z} = \frac{\lambda_0}{\Lambda} \tag{8.20}$$

which means that inside a waveguide, the wavelength in the direction of propagation differs from the one in free space for a given frequency. This indicates that the propagation speed in z-direction differs from the speed of light c_0 and is also frequency-dependent since the factor Λ shows strong variations over frequency. This difference in the propagation speed is the main reason for the *waveguide dispersion*, which means that different frequencies travel at different velocities and thus

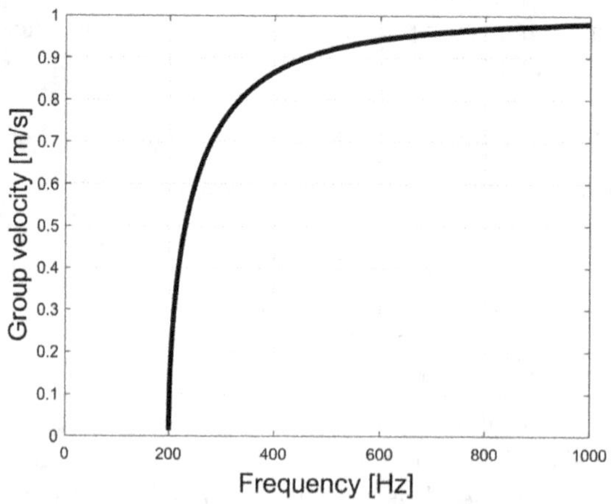

Figure 8.4 *Group velocity for a mode with a cutoff frequency of 200 Hz*

a bandpass signal gets spread in time and the shape of the envelope gets distorted. In general, these effects are analyzed by the *group velocity*

$$v_G = \frac{\partial \omega}{\partial k_z} = \left(\frac{\partial k_z}{\partial \omega}\right)^{-1} = \left(\frac{\partial}{\partial \omega}\sqrt{\frac{\omega^2}{c^2} - \kappa_c^2}\right)^{-1} = c_0 \Lambda \tag{8.21}$$

which gives the propagation speed for the signal contained in the wave [19]. From (8.21) follows that information, which propagates with the group velocity, inside of waveguides travels slower than the speed of light and depends on frequency. An exemplary plot of the group velocity of a toy example, which will be used later again, is shown in Figure 8.4.

These are the main results that will be used in the scattering model below. But first, a short summary of cylindrical waveguides will be given. In cylindrical waveguides the electric and magnetic fields follow different shapes given by

$$E_z(r, \varphi) = \hat{E} J_m(\kappa_c r)\cos(m\varphi)e^{-jk_z z} \tag{8.22}$$

and

$$H_z(r, \varphi) = \hat{H} J_m(\kappa_c r)\cos(m\varphi)e^{-jk_z z}. \tag{8.23}$$

In both expressions, the term $J_m(\cdot)$ refers to the mth order *Bessel function of the first kind.* The calculation of κ_c differs for both cases, in case of (8.22) the cutoff wavenumber is given by the nth zeros of the mth order Bessel function P_{mn} divided by the radius R of the waveguide

$$\kappa_c = \frac{P_{mn}}{R}. \tag{8.24}$$

The cutoff wavenumbers for the mode of the magnetic field are calculated with the nth zero of the derivative of the mth Bessel function P'_{mn} by

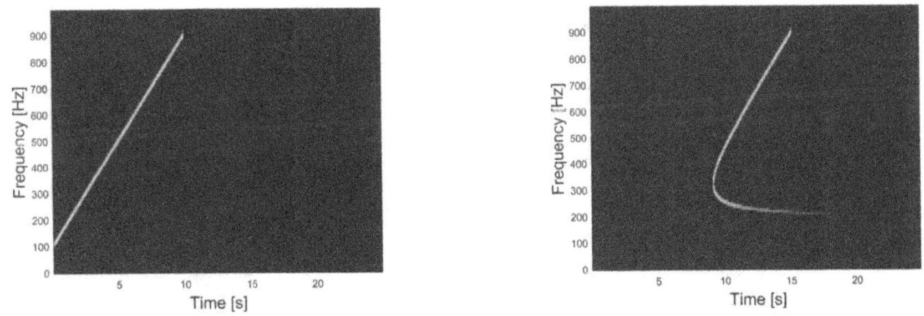

(a) Spectrogram input signal

(b) Spectrogram output signal

Figure 8.5 Time-frequency representation of an LFM pulse before and after propagation through a hypothetic waveguide with a single propagating mode

$$\kappa_c = \frac{P'_{mn}}{R}. \tag{8.25}$$

These values of κ_c define the different values of k_z for the different propagating modes in the following model using (8.19). This calculation using the zeros of the Bessel function implies that the number of propagating modes differs for each Bessel function. For a better understanding of the effect that appears in the waveguide, an illustrative example with a linear frequency modulation (LFM) waveform is given next.

8.3.2 Example of LFM pulse propagation inside a waveguide

A convenient representation of wave propagation is the model of a linear system. An input waveform $\tilde{\mathbf{E}}(0, \omega)$ is transformed with the transfer function $H(z, \omega)$, which is given by [18]

$$H(z, \omega) = e^{-jk_z(\omega)z}. \tag{8.26}$$

The expression for $k_z(\omega)$ was already given in (8.19). To visualize the dispersion effect that appears in waveguides a toy example with an LFM waveform from 100 Hz to 900 Hz is shown in Figure 8.5. The propagating mode has a cutoff frequency of 200 Hz, which corresponds to the group velocity shown in Figure 8.4.

This example is not supposed to be realistic, but clearly shows the slow propagation of the lower frequencies, which are actually passed by the high frequencies. Frequencies below the cutoff are damped and are not visible anymore. A similar example with the same waveform, but with three propagating modes is shown in Figure 8.6. For this example, the transfer function must be adapted, since (8.26) supports only a single mode. With the assumption of linear systems, the principle of superposition can be applied, which means that the transfer functions of different modes can be added. This implies that the signal is propagating as a sum of different modes, but not necessarily equally distributed. To allow a weighting of the modes, a factor $S^{(m,n)}$ is introduced to achieve the final transfer function

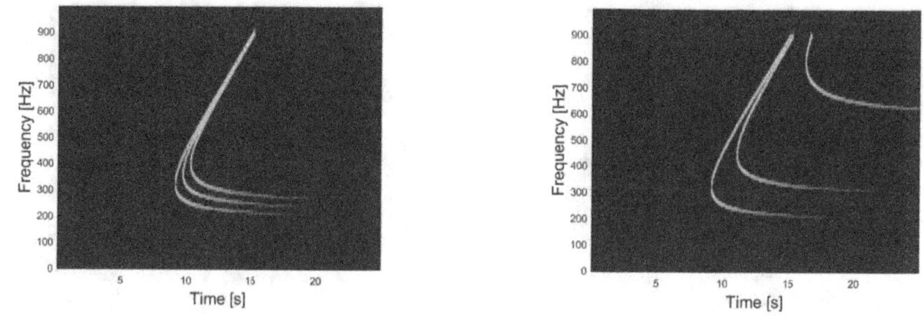

(a) Spectrogram Output signal with f_{c1} = 200 Hz, f_{c2} = 230 Hz and f_{c3} = 260 Hz

(b) Spectrogram Output signal with f_{c1} = 200 Hz, f_{c2} = 300 Hz and f_{c3} = 600 Hz

Figure 8.6 Time-frequency representation of an LFM pulse after propagation through a hypothetic waveguide with three propagating modes

$$H(z,\omega) = \sum_{m=0}^{M} \sum_{n=0}^{N(m)} S^{(m,n)} e^{-jk_z^{(m,n)} z} \tag{8.27}$$

where M and $N(m)$ are the upper bounds of the mode indices m and n, which allow a propagating mode for the given frequency. For circular waveguides, the dependency of N on m in (8.27) follows from the calculation using the nth zero of the mth order Bessel function and the derivative of the mth order Bessel function, respectively, as it was described above. It follows that $N(m)$ is reduced with an increasing m for circular waveguides. In case of rectangular waveguides, this dependency does not exist, since the cutoff frequencies are calculated using the waveguide dimension directly. Subsequently, the notation of the dependency of k_z on ω is omitted to simplify the expressions.

The example in Figure 8.6(a) shows the propagation with three modes, which have a closely spaced cutoff frequency. The high frequencies seem to add up to a single LFM pulse, but if the cutoff frequencies are more apart from each other, as in Figure 8.6(b), the traveling time of the separate modes differs significantly and produces three clearly visible LFM pulses. This delay and dispersion in the waveguide are the reasons for range extended returns, which are visible in ISAR images. The next step is a formulation of a more realistic model that can be applied to realistic jet engines.

8.3.3 Jet engine scattering model

Modeling jet engines as waveguides and using modal decompositions of the incident wave is a common way to determine the RCS of engines and aircraft [20]. Some applications that use this kind of model in the imaging domain can be found in the literature, e.g. in [21], where it was used to remove range extended returns from an ISAR image and increase the matching score of a template matching target recognition system. Another example can be found in [22], which uses a time-frequency transformation-based ISAR imaging algorithm to produce several ISAR images that differ in the used frequency. Furthermore, parts that are unstable across the chosen frequency range are removed from the image. In this way, an image that only consists of stable point scatterers is created. It was further shown that the parts, which are removed from the image, show the highest variation in frequencies, that are very near to the theoretical modes of the inlet structure. The approach in this chapter is different, since the range extended return is not considered as an artifact that should be removed, but as a feature that should be used for classification.

The basic idea is to model the engine structure as a terminated waveguide and thus allow a modal decomposition of the waveform inside the engine. Using this approach, the modal sum is calculated only once for a given jet engine and can be stored for the classification process. Further, an analytical model is available and, therefore, the computational demands are reduced. The termination of the waveguide is realized by the fan or the first stage of the compressor, dependent on the engine. To find a mathematical formulation of the model, the normal form of point scatterers

$$s(k_r, R) = e^{-j2k_r R} \tag{8.28}$$

which only depends on the range R of the target and the wavenumber k_r of the transmitted waveform is used here. This expression must be extended by the transfer function (8.27) of a waveguide with length L

$$H(L, \omega) = \sum_{m=0}^{M} \sum_{n=0}^{N(m)} S^{(m,n)} e^{-j2k_{wg}^{(m,n)} L} \tag{8.29}$$

where the term k_z is replaced by k_{wg}, since the direction of propagation is not z anymore. Instead of z, the length L of the waveguide is used as the propagation distance of the waveguide, the factor 2 stands for the two-way propagation inside the engine. Combining the terms above gives the normal form of waveguide scatterers

$$\begin{aligned} s\left(k_r, k_{wg}, R, L\right) &= \sum_{m=0}^{M} \sum_{n=0}^{N(m)} S^{(m,n)} e^{-j2\left(k_r R + k_{wg}^{(m,n)} L\right)} \\ &= e^{-j2k_r R} \sum_{m=0}^{M} \sum_{n=0}^{N(m)} S^{(m,n)} e^{-j2k_{wg}^{(m,n)} L}. \end{aligned} \tag{8.30}$$

The first term in the exponent of (8.30) represents the distance of the radar to the entrance of the engine, where the opening structure also gives a strong point-like scattering effect. The different scattering effects of an engine are depicted in Figure 8.7, which shows three different types of scattering for an incident electric field $\mathbf{E}^{in}$ [23]. The first one is the point-like scattering at the opening structure $\mathbf{E}^{s}_{rim}$, the second one is the waveguide scattering from the fan $\mathbf{E}^{s}_{wg}$ and the third one is the point-like scattering from the exterior of the engine $\mathbf{E}^{s}_{ext}$.

The first two of the scattering mechanisms above are relevant for the model used here, the scattering from the exterior is ignored, since this scattering can be very diffuse and might not contribute in the backscattering direction. The scattering of the outer rim and the cavity should always be present for this kind of engine. An example of a real ISAR image with a zoomed plot of the jet engine is shown in

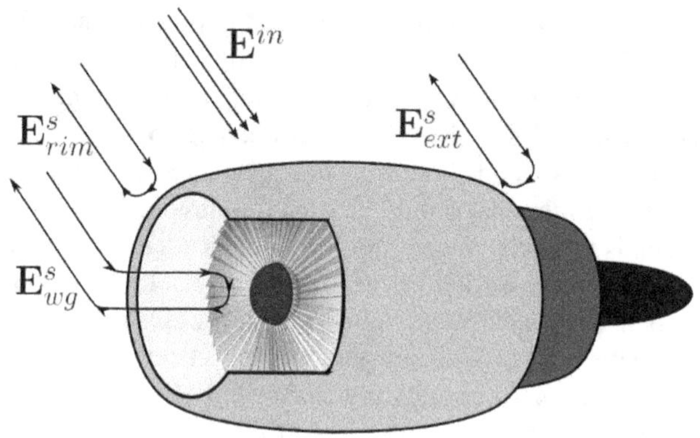

Figure 8.7 Scattering mechanisms of a jet engine. ©Cambridge University Press, 2017, reprinted with permission from [7].

Figure 8.8. The data was measured with the TIRA system of Fraunhofer FHR, the bandwidth is 800 MHz and the processing interval was selected to achieve a corresponding cross-range resolution.

In the zoomed image in Figure 8.8 two strong point-like scattering centers can be seen close to each other in the down-range direction. The first one corresponds to the opening of the jet engine, the second one is the result of the fast propagating modes that generate an echo from the fan. The higher modes with a slower group velocity produce the ponytail behind the physical position of the engine. Since the matched filter in the processing chain is fitted to a wave that is propagating with the speed of light, the slow group velocity of the higher modes inside a waveguide has the same effect as a longer distance of a wave traveling with the speed of light.

A point that was not yet considered is the calculation of the weighting parameters $S^{(m,n)}$ in the model, which can be done, e.g. with the method in [24]. The results presented there can principally be used to determine the coefficients of the jet engine scattering model, but this method has a major drawback, since it is based on the assumption of a perfect waveguide structure, which is not true for jet engines. It is known from waveguide theory that any discontinuity along the inner surface of a waveguide will excite all modes, also the evanescent ones [25, 26]. From this point

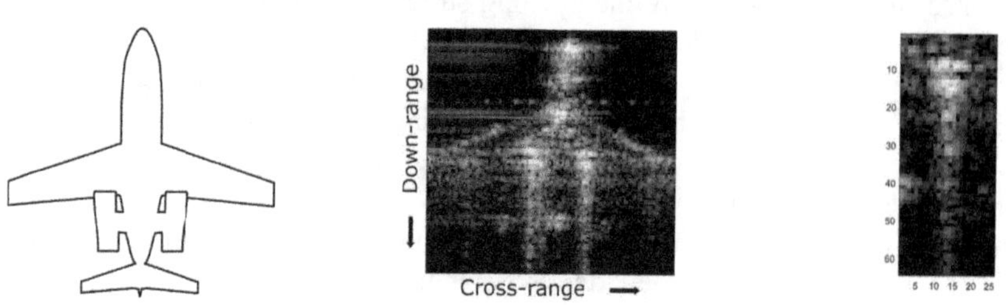

(a) Target shape (b) ISAR image (c) Zoom

Figure 8.8 Example of a real ISAR image with range extended returns of view, it is very likely that inside a jet engine the propagating modes are equally distributed, since a large number of discontinuities can be found inside the air intake of engines, e.g. small things like screws or large ones like the air bypass.

Due to the just mentioned reason, the method to calculate the coefficients $S^{(m,n)}$, which was presented for completeness, will not be used in the further presented work. Instead an equal distribution of the propagating modes is assumed and evanescent modes are ignored. To validate the model and the given assumptions, a comparison of the example in Figure 8.8 and the model produced with the TIRA parameters is presented in the next part.

8.3.4 Scattering of an LFM waveform

An example of an ISAR image that was measured with the TIRA system can be seen in Figure 8.8. Also in the previous section, a comparison between this example and the output of the model with TIRA parameters is given. The parameters of the TIRA system are repeated here, as they are given in [27]

- bandwidth 800 MHz
- pulse duration 256 μs
- carrier frequency 16.7 GHz
- sampling frequency 4 MHz
- N samples 1 024.

These numbers give an unambiguous range of 192 m and a range resolution of 18.75 cm. The parameters of the opening geometry, i.e. the diameter and the length of the jet engine, are estimated from pictures of the target in Figure 8.8 because ground truth data was not available. The estimated length of the waveguide is 50cm, which gives a total propagation distance of 1 m. The transfer function of this model and thus the point spread function (PSF) for this kind of scattering is shown in Figure 8.9, where on the left side a plot with the complete unambiguous range is

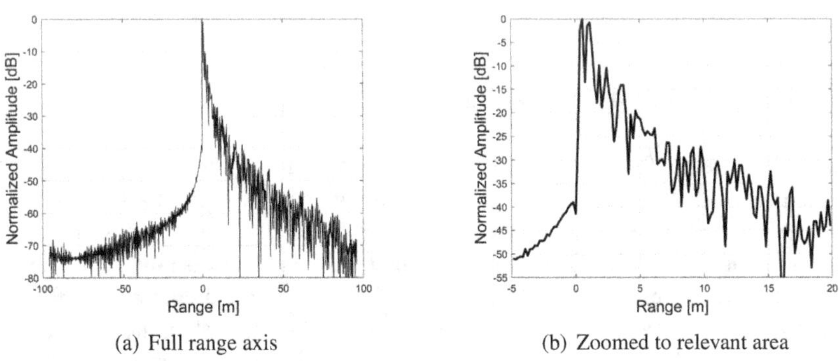

(a) Full range axis (b) Zoomed to relevant area

Figure 8.9 Calculated PSF for jet engine scattering

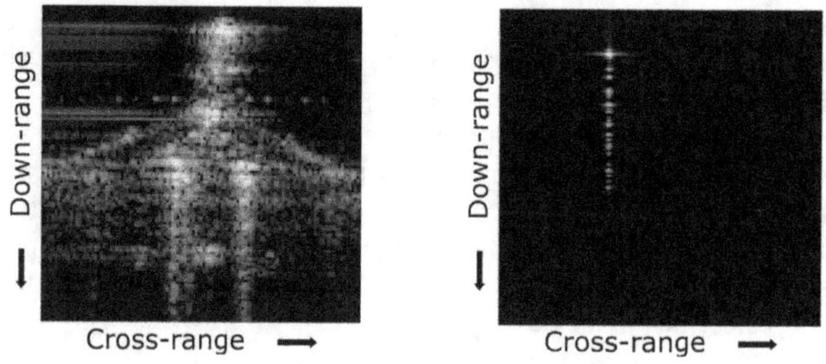

(a) Real ISAR image (b) Jet engine PSF

Figure 8.10 Comparison of an ISAR image and a 2D-PSF of a jet engine

depicted and on the right a zoomed plot shows the decay of the amplitude over range. The maximum amplitude is in both plots normalized to 0 dB.

The contribution of the propagating modes in this example is equally distributed, as it was already mentioned at the end of Section 8.3.3. This equal distribution explains the steady decay of the visible echos of the simulated example in Figure 8.9. The values of the plot show that 5 m behind the physical entrance, the amplitude is roughly 20 dB below the maximum 10 m behind the entrance the amplitude is still 30 dB below the maximum. Since most modes of the model propagate close to the speed of light, the amplitudes close to the entrance of the engine are the highest ones along the PSF. The slower propagating modes result in the smaller amplitudes behind the physical position of the engine. Similar to the real echo in Figure 8.8(b), which is depicted again in Figure 8.10 together with the 2D-version of the jet engine PSF in the same dynamic range. The two-dimensional PSF is generated with a standard sinc-function in the cross-range domain, which is defined by the movement of the target.

The similarity between the real ISAR image and the calculated jet engine PSF is clearly visible in Figure 8.10, but for an automatic recognition of this kind of scattering, an algorithm that is able to distinguish these scatterers from ordinary point scatterers is required. The theoretical background for this algorithm will be presented in the next section and the application to radar data will be shown in Section 8.5.

8.4 Sparse decomposition framework

Over the last years, the topic of *sparse representation* (SR) became very popular since many applications showed that this technique can be used in a wide variety of areas. Together with these applications, a lot of attention was given to the theoretical foundations and therefore, a well-understood framework is now available. In this chapter the basics of SR will be described as far as they are necessary for the decomposition framework used later in this chapter. As a first step, the basic terms and notations will be given in Sections 8.4.1 and 8.4.2, the algorithm that is chosen for the application is described in detail.

8.4.1 Terminology and problem description

The aim of the SR framework is to represent an input signal by a linear combination of elementary signals. This input signal is limited here to a finite-dimensional vector $\boldsymbol{x} \in \mathbb{C}^N$, representing a complex-valued sampled signal. The elementary signals that are used to approximate the signal are collected as columns of a Matrix $\boldsymbol{\Phi} \in \mathbb{C}^{N\times M}$, which is called *Dictionary*. The weighting of these elementary signals is represented by the coefficients of the vector $\boldsymbol{\alpha} \in \mathbb{C}^M$

$$\boldsymbol{x} = \boldsymbol{\Phi}\boldsymbol{\alpha}. \tag{8.31}$$

The single columns $\boldsymbol{\varphi}_i$ of the dictionary are called *atoms* and to represent an arbitrary signal, these atoms must span the $\mathbb{C}^N$, i.e. the dictionary $\boldsymbol{\Phi}$ must be a full rank matrix. If the number of atoms M is the same as the dimension N, the dictionary is a basis of the signal space. If $M > N$, the matrix $\boldsymbol{\Phi}$ is called an *overcomplete dictionary*. These overcomplete dictionaries introduce the problem of linear dependency in (8.31) but hopefully allow a representation of the input signal with fewer elements.

If the input signal $\boldsymbol{x}$ can be exactly reproduced with $s \ll N$ atoms of the dictionary, the coefficient vector $\boldsymbol{\alpha}$ of this representation is called *s-sparse*. In most practical applications, this ideal case of an exactly s-sparse coefficient vector is not possible due to noise and model imperfections and therefore, an error $\boldsymbol{x}$ between the input signal and the sparse approximation is allowed

$$\|\boldsymbol{x} - \boldsymbol{\Phi}\boldsymbol{\alpha}\|_2 \leq \eta. \tag{8.32}$$

With this error term the expression $\boldsymbol{\Phi}\boldsymbol{\alpha}$ is strictly speaking not a sparse representation anymore, since the signal is not exactly represented, it is only approximated. The approximation with the minimum η for a given s, i.e. the number of used atoms, is called *best s-term approximation*. For this approximation problem, the expression compressible representation is introduced and vector $\boldsymbol{x} \in \mathbb{C}^N$ is called compressible if its best s-term approximation decays quickly in s [28]. Although the expression sparse representation is strictly speaking not correct for the mentioned best s-term approximation that is calculated normally, it is used synonymously most times and also in this work SR and best s-term approximation are used equivalently.

To formulate the mathematical optimization problem that must be solved to find the best s-term approximation, a common mathematical tool, namely the ℓ_p norm $\|\cdot\|_p$, is introduced here. In (8.32) the ℓ_2 norm $\|\cdot\|_2$, which is also known as the Euclidean norm, was already used. This norm is a special case of the more general *ℓ_p norm*, which is defined for $p \in [1, \infty]$ as [29]

$$\|\boldsymbol{x}\|_p = \begin{cases} \left(\sum_{i=1}^{N} |x_i|^p\right)^{\frac{1}{p}} & \text{for } p \in [1, \infty) \\ \max\limits_{i=1,2,\ldots,N} |x_i| & \text{for } p = \infty \end{cases} \tag{8.33}$$

with the Euclidean norm as a special case for $p = 2$. The ℓ_p norm can also be extended to the case $0 < p < 1$, which leads to a violation of the triangle inequality and thus to a quasinorm. For the SR framework the case $p \rightarrow 0$ is very important, since it can be shown that this case gives the number of nonzero elements of the vector $\boldsymbol{x}$.

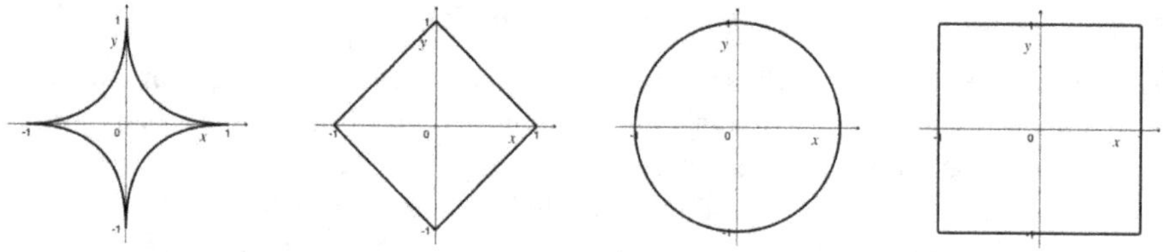

(a) $\ell_{0.5} - norm$ (b) $\ell_1 - norm$ (c) $\ell_2 - norm$ (d) $\ell_{10000} - norm$

Figure 8.11 Unit circles in different ℓ_p-norms

The index of the nonzero elements in $\boldsymbol{x}$ is called *support* of $\boldsymbol{x}$ and is written as supp($\boldsymbol{x}$). As already mentioned, it can be shown that

$$\lim_{p\to 0} \|\boldsymbol{x}\|_p^p = |\text{supp}(\boldsymbol{x})| \tag{8.34}$$

and therefore, the notation $\|\boldsymbol{x}\|_0 := |\text{supp}(\boldsymbol{x})|$ is used to calculate the cardinality of the support of $\boldsymbol{x}$. Although this is not even a quasinorm, it is common to use the expression ℓ_0 *norm* for the number of nonzero elements of $\boldsymbol{x}$. In Figure 8.11 a visualization for different values of p is shown. These images show unit circles in different norms and it can be seen that for the case $p \to \infty$ the maximum norm is achieved and that for $p \to 0$ the unit circle changes to a cross that indicates the presence of a single value.

Therefore, the mathematical problem that must be solved to determine the sparsest solution of (8.31) can be written as

$$\min_{\boldsymbol{\alpha}} \|\boldsymbol{\alpha}\|_0 \quad \text{s.t.} \quad \boldsymbol{x} = \boldsymbol{\Phi\alpha} \tag{8.35}$$

and for (8.32), which is the more relevant case from a practical point of view, the corresponding problem is

$$\min_{\boldsymbol{\alpha}} \|\boldsymbol{\alpha}\|_0 \quad \text{s.t.} \quad \|\boldsymbol{x} - \boldsymbol{\Phi\alpha}\|_2 < \eta. \tag{8.36}$$

Both problems, i.e. (8.35) and (8.36), are generally intractable, since both problems are NP-hard [28]. To solve these problems with tools from mathematical optimization, the initial formulations of (8.35) and (8.36) are replaced by

$$\hat{\boldsymbol{\alpha}} = \min_{\boldsymbol{\alpha}} \|\boldsymbol{\alpha}\|_1 \quad \text{s.t.} \quad \boldsymbol{x} = \boldsymbol{\Phi\alpha} \tag{8.37}$$

and

$$\hat{\boldsymbol{\alpha}} = \min_{\boldsymbol{\alpha}} \|\boldsymbol{\alpha}\|_1 \quad \text{s.t.} \quad \|\boldsymbol{x} - \boldsymbol{\Phi\alpha}\|_2 < \eta. \tag{8.38}$$

The particular choice of $p = 1$ in the minimization problem has been done on purpose, since this is the so-called *convex relaxation* of (8.35) and (8.36) and it can be shown that this choice will lead to sparse solutions.

An alternative formulation for the problems in (8.36) and (8.38) can be found by combining the ℓ_0 or ℓ_1 term and the constraint using a multiplicative Lagrange factor λ to

$$\boldsymbol{\alpha}^* = \min_{\boldsymbol{\alpha}} \lambda\|\boldsymbol{\alpha}\|_p + \tfrac{1}{2}\|\boldsymbol{x} - \boldsymbol{\Phi\alpha}\|_2^2 \tag{8.39}$$

where $\boldsymbol{\alpha}^*$ is the optimal solution. This notation will, for some choice of the parameter λ, yield the same result as (8.36) and (8.38), but since these values of λ are generally unknown, the notations of (8.36) and (8.38) are a more natural parameterization using, e.g. an estimated noise level for the parameter η. The second term in (8.39), i.e. the squared ℓ_2 norm of the difference between $\boldsymbol{x}$ and $\boldsymbol{\Phi\alpha}$, is called *data fidelity* term, while the first term, i.e. the ℓ_p norm with the weight λ, is called *penalty term*.

Several algorithms have been presented over the past years to solve the problems above approximately and in some cases exactly. The success of these algorithms depends heavily on the structure of the dictionary $\boldsymbol{\Phi}$. Therefore, several different characterizations of the dictionary that should assess the success of the SR algorithm in advance were introduced over the last years, e.g. the *restricted isometry property*, the *null space property*, or the *spark*. These different characteristics are used to derive sparsity levels that are guaranteed to be recovered from an arbitrary vector $\boldsymbol{x}$ with a given dictionary $\boldsymbol{\Phi}$ [29]. The drawback is that these theoretical bounds are very pessimistic compared to empirical results and, therefore, the characteristics mentioned above are not used here. The only dictionary characteristic that is used within this chapter is the so-called *mutual coherence* $\mu(\boldsymbol{\Phi})$, which is more intuitive than the other mentioned so far. It is defined by

$$\mu(\boldsymbol{\Phi}) = \max_{i \neq j} \frac{|\boldsymbol{\varphi}_i^H \boldsymbol{\varphi}_j|}{\|\boldsymbol{\varphi}_i\|_2 \|\boldsymbol{\varphi}_j\|_2} \tag{8.40}$$

which gives a maximum value of 1 for two linear dependent atoms in the dictionary and a minimum value of

$$\mu(\boldsymbol{\Phi}) \geq \sqrt{\frac{M-N}{N(M-1)}}. \tag{8.41}$$

The mutual coherence characterizes the dependence between the columns of the dictionary $\boldsymbol{\Phi}$ and in the design process of a dictionary a small value is desired, since a lower value guarantees a better result of the SR [30]. Precisely, if the sparsity s of $\hat{\boldsymbol{\alpha}}$, which represents the output of an arbitrary algorithm, is below an upper bound given by

$$s < \frac{1}{2}\left(1 + \frac{1}{\mu(\boldsymbol{\Phi})}\right) \tag{8.42}$$

it is necessarily the unique solution of (8.37). This means that a dictionary with a low coherence allows a less sparse signal to be recovered exactly [31]. This is one example of a theoretical convergence guarantee and although many of these results exist, most of them are still too pessimistic compared to empirical results [30] and are therefore not considered further here.

8.4.2 Morphological component analysis

Over the last years, many algorithms were introduced to solve the problem of finding the sparsest representation using an overcomplete dictionary, as it was presented in Section 8.4.1. The purpose of this part is not to give a full overview of all available algorithms but to show the principle of the *iterative thresholding algorithms* and an algorithm called *morphological component analysis* (MCA). A more extensive review of SR algorithms can be found, e.g. in [32].

For the description of the algorithms, the following notation is used. The common output of all algorithms is an estimate $\hat{\boldsymbol{\alpha}}$ of the true sparse vector $\boldsymbol{\alpha}$. If this estimate is build up in an iterative way, a superscript with square brackets is used to determine the estimate of the current iteration, i.e. $\hat{\boldsymbol{\alpha}}^{[\nu]}$ is the estimate of the sparse vector $\boldsymbol{\alpha}$ at iteration ν. The difference between the signal $\boldsymbol{x}$ and the reconstruction at iteration ν is given by the current *residual*

$$\boldsymbol{r}^{[\nu]} = \boldsymbol{x} - \boldsymbol{\Phi}\hat{\boldsymbol{\alpha}}^{[\nu]}. \tag{8.43}$$

Finally, the support of the estimated sparse coefficient vector at iteration ν is written as $\text{supp}(\hat{\boldsymbol{\alpha}}^{[\nu]})$.

Before the iterative thresholding algorithms will be shown, two basic thresholding operations are introduced. The first one $H_T(x)$ is called *hard thresholding* and keeps the argument x unchanged if it is above a certain threshold$T$

$$H_T(x) = \begin{cases} x & \text{for } |x| \geq T \\ 0 & \text{for } |x| < T. \end{cases} \tag{8.44}$$

The second thresholding operation $S_T(x)$ is called *soft thresholding* or *shrinkage* and is defined by

$$S_T(x) = \begin{cases} x - T & \text{for } x \geq T \\ 0 & \text{for } |x| < T \\ x + T & \text{for } x \leq -T. \end{cases} \tag{8.45}$$

Both thresholding operations are defined here for real-valued numbers; in case of complex numbers, the thresholding operation is applied on the magnitude and the phase is kept unchanged. Further both operations can also be applied to vectors, and in this case each element of the vector is treated independently. A visualization of both thresholding methods is shown in Figure 8.12. The output on the y-axis of Figure 8.12(b) clearly shows the shrinkage effect of the soft thresholding operation.

These thresholding operations possess a relation to SR that will be shown without proof before the algorithms are formulated. To show this relation, a reduced form of (8.39) with a unitary matrix as dictionary is used. In this case, the problem becomes a data approximation with ℓ_p constraint and with the definition $\tilde{\boldsymbol{\alpha}} = \boldsymbol{\Phi}^{-1}\boldsymbol{x}$ the reduced form becomes

$$\boldsymbol{\alpha}^* = \min_{\boldsymbol{\alpha}} h(\boldsymbol{\alpha}) = \min_{\boldsymbol{\alpha}} \lambda\|\boldsymbol{\alpha}\|_p + \tfrac{1}{2}\|\tilde{\boldsymbol{\alpha}} - \boldsymbol{\alpha}\|_2^2 \tag{8.46}$$

To reduce the problem, both terms of (8.46) are written as sum, which gives for$p \in \{0, 1\}$

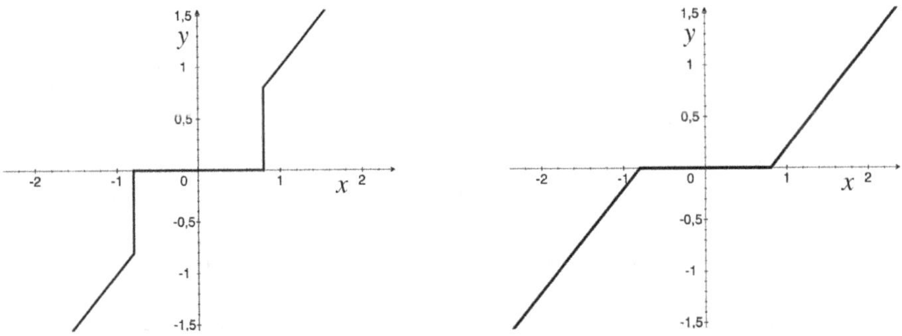

(a) Hard thresholding (b) Soft thresholding

Figure 8.12 Hard and soft thresholding

$$\begin{aligned}\alpha^* &= \min_{\alpha} h(\alpha) = \min_{\alpha} \lambda \|\alpha\|_p + \frac{1}{2}\|\tilde{\alpha} - \alpha\|_2^2 \\ &= \sum_{n=1}^{N} \lambda\, |\alpha_n|^p + \frac{1}{2}\left(\tilde{\alpha}_n - \alpha_n\right)^2\end{aligned} \tag{8.47}$$

which gives N independent problems that can be optimized separately [33]. In case of the ℓ_0 norm, the optimal solution α^* is given by

$$\alpha_n^* = \begin{cases} \tilde{\alpha}_n & \text{for } |\tilde{\alpha}_n| \geq \sqrt{2\lambda} \\ 0 & \text{for } |\tilde{\alpha}_n| < \sqrt{2\lambda} \end{cases} \tag{8.48}$$

which is the hard thresholding operation with the threshold $\sqrt{2\lambda}$. This equivalence is used, e.g. in image denoising with wavelets [33], which assumes that the signal energy can be represented by a few coefficients and the remainder contains only noise.

The solution of the reduced problem with ℓ_1 constraint can be summarized as follows:

$$\alpha_n^* = \begin{cases} \tilde{\alpha}_n - \lambda & \text{for } \tilde{\alpha}_n \geq \lambda \\ 0 & \text{for } |\tilde{\alpha}_n| < \lambda \\ \tilde{\alpha}_n + \lambda & \text{for } \tilde{\alpha}_n \leq -\lambda \end{cases} \tag{8.49}$$

which corresponds to the definition of the shrinkage operator in (8.45) with the threshold λ.

The general idea of the iterative thresholding algorithms is to find a fixed-point equation with the overcomplete dictionary [28], which results in the basic scheme of Algorithm 1. The step size τ in the algorithm depends on the chosen dictionary and an adequate choice is an upper bound of

$$\tau \leq \frac{1}{\lambda_{max}\left(\mathbf{\Phi}^H \mathbf{\Phi}\right)} \tag{8.50}$$

where $\lambda_{max}(\boldsymbol{A})$ is the largest eigenvalue of the matrix $\boldsymbol{A}$ [30].

The operation threshold$_T$ in Algorithm 1 represents either the hard thresholding $H_T(\cdot)$ or the shrinkage operation $S_T(\cdot)$ dependent on the choice of p. In case of hard thresholding $(p = 0)$, the algorithm is called *iterative hard thresholding* (IHT) and with soft thresholding $(p = 1)$, it is called *iterative shrinkage thresholding* (IST).

The already mentioned MCA is based on these iterative thresholding algorithms and can be used with both the IHT and IST. In this section, the algorithm is specifically designed to decompose signals into different components and relies on the principle of *morphological diversity*.

Algorithm 1: Iterative thresholding

Input: x, Φ, η, S, T

Init: $\hat{\alpha}^{[0]} = 0$, $r^{[0]} = x$, $v = 1$, $\tau = \frac{1}{\lambda_{max}(\Phi^H\Phi)}$;

while *stopping criteria is not met* **do**

  $\hat{\alpha}^{[v]} = \text{Threshold}_T\left(\alpha^{[v-1]} + \tau\Phi^H r^{[v-1]}\right)$;

  $r^{[v]} = x - \Phi\hat{\alpha}^{[v]}$;

  if $||r^{[v]}||_2^2 \leq \eta$ *or* $v = S$ **then**

    stop thresholding;

  else

    $v = v + 1$;

  end

end

Output: $\hat{\alpha}^{[v]}$, $r^{[v]}$

The model of morphological diversity assumes that the input data $\boldsymbol{x}$ can be modeled as the sum of K different components $\boldsymbol{x}_{(k)}$ [31], i.e.

$$\boldsymbol{x} = \sum_{k=1}^{K} \boldsymbol{x}_{(k)}. \tag{8.51}$$

Each $\boldsymbol{x}_{(k)}$ is called a *morphological component* and should be sparsely represented by a given dictionary $\boldsymbol{\Phi}_{(k)}$ with the coefficient vector $\boldsymbol{\alpha}_{(k)}$. Up to this point, the dictionary was assumed to be one large matrix $\boldsymbol{\Phi}$, which is still valid here, but this matrix can now be divided into several subdictionaries $\boldsymbol{\Phi}_{(k)}$ and each component can be represented as

$$\boldsymbol{x}_{(k)} = \boldsymbol{\Phi}_{(k)}\boldsymbol{\alpha}_{(k)} \quad \forall\, k \in \{1, \ldots, K\}. \tag{8.52}$$

Each dictionary $\boldsymbol{\Phi}_{(k)}$ must provide a sparse representation of the component it is serving, while being inefficient in representing the other contents. One possibility to ensure this behavior is to use dictionaries, which are *mutually incoherent*, i.e. the combined dictionary possesses a low mutual coherence.

The idea of MCA is the estimation of the components $\boldsymbol{\alpha}_{(k)}$ by solving the constraint optimization

$$\hat{\boldsymbol{\alpha}}_{(1)}, \ldots, \hat{\boldsymbol{\alpha}}_{(K)} = \min_{\boldsymbol{\alpha}_{(1)}, \ldots, \boldsymbol{\alpha}_{(K)}} \sum_{k=1}^{K} \|\boldsymbol{\alpha}_{(k)}\|_p \quad \text{s.t.} \quad \|\boldsymbol{x} - \sum_{\mathrm{k}=1}^{\mathrm{K}} \boldsymbol{\Phi}_{(\mathrm{k})}\boldsymbol{\alpha}_{(\mathrm{k})}\|_2^2 < \eta. \tag{8.53}$$

The parameter η in the constraint function is used to take noise into account. Assuming that each dictionary gives a sparse representation of its specific content, (8.53) gives the coefficient vectors of the different components. It should be mentioned that this algorithm should not be mistaken with the morphological operations known from image processing [34]. Although MCA was also introduced in the context of image processing [33], the term *morphology* refers to the general study of structures underlying measured data in this case.

A solution of (8.53) can be achieved by iterative hard thresholding ($p = 0$) or iterative soft thresholding ($p = 1$) [31]. The difference compared to the general scheme of the algorithms presented above is now the given structure of the dictionary. To solve this problem, the so-called *block coordinate relaxation* (BCR) method is exploited. This method was introduced for unitary subdictionaries, i.e. orthonormal transformations as in the optimal solutions of (8.46), and the method is applied iteratively with all but one $\boldsymbol{\alpha}_{(k)}$ fixed [35]. In this case, (8.53) can be written for $p = 1$ as

$$\hat{\boldsymbol{\alpha}}_{(l)} = \min_{\boldsymbol{\alpha}_{(l)}} \|\boldsymbol{\alpha}_{(l)}\|_1 \quad \text{s.t.} \quad \left\| \left(\boldsymbol{x} - \sum_{k=1, k \neq l}^{K} \boldsymbol{\Phi}_{(k)}\boldsymbol{\alpha}_{(k)} \right) - \boldsymbol{\Phi}_{(l)}\boldsymbol{\alpha}_{(l)} \right\|_2^2 < \eta \tag{8.54}$$

which is of the same type as (8.46) that is known to possess the closed-form solution in (8.49). The iterative application of (8.54) will converge to the optimum of (8.53) for $p = 1$ if the dictionary $\boldsymbol{\Phi}$ is built of unitary subdictionaries $\boldsymbol{\Phi}_{(k)}$, which guarantee a unique reconstruction for each subdictionary [33]. The idea of MCA is to apply the scheme of (8.54) under the assumption that it will also work for redundant subdictionaries, which is similar to the iterative thresholding algorithms in Algorithm 1. An application of this algorithm to SAR images was presented in [36], which was focusing on the separation of SAR images into target and clutter content.

A point that is crucial for MCA is the desired coarse to fine behavior of the decomposition. This means, the algorithm is applied with a decreasing threshold from one iteration to the next. To achieve this behavior, an initial maximum and a minimal threshold must be determined and the actual threshold is varied between these two points. The two standard schemes are either a linear or an exponential decrease determined by the predefined maximum number of iterations. To determine the initial threshold, a common strategy is to set it to a certain value that only coefficients from a single subdictionary are chosen at the first iteration. The algorithm can also be stopped early if the targeted error bound is achieved before the minimal threshold is reached. Additionally to these standard schemes, adaptive methods can be used to determine the threshold in each iteration [37]. The structure of MCA is shown in Algorithm 2 using the shrinkage operation.

This iterative behavior with a decreasing threshold also allows to correct separations from early iterations, i.e. content of the data $\boldsymbol{x}$ is allowed to change from one morphological component to another during the iterations. This is a result of the block-wise application of the thresholding for each subdictionary with the residual of the previous iteration. Further, the algorithm can also be used with the hard instead of the soft thresholding, which would lead to the minimization of the ℓ_0 norm in (8.53). Also a combination of both thresholding methods is possible; in [33], it was mentioned that the best results are often reached if the soft thresholding is replaced by hard thresholding towards the end of the decomposition.

Algorithm 2: Morphological Component Analysis

Input: x, $\Phi = [\Phi_{(1)}, \Phi_{(2)}, \ldots, \Phi_{(K)}]$, η, S

Init: $\hat{\alpha}_{(k)}^{[0]} = 0\ \forall k$, $r^{[0]} = x$, $\nu = 1$, $\tau = \frac{1}{\lambda_{max}(\Phi^H \Phi)}$;

let $k^* = \max_k ||\Phi_{(k)}^H x||_\infty$, then $T^{[1]} = \max_{k \neq k^*} ||\Phi_{(k)}^H x||_\infty$;

while *stopping criteria is not met* **do**

  for $k = 1$ **to** K **do**

    $\hat{\alpha}_{(k)}^{[\nu]} = S_T\left(\alpha_{(k)}^{[\nu-1]} + \tau \Phi_{(k)}^H r^{[\nu-1]}\right)$;

  end

  $r^{[\nu]} = x - \sum_{k=1}^{K} \Phi_{(k)} \hat{\alpha}_{(k)}^{[\nu]}$;

  if $||r^{[\nu]}||_2^2 \leq \eta$ *or* $\nu = S$ **then**

    stop MCA;

  else

    $\nu = \nu + 1$;

    Set $T^{[\nu]}$ according to chosen updating scheme;

  end

end

Output: $\hat{\alpha}_{(1)}^{[\nu]}, \ldots, \hat{\alpha}_{(K)}^{[\nu]}, r^{[\nu]}$

A point that is not described closer here is a variation of the IST algorithm called *fast iterative shrinkage thresholding algorithm* (FISTA), which is used in the results below. This variation adds a backtracking linear research element to the algorithm, which uses a specific linear combination of the previous two estimates $\hat{\boldsymbol{\alpha}}$ and not only the last one. Furthermore, the determination of τ is dependent on the dictionary $\boldsymbol{\Phi}$. Details about this variation can be found in [32] and references therein.

8.5 Results

The results obtained with the algorithm of the previous chapter are separated into two parts. The first one is a simulation with scatterers of different shapes with the aim to decompose the complex-valued raw data according to the scattering centers. Results for this simulation are presented in Section 8.5.1.

The second part is dedicated to the recognition and localization of jet engines in exemplary ISAR images measured by the Fraunhofer FHR TIRA system and the results are shown in Section 8.5.2.

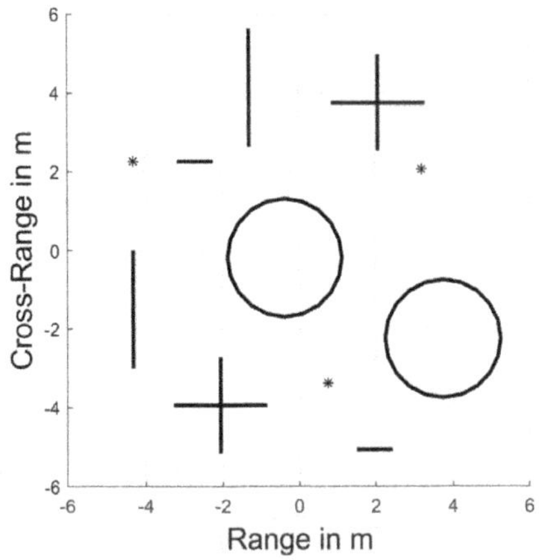

Figure 8.13 *Plot of the scene used for the simulation*

8.5.1 Simulation of objects with different shapes

Here, the sparse decomposition is applied to complex-valued phase histories. These phase histories are created using a discretized version of

$$y(k_x, k_y) = \iint e^{-j(k_x x + k_y y)} a(x, y) dx dy \tag{8.55}$$

which is a Fourier transform between the spatial coordinates in x and y and the spatial wavenumber domain in k_x and k_y. This bandlimited Fourier transform represents the basic ISAR imaging relation between the scene $a(x, y)$ and the measured signal $y(k_x, k_y)$. The resolution of the resulting image is defined by the span of k_x and k_y, which depend on the bandwidth of the transmitted waveform and the integration angle during the measurement. The sampling points of the discretized version are defined by the discrete sampling times t_i of the radar receiver and the observation angles φ_i of the measurement. For details on ISAR imaging, the reader is referred to the literature, e.g. [38] or [39].

The radar parameters used in this section are the same as in Section 8.3.4 and are chosen according to the TIRA parameters given in [27]. The integration angle was chosen to achieve the same resolution in cross range as in range. The center integration angle is set to zero degrees and in both dimensions 64 bins are used, i.e. the simulated phase histories are given as a complex-valued 64×64 matrix. A plot of the scene that is used for the simulations is shown in Figure 8.13.

This plot shows two circles, two crosses, two vertical lines, two horizontal lines and three point scatterers, which are marked by the $*$ sign. For this simulation these targets are modeled as a collection of point scatterers with a difference in range and cross range that fits the size of the corresponding resolution cells. For a more challenging decomposition a small random shift value is added to the position to create an off-grid problem.

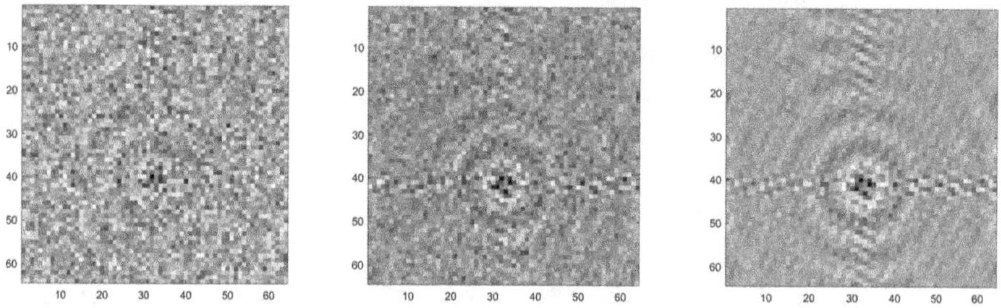

(a) Phase history (-4dB SNR) (b) Phase history (0dB SNR) (c) Phase history (10dB SNR)

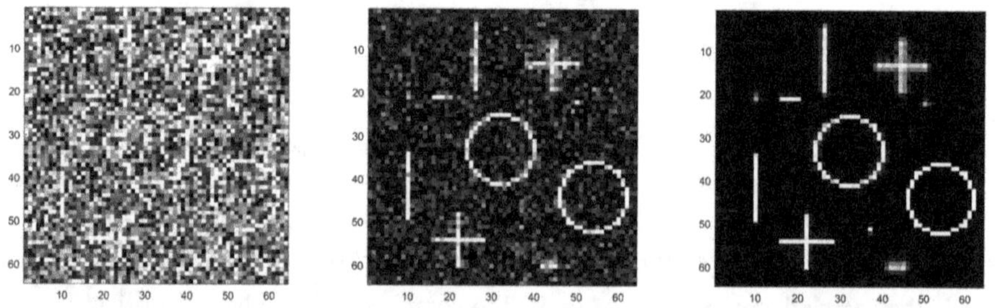

(d) Image (-4dB SNR) (e) Image (0dB SNR) (f) Image (10dB SNR)

Figure 8.14 Simulated phase histories and reconstruction with different SNR values

A separate phase history is calculated for each of the shapes in Figure 8.13 and stored for later comparison with the decomposition results. The overall received signal is calculated as the sum of all shapes and a noise term is added to achieve a predefined signal-to-noise ratio (SNR), which is varied here between -10 and 20 dB. Since the modeled signal represents the complex-valued phase history of a radar receiver, it cannot be plotted directly. For this reason the real part of the signal is shown in Figure 8.14 for three different SNR levels together with the reconstruction of the scene. This reconstruction is calculated with the standard signal model, i.e. a matched filter designed with (8.55) and the parameters that have been used for the simulation. This matched filter will be also used as the point scatterer dictionary.

The aim of this simulation is the decomposition of the data shown in Figure 8.14 using the MCA algorithm. In this simulation, the main parameters of the algorithm are an exponential decrease of the threshold T within 100 iterations. The minimal threshold is calculated as three times the noise variance. Further, the scene is simulated with a random shift and three different dictionary sizes are utilized, without oversampling and with oversampling factors of two and four.

The decomposition is performed for each dictionary with 15 different SNR values from -10 to 20 dB in steps of 2 dB. Therefore, only exemplary results using the

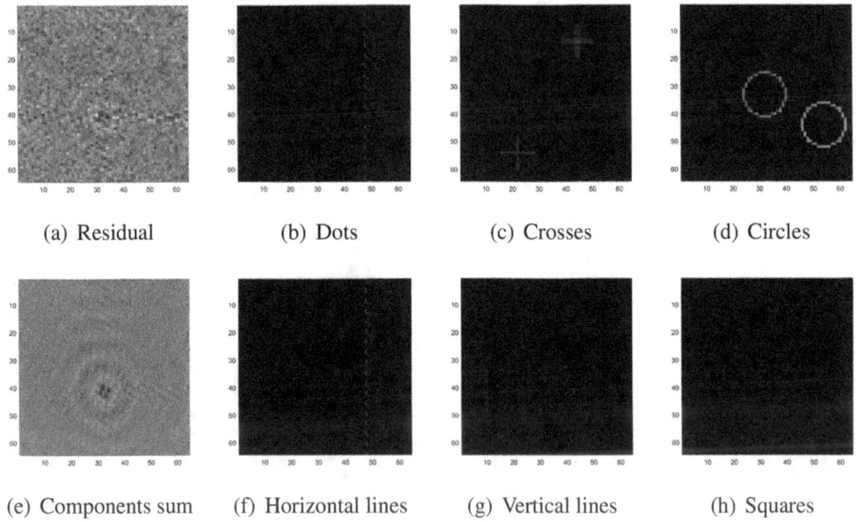

(a) Residual (b) Dots (c) Crosses (d) Circles

(e) Components sum (f) Horizontal lines (g) Vertical lines (h) Squares

Figure 8.15 MCA of radar simulation after 20 iterations

0 dB data from Figure 8.14(b) and (e) are shown here. The dictionaries used in these examples possess an oversampling factor of two.

The decomposition is shown from Figures 8.15 to 8.17 for 20, 60, and 100 iterations. For the visualization of the decomposition, not the coefficient vectors but the reconstruction $\mathbf{\Phi}_{(k)}\hat{\boldsymbol{\alpha}}_{(\boldsymbol{k})}$ in the image domain are shown together with the remaining residual and the sum of the single components in the raw data domain.

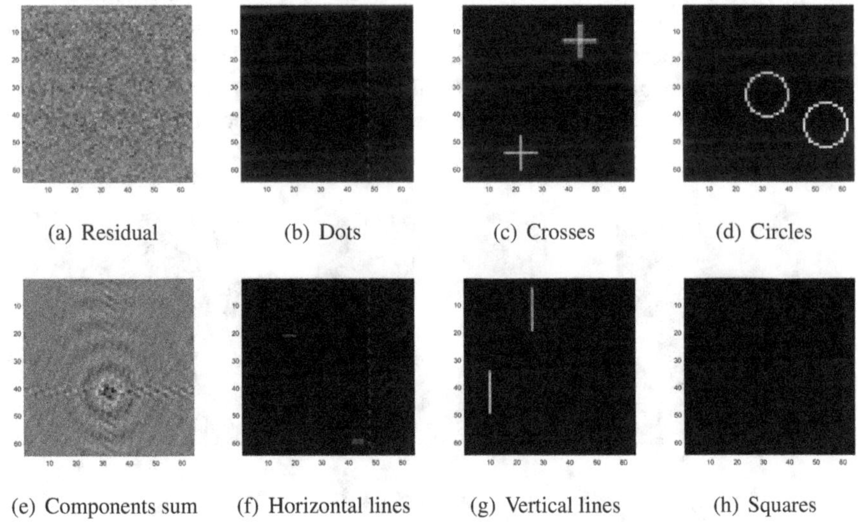

(a) Residual (b) Dots (c) Crosses (d) Circles

(e) Components sum (f) Horizontal lines (g) Vertical lines (h) Squares

Figure 8.16 MCA of radar simulation after 60 iterations

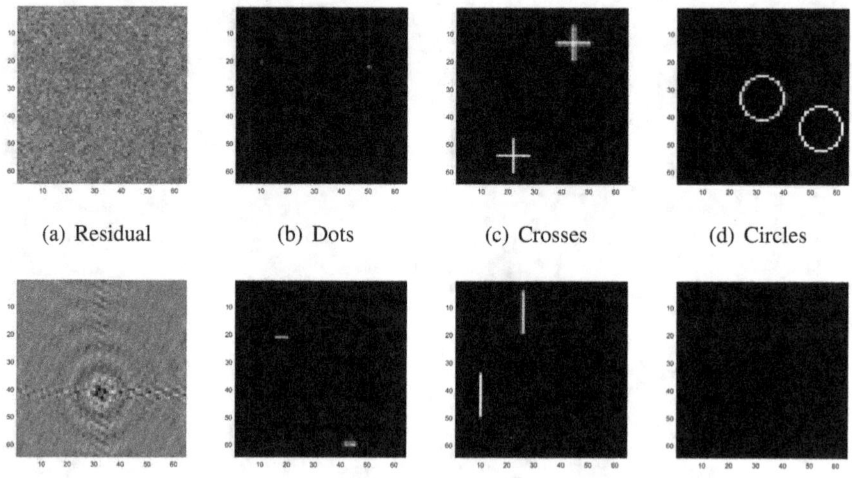

(a) Residual (b) Dots (c) Crosses (d) Circles

(e) Components sum (f) Horizontal lines (g) Vertical lines (h) Squares

Figure 8.17 MCA of radar simulation after 100 iterations

The decomposition after 100 iterations is the final result. All plots are in the same dynamic range to allow a comparison between the different figures.

This example already shows that this decomposition has a strong denoising effect and that the images of the individual shapes only contain the corresponding shapes from the simulation. The image related to the squares dictionary does not show any content, since squares are not present in the scene. Although the amplitude of the scattering centers is not reconstructed perfectly, the result clearly indicates the recognition of the different shapes. For example, the difference in the amplitude of the cross shape scatterers is probably due to the off-grid simulation. The cross in the lower part of the image is more clearly visible than the cross in the upper part, this indicates that the lower cross was closer to a dictionary function than the upper one.

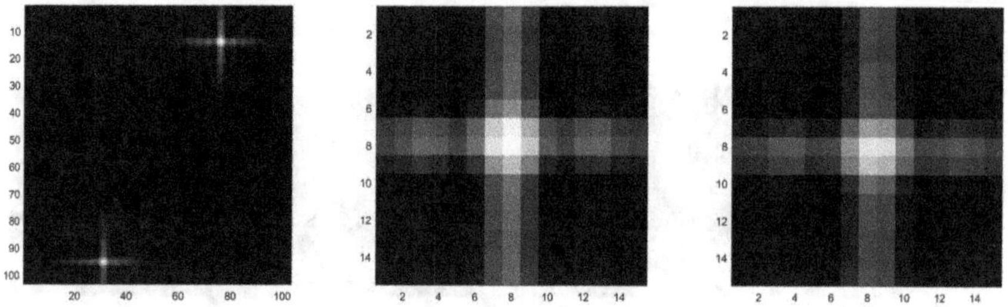

(a) Full Dictionary (b) Blurry Cross (c) Sharp Cross

Figure 8.18 Magnitudes of the selected cross atoms

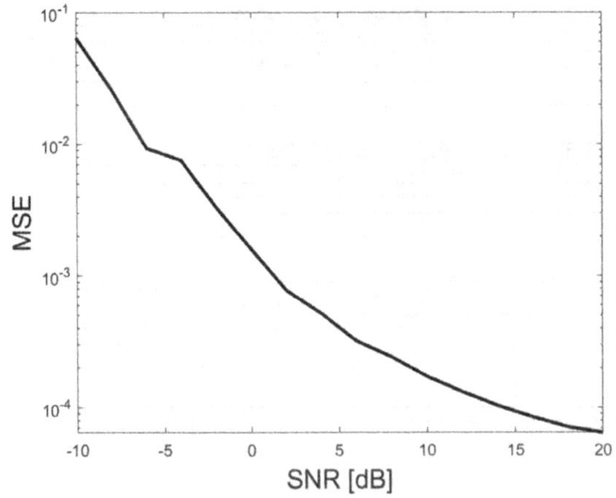

Figure 8.19 MSE in the raw data domain over SNR

For a closer look at this specific example, the magnitudes of the chosen cross atoms are depicted in Figure 8.18. These images show that for both crosses several dictionary functions are selected, and that the blurry cross in the upper part actually possesses an atom with a larger magnitude than the sharp cross at the bottom. Nevertheless, if the magnitudes of the selected atoms around the central point of the crosses are summed up, the blurry cross results in a higher value. In fact, the cumulative sum is 12.3% higher in the depicted 15×15 area for the blurry cross. This means that more atoms with a larger magnitude have been selected to represent this cross compared to the lower one. All plots in Figure 8.18 are in the same dynamic range.

Since it is not practical to show all results in this manner, a more compact presentation of the results is used to summarize these. In Figure 8.19, the mean square error (MSE) of the phase histories compared to the stored noise-free reference values is shown for the selected range of SNR values. This curve shows that the decomposition and the denoising work very well with an almost exponential decrease over the selected SNR range. From the decomposition example above, it becomes clear that the algorithm works already very well for an SNR of 0 dB and this curve shows that the performance is increasing further with increasing SNR.

This simulation was used as a proof of concept that this algorithm is able to separate different sources if an appropriate model is available. The next step is the application of this algorithm with the model from Section 8.3 to real ISAR images.

8.5.2 Recognition of jet engines in ISAR images

The results of this section are achieved with the data shown in Figure 8.8 and the scattering model of Section 8.3.4. The decomposition is applied in the complex-valued raw data domain and the jet engine dictionary was built in the way that the main part of the scattering centers fit into the image. This means that the last 22

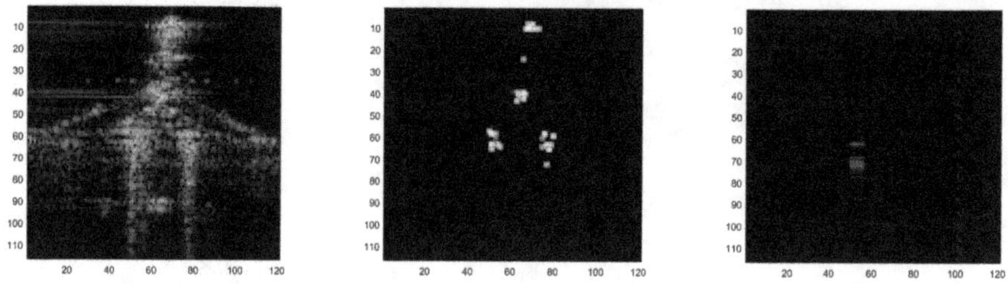

(a) Residual (b) point scatterer (c) waveguide scatterer

Figure 8.20 MCA of ISAR image after 200 iterations

range cells cannot be recognized as jet engine, since no atom was calculated for these positions in the image. The point scatterer dictionary was built over the full image and uses the standard isometric point scatterer assumption, which leads to two-dimensional sinc as point spread function in the image. As it was mentioned in Section 8.3.4, the cross range dimension of the jet engine atoms is also calculated using this standard model.

The decomposition is done with the MCA algorithm, but the parameters have changed compared to the example above. The number of iterations is set to 400 and the hard thresholding is chosen in the iterative algorithm. The lower bound of the threshold is calculated with the median absolute value deviation (MAD) from statistics. The MAD value should be an estimate of the additive white Gaussian noise in the image, which is not known anymore. The problem with this estimate is that imaging artifacts due to imperfect focusing and range compensation have not the same effect as additive white noise and therefore, the minimum value of the threshold is calculated arbitrarily by eight times the MAD estimate. This lower bound and the other parameters are chosen empirically in this case.

The decomposition is shown in Figures 8.20 and 8.21 after 200 iterations and 400 iterations, i.e. the final result, respectively.

These results show that the engines have been detected in the image, although the position of the right engine is shifted by some range cells. Most of the image is

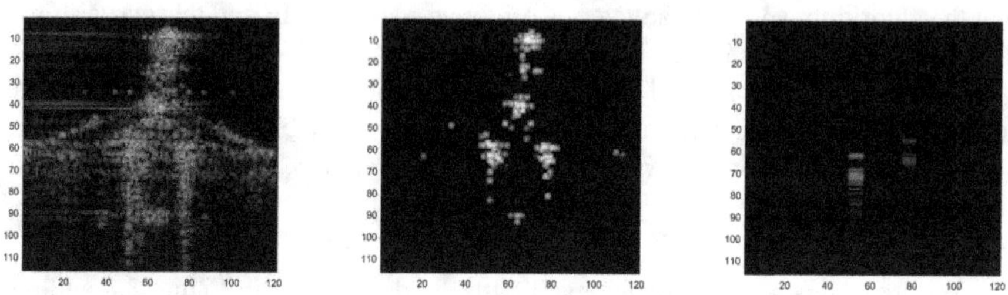

(a) Residual (b) point scatterer (c) waveguide scatterer

Figure 8.21 MCA of ISAR image after 400 iterations

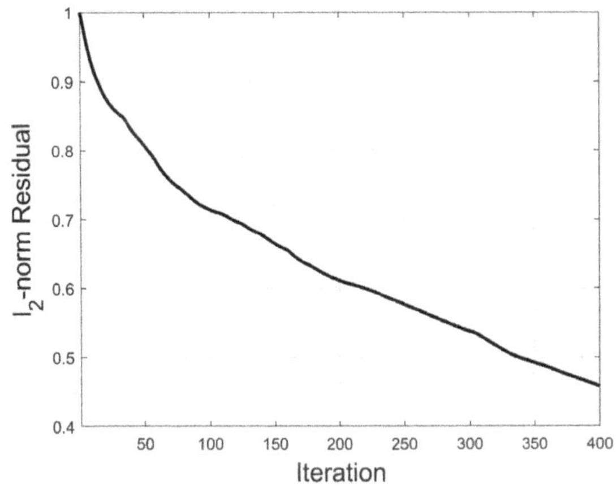

Figure 8.22 Norm of residual during the ISAR decomposition

actually covered by the point scatterer dictionary, also at the position of the engines, which is due to the strong reflection at the opening rim of the engine. Some point scatterers appear also behind the engine along the ponytail of the duct delayed returns, which are not covered by the model of the engine scatterer. Besides these strong scatterers, the residual seems still to contain most of the energy of the image and gives the contour of the target.

To evaluate how much of the energy is covered by the selected atoms, the Frobenius norm of the residual during the decomposition is shown in Figure 8.22. This curve shows that more than 50% of the energy is covered by this decomposition, although the number of scattering centers is rather low in the component images. This is an indicator of the common assumption that the main energy is concentrated in a few scattering centers, but the contour of the target becomes visible with a large number of minor scattering centers around the target.

The curve of Figure 8.22, together with the decomposition results, confirms that the main energy of the image is contained in point-like scattering and only a small part of the energy is covered by the duct delayed returns. Further, the modeled point scattering at the opening of the engine seems not enough to compete with the point scatterer dictionary at this point, but it should be mentioned that an increase of the point scatterer in the engine model would also lead to a higher mutual coherence between the two dictionaries, which might lead to higher confusion between the two scattering mechanisms. Therefore, even if the model applied here is far from being perfect, it should give more reliable results due to the higher mutual incoherence.

8.6 Conclusion and future work

These results show a promising tool for scattering identification in radar images if appropriate models are available. Although the magnitude of the detected jet engines is rather low compared to the point scatterers, which are still the main scattering effect at the engine, the results are satisfying, since the engines have clearly been detected. The ratio between the rim opening reflection and the waveguide effect is an open point for this specific model, which might strongly depend on the actual engine and the aspect angle of the target. Nevertheless, the simulation using different shapes supports this procedure, since, with the chosen dictionaries, which actually fit the data besides the random position shift, the decomposition into several scattering mechanisms is possible.

Future work should focus on the improvement of the model and the possibility to include this kind of decomposition into the imaging process directly and thus perform the autofocus with the different kinds of assumed scattering mechanisms.

References

[1] Nguyen D.H., Kay J.H., Orchad B.J., *et al.* 'Classification and tracking of moving ground vehicles'. *Lincoln Laboratory Journal.* 2002, vol. 13(2), pp. 275–308.

[2] Nebabin V.G. 'Methods and techniques of radar recognition' in Barton D.K. (ed.). Norwood, MA: Artech House; 1995.

[3] Bowden L. 'The story of IFF (identification friend or foe'. *IEE Proceedings A: Physical Science, Measurement and Instrumentation, Management and Education, Reviews.* 1985, vol. 132(6), pp. 435–7.

[4] Fisher M. *The forgotten story of Iran Air Flight 655 [online].* 2013. Available from https://www.washingtonpost.com/news/worldviews/wp/2013/10/16/the-forgotten-story-of-iran-air-flight-655/?noredirect=on&utm_term=.10066a543c30 [Accessed 6 Apr 2019].

[5] BBC. *Iran plane crash: What we know about flight PS752 [online].* 2020. Available from www.bbc.com/news/world-middle-east-51047006.

[6] Wagner S., Dommermuth F., Ender J. 'Detection of jet engines via sparse decomposition of isar images for target classification purposes'. Proceedings of the 13th European Radar Conference; London, UK; 2016. pp. 77–80.

[7] Wagner S., Ender J. 'Recognition of jet engines via sparse decomposition of ISAR images using a waveguide scattering model'. *International Journal of Microwave and Wireless Technologies.* 2017, vol. 9(6), pp. 1339–43.

[8] Chauveau J., de Beaucoudrey N., Saillard J. 'Use of resonance parameters of air-intakes for the identification of aircrafts' in Kouemou G. (ed.). *Radar Technology.* Rijeka: IntechOpen; 2010. Available from https://doi.org/10.5772/7183.

[9] Cloude S.R., Pottier E. 'A review of target decomposition theorems in radar polarimetry'. *IEEE Transactions on Geoscience and Remote Sensing.* 1996, vol. 34(2), pp. 498–518.

[10] Krogager E., Lee J., Boerner W. 'Polarimetric analysis of radar signature of a manmade structure'. 2006 International Radar Symposium; 2006. pp. 1–4.

[11] Potter L.C., Moses R.L. 'Attributed scattering centers for SAR ATR'. *IEEE Transactions on Image Processing*. 1997, vol. 6(1), pp. 79–91.

[12] Hung-Chih Chiang., Moses R.L., Potter L.C. 'Model-based classification of radar images'. *IEEE Transactions on Information Theory*. 2000, vol. 46(5), pp. 1842–54.

[13] Ding B., Wen G., Huang X., Ma C., Yang X. 'Target recognition in synthetic aperture radar images via matching of attributed scattering centers'. *IEEE Journal of Selected Topics in Applied Earth Observations and Remote Sensing*. 2017, vol. 10(7), pp. 3334–47.

[14] Ding B., Wen G., Ma C., Yang X. 'Decision fusion based on physically relevant features for SAR ATR'. *IET Radar, Sonar & Navigation*. 2017, vol. 11(4), pp. 682–90.

[15] Jackson J.A. *Three-Dimensional Feature Models for Synthetic Aperture Radar and Experiments in Feature Extraction*. The Ohio State University; 2009.

[16] de Wit J.J.M., van Rossum W.L. 'Extraction of building features from stand-off measured through-wall radar data'. *IEEE Journal of Selected Topics in Applied Earth Observations and Remote Sensing*. 2016, vol. 9(1), pp. 149–58.

[17] Samadi S., Çetin M., Masnadi-Shirazi M.A. 'Sparse representation-based synthetic aperture radar imaging'. *IET Radar, Sonar & Navigation*. 2011, vol. 5(2), pp. 182–93.

[18] Orfanidis S.J. *Electromagnetic waves and antennas [online]*. 2016. Available from https://www.ece.rutgers.edu/~orfanidi/ewa/.

[19] Stratton J.A., Theory E. *International Series in Pure and Applied Physics*. New York, NY, USA: McGraw Hill; 1941.

[20] Anastassiu H.T. 'A review of electromagnetic scattering analysis for inlets, cavities, and open ducts'. *IEEE Antennas and Propagation Magazine*. 2003, vol. 45(6), pp. 27–40.

[21] Borden B. 'An observation about radar imaging of re-entrant structures with implications for automatic target recognition'. *Inverse Problems*. 1997, vol. 13(6), pp. 1441–52.

[22] Trintinalia L.C., Hao Ling, Ling H. 'Extraction of waveguide scattering features using joint time-frequency ISAR'. *IEEE Microwave and Guided Wave Letters*. 1996, vol. 6(1), pp. 10–12.

[23] Moghaddar A., Walton E.K. 'Time-frequency distribution analysis of scattering from waveguide cavities'. *IEEE Transactions on Antennas and Propagation*. 1993, vol. 41(5), pp. 677–80.

[24] Huang C.C. 'Simple formula for the RCS of a finite hollow circular cylinder'. *Electronics Letters*. 1983, vol. 19(20), pp. 854–6.

[25] Wexler A. 'Solution of waveguide discontinuities by modal analysis'. *IEEE Transactions on Microwave Theory and Techniques*. 1967, vol. 15(9), pp. 508–17.

[26] Cooper D.N. 'Complex propagation coefficients and the step discontinuity in corrugated cylindrical waveguide'. *Electronics Letters*. 1971, vol. 7(5), pp. 135–6.

[27] Mehrholz D. 'Ein Verfolgungs- und Abbildungsradarsystem Zur Beobachtung von Weltraumobjekten'. *Frequenz*. 1996, vol. 50(7–8), pp. 138–46.
[28] Foucart S., Rauhut H.A. 'Mathematical introduction to compressive sensing'. *Applied and Numerical Harmonic Analysis*. New York: Springer Science+Business Media; 2013.
[29] Davenport M.A., Duarte M.F., Eldar Y.C. 'Introduction to compressed sensing' in Eldar Y.C., Kutyniok G. (eds.). *Compressed Sensing – Theory and Applications*. New York: Cambridge University Press; 2012. pp. 1–64.
[30] Elad M. *Sparse and Redundant Representations*. New York, NY: Springer Science+Business Media; 2010.
[31] Starck J.L., Murtagh F., Fadili J.M. *Sparse Image and Signal Processing*. New York, NY: Cambridge University Press; 2010.
[32] Zhang Z., Xu Y., Yang J., Li X., Zhang D. 'A survey of sparse representation: algorithms and applications'. *IEEE Access*. 2015, vol. 3, pp. 490–530.
[33] Starck J.L., Elad M., Donoho D. 'Redundant multiscale transforms and their application for morphological component separation'. *in Advances in Imaging and Electron Physics vol. 132 of Advances in Imaging and Electron Physics*. Elsevier; 2004. pp. 287–348. Available from http://www.sciencedirect.com/science/article/pii/S1076567004320069.
[34] Pratt W.K. *Digital Image Processing*. 4th ed. John Wiley & Sons; 2007.
[35] Bruce A.G., Sardy S., Tseng P. 'Block coordinate relaxation methods for non-paramatric signal denoising'. Proceedings of SPIE; 1998. vol. 3391.
[36] Wagner S. 'Morphological component analysis in SAR images to improve the generalization of ATR systems'. 2015 3rd International Workshop on Compressed Sensing Theory and its Applications to Radar, Sonar and Remote Sensing (CoSeRa); 2015. pp. 46–50.
[37] Bobin J., Starck J.-L., Fadili J.M., Moudden Y., Donoho D.L. 'Morphological component analysis: an adaptive thresholding strategy'. *IEEE Transactions on Image Processing*. 2007, vol. 16(11), pp. 2675–81.
[38] Borden B. *Radar Imaging of Airborne Targets*. Bristol, UK and Philadelphia, PA, USA: Institute of Physics Publishing; 1999.
[39] Chen V.C., Martorella M. *Inverse Synthetic Aperture Radar Imaging*. Edison, NJ: SciTech Publishing; 2014.

Chapter 9

Waveform design for efficient radar signal recovery

Mohammad Alaee-Kerahroodi[1]*, Bhavani Shankar M. R*[1]*, and Björn Ottersten*[1]

In this chapter, we focus on the design of transmit waveform and the resulting signal recovery. Waveform design plays a key role in determining the useful properties of the targets and environment. In particular, optimizing waveforms to have good correlation properties may improve the detection performance of weak targets. An emitted probing signal with low auto-correlation sidelobes maximizes the signal-to-interference-plus-noise-ratio (SINR), when complemented by a matched filter at the receiver, while significantly mitigating the signals from adjacent range bins. In the classical pulsed radar systems, a short code is transmitted per pulse repetition interval (PRI), followed by a long *silence*. In this case, the aperiodic auto-correlation function of the transmitted code should be optimized to have low peak sidelobe level (PSL) and integrated sidelobe level (ISL) for pulse compression. In the emerging phase-modulated continuous wave (PMCW) automotive radar applications, the radar sensor can have zero range sidelobes, provided that they use particular phase sequences with perfect periodic auto-correlation function. Additionally, waveform optimization is coupled with the advent of multiple input, multiple output (MIMO) radar systems which employ several probing signals at the transmitter; designing set of sequences with good correlation properties is an example. Furthermore, imposing practical constraints like unimodularity (or constant-modulus) or finite or discrete-phase (potentially binary) alphabet at the design stage enables the designs from the waveform optimization favorable for implementation. These constraints are important since the emergent applications tend to be power constrained necessitating power-efficient operation of the sensing modules. A popular family of optimization frameworks that have found success in the design of radar waveforms considering these constraints is the coordinate descent (CD) method; its use in waveform design for efficient recovery will be discussed in this chapter.

[1]Interdisciplinary Centre for Security, Reliability and Trust (ST), University of Luxembourg, Luxembourg

9.1 Signal model

Let $\mathbf{x} = [x_1, x_2, \ldots, x_N]^T \in \mathbb{C}^N$ be the transmitted fast-time radar code vector with N being the number of coded sub-pulses (code length)[a]. The aperiodic auto-correlation of $\mathbf{x}$ at lag k (e.g., matched filter output in pulse radars) is defined as[b]

$$(\mathbf{x} \circledast \mathbf{x})_k^{\mathscr{A}} \equiv r_k^{\mathscr{A}} = \sum_{n=1}^{N-k} x_n x_{n+k}^{*} = r_{-k}^{\mathscr{A}*}. \tag{9.1}$$

Similarly, the periodic auto-correlation (matched filter output in PMCW radars) of $\mathbf{x}$ at lag k is

$$(\mathbf{x} \circledast \mathbf{x})_k^{\mathscr{P}} \equiv r_k^{\mathscr{P}} = \sum_{n=1}^{N} x_n x_{n+k \bmod(N)}^{*} = r_{-k}^{\mathscr{P}*}, \tag{9.2}$$

where $-N+1 \leq k \leq N-1$ and $(.)^*$ denotes complex conjugate. In this chapter, for ease of notations, we denote

$$(\mathbf{x} \circledast \mathbf{x})_k \equiv r_k = \begin{cases} r_k^{\mathscr{A}}, & \text{for aperiodic case} \\ r_k^{\mathscr{P}}, & \text{for periodic case.} \end{cases} \tag{9.3}$$

Finding sequences with good aperiodic auto-correlation properties is crucial since transmit waveforms having non-zero auto-correlation sidelobes cause missed or false targets at the output of the matched filter (see Figure 9.1), thereby degrading the system performance. In PMCW radar systems, the waveforms are continuously transmitted, while the receivers perform matched filtering to separate the transmitted signal and the detect target. An attractive example for these systems is the emerging automotive radar sensors which require the ideal periodic auto-correlation function when processed by a matched filter [1]. In many practical systems, the synthesizer requires an idle time between every transmission, and consequently, the periodic

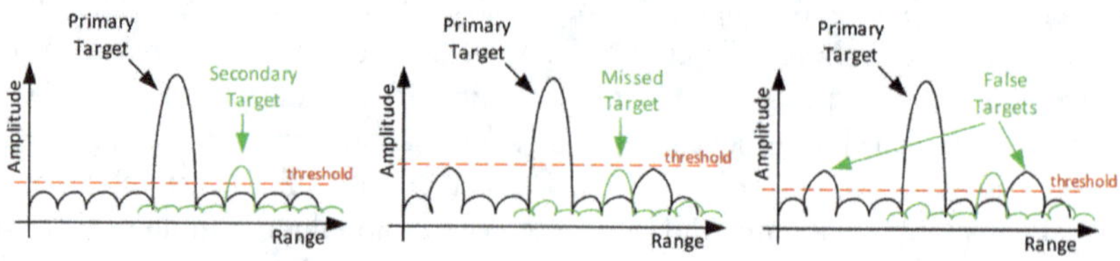

(a) Correct detection. (b) Missed target. (c) False targets.

Figure 9.1 *Sketch of auto-correlation function, displaying the effects of waveform optimization in detecting weak signals. In (a), two targets are correctly detected when a waveform with low PSL is adopted as the transmitting signal. In (b), the secondary target is missed, and in (c), false targets are detected due to the large sidelobes of the transmitted waveform.*

[a] $\mathbb{C}^N$ depicts set of N-dimensional vectors of complex numbers.
[b] In this chapter, the operator ($\circledast$) will be used to denote correlation.

auto-correlation cannot be typically performed in the receive matched filter; this brings back the focus on aperiodic correlation.

In radar systems, the goodness of the transmit waveforms is usually evaluated by two metrics: (i) PSL and (ii) ISL. These are formally defined as [2].

$$\begin{aligned} \text{PSL} &= \max\{|r_k|\}_{k=1}^{k=N-1}, \\ \text{ISL} &= \textstyle\sum_{k=1}^{N-1} |r_k|^2. \end{aligned} \tag{9.4}$$

The sequence length, N, by the definition of correlation, affects the PSL and ISL. Towards explicitly incorporating this dependence on N, the normalized correlation level, PSL and ISL in dB are defined as

$$\begin{aligned} &\text{Correlation level (dB)} = 10\log_{10}\frac{|r_k|^2}{N^2}, \quad k = 0, \ldots, N-1, \\ &\text{PSLR (dB)} = 10\log_{10}\frac{\max\{|r_k|^2\}_{k=1}^{k=N-1}}{N^2}, \\ &\text{ISLR (dB)} = 10\log_{10}\frac{\sum_{k=1}^{N-1}|r_k|^2}{N^2}. \end{aligned} \tag{9.5}$$

The aforementioned ISL metric is inversely proportional to the MF, where

$$\text{MF} = \frac{N^2}{2 \times \text{ISL}}. \tag{9.6}$$

Minimizing the ISL is equivalent to maximizing the MF. In classical pulse compression, given the hardware limitations, a direct design of discrete phase (potentially binary) sequences is of special interest. Some strategies for the construction of such sequences are reviewed in the following section.

9.1.1 Binary sequences

The binary sequences consist of the codes with elements $\{0, 1\}$ or $\{-1, +1\}$. Accordingly, the phase of the transmitted signal alternates between $0°$ and $180°$. Designing binary sequences is practically important since matched filter processing can be performed by a simple shift and add operation at the receive side; this reduces the complexity and cost of implementation compared to the sequences employing multi-phase signals. The well-known binary sequences are

- *Barker Codes.* A class of binary codes (known up to length 13 [3]) whose peak sidelobe of their aperiodic/periodic auto-correlation function is not >1 [4]. Unfortunately, there are only a few known binary Barker codes at lengths 1, 2, 3, 4, 5, 7, 11 and 13 sharing this unique property. The longest known Barker code length is 13, so the pulse compression radar using these Barker codes will be limited to the maximum compression ratio of 13. However, for many radar applications, it is desirable to generate pulse compression ratios in excess of this ratio.
- *MPS Codes.* Best known binary sequences (known up to length 105 [5]) obtained via global search whose peak sidelobe of their aperiodic auto-correlation function is the minimum possible for a given code length. Finding MPS codes for long sequence lengths is a time-consuming task and cannot be performed without using high-performance processors.

- *Almost Perfect Auto-correlation Sequences (APAS)* are defined such that only two of the out-of-phase periodic correlation coefficients are non-zero [6–8]. These two peaks are offset by $\frac{N}{2}$ to each other, where N is the code length. By adding these two peaks constructively in the receiver, a perfect periodic auto-correlation profile is obtained. This implies the absence of range sidelobes; hence, even very strong radar reflections do not cover very weak signals in other range bins [8]. However, these sequences are limited to the specified lengths and also have been found in a limited number at different lengths.
- *Legendre Codes (quadratic residue sequences)* are codes with -1 and $+1$ entries except for an initial zero element. They are defined for odd prime lengths and have near-perfect periodic auto-correlations: the peak is equal to the number of non-zero sequence elements (one less than the sequence length, due to the initial zero), and all the sidelobes are negative one [9].
- *m-sequence.* These pseudo-random sequences[c] (called also maximum length codes) are defined for length $N = 2^n - 1$, $n = 1, 2, \ldots$ One of the prominent features of an m-sequence is that its periodic auto-correlation sidelobes are always equal to -1. In terms of aperiodic auto-correlation, they can achieve PSLs on the order of $\sqrt{N}$. Usually, they are generated using shift registers with different stages and selected outputs used for feedback. If the connections are well chosen, the resulting sequence will have the maximum length (i.e., without repetition) i.e., $N = 2^n - 1$, and being n the number of stages of the shift register.
- *Gold Codes* are sequences derived from the result of modulo-2 addition of two different m-sequences of length $N = 2^n - 1$ generated by two distinct polynomials with registers of the same length and n being a positive integer not multiple of 4 [10, 11].
- *Kasami Codes* are sets of sequences constructed for all even degrees of m-sequences with length $N = 2^n - 1$ and exhibit good periodic auto- and cross-correlation functions. Kasami sequences are classified in *small* and *large* sets. The collection of sequences that are part of the large set of Kasami sequences, also contains the Gold sequences and the small set [11].

Legendre codes, m-sequences, Gold codes and Kasami sequences have ideal periodic auto-correlation functions but have no constraints on the sidelobes of their aperiodic autocorrelation functions, i.e., their PSL value is not pre-determined. Besides, they are defined for specific code lengths and are not available for arbitrary lengths. Figure 9.2 shows the auto-correlation function r_k of a random binary code and m-sequence of length $N = 127$.

[c]These codes are called pseudo-random because the statistics associated with their occurrence is similar to that associated with the coin-toss sequences.

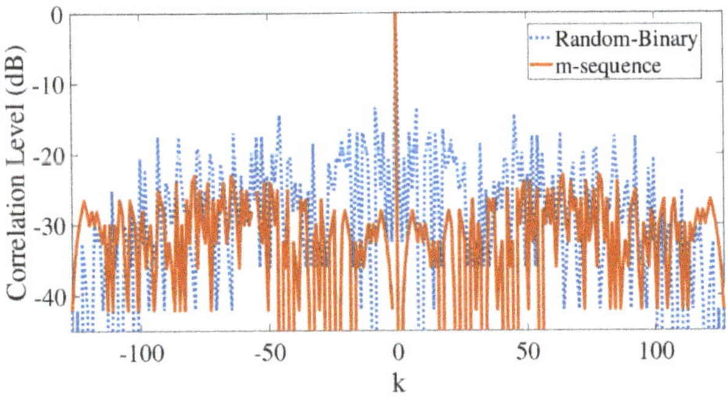

(a) Aperiodic.

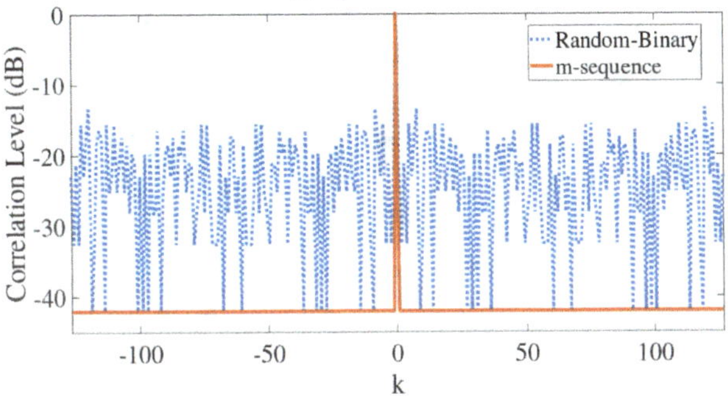

(b) Periodic.

Figure 9.2 *Auto-correlation function of random binary and m-sequence at length N=127. (a) indicates aperiodic auto-correlation functions and (b) depicts periodic auto-correlation functions of random-binary and m-sequence.*

9.1.2 *Polyphase sequences*

Waveforms consisting of more than two phases are called polyphase sequences. These sequences have more degrees of freedom for selecting the phases of the code entries and therefore typically provide lower sidelobe levels compared to the binary phase.

- *Generalized Barker Codes*. A class of polyphase codes (known up to length 77 [3]) whose peak aperiodic/periodic auto-correlation sidelobe is no larger than one [4].
- *Golomb Codes*. These offer perfect (impulse-like) periodic auto-correlation and can be constructed by

$$x_n = \exp\left[j\pi \frac{n(n-1)}{N}\right], \quad n = 1, \ldots, N. \tag{9.7}$$

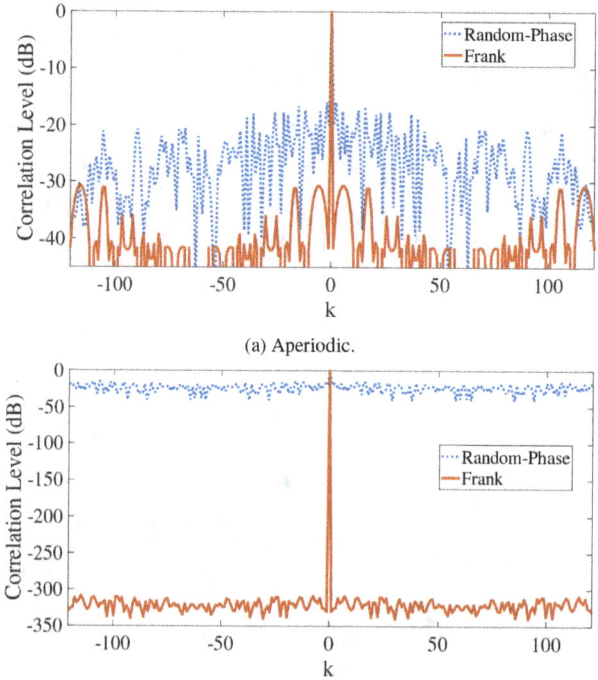

(a) Aperiodic.

(b) Periodic.

Figure 9.3 *Auto-correlation function of random-phase and Frank at length N=121. (a) indicates aperiodic auto-correlation functions and (b) depicts periodic auto-correlation functions of random-binary and Frank.*

- *Frank Codes*. A class of codes with perfect periodic and good aperiodic auto-correlation function, which are available when the code length is a perfect square, i.e., $N = L^2$. They are mathematically expressed as

$$x_{(n-1)L+\widetilde{n}} = \exp\left[j2\pi\frac{(n-1)(\widetilde{n}-1)}{L}\right], \quad n, \widetilde{n} = 1, \ldots, L. \tag{9.8}$$

Further, it is numerically shown that the Frank sequence offers lower aperiodic PSL/ISL values than either the Golomb, random-phase or *m*-sequence [12]. Figure 9.3 shows the auto-correlation function r_k of a random-phase sequence and Frank code of length $N = 121$.

In this chapter, the term *random-phase* is used to indicate the family of the uni-modular sequences that are defined by

$$x_n = e^{j\phi_n}, \quad n = 1, \ldots, N \tag{9.9}$$

where ϕ_n, $n = 1, \ldots, N$, are independent and identically distributed random variables with a uniform probability density function over $[0, 2\pi)$. Also, the quantization over the waveforms with 2^q quantization level ($q \geq 1$ is an integer) is obtained by

$$\widehat{x}_n = \exp\left(j \left\lfloor \frac{\arg\{x_n\}}{2\pi/2^q} \right\rfloor \frac{2\pi}{2^q}\right), \quad n = 1, \ldots N. \tag{9.10}$$

9.2 Design methodology

Research in the area of waveform design for radar systems is focused on designing sequences with appropriate correlation properties to enhance the detection performance by avoiding false targets. To this end, the waveform design is often formulated as an optimization problem with a certain metric that represents the quality objective, along with the set of constraints on the transmit signals. Among others, the most widely used signal quality objectives are the PSL and the ISL. Moreover, due to implementation and technological considerations, the transmit signals should also comply with certain constraints, e.g., total power, constant modulus, etc. Recently, iterative algorithms have been proposed to design both periodic and aperiodic sequences, including binary sequences. In this section, we introduce the CD framework that can be used to design radar transmit waveform of arbitrary lengths.

The CD algorithm is a derivative-free optimization method that has a powerful performance in optimizing large-scale signal processing problems in machine-learning, compressed sensing and radar systems [2, 13–16]. The principle behind the CD framework is to iteratively optimize a multi-variable objective function considering one variable (coordinate) at a time while keeping the other variables (coordinates) fixed. This potentially involves solving a set of simpler single variable sub-problems in a loop [17]. Recall that $\mathbf{x} = [x_1, x_2, \ldots, x_N]^T \in \mathbb{C}^N$ denotes a transmit code vector of length N and let $f(.)$ be a multi-variable continuous real function (possibly non-convex, non-smooth). The following optimization problem

$$\mathscr{P}_{\mathbf{x}} \begin{cases} \min\limits_{\mathbf{x}} & f(x_1, x_2, \ldots, x_N), \\ \text{s. t.} & x_n \in \mathscr{X}_n,\ n = 1, 2, \ldots, N, \end{cases} \tag{9.11}$$

can be iteratively solved using the CD technique, by finding the solutions of the following sub-problems for $i = 1, 2, 3, \ldots$:

$$\begin{aligned} x_1^{(i)} &= \arg\min\nolimits_{x_1 \in \mathcal{X}_1} f(x_1, x_2^{(i-1)}, x_3^{(i-1)}, \ldots, x_N^{(i-1)}), \\ x_2^{(i)} &= \arg\min\nolimits_{x_2 \in \mathcal{X}_2} f(x_1^{(i)}, x_2, x_3^{(i-1)}, \ldots, x_N^{(i-1)}), \\ x_3^{(i)} &= \arg\min\nolimits_{x_3 \in \mathcal{X}_3} f(x_1^{(i)}, x_2^{(i)}, x_3, \ldots, x_N^{(i-1)}), \\ &\vdots \\ x_N^{(i)} &= \arg\min\nolimits_{x_N \in \mathcal{X}_N} f(x_1^{(i)}, x_2^{(i)}, x_3^{(i)}, \ldots, x_N), \end{aligned}$$

where $\mathscr{X}_k$ is the constraint set for the kth variable $x_k^{(})$. Thus, by applying the classical CD method to solve $\mathscr{P}_x$, a single variable is optimized at every iteration i. If at some iteration, say i_K, the objective is not decreasing at every coordinate direction, then we have obtained the local optimum $\mathbf{x}^\star \equiv \mathbf{x}^{(i_K)} = [x_1^{(i_K)}, x_2^{(i_K)}, ..., x_N^{(i_K)}]^T$.

Coordinate descent

Result: Optimized code vector $\mathbf{x}^\star$
initialization;
for $i = 0, 1, 2, \ldots$ **do**
  Pick an index n from $n = \{1, 2, \ldots, N\}$;
  Optimize the n-th coordinate
  $$x_n^{(i+1)} = \arg\min_{x_n \in \mathscr{X}_n} f(\underbrace{x_1^{(i+1)}, \ldots, x_{n-1}^{(i+1)}}_{\text{previously optimized}}, \underbrace{x_n}_{\text{current}}, \underbrace{x_{n+1}^{(i)}, \ldots, x_N^{(i)}}_{\text{to be optimized}});$$
end

Algorithm 1: Sketch of the CD Method

The aforementioned framework is rather general and offers us the freedom of update, i.e., choice of the index n at k-th iteration. Following are several update rules on choosing the coordinates at each iteration:

- **Randomized**: Randomly select some coordinate to update (sample with replacement).
- **Cyclic**: Run all coordinates in cyclic order, $n = 1 \longrightarrow 2 \ldots \longrightarrow N \longrightarrow 1 \longrightarrow \ldots$
- **Greedy** (called also MBI or Gauss-Southwell): Evaluate the new objective value by making an update to each x_n separately and choose the best one.

It is worth noting that the greedy selection rule could be costly with a large number of variables but typically has faster convergence than randomized/cyclic selection of the variables [18].

Remark 1 (Convergence of CD): Regarding the convergence of CD method [19]:

- The objective function values are non-decreasing, i. e., $f(\mathbf{x}^0) \geq f(\mathbf{x}^1) \geq \ldots$.
- If f is strictly convex and smooth, the algorithm converges to a global minimum (optimal solution).
- If f (i) is continuously differentiable over the feasible set, (ii) has separable constraints and (iii) has a unique minimizer at each step, then CD method will converge to a stationary point.

9.3 Waveform design using CD optimization framework

In this section, we introduce a couple of examples indicating the utility of the CD design approach in waveform design for radar systems. Subsequently, we demonstrate the superior signal recovery aspects of these novel designs. For the waveform design, we consider constant modulus and discrete phase constraints in the design stage; these result in a significantly more complex problem than the unconstrained version. The constant modulus constraint can be mathematically written as $|x_n| = 1$, which shows the continuous phase constraint, indicating that the phase of n-th code entry of the transmit waveform can have arbitrary precision $\phi_n \in [0, 2\pi)$, where

$$x_n = e^{j\phi_n}, \quad n = 1, \ldots, N. \tag{9.12}$$

On the other hand, discrete phase constraint can be written by defining the set

$$\Omega_M = \left\{1, e^{j\frac{2\pi}{M}}, \ldots, e^{j\frac{2\pi(M-1)}{M}}\right\}, \tag{9.13}$$

which implies the problem considers the direct design of quantized phased sequences, where M indicates the alphabet size (MPSK sequences). We begin with an application dealing only with discrete phase design, then gradually consider the continuous phase design and finally present a system with enhanced dimensions.

9.3.1 Application 1: spectrum-sharing

The interest in spectrum-sharing systems arises predominantly from the fact that the spectrum required by the wireless media is a scarce resource, and both communications and remote sensing systems exhibit the enhanced performance from utilizing a wider spectrum [20]. In the communication systems, the data rate is $B\log_2(1 + \text{SINR})$, where SINR denotes the signal-to-noise ratio; clearly, the rate increases with the bandwidth B. In radar systems, the range resolution is proportional to the transmit signal bandwidth B, i.e., range resolution $= \frac{c}{2B}$, where c is the speed of light. Thus, efficient methods for spectrum-sharing need to be considered by these systems. In the following example, we assume that the communication system has the license to transmit in a particular band, and the radar wants to use this band opportunistically without causing elevated interference to communications; this is reflective of the current trends in spectrum allocation.

9.3.1.1 Multi-objective optimization and scalarization

In the aforementioned spectrum sharing context, we present the design of discrete-phase sequences avoiding reserved frequency bands which are occupied by communications (or narrowband interferers) while simultaneously considering the PSL as a radar system metric. This is a multi-objective optimization problem, not dealt in the CD framework presented above. Towards addressing this challenge, we first detail the constituent problems.

First, the problem of PSL minimization is

$$\mathscr{P}_1 \begin{cases} \min\limits_{\mathbf{x}} \quad \max\limits_{k\in 1,\ldots N-1} \left|(\mathbf{x} \circledast \mathbf{x})_k\right| \\ s.t. \quad x_n \in \Omega_M, \;\; n = 1, \ldots, N. \end{cases} \tag{9.14}$$

Note that we exclude $k = 0$ lag from the auto-correlation because it corresponds to the mainlobe rather than the sidelobe.

Second, with regards to suppression of certain spectral bands, let

$$\Gamma = \bigcup_{m=1}^{\widetilde{N}} (w_{m_1}, w_{m_2}), \tag{9.15}$$

be the set of frequencies which the radar transmission should avoid[d]. Corresponding to Γ, we choose a number $\widetilde{N}$ large enough so that points of the DFT can cover Γ densely. Let $\mathbf{F}_{\widetilde{N}}$ denote the $\widetilde{N} \times \widetilde{N}$ DFT matrix, and $\mathbf{S}$ be the matrix which contains the columns of $\mathbf{F}_{\widetilde{N}}$ corresponding to the frequencies in Γ. Also, let $\mathbf{G}$ denote the matrix comprising the remaining columns in $\mathbf{F}_{\widetilde{N}}$, and

$$\widetilde{\mathbf{x}} = [x_1, x_2, \ldots, x_N, 0, \ldots, 0]^T \in \mathbb{C}^{\widetilde{N}}. \tag{9.16}$$

Note that $\mathbf{G}^H\mathbf{G} = \mathbf{I}$ and $\mathbf{S}^H\mathbf{S} = \mathbf{I}$. Then, the spectral power of $\mathbf{x}$ in Γ can be suppressed by minimizing [21], $\|\mathbf{S}^H\widetilde{\mathbf{x}}\|$, which is equivalent to

$$\mathscr{P}_2 \begin{cases} \min\limits_{\mathrm{x},\boldsymbol{\alpha}} \quad \|\widetilde{\mathbf{x}} - \mathbf{G}\boldsymbol{\alpha}\| \\ \text{s.t.} \quad \mathbf{x} \in \Omega_M \end{cases} \tag{9.17}$$

where $\boldsymbol{\alpha}$ is auxiliary variable. Let us define

$$f_1(\mathbf{x}) = \max_{k\in 1,\ldots N-1} \left|(\mathbf{x} \circledast \mathbf{x})_k\right|, \quad f_2(\mathbf{x}, \boldsymbol{\alpha}) = \|\widetilde{\mathbf{x}} - \mathbf{G}\boldsymbol{\alpha}\|\,. \tag{9.18}$$

The problem of designing discrete-phase sequences by minimizing the PSL and shaping the spectral powers is

$$\mathscr{P}_{\mathbf{x},\boldsymbol{\alpha}} \begin{cases} \min\limits_{\mathbf{x},\boldsymbol{\alpha}} \quad f_1(\mathbf{x}), f_2(\mathbf{x}, \boldsymbol{\alpha}) \\ \text{s.t.} \quad x_n \in \Omega_M \end{cases} \tag{9.19}$$

which is constrained, bi-objective, multi-variable ($\mathbf{x}$ is a vector) and non-convex (due to the constraint set being non-convex). In multi-objective optimization problems, a feasible solution that minimizes all the objectives simultaneously does not exist usually [22]. Accordingly, one solution is to use *Pareto-optimization* by resorting to scalarization which exploits a specific weighted sum of $f_1(.)$ and $f_2(.)$ as the objective. Specifically, defining the function $f_\theta(\mathbf{x})$, parameterized in the Pareto weight $\theta \in [0, 1]$,

[d]Without loss of generality, only normalized frequencies (from 0 to 1 Hz) are considered for notational simplicity.

$$f_\theta(\mathbf{x}, \boldsymbol{\alpha}) = \max_{k=1,\ldots,N-1} \left[\theta \left|(\mathbf{x} \circledast \mathbf{x})_k\right| + (1-\theta)\,\|\widetilde{\mathbf{x}} - \mathbf{G}\boldsymbol{\alpha}\|\right], \tag{9.20}$$

and the scalarization leads to the following design problem

$$\mathscr{P}^\theta_{\mathbf{x},\boldsymbol{\alpha}} \begin{cases} \min\limits_{\mathbf{x},\boldsymbol{\alpha}} & f_\theta(\mathbf{x}, \boldsymbol{\alpha}) \\ \text{s.t.} & x_n \in \Omega_M, \end{cases} \tag{9.21}$$

which reduces to classical PSL minimization by setting $\theta = 1$ or traditional spectral shaping when $\theta = 0$. Moreover, for any θ, an optimal solution to (9.21) is a Pareto-optimal point to problem (9.19) (see [2] and references therein for details).

9.3.1.2 Utilization of the CD framework

Using the CD framework, the optimization problem at $(i+1)$-th iteration is

$$\mathscr{P}^\theta_{\boldsymbol{\alpha}} \ \min_{\boldsymbol{\alpha}} \left\|\widetilde{\mathbf{x}}^{(i)} - \mathbf{G}\boldsymbol{\alpha}\right\| \tag{9.22}$$

where by using $\mathbf{G}^H\mathbf{G} = \mathbf{I}$ and finding the least square solution, we obtain

$$\boldsymbol{\alpha}^{(i+1)} = \mathbf{G}^H \widetilde{\mathbf{x}}^{(i)}. \tag{9.23}$$

For fixed $\boldsymbol{\alpha}$, the optimization problem at the same iteration is

$$\mathscr{P}^\theta_{\mathbf{x}} \begin{cases} \min\limits_{x} & f_\theta(\mathbf{x}, \boldsymbol{\alpha}^{(i+1)}) \\ \text{s.t.} & x_n \in \Omega_M \end{cases} \tag{9.24}$$

which is non-convex and constrained. Let us assume that x_d is the only variable in the code vector $\mathbf{x}$ (other entries being fixed). The objective function in $\mathscr{P}^\theta_{\mathbf{x}}$ with dependence to $x_d = e^{j\phi_d}$ is

$$\begin{aligned} f_\theta(\phi_d, \boldsymbol{\alpha}^{(i+1)}) = \max_{k=1,\ldots,N-1} \Big[& \theta \left| a_{1k} e^{j\phi_d} + a_{2k} e^{-j\phi_d} + a_{3k} \right| + \\ & (1-\theta)\left(\left|e^{j\phi_d} - \mathbf{g}_d \boldsymbol{\alpha}\right| + b_1\right) \Big], \end{aligned} \tag{9.25}$$

where $\mathbf{g}_d$ is the dth row of the matrix $\mathbf{G}$. In the case of aperiodic auto-correlation,[e]

$$a_{1k} = x^*_{d+k}\mathbf{I}_A(d+k), \quad a_{2k} = x_{d-k}\mathbf{I}_A(d-k), \quad a_{3k} = (\mathbf{x}_{\bar{d}} \circledast \mathbf{x}_{\bar{d}})^A_k, \tag{9.26}$$

and in case of periodic auto-correlation,

[e]Notice that $\mathbf{I}_A(v)$ is the indicator function of the set $A = \{1,2,\ldots,N\}$, i.e., $\mathbf{I}_A(v)$ if $v \in A$, otherwise $\mathbf{I}_A(v) = 0$.

$$\begin{aligned} a_{1k} &= x^*_{d+k}\mathbf{I}_A(\widetilde{d}+k) + x^*_{d+k-N}\mathbf{I}_A(d+k-N),\\ a_{2k} &= x_{d-k}\mathbf{I}_A(d-k) + x_{d-k+N}\mathbf{I}_A(d-k+N),\\ a_{3k} &= (\mathbf{x}_{\bar{d}} \circledast \mathbf{x}_{\bar{d}})^{\mathscr{P}}_k, \end{aligned} \tag{9.27}$$

In both cases, $\mathbf{x}_{\bar{d}}$ represents the $\mathbf{x}$ code vector whose dth entry is zero, i.e.

$$\begin{aligned} \mathbf{x}_{\bar{d}} &= [x_1, \ldots, x_{d-1}, 0, x_{d+1}, \ldots x_N]^T \in \mathbb{C}^N,\\ \bar{\mathbf{x}} &= [\mathbf{x}^T_{\bar{d}}, 0, \ldots, 0]^T \in \mathbb{C}^{\widetilde{N}}, \end{aligned} \tag{9.28}$$

and $b_1 = \|\bar{\mathbf{x}} - \mathbf{G}\boldsymbol{\alpha}\|$. Thus, the scalar optimization problem with respect to ϕ_d is

$$\mathscr{P}^{\theta}_{\phi_d} \begin{cases} \min\limits_{\phi_d} & f_\theta(\phi_d, \boldsymbol{\alpha}^{(i+1)}) \\ \text{s.t.} & \phi_d \in \arg(\Omega_M) \end{cases} \tag{9.29}$$

where the optimal phase code entry $\phi_d{}^\star$ can be obtained by exhaustive search over the feasible set Ω_M. However, the following proposition provides an efficient method by using the FFT.

Proposition 1: Let us define[f]

$$\begin{aligned} \widetilde{\xi}_{dk} &= \left|\mathrm{FFT}[a_{1k}, a_{3k}, a_{2k}, \mathbf{0}_{1\times(M-3)}]^T\right| \in \mathbb{R}^M,\\ \bar{\xi}_d &= \left|\mathrm{FFT}[1, \mathbf{g}_d\boldsymbol{\alpha}, \mathbf{0}_{1\times(M-2)}]^T\right| \in \mathbb{R}^M, \end{aligned} \tag{9.30}$$

and $\boldsymbol{\nu}_{dk} = \max\limits_{k=1,\ldots,N-1} \left[\theta\widetilde{\xi}_{dk} + (1-\theta)\bar{\xi}_d\right]$. After removing the constant terms in the objective function $f_\theta(\phi_d, \boldsymbol{\alpha}^{(i+1)})$, the optimal solution to problem $\mathscr{P}^{\theta}_{\phi_d}$ can be obtained by finding $\widetilde{i}^\star = \arg\min \boldsymbol{\nu}_{dk}$, and setting $\widetilde{\phi}^\star = \frac{2\pi(\widetilde{i}^\star - 1)}{M}$.

Proof: The proof is straight forward and can be obtained using Appendix C in [2].

Hence, the optimal phase code entry will be obtained by $x^\star_d = e^{j\phi^\star_d}$ using discrete Fourier transform/fast Fourier transform (DFT/FFT). Based on **Algorithm** 9.2 in Section 9.2, once the optimal code entry $x^\star_d$ is obtained, we repeat the procedure until $d = N$, and all the entries of the code vector $\mathbf{x}$ are updated. Subsequently, we fix $\mathbf{x}$ and optimize the problem with respect to $\boldsymbol{\alpha}$ again, and the procedure continues until a stopping criteria be met.

9.3.1.3 Design examples

Figure 9.4 shows the spectral density and Figure 9.5 depicts the periodic auto-correlation function[g] of the obtained sequences at length $N = 256$, when alphabet size $M = 128$.

[f] The absolute and the square power are element wise.

[g] The behavior of the optimized sequence in case of considering the aperiodic auto-correlation sidelobes as $f_1(x)$ is also similar to that of reported in Figures 9.4 and 9.5 for the periodic case.

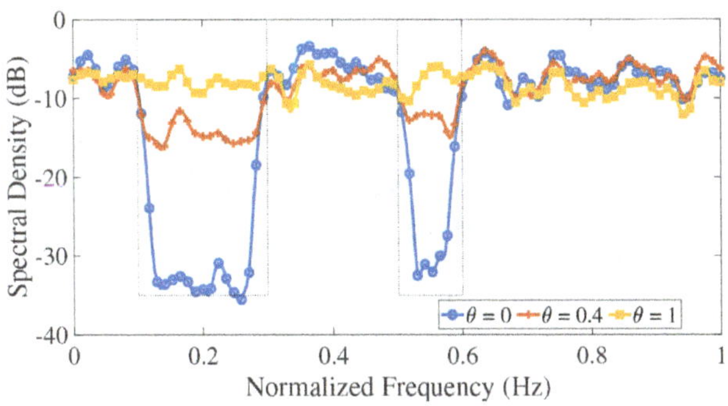

Figure 9.4 *The spectral power of the optimized sequences at $\theta = \in \{0, 0.4, 1\}$ with $N = 256$ and $M = 128$*

We adopt $\Gamma = \{\{0.1,\ 0.3\},\ \{0.5,\ 0.6\}\}$, which means that set of frequencies which $\boldsymbol{x}$ should avoid lie in the intervals $\{0.1, 0.3\}$ and $\{0.5, 0.6\}$. To obtain the above figures, we further initialize the CD algorithm with random-phase sequences, i.e.,

$$\mathbf{x}^{(0)} = [x_1^{(0)}, x_2^{(0)}, \ldots, x_N^{(0)}]^T \in \mathbb{C}^N, \tag{9.31}$$

where $x_n^0 = e^{j\phi_n}$, $n = 1, \ldots, N$, $\phi_n \in [0, 2\pi)$ and set the stopping threshold 10^{-5}, which examines the differences between the objective function at every iteration of CD (after updating all the code entries). As expected, the nulls in the spectrum are higher when the weighting factor, θ, is closer to 0. But, they suffer from a higher auto-correlation sidelobes comparing with the cases when θ values are close to 1. In this example, $\theta = 0.4$ is depicted as a representative value that depicts a trade-off between the PSLR and the spectral power; the actual value of θ depends on the application requirements (e.g., PSLR and spectral interference values that can be accommodated).

9.3.2 Application 2: l_p-norm minimization and designing waveforms with good auto-correlation properties

A classical problem in radar waveform design is to obtain sequences with good auto-correlation properties, i.e., possessing small PSL and ISL. To this end, it is known that the minimization of the l_p-norm of the auto-correlation vector $[r_1, r_2, \ldots, r_{N-1}]$ allows to trade-off ISL and PSL values of the devised sequence as the value of p increases [2, 23]. The motivation of the l_p-norm minimization instead of directly PSL minimization lies in the fact that l_p-norm is a smooth function; by increasing p, we gradually move from ISL minimization to PSL minimization thereby providing the designer with options for dimensioning the system based on requirements. Specifically, as $p \to +\infty$, the l_p-norm metric tends to the l_∞-norm of the autocorrelation sidelobes, which coincides with the PSL.

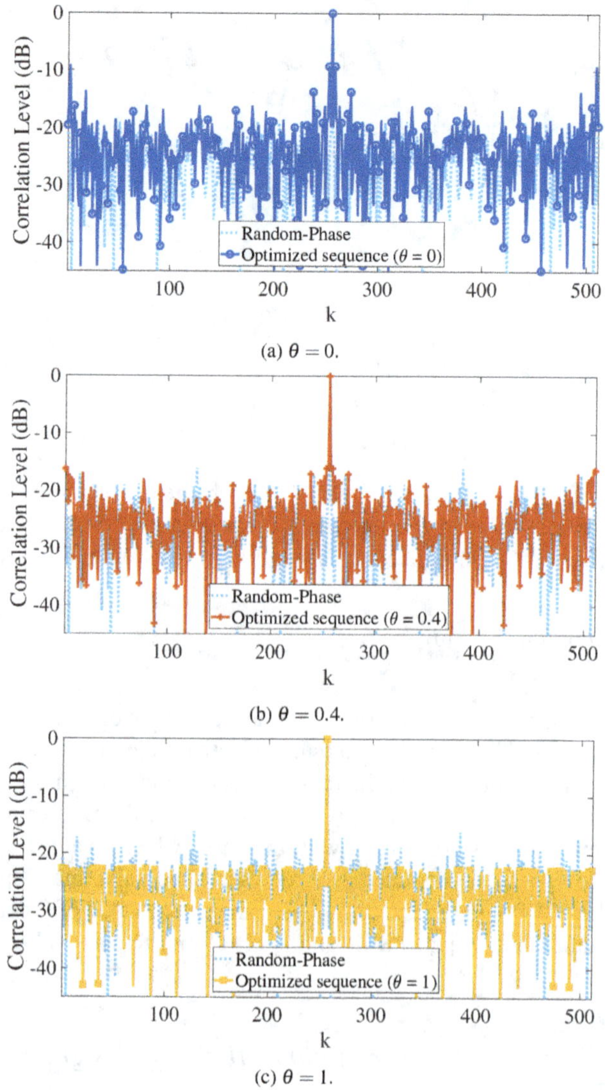

Figure 9.5 The periodic auto-correlation function of the optimized waveforms at N=256 and M=128 for different values of Pareto weight. (a) indicates θ = 0, and (b) depicts θ = 0.4, and (c) shows θ = 1.

9.3.2.1 BSUM approach

In general, the l_p-norm minimization problem for continuous and discrete phase cases can be formulated as

$$\mathcal{H}^{\infty}\begin{cases}\min\limits_{x} & \sum\limits_{k=1}^{N-1}|(\mathbf{x}\circledast\mathbf{x})_k|^p\\ \text{s.t.} & |x_n|=1\end{cases},\qquad \mathcal{H}^{M}\begin{cases}\min\limits_{x} & \sum\limits_{k=1}^{N-1}|(\mathbf{x}\circledast\mathbf{x})_k|^p\\ \text{s.t.} & x_n\in\Omega_M\end{cases}. \tag{9.32}$$

Following the CD approach to solve $\mathcal{H}^{\infty}$ and $\mathcal{H}^{M}$, the straightforward way is to write out the cost functions in (9.32) as a function of the single variable followed by its optimization. However, the single variable objectives tend to be difficult to solve. To tackle this, the iterative Block Successive Upper-bound Minimization (BSUM) algorithm [24] can be exploited, where each variable block corresponds to one code entry and the original cost function is replaced by a surrogate function having specific properties (i.e., being an upper bound at solution of the previous iteration) that ease the problem solution. The CD approach is then used on the surrogate function. While several surrogate functions can be considered, we utilize the one presented in [23] for exposition.

For the specific problem considered above, at step $(i+1)$ of the iterative procedure, the following optimization problems are considered:

$$\mathcal{H}^{\infty}_{x_d}\begin{cases}\min\limits_{x_d} & \sum\limits_{k=1}^{N-1}\widetilde{\tau}_k|r_k(x_d)|^2+\widetilde{\lambda}_k\Re\left\{r_k^*(x_d)\dfrac{r_k^{(i)}}{\left|r_k^{(i)}\right|}\right\}\\ \text{s.t.} & |x_d|=1\end{cases} \tag{9.33}$$

$$\mathcal{H}^{M}_{x_d}\begin{cases}\min\limits_{x_d} & \sum\limits_{k=1}^{N-1}\widetilde{\tau}_k|r_k(x_d)|^2+\widetilde{\lambda}_k|r_k(x_d)|\\ \text{s.t.} & x_d\in\Omega_M\end{cases} \tag{9.34}$$

where

$$\widetilde{\tau}_k=\frac{t_n^p-\left|r_k^{(i)}\right|^p-p\left|r_k^{(i)}\right|^{p-1}\left(t_n-\left|r_k^{(i)}\right|\right)}{\left(t_n-\left|r_k^{(i)}\right|\right)^2}, \tag{9.35}$$

$$\widetilde{\lambda}_k=p\left|r_k^{(i)}\right|^{p-1}-2\widetilde{\tau}_k\left|r_k^{(i)}\right|, \tag{9.36}$$

$$t_n=\left(\sum_{k=1}^{N-1}\left|r_k^{(i)}\right|^p\right)^{\frac{1}{p}}, \tag{9.37}$$

with $r_k^{(i)}$ being the optimized auto-correlation lag at step (i). Following proposition is a key to solve $\mathcal{H}^{\infty}_{x_d}$.

Proposition 2: Performing the change of variable $x_d = e^{j\phi_d}$, $\beta_d \triangleq \tan\left(\frac{\phi_d}{2}\right)$, we obtain

$$
\begin{aligned}
|\widetilde{r}_k(\beta_d)|^2 &= \frac{\widetilde{p}_k(\beta_d)}{q(\beta_d)}, \\
\Re\left\{\widetilde{r}_k^{*}(\beta_d)\frac{r_k^{(i)}}{\left|r_k^{(i)}\right|}\right\} &= \frac{\bar{p}_k(\beta_d)}{q(\beta_d)},
\end{aligned}
\tag{9.38}
$$

where

$$
\begin{aligned}
\widetilde{p}_k(\beta_d) &= \mu_{1k}\beta_d^4 + \mu_{2k}\beta_d^3 + \mu_{3k}\beta_d^2 + \mu_{4k}\beta_d + \mu_{5k}, \\
\bar{p}_k(\beta_d) &= \kappa_{1k}\beta_d^4 + \kappa_{2k}\beta_d^3 + \kappa_{3k}\beta_d^2 + \kappa_{4k}\beta_d + \kappa_{5k}, \\
\widetilde{q}(\beta_d) &= (1+\beta_d^2)^2,
\end{aligned}
\tag{9.39}
$$

with μ_{ik}, and κ_{ik}, $i = 1, 2, \ldots, 5$, as real value coefficients defined in the Appendix.

Proof: See the Appendix.

According to the above proposition, the optimal solution of problem $\mathscr{H}_{x_d}^{\infty}$ can be obtained by finding $\beta_d^{\star}$ from the optimization problem

$$
\widetilde{\mathscr{H}}_{\beta_d}^{\infty}
\begin{cases}
\min\limits_{\beta_d} & \dfrac{1}{\widetilde{q}(\beta_d)} \sum\limits_{k=1}^{N-1} \widetilde{\tau}_k\widetilde{p}_k(\beta_d) + \widetilde{\lambda}_k\bar{p}_k(\beta_d) \\
\text{s.t.} & \beta_d \in \mathbb{R}
\end{cases}
\tag{9.40}
$$

This problem can be solved by finding the real roots of a quartic function (related to the first-order derivative of the objective) and evaluating the objective function in these points as well as infinity. Hence, the optimal phase code entry will be obtained by $x_d^{\star} = e^{j\phi_d^{\star}}$ where $\phi_d^{\star} = 2\tan^{-1}(\beta_d^{\star})$.

To optimize the sequence with discrete phase constraint, we solve the following optimization problem:

$$
\begin{cases}
\min\limits_{\phi_d} & \sum\limits_{k=1}^{N-1} \widetilde{\tau}_k|a_{1k}e^{j\phi_d} + a_{2k}e^{-j\phi_d} + a_{3k}|^2 + \widetilde{\lambda}_k|a_{1k}e^{j\phi_d} + a_{2k}e^{-j\phi_d} + a_{3k}| \\
\text{s.t.} & \phi_d \in \arg(\Omega_M)
\end{cases}
\tag{9.41}
$$

where a_{1k}, a_{2k} and a_{3k} are defined in (9.26). Similar to (9.30), by defining

$$
\begin{aligned}
\zeta_{dk} &= \left|\text{FFT}\left[a_{1k}, a_{3k}, a_{2k}, \mathbf{0}_{1\times(M-3)}\right]^T\right| \in \mathbb{R}^M, \\
\boldsymbol{\nu} &= \sum_{k=1}^{N-1}\left(\widetilde{\tau}_k\zeta_{dk}^2 + \widetilde{\lambda}_k\zeta_{dk}\right),
\end{aligned}
\tag{9.42}
$$

the optimal $x_d^{\star}$ can be efficiently obtained by finding $\widetilde{i}^{\star} = \arg\min \boldsymbol{\nu}_{dk}$, and setting $\widetilde{\phi}^{\star} = \frac{2\pi(\widetilde{i}^{\star}-1)}{M}$.

Once the optimal code entry $x_d^{\star}$ is obtained (for either continuous phase or discrete phase cases), we repeat the procedure based on **Algorithm 9.2**, until all the entries of the code vector $\mathbf{x}$ are update and a stopping criterion be met.

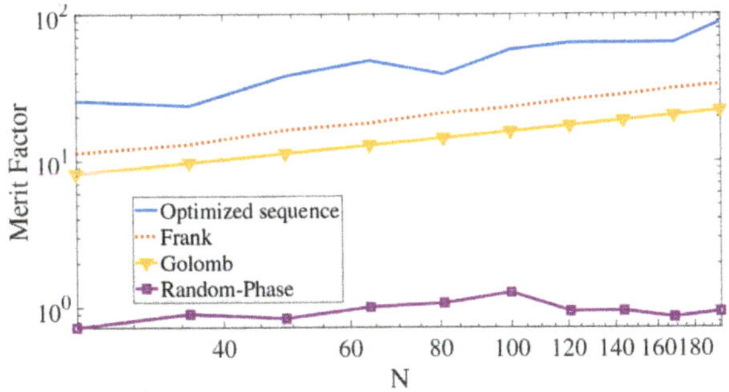

(a) Aperiodic MF values of the optimized sequences at different lengths in comparison with Frank, Golomb, and Random-Phase.

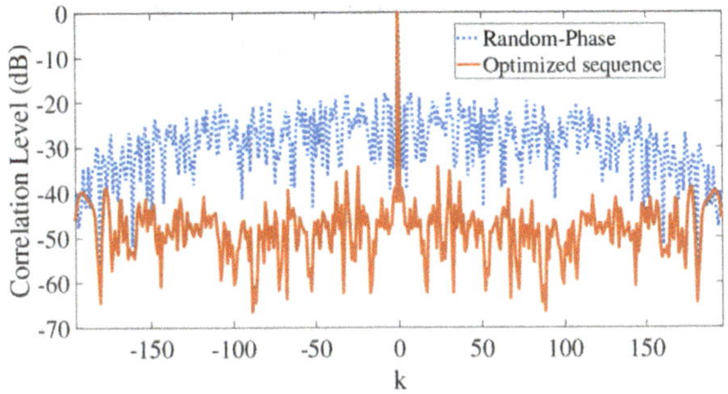

(b) Aperiodic auto-correlation function of random-phase and the optimized sequence at length $N = 196$.

Figure 9.6 *Performance of the optimized waveform in terms of aperiodic MF. (a) indicates MF values of optimized sequence in comparison with Frank, Golomb, and Random-Phase at different lengths and (b) depicts aperiodic auto-correlation functions of random-Phase and optimized sequence.*

9.3.2.2 Design examples

To analyse the performance of the algorithm in terms of ISL minimization, we set $p = 2$ in the l_p-norm optimization framework. In Figure 9.6, we initialize Algorithm 9.2 by Frank sequences and compare the merit factors of Frank, Golomb, random-phase and optimized sequences under continuous phase constraint. The code lengths are $N = [5, 6, \ldots, 14]^2$, since the length of Frank sequences is constrained to a perfect square. For all the considered code lengths, the optimized waveforms give the largest merit factors. In this figure, the correlation levels of the optimized waveform and a random-phase sequence at $N = 196$ is also depicted.

In [23], the effect of the parameter p on l_p-norm minimization is analysed. Assessing the results, it has been shown that gradually increasing the p value is probably the best approach to obtain the lowest PSL. Consequently, we use a similar increasing scheme to minimize the PSL and set $p = 2, 4, 6, 8, 10, 12$; the sequence optimized for $p = 2$ is used as an initial sequence for $p = 4$ and so on. In Figure 9.7, we compare the PSLR (dB) values of the aperiodic auto-correlation function of

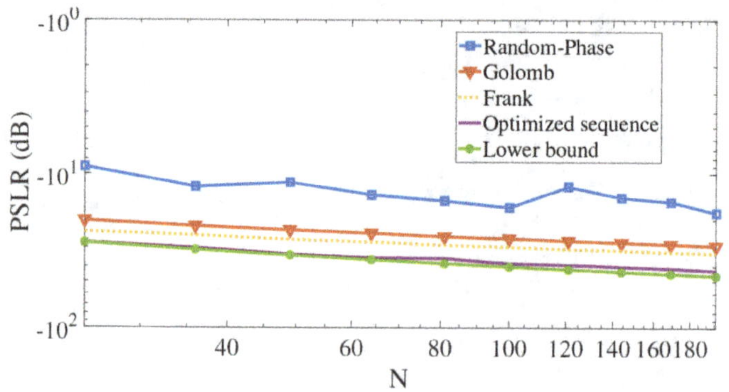

(a) Aperiodic PSLR (dB) values of the optimized sequences at different lengths in comparison with Frank, Golomb, Random-Phase, and the lower bound.

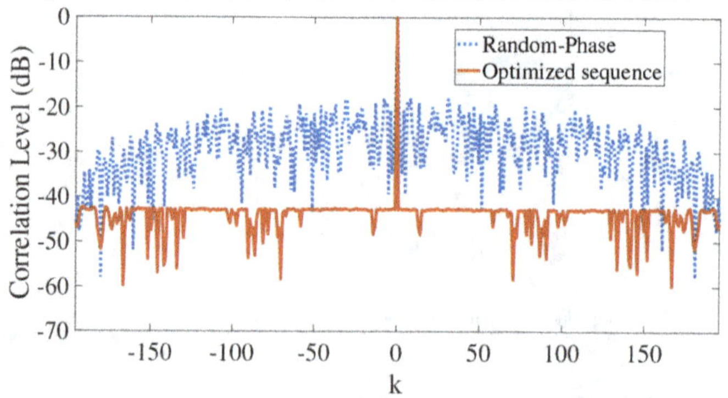

(b) Aperiodic auto-correlation function of Random-Phase and the optimized sequence at length $N = 196$.

Figure 9.7 Performance of the optimized waveform in terms of aperiodic PSL.

Frank, Golomb, random-phase and the optimized sequences under constant modulus constraint (continuous phase) on a log–log scale. Additionally, in Figure 9.7 the lower bound for the aperiodic PSLR values is also depicted. Note that for a unimodular sequence, the lowest possible aperiodic PSL is 1, since $|r_{N-1}| = |x_N x_1^*| = 1$. Interestingly, the PSLR values of the optimized sequences that are obtained when **Algorithm 9.2** is initialized by Frank, are very close to the lower bound. Figure 9.7 also represents the correlation levels of the optimized sequence and a random-phase code at length $N = 196$. We note that the correlation sidelobes of the random-phase sequence are comparatively large, while the optimized sequence exhibits relatively uniform correlation sidelobes at different lags.

9.3.3 *Application 3: waveform design with good correlation properties in hybrid phased mimo radar*

The realization of the potential of MIMO radars, which employ several antennas at the radar station, has attracted a significant interest to waveform design and diversity. Unlike a standard phased-array radar, a MIMO radar can transmit multiple probing signals that can be distinct in the receiver. The resulting waveform and spatial diversity provides MIMO radar with superior capabilities in comparison to the traditional radar settings. However, phased-array systems offer higher SNR. To exploit the benefits of MIMO and phased-array systems, HPM systems have been conceived. In the sequel, we briefly describe the system model for HPM and provide a waveform optimization mechanism.

9.3.3.1 HPM: A hybrid system

Similar to a MIMO radar, a HPM radar has separate transmit and receive arrays consisting of N_T and N_R elements, respectively. However, the HPM transmit array is partitioned into D subarrays ($1 \leq D \leq N_T$) which are, in general, allowed to overlap and have different sizes. The number of elements in the dth subarray are N_d such that $1 \leq N_d \leq N_T$ and $\sum_{d=1}^{D} N_d = N_T$.

All elements of the same subarray transmit scaled versions of a single waveform that is mutually orthogonal to the waveforms transmitted by other subarrays. The equivalent virtual ULA of HPM array, which is a generalization of MIMO array, consists of DN_R elements while the virtual array corresponding to a MIMO radar has N_TN_R elements. Let each antenna transmit a code vector composed of N sub-pulses (intra-pulse coding). The m-th transmit antenna sends the signal

$$\mathbf{x}_m = [x_m(1), x_m(2), \ldots, x_m(N)]^T \in \mathbb{C}^N, \tag{9.43}$$

where $x_m(n)$ is the n-th sub-pulse of the transmit code vector $\mathbf{x}_m$ and the sequence energy satisfies $\mathbf{x}_m^H\mathbf{x}_m = N$. The aperiodic cross-correlation of $\{x_m(n)\}_{n=1}^N$ and $\{x_l(n)\}_{n=1}^N$ at lag k is defined as,

$$(\mathbf{x}_m \circledast \mathbf{x}_l)_k^{\mathscr{A}} \equiv r_{ml}^{\mathscr{A}}(k) = \textstyle\sum_{n=1}^{N-k} x_m(n)x_l^*(n+k) = r_{lm}^{\mathscr{A}*}(-k), \tag{9.44}$$

Similarly, the periodic cross-correlation of $\{x_m(n)\}_{n=1}^N$ and $\{x_l(n)\}_{n=1}^N$ at lag k is,

$$(\mathbf{x}_m \circledast \mathbf{x}_l)_k^{\mathscr{P}} \equiv r_{ml}^{\mathscr{P}}(k) = \textstyle\sum_{n=1}^{N} x_m(n)x_l^*(n+k)_{\mathrm{mod}(N)} = r_{lm}^{\mathscr{P}*}(-k), \tag{9.45}$$

where $m, l = 1, \ldots, N_T$, $-N+1 \leq k \leq N-1$ and $(.)^*$ denotes complex conjugate. Notice that, in both cases of aperiodic and periodic, when $m = l$, the cross-correlation becomes the auto-correlation function of $\{x_m(n)\}_{n=1}^N$. Similar to the previous section, we assume

$$(\mathbf{x}_m \circledast \mathbf{x}_l)_k \equiv r_{ml}(k) = \begin{cases} r_{ml}^{\mathscr{A}}(k), & \text{aperiodic} \\ r_{ml}^{\mathscr{P}}(k), & \text{periodic} \end{cases} \tag{9.46}$$

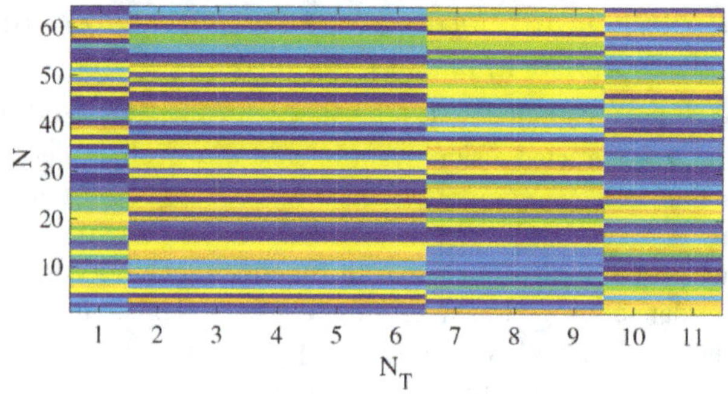

Figure 9.8 An example of non-overlapping subarrays. The similar colors show same waveform is transmitting from the antenna elements.

The problem of designing transmit waveforms for HPM radars can be considered in two cases, (a) non-overlapping subarrays and (b) overlapping subarrays. In this chapter, we only concentrate on the non-overlapping subarrays. In this case, let $\{\mathbf{x}_d\}_{d=1}^{D}$ be the independent waveforms forming the transmit code matrix $\mathbf{X} \in \mathbb{C}^{N\times N_T}$, such that

$$\mathbf{X} = \Big[\underbrace{\mathbf{x}_1, \mathbf{x}_1, \ldots,}_{N_1 \text{ times}} \underbrace{\mathbf{x}_2, \mathbf{x}_2, \ldots,}_{N_2 \text{ times}} \underbrace{\ldots, \mathbf{x}_D, \mathbf{x}_D, \ldots,}_{N_D \text{ times}} \Big], \tag{9.47}$$

which indicates that every subarray transmits N_D times the waveform $\mathbf{x}_d$. The ISL for the non-overlapping arrays is

$$f_o(\mathbf{X}) = \sum_{d=1}^{D} N_d \sum_{\substack{k=-N+1 \\ k\neq 0}}^{N-1} |r_{dd}(k)|^2 + \sum_{d=1}^{D}\sum_{\substack{l=1 \\ l\neq d}}^{D} N_d N_l \sum_{k=-N+1}^{N-1} |r_{dl}(k)|^2, \tag{9.48}$$

which is equivalent to the ISL defined for conventional MIMO radar systems where $N_d = 1$ for $d = 1, 2, \ldots, D$, and hence, $D = N_T$ [25].

Figure 9.8 depicts a graphical representation of a set of random-phase sequences when $N_T = 11$, $D = 4$, and the sequence length $N = 64$, when N_d is adopted (in order) from $\{1, 5, 3, 2\}$. In this figure, we use the same color when a similar waveform is used in the antenna elements.

This figure indicates intuitively the scheme of the transmit waveforms from every antenna elements.

9.3.3.2 CD adaptation in HPM

Under continuous and discrete phase constraints, the optimization problems are

$$\mathscr{P}_{\mathbf{X}}^{\infty} \begin{cases} \underset{\mathbf{X}}{\text{minimize}} & f_o(\mathbf{X}) \\ \text{subject to} & |x_d(n)| = 1 \end{cases}, \quad \mathscr{P}_{\mathbf{X}}^{M} \begin{cases} \underset{\mathbf{X}}{\text{minimize}} & f_o(\mathbf{X}) \\ \text{subject to} & x_d(n) \in \Omega_M \end{cases} \tag{9.49}$$

where $f_o(\mathbf{X})$ is defined in (9.46) for both aperiodic and periodic auto- and cross-correlation functions. The constraint $|x_d(n)| = 1$ in problem $\mathscr{P}_{\mathbf{X}}^{\infty}$ (continuous phase constraint) means that n-th subpulse at the d-th transmit waveform can be written as

$$x_d(n) = e^{j\phi_d(n)}, \quad d = 1, \ldots, D \text{ and } n = 1, \ldots, N, \tag{9.50}$$

with $\phi_d(n)$ being the phase of the n-th subpulse of the transmit code vector $\mathbf{x}_d$. Let $\widetilde{x}$ be n-th entry of the code vector $\mathbf{x}_d$, when $d = \widetilde{d}$ and $n = \widetilde{n}$, i. e., $\widetilde{x} = x_{\widetilde{d}}(\widetilde{n})$, and

$$\begin{aligned} \mathbf{x}_{\widetilde{d}} &= [x_{\widetilde{d}}(1), x_{\widetilde{d}}(2), \ldots, x_{\widetilde{d}}(\widetilde{n}), \ldots, x_{\widetilde{d}}(N)]^T \in \mathbb{C}^N, \\ \bar{\mathbf{x}}_{\widetilde{d}} &= [x_{\widetilde{d}}(1), x_{\widetilde{d}}(2), \ldots, 0, \ldots, x_{\widetilde{d}}(N)]^T \in \mathbb{C}^N. \end{aligned} \tag{9.51}$$

We assume $\widetilde{x}$ to be the only variable in code matrix $\mathbf{X}$. The auto- and cross-correlation functions with respect to $\widetilde{x}$ can be respectively obtained by

$$\begin{aligned} \widetilde{r}_{dd}(k) &= a_1(k)\widetilde{x} + a_2(k)\widetilde{x}^* + a_3(k), \\ \widetilde{r}_{dl}(k) &= b_1(k)\widetilde{x} + b_2(k), \end{aligned} \tag{9.52}$$

where[h], in case of aperiodic auto- and cross-correlations,

$$\begin{aligned} a_1(k) &\triangleq a_1^{\mathcal{A}}(k) = x_{\widetilde{d}}^*(\widetilde{n} + k)\mathbf{I}_A(\widetilde{n} + k), \\ a_2(k) &\triangleq a_2^{\mathcal{A}}(k) = x_{\widetilde{d}}(\widetilde{n} - k)\mathbf{I}_A(\widetilde{n} - k), \\ a_3(k) &\triangleq a_3^{\mathcal{A}}(k) = (\bar{\mathbf{x}}_{\widetilde{d}} \circledast \bar{\mathbf{x}}_{\widetilde{d}})_k^{\mathcal{A}}, \end{aligned} \tag{9.53}$$

and

$$\begin{aligned} b_1(k) &\triangleq b_1^{\mathscr{A}}(k) = x_l^*(\widetilde{n} + k)\mathbf{I}_A(\widetilde{n} + k), \\ b_2(k) &\triangleq b_2^{\mathscr{A}}(k) = (\bar{\mathbf{x}}_{\widetilde{d}} \circledast \mathbf{x}_l)_k^{\mathscr{A}}. \end{aligned} \tag{9.54}$$

Also, in case of periodic auto- and cross-correlations,

$$\begin{aligned} a_1(k) &\triangleq a_1^{\mathcal{P}}(k) = x_{\widetilde{d}}^*(\widetilde{n} + k)\mathbf{I}_A(\widetilde{n} + k) + x_{\widetilde{d}}^*(\widetilde{n} + k - N)\mathbf{I}_A(\widetilde{n} + k - N), \\ a_2(k) &\triangleq a_2^{\mathcal{P}}(k) = x_{\widetilde{d}}(\widetilde{n} - k)\mathbf{I}_A(\widetilde{n} - k) + x_{\widetilde{d}}(\widetilde{n} - k + N)\mathbf{I}_A(\widetilde{n} - k + N), \\ a_3(k) &\triangleq a_3^{\mathcal{P}}(k) = (\bar{\mathbf{x}}_{\widetilde{d}} \circledast \bar{\mathbf{x}}_{\widetilde{d}})_k^{\mathcal{P}}, \end{aligned} \tag{9.55}$$

and

$$\begin{aligned} b_1(k) &\triangleq b_1^{\mathscr{P}}(k) = x_l^*(\widetilde{n} + k)\mathbf{I}_A(\widetilde{n} + k) + x_l^*(\widetilde{n} + k + N)\mathbf{I}_A(\widetilde{n} + k + N) \\ &+ x_l^*(\widetilde{n} + k - N)\mathbf{I}_A(\widetilde{n} + k - N), \\ b_2(k) &\triangleq b_2^{\mathscr{P}}(k) = (\bar{\mathbf{x}}_{\widetilde{d}} \circledast \mathbf{x}_l)_k^{\mathscr{P}}. \end{aligned} \tag{9.56}$$

[h]The dependence of the coefficients $a_1(k)$, $a_2(k)$, $a_3(k)$, $b_1(k)$ and $b_2(\mathbf{k})$ to d and n are omitted due to the simplicity of the notation.

Notice that $\mathbf{I}_A(v)$ is the indicator function of the set $A = \{1, 2, \ldots, N\}$, i.e., $\mathbf{I}_A(v) = 1$ if $v \in A$, otherwise $\mathbf{I}_A(v) = 0$. Using the CD framework, the idea is to iteratively minimize $f_o(\widetilde{x})$ by updating d and n while covering all the entries of the code matrix $\mathbf{X}$. Every iteration means that all entries of the code matrix $\mathbf{X}$ are updated, which needs $N \times D$ times optimization of

$$\mathcal{P}_x^{\infty} \begin{cases} \underset{\widetilde{x}}{\text{minimize}} & f_o(\widetilde{x}) \\ \text{subject to} & |\widetilde{x}| = 1 \end{cases}, \quad \mathcal{P}_x^{M} \begin{cases} \underset{\widetilde{x}}{\text{minimize}} & f_o(\widetilde{x}) \\ \text{subject to} & \widetilde{x} \in \Omega_M \end{cases}, \tag{9.57}$$

for $n = 1, \ldots, N$, and $d = 1, \ldots, D$. Using proposition 9.37 at the previous section, it would be easy to show by the change of variable $\widetilde{x} = e^{j\widetilde{\phi}}$, and $\widetilde{\beta} = \tan\frac{\widetilde{\phi}}{2}$, we obtain

$$\begin{aligned} &\sum_{\substack{k=-N+1 \\ k\neq 0}}^{N-1} \left| a_1(k)e^{j\widetilde{\phi}} + a_2(k)e^{-j\widetilde{\phi}} + a_3(k) \right|^2 = \frac{p_a(\widetilde{\beta})}{q(\widetilde{\beta})}, \\ &\sum_{k=-N+1}^{N-1} |b_1(k)e^{j\widetilde{\phi}} + b_2(k)|^2 = \frac{p_b(\widetilde{\beta})}{q(\widetilde{\beta})}, \end{aligned} \tag{9.58}$$

where

$$\begin{aligned} p_a(\widetilde{\beta}) &= \mu_1\widetilde{\beta}^4 + \mu_2\widetilde{\beta}^3 + \mu_3\widetilde{\beta}^2 + \mu_4\widetilde{\beta} + \mu_5, \\ p_b(\widetilde{\beta}) &= \kappa_1\widetilde{\beta}^4 + \kappa_2\widetilde{\beta}^3 + \kappa_3\widetilde{\beta}^2 + \kappa_4\widetilde{\beta} + \kappa_5, \\ q(\widetilde{\beta}) &= (1 + \widetilde{\beta}^2)^2, \end{aligned} \tag{9.59}$$

with μ_i, and κ_i, $i = 1, 2, \ldots, 5$, as real value coefficients and can be obtained from (9.39). By ignoring Δ in $f_o(\widetilde{x})$, the objective function can be simplified as a fraction of four degrees polynomials

$$\frac{\lambda_1\widetilde{\beta}^4+\lambda_2\widetilde{\beta}^3+\lambda_2\widetilde{\beta}^2+\lambda_4\widetilde{\beta}+\lambda_5}{(1+\widetilde{\beta}^2)^2}, \tag{9.60}$$

where λ_i, $i = 1, \ldots, 5$ are real-valued coefficients specified as

$$\lambda_i = N_d\mu_i + \kappa_i \sum_{\substack{l=1 \\ l\neq d}}^{D} N_d N_l.$$

Finally, the optimization problem with respect to the variable $\widetilde{\beta}$ can expressed as

$$\mathcal{P}_{\beta}^{\infty} \begin{cases} \underset{\widetilde{\beta}}{\text{minimize}} & \dfrac{\lambda_1\widetilde{\beta}^4 + \lambda_2\widetilde{\beta}^3 + \lambda_3\widetilde{\beta}^2 + \lambda_4\widetilde{\beta} + \lambda_5}{(1 + \widetilde{\beta}^2)^2} \\ \text{subject to} & \widetilde{\beta} \in \mathbb{R} \end{cases} \tag{9.61}$$

Consequently, the optimal $\widetilde{\beta}^\star$ can be obtained by finding the real roots of the first-order derivative of the objective function in P_{β}^{∞} and evaluating the objective function in these points as well as the boundaries. After finding the optimal value $\widetilde{\beta}^\star$, we obtain $\widetilde{\phi}^\star = 2\,\text{atan}(\widetilde{\beta}^\star)$ and $\widetilde{x}^\star = e^{j\widetilde{\phi}^\star}$.

In problem $\mathscr{P}^M_{\widetilde{x}}$, the objective function in terms of $\widetilde{\phi} = \arg(\widetilde{x}) \in [0, 2\pi)$ can be expanded as

$$
\begin{aligned}
f_o(\widetilde{\phi}) &= N_d \sum_{\substack{k=-N+1 \\ k \neq 0}}^{N-1} \left| a_1(k)e^{j\widetilde{\phi}} + a_2(k)e^{-j\widetilde{\phi}} + a_3(k)\right|^2 \\
&+ \sum_{\substack{l=1 \\ l \neq d}}^{D} N_d N_l \sum_{k=-N+1}^{N-1} |b_1(k)e^{j\widetilde{\phi}} + b_2(k)|^2 + \Delta,
\end{aligned}
\tag{9.62}
$$

and the optimization problem would be

$$
\mathscr{P}^M_{\widetilde{\phi}} \begin{cases} \underset{\widetilde{\phi}}{\text{minimize}} & f_o(\widetilde{\phi}) \\ \text{subject to} & \widetilde{\phi} \in \arg(\Omega_M) \end{cases}
\tag{9.63}
$$

In this problem, the optimal phase code entry $\widetilde{\phi}^\star$ can be obtained by exhaustive search among the feasible set, i.e., all the possible values in $\arg(\Omega_M)$. However, following proposition 9.29, the solution of $\mathscr{P}^M_{\widetilde{\phi}}$ can be obtained using FFT. Let us define[i]

$$
\begin{aligned}
\zeta_d(k) &= \left|\text{FFT}\left[a_1(k), a_3(k), a_2(k), \mathbf{0}_{1\times(M-3)}\right]^T\right|^2 \in \mathbb{R}^M, \\
\zeta_l(k) &= \left|\text{FFT}\left[b_1(k), b_2(k), \mathbf{0}_{1\times(M-2)}\right]^T\right|^2 \in \mathbb{R}^M,
\end{aligned}
\tag{9.64}
$$

and

$$
\widetilde{\boldsymbol{\nu}} = N_d \sum_{\substack{k=-N+1 \\ k \neq 0}}^{N-1} \zeta_d(k) + \sum_{\substack{l=1 \\ l \neq d}}^{D} N_d N_l \sum_{k=-N+1}^{N-1} \zeta_l(k) \in \mathbb{R}^M.
\tag{9.65}
$$

The optimal solution to problem $\mathscr{P}^M_{\widetilde{\phi}}$ can be obtained by finding $\widetilde{i}^\star = \arg\min \widetilde{\boldsymbol{\nu}}$ and setting $\widetilde{\phi}^\star = \frac{2\pi(\widetilde{i}^\star - 1)}{L}$. Hence, the optimal phase code entry can be efficiently computed as $\widetilde{x}^\star = e^{j\widetilde{\phi}^\star}$ using DFT/FFT.

9.3.3.3 Design examples

Following, we present some numerical examples to show the performance of Algorithm 9.2 in optimizing waveforms with good ISLR values at HPM radars. We set $N_T = 11$, $D = 4$ and the sequence length $N = 64$. In case of discrete phase, we set alphabet size $M = 16$. In Figure 9.9, the convergence behavior of the CD

[i]The absolute and the square power are element wise.

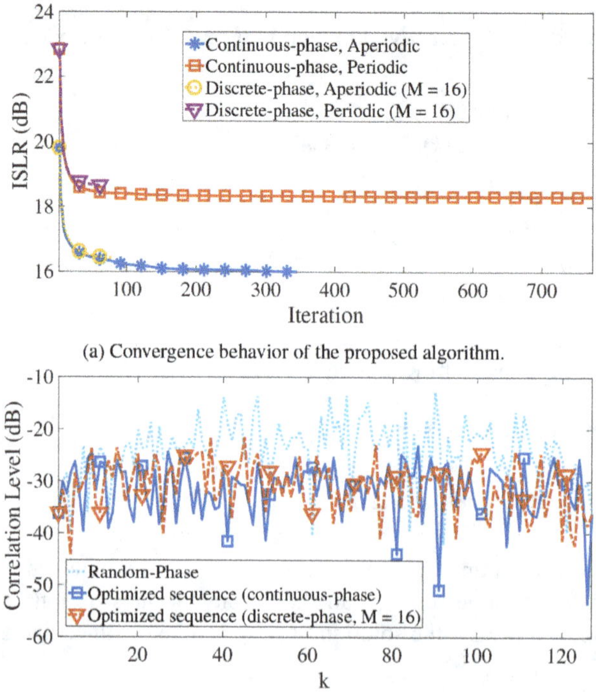

(a) Convergence behavior of the proposed algorithm.

(b) Aperiodic cross-correlation of the optimized sequences.

Figure 9.9 *Convergence behavior of the proposed algorithm and cross-correlations of the optimized sequences*

method in case of minimizing ISLR, and aperiodic cross-correlation of 2nd and 4 th optimized sequences are depicted. Indeed, the ISLR values decrease monotonically and converge to a stationary point, when the iteration increases. Also, very small aperiodic cross-correlation values of the obtained sequences indicate the competence of the proposed method for designing set of sequences with enhanced ISL values.

9.4 Signal recovery and prototype development

In the previous section, many applications of optimization for waveform design have been presented. However, the performance metric in the design relates to the waveform properties and not to its impact on the overall performance. In this section, we consider this aspect and present two system analyses. At first, we consider a simulation framework to evaluate the efficiency of signal recovery. Subsequently, we move beyond the simulation framework to demonstrate the feasibility of the proposed methods through a custom-built SDR-based prototype.

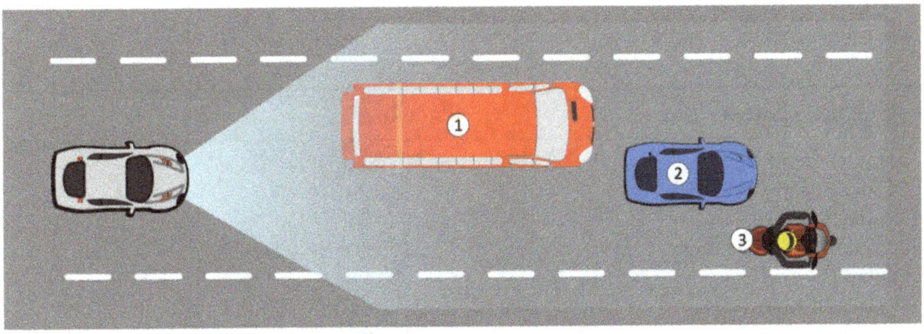

Figure 9.10 *An illustration of mmWave radar sensing in LRR mode. The automotive radar deployed on a vehicle on the left and could constantly estimate the distance between the vehicle it is mounted on and the vehicles in front of it, and alert the driver when the two become too close. Three targets are observable in the scene, provided that appropriate waveform been selected in the radar transmitter unit.*

9.4.1 Signal recovery

Signal recovery is essential in a wide range of applications across active sensing and communication systems. To implement different applications, we often need to pre-process and compare signals, design digital filters, transform signals, perform measurements and detect events. In this section, we indicate the effect of choosing appropriate waveforms in signal recovery for radar application, precisely by detecting a weak signal in the vicinity of a strong signal.

Consider a mmWave automotive Long Range Radar (LRR) used for adaptive cruise control (ACC). The radar occupies a portion of the band around 79 GHz and constantly estimates the distance between the autonomous vehicle and the objects in front of it, and reacts appropriately when an object become too close to the vehicle (see Figure 9.10). A typical parameters of such a radar system is reported in Table 9.1.

The received signal is a time-delayed copy of the transmitted signal where the delay is related to the range of target. The maximum range of the radar is 225 m, and the system needs to be able to distinguish two different targets that are around 0.5 m apart. Measurements in the mmWave band show that bus can have an RCS up to 30 dBsm, whereas small targets can have an RCS < -20 dBsm when seen from an unfortunate angle [8]. This large dynamic range of RCS may cause to a miss/false detection as sketched in the Figure 9.1. A worst case scenario can happen when the other vehicles have similar relative speed with the automotive vehicle and are located in a similar angular beam. Considering a typical angular resolution 5 degree for the automotive radar, in a distance >50 m, the spacing between the vehicles should be >4 m to be possibly discriminated in angle. In this case, to precisely recover the signal, the transmitting waveform should have low PSL and ISL.

Table 9.1 PMCW radar parameters

System parameters	Value
Operating frequency	79 (GHz)
Transmitting power	12 (dBm)
Antenna gain	10 (dB)
Maximum detection range	225 (m)
Bandwidth	300 (MHz)
Range resolution	0.5 (m)
Receiver noise figure	15 (dB)
Transmission time	27.3 (μs)
Inter-pulse duration	10.7 (μs)
PMCW code length	8 191
Number of pulses	256
Doppler FFT size	512
Max unambiguous relative velocity	89 (km/h)
Total active frame time	7.68 (ms)

Let us assume that the three targets that are depicted in Figure 9.10 are located at a distance of 50, 70 and 90 m from the automotive PMCW radar, with RCS 30, 10 and 0 dBsm, respectively. We assume that the relative speed of all the targets and the radar is equal to 50 kmph. In the processing chain of the PMCW radar, we pass the data cube through a matched filter and perform the Doppler FFT.

In Figure 9.11, we visualize the recovered signal in the range-Doppler domain, which can help to intuitively understand connections between the adopted waveform and the targets. In Figure 9.11, we use random-phase as the transmit sequence in the automotive radar, while in Figure 9.11 we use the optimized sequence obtained from solving the optimization problem $\mathscr{H}^{\infty}$ in Application 2 of Section 9.3. The true locations of the three targets are indicated with squares (and numbers) in this figure.

The range-Doppler map is acted upon by a 2D-CFAR detector considering 4 samples as guard and 8 samples as training cells. The false alarm probability is adjusted to 10^{-3} for the CFAR detector. In Figure 9.11, first and second targets are detected, but the third target is missed. Also, many false targets are reported due to the high energy of the sidelobes of transmitting sequence. In Figure 9.11, using same CFAR detector and false alarm probability, the three targets are precisely detected and there is no false alarm, which shows the advantage of transmitting the optimized waveform.

To better visualize the sidelobes of the transmitting waveform which masked the weak target in Figure 9.12, we plot the Doppler cut of the range-Doppler map when the relative speed is 50 kmph. The third target which has a lower RCS and is located at a larger distance from the automotive radar was missed when random-phase is employed as the transmitting sequence because of the CFAR threshold. Due to better waveform design, the three targets can be precisely detected if the optimized sequence be employed as the transmitting waveform.

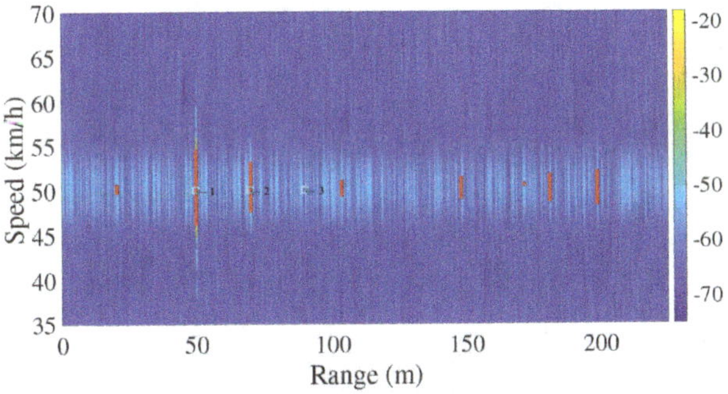

(a) Random-Phase.

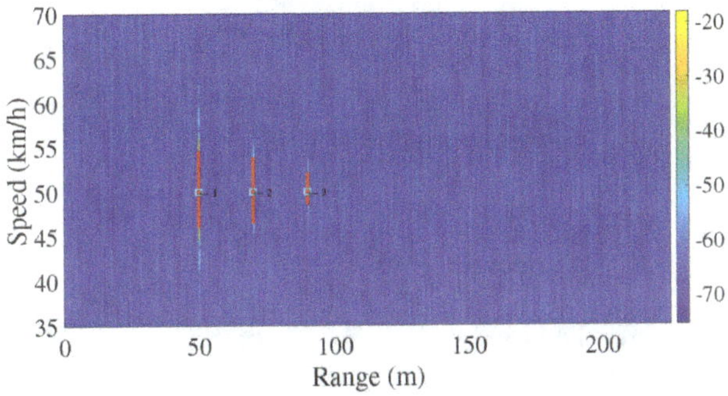

(b) Optimized sequence.

Figure 9.11 Comparing range-Doppler map by using random-phase and the optimized sequence as the transmit waveform in a PMCW radar

9.4.2 Prototype

In this section, we verify the operational feasibility of the designed algorithm by implementing it using a custom-built SDR-based prototype. We consider a spectrum sharing scenario as illustrated in Figure 9.13, where a vehicle comprising a PMCW radar coexists with communications as the source of interference. To avoid creating interference to the communications and also enhancing the radar performance, we adopt the solution of problem $\mathscr{P}_{\mathbf{x},\boldsymbol{\alpha}}$ with $\theta = 0.1$, $N = 400$, $M = 128$ as the optimized transmitting sequence for the radar system.

In Figure 9.14 (a), the block diagram of the developed prototype is depicted. To generate the communications signal, we used GNU Radio OFDM block and emulated the presence/absence of multiple users by suppressing selected OFDM subcarriers. To perform the cognition and sense the radar environment on a continuous basis, we developed the depicted application in Figure 9.14–(b) in LabView NXG 3.1, which was connected to ettus USRP B210. We further developed a cognitive radar application depicted in Figure 9.14–(c) in LabView NXG 3.1 and connected it to NI-USRP 2944R. To verify the performance of the optimization algorithm in a controlled environment, we established all the connections shown in Figure 9.14–(a) using RF cables and splitters/combiners.

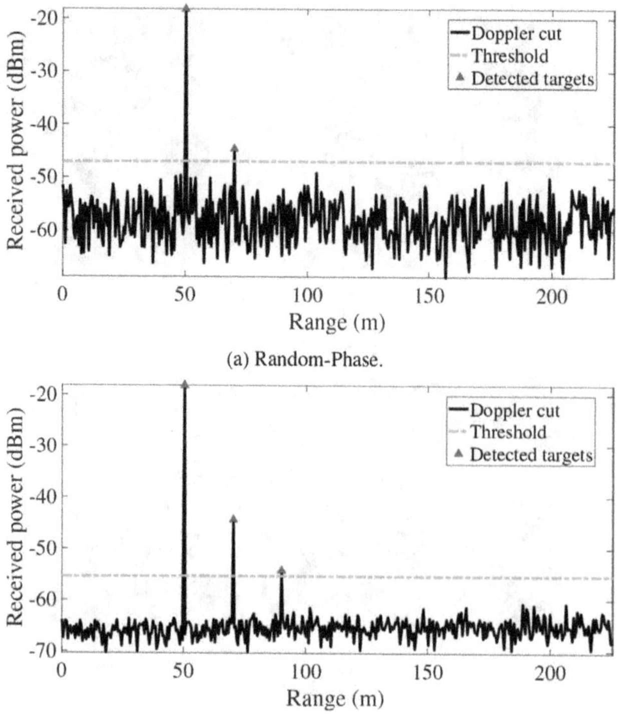

(a) Random-Phase.

(b) Optimized Sequence.

Figure 9.12 *Doppler cut of the range-Doppler map when relative speed is 50 kmph. The weak target is missed when random-phase is employed as the transmitting sequence. The three targets can be precisely detected if the optimized sequence be employed as the transmitting waveform.*

In the experiment of Figure 9.14, the communications signal occupies 10 MHz bandwidth at the center frequency 3.775 GHz, and radar occupies 20 MHz in the same center frequency. The spectrum sensing application estimates the vacant bands on-the-fly using an energy detector and sends the list of occupied frequencies to the cognitive radar through a network connection. The cognitive radar optimizes its waveform based on the proposed algorithm in Section 9.3.1 and performs standard radar the receive processing blocks, including matched filtering, windowing and range-Doppler processing.

9.5 Summary

The synthesis of phase sequences exhibiting good correlation features has been addressed in this chapter. Specifically, PSL, ISL and spectrum shaping have been adopted as performance metrics and either a continuous or a discrete phase

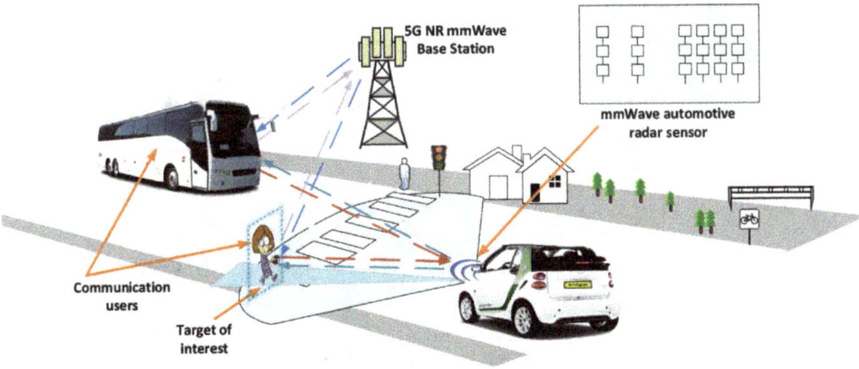

Figure 9.13 *An illustration of PMCW automotive radar sensing in coexistence with the communications. The kid is the target-of-interest for the automotive radar deployed on a vehicle on the right and could cross the street in front of the car, whereas the bus is approaching the car from the opposite direction in a neighboring lane. The buildings and vegetation comprise the clutter. At the same time, the pedestrian and the bus are equipped with communications/radar transmitters that make interference the autonomous vehicle.*

constraint is imposed at the design stage. The non-convex and, in general, NP-hard problems resulting from scalarization are handled via an iterative procedure through the CD optimization framework. Each iteration of the algorithm requires the solution of a non-convex problem which is handled in this chapter. At the analysis stage, some interesting case studies have been provided to illustrate the effectiveness of the developed CD design approach. These results highlight the ability of the new procedure to design constant modulus sequences with enhanced correlation properties towards efficient signal recovery.

Appendix: Proof of Proposition 9.37

Let

$$\begin{aligned} |r_k(e^{j\phi_d})|^2 &= \left| a_{1k}e^{j\phi_d} + a_{2k}e^{-j\phi_d} + a_{3k} \right|^2 \\ &= \left| \left(a_{dkr}e^{j\phi_d} + b_{dkr}e^{-j\phi_d} + c_{dkr} \right) + j\left(a_{dki}e^{j\phi_d} + b_{dki}e^{-j\phi_d} + c_{dki} \right) \right|^2 \\ &= A_{dk} + B_{dk} \end{aligned} \tag{9.66}$$

where $a_{dkr} = \Re(a_{1k})$, $b_{dkr} = \Re(a_{2k})$, $c_{dkr} = \Re(a_{3k})$, $a_{dki} = \Im(a_{1k})$, $b_{dki} = \Im(a_{2k})$ and $c_{dki} = \Im(a_{3k})$. Also,

$$A_{dk} = \left((a_{dkr} + b_{dkr})\cos(\phi_d) + (b_{dki} - a_{dki})\sin(\phi_d) + c_{dkr} \right)^2 \tag{9.67}$$

$$B_{dk} = \left((a_{dki} + b_{dki})\cos(\phi_d) + (a_{dkr} - b_{dkr})\sin(\phi_d) + c_{dki} \right)^2 \tag{9.68}$$

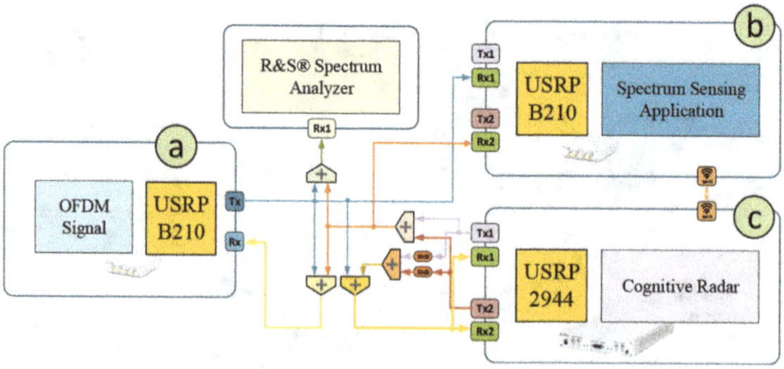

(a) Connection diagram of the developed prototype.

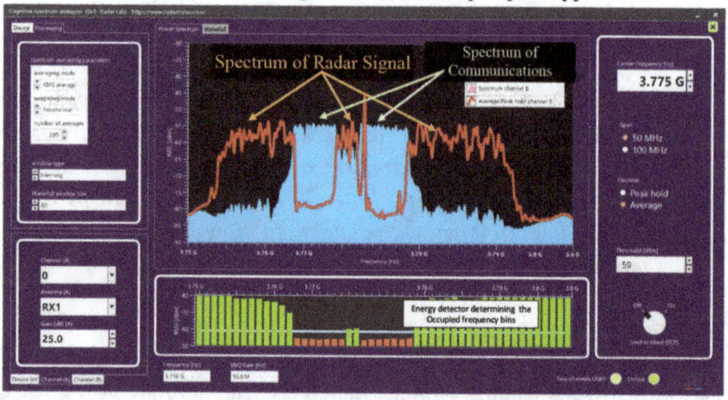

(b) Developed spectrum sensing application

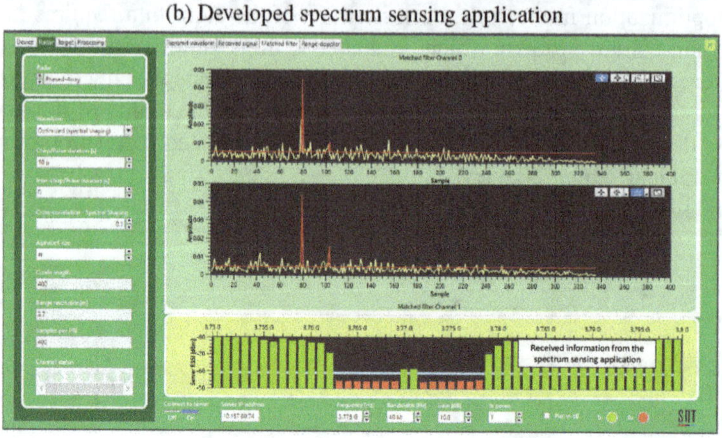

(c) Developed cognitive radar application

Figure 9.14 *Real-time experiment of the developed cognitive radar software. The optimized waveform is transmitted through the USRP 2944R. In a same time, the radar receives information from the spectrum sensing application to adapts its transmitting waveform in case any interference happens.*

Expanding (9.64) and using the trigonometric relationships $\sin\phi_d = \frac{2\tan\left(\frac{\phi_d}{2}\right)}{1+\tan^2\left(\frac{\phi_d}{2}\right)}$, $\cos\phi_d = \frac{1-\tan^2\left(\frac{\phi_d}{2}\right)}{1+\tan^2\left(\frac{\phi_d}{2}\right)}$ and defining $\beta_d = \tan\left(\frac{\phi_d}{2}\right)$ we obtain

$$A_{dk} = \frac{\mu'_{dk}\beta_d^4+\kappa'_{dk}\beta_d^3+\xi'_{dk}\beta_d^2+\eta'_{dk}\beta_d+\rho'_{dk}}{(1+\beta_d^2)^2} \tag{9.69}$$

with

$$\begin{aligned}
\mu'_{dk} &= (a_{dkr}+b_{dkr})^2 - 2c_{dkr}(a_{dkr}+b_{dkr}) + c_{dkr}^2 \\
\kappa'_{dk} &= -4(a_{dkr}+b_{dkr})(b_{dki}-a_{dki}) + 4c_{dkr}(b_{dki}-a_{dki}) \\
xi'_{dk} &= -2(a_{dkr}+b_{dkr})^2 + 2c_{dkr}^2 + 4(b_{dki}-a_{dki})^2 \\
\eta'_{dk} &= 4(a_{dkr}+b_{dkr})(b_{dki}-a_{dki}) + 4c_{dkr}(b_{dki}-a_{dki}) \\
\rho'_{dk} &= (a_{dkr}+b_{dkr})^2 + 2(a_{dkr}+b_{dkr})c_{dkr} + c_{dkr}^2
\end{aligned} \tag{9.70}$$

A similar procedure on B_{dk} yields

$$B_{dk} = \frac{\mu''_{dk}\beta_d^4+\kappa''_{dk}\beta_d^3+\xi''_{dk}\beta_d^2+\eta''_{dk}\beta_d+\rho''_{dk}}{(1+\beta_d^2)^2} \tag{9.71}$$

where

$$\begin{aligned}
\mu''_{dk} &= (a_{dki}+b_{dki})^2 - 2c_{dki}(a_{dki}+b_{dki}) + c_{dki}^2 \\
\kappa''_{dk} &= -4(a_{dki}+b_{dki})(a_{dkr}-b_{dkr}) + 4c_{dki}(a_{dkr}-b_{dkr}) \\
\xi''_{dk} &= -2(a_{dki}+b_{dki})^2 + 2c_{dki}^2 + 4(a_{dkr}-b_{dkr})^2 \\
\eta''_{dk} &= 4(a_{dki}+b_{dki})(a_{dkr}-b_{dkr}) + 4c_{dki}(a_{dkr}-b_{dkr}) \\
\rho''_{dk} &= (a_{dki}+b_{dki})^2 + 2(a_{dki}+b_{dki})c_{dki} + c_{dki}^2
\end{aligned} \tag{9.72}$$

Finally,

$$|\widetilde{r}_k(\beta_d)|^2 = \frac{\mu_{1k}\beta_d^4+\mu_{2k}\beta_d^3+\mu_{3k}\beta_d^2+\mu_{4k}\beta_d+\mu_{5k}}{(1+\beta_d^2)^2} \tag{9.73}$$

where $\mu_{1k} = \mu'_{dk} + \mu''_{dk}$, $\mu_{2k} = \kappa'_{dk} + \kappa''_{dk}$, $\mu_{3k} = \xi'_{dk} + \xi''_{dk}$, $\mu_{4k} = \eta'_{dk} + \eta''_{dk}$, $\mu_{5k} = \rho'_{dk} + \rho''_{dk}$.
Similarly, let

$$\Re\left\{r_k^*(e^{-j\phi_d})\frac{r_k^{(n)}}{\left|r_k^{(n)}\right|}\right\} = \Re\left\{\left[a_{1k}^*e^{-j\phi_d} + a_{2k}^*e^{j\phi_d} + a_{3k}^*\right]\frac{r_k^{(n)}}{\left|r_k^{(n)}\right|}\right\} \tag{9.74}$$
$$= (\widetilde{a}_{dkr} + \widetilde{b}_{dkr})\cos(\phi_d) + (\widetilde{a}_{dki} - \widetilde{b}_{dki})\sin(\phi_d) + \widetilde{c}_{dkr}$$

where
and $\widetilde{a}_{dk} = \widetilde{a}_{dkr} + j\widetilde{a}_{dki}$, $\widetilde{b}_{dk} = \widetilde{b}_{dkr} + j\widetilde{b}_{dki}$.

Similar to Appendix: Proof of proposition 9.37, using the change of variable $\cos(\phi_d) = \frac{1-\beta_d^2}{1+\beta_d^2}$ and $\sin(\phi_d) = \frac{2\beta_d}{1+\beta_d^2}$,

$$\Re\left\{\widetilde{r}_k^*(\beta_d)\frac{r_k^{(n)}}{\left|r_k^{(n)}\right|}\right\} = \frac{1}{(1+\beta_d^2)^2}\{(\widetilde{a}_{dkr}+\widetilde{b}_{dkr})(1-\beta_d^4) + (\widetilde{a}_{dki}-\widetilde{b}_{dki})(2\beta_d)(1+\beta_d^2) + \widetilde{c}_{dkr}(1+\beta_d^2)^2\}. \tag{9.75}$$

Finally, defining $\kappa_{1k} = \widetilde{c}_{dkr} - \widetilde{a}_{dkr} - \widetilde{b}_{dkr}$, $\kappa_{2k} = 2(\widetilde{a}_{dki} - \widetilde{b}_{dki})$, $\kappa_{3k} = 2\widetilde{c}_{dkr}$, $\kappa_{4k} = 2(\widetilde{a}_{dki} - \widetilde{b}_{dki})$, $\kappa_{5k} = \widetilde{a}_{dkr} + \widetilde{b}_{dkr} + \widetilde{c}_{dkr}$ yields

$$\Re\left\{\widetilde{r}_k^*(\beta_d)\frac{r_k^{(n)}}{\left|r_k^{(n)}\right|}\right\} = \frac{\kappa_{1k}\beta_d^4 + \kappa_{2k}\beta_d^3 + \kappa_{3k}\beta_d^2 + \kappa_{4k}\beta_d + \kappa_{5k}}{(1+\beta_d^2)^2}. \tag{9.76}$$

References

[1] Bourdoux A., Ahmad U., Guermandi D. 'PMCW waveform and MIMO technique for a 79 GHz CMOS automotive radar'. 2016 IEEE Radar Conference (RadarConf); 2016. pp. 1–5.

[2] Alaee Kerahroodi M., Aubry A., De Maio A., Naghsh M.M., Modarres-Hashemi M. 'A coordinate-descent framework to design low PSL/ISL sequences'. *IEEE Transactions on Signal Processing*. 2017, vol. 65(22), pp. 5942–56.

[3] Schmidt K.-U. 'Binary sequences with small peak sidelobe level'. *IEEE Transactions on Information Theory*. 2012, vol. 58(4), pp. 2512–15.

[4] Barker R. 'Group synchronizing of binary digital systems'. *Communication Theory*. 1953, pp. 273–87.

[5] Nunn C.J., Coxson G.E. 'Best-known autocorrelation peak sidelobe levels for binary codes of length 71 to 105'. *IEEE Transactions on Aerospace and Electronic Systems*. 2008, vol. 44(1), pp. 392–5.

[6] Wolfmann J. 'Almost perfect autocorrelation sequences'. *IEEE Transactions on Information Theory*. 1992, vol. 38(4), pp. 1412–18.

[7] Cai K., Weng G., Cheng X. 'Binary almost-perfect sequence sets'. *IEEE Transactions on Information Theory*. 2010, vol. 56(7), pp. 3594–604.

[8] Van Thillo W., Gioffré P., Giannini V., *et al.* 'Almost perfect auto-correlation sequences for binary phase-modulated continuous wave radar'. 2013 European Microwave Conference; 2013. pp. 1803–6.

[9] Baden J.M. 'Legendre sequences for periodic non-coherent pulse compression'. *IET Radar, Sonar & Navigation*. 2016, vol. 10(1), pp. 225–6.

[10] Pursley M., Roefs H. 'Numerical evaluation of correlation parameters for optimal phases of binary shift-register sequences'. *IEEE Transactions on Communications*. 1979, vol. 27(10), pp. 1597–604.

[11] Velazquez-Gutierrez J.M., Vargas-Rosales C. 'Sequence sets in wireless communication systems: a survey'. *IEEE Communications Surveys & Tutorials*. 2017 Second quarter, vol. 19(2), pp. 1225–48.

[12] He H., Li J., Stoica P. '*Waveform Design for Active Sensing Systems*'. *Cambridge University Press*; 2012.

[13] Donoho D.L. 'Compressed sensing'. *IEEE Transactions on Information Theory*. 2006, vol. 52(4), pp. 1289–306.

[14] Hong M., Razaviyayn M., Luo Z.-Q., Pang J.-S. 'A unified algorithmic framework for block-structured optimization involving big data: with applications in machine learning and signal processing'. *IEEE Signal Processing Magazine*. 2016, vol. 33(1), pp. 57–77.

[15] Borwein P., Ferguson R. 'Polyphase sequences with low autocorrelation'. *IEEE Transactions on Information Theory*. 2005, vol. 51(4), pp. 1564–7.

[16] Huang W., Naghsh M.M., Lin R., Li J. 'Doppler sensitive discrete-phase sequence set design for MIMO radar'. *IEEE Transactions on Aerospace and Electronic Systems*. 2020, vol. 56(6), pp. 4209–23.

[17] Wright S.J. 'Coordinate descent algorithms'. *Mathematical Programming*. 2015, vol. 151(1), pp. 3–34.

[18] Nutini J., Schmidt M., Laradji I., *et al.* 'Coordinate descent converges faster with the Gauss-Southwell rule than random selection'. International Conference on Machine Learning; 2015. pp. 1632–41.

[19] Dimitri P. 'Nonlinear programming'. *The Journal of the Operational Research Society*. 1999, vol. 48, p. 334.

[20] Mishra K.V., Bhavani Shankar M.R., Koivunen V., Ottersten B., Vorobyov S.A. 'Toward millimeter-wave joint radar communications: a signal processing perspective'. *IEEE Signal Processing Magazine*. 2019, vol. 36(5), pp. 100–14.

[21] He H., Stoica P., Li J. 'Waveform design with stopband and correlation constraints for cognitive radar'. *2010 2nd International Workshop on Cognitive Information Processing*. 2010, pp. 344–9.

[22] Deb K. *Multi-Objective Optimization Using Evolutionary Algorithms*. ed. 16. John Wiley & Sons; 2001.

[23] Song J., Babu P., Palomar D.P. 'Sequence design to minimize the weighted integrated and peak sidelobe levels'. *IEEE Transactions on Signal Processing*. 2016, vol. 64(8), pp. 2051–64.

[24] Razaviyayn M., Hong M., Luo Z.-Q. 'A unified convergence analysis of block successive minimization methods for nonsmooth optimization'. *SIAM Journal on Optimization*. 2013, vol. 23(2), pp. 1126–53.

[25] Alaee-Kerahroodi M., Modarres-Hashemi M., Naghsh M.M. 'Designing sets of binary sequences for MIMO radar systems'. *IEEE Transactions on Signal Processing*. 2019, vol. 67(13), pp. 3347–60.

Chapter 10

Sea ice concentration estimation techniques using machine learning: an end-to-end workflow for estimating concentration maps from SAR images

Stefan Dominicus[1], Amit Kumar Mishra[2], and Christo Rautenbach[3]

Sea ice concentration (SIC) is an important metric used to characterize polar sea ice behavior. Understanding this behavior and accurately representing it is of critical importance for climate science research and also has important uses in the context of maritime navigation. An end-to-end workflow for generating learned concentration estimation models from synthetic aperture radar (SAR) data, trained on existing passive microwave (PMW) data, is presented here. A novel objective function was introduced to account for uncertainty in the PMW measurements, which can be extended to account for arbitrary sources of error in the training data, and a recent set of in situ observations was used to evaluate the reliability of the chosen PMW concentration estimation model. Google Colaboratory was used as the development platform, and all notebooks, training data, and trained models are available on GitHub. This chapter is an overview of the most interesting aspects of this investigation, and a detailed report is also available on GitHub.

GitHub Repository Link: https://github.com/stefandominicus/FYP_ML_SIC

[1]Honors Student, Electrical Engineering Department, University of Cape Town, South Africa
[2]Radar Remote Sensing Group, University of Cape Town, South Africa, akmishra@ieee.org
[3]Coastal and Estuarine Processes Group, Engineer at NIWA, New Zealand, christo.rautenbach@niwa.co.nz

10.1 Introduction

Polar sea ice coverage, and the phenomena that drive changes in this coverage, forms a critical link in the Earth's climate system. Sea ice is located right at the ocean–atmosphere interface and is thus intricately linked to climate system evolution [1]. Current models used to estimate SIC are derived from PMW radar satellite data, but these instruments offer poor spatial resolutions and are susceptible to interference due to atmospheric variations. Additionally, these models are usually calibrated for Arctic conditions, and evidence from recent in situ observations in the Southern Ocean suggests that their estimates are significantly biased and are not reliable in the Antarctic marginal ice zone (MIZ). Sea ice behavior is characterized by a number of important variables (including SIC) that have been tied to particular aspects of climate system evolution. It is clear that understanding the interplay between these variables is an important requirement for accurately modeling large-scale climate trends [2].

10.1.1 Context

Ice edge location has been the traditional choice for quantifying ice coverage and is defined as the point where the ice concentration first exceeds 15% [2]. However, recent investigations in the Southern Ocean have chosen to focus on the much broader MIZ, which is defined as the region for which the SIC lies between 15% and 80%. Studying concentration rather than edge location has highlighted many of the shortcomings in the current modeling techniques [3].

There are a few factors that make this a difficult problem to solve. First, consider that much of what is known in the Arctic in terms of climate dynamics, and sea ice in particular, has been shown to be of little use in the Antarctic region due to the significant differences in the underlying driving forces influencing sea ice dynamics in each region. There is currently no consensus on what factors may be the primary driving forces of sea ice dynamics in the Antarctic [2]. Additionally, much of the current understanding in this field regarding sea ice characteristics was derived from old Arctic sea ice, which is of questionable relevance to the present-day Arctic environment, let alone the Antarctic [4].

The second and, perhaps, more critical factor hindering the development of accurate Antarctic sea ice models is the chronic lack of validated data. Although more in situ measurement data is slowly becoming available, the need still exceeds what existing datasets have to offer. The accuracy of existing models may be sufficient for some long-term synoptic analysis, but coupled models are still not able to capture large-scale sea ice coverage features due to incomplete parameterizations of the system dynamics [4].

Satellite-based remote sensing has been proposed as a way to obtain more data from the polar regions, but perhaps the biggest obstacle when using satellite data is that some understanding of the sensed environment is still required in order to properly interpret the data provided by the remote platform [5]. Interpretation models have been developed for the Arctic region, aided by the superior availability of validation data. As before, these models can be transferred for use in the Antarctic but will not capture many important details [3], which renders the model all but useless when a high temporal resolution is desired [1].

Another factor that has, to some extent, precluded the use of satellite observations in Antarctic sea ice analysis is that so much of the characteristic behavior takes place during intense stormy conditions, where clouds would obscure the scene from view by a satellite [3]. Even if PMW instruments are used, the increased water and water vapor content in the atmosphere can significantly degrade the quality and accuracy of the reported information [6]. However, a promising alternative is to use SAR instruments that are not affected by cloud cover or sunlight and are much less sensitive to atmospheric water vapor content. The use of SAR platforms also presents the advantage of increased spatial resolution [4]. This improvement in spatial resolution will allow the interpretive models being developed to achieve greater intricacy [1].

10.1.2 Existing work

In this context, there is a clear desire for an improved modeling approach that is not reliant on existing numerical models, can offer improved spatiotemporal resolutions, is flexible and can be adapted to suit the region of interest, and takes into account the uncertainty present in any datasets used so as not to be hindered by the high variability of some existing measurements. A recent paper presented a novel method of SIC estimation using deep learning [6], which was the most comprehensive effort found in addressing these desires.

Traditionally, when higher spatial resolution information is required, it can be obtained by the manual interpretation of SAR images, which also have the advantage of being less susceptible to interference from atmospheric water content (compared to the more widely used PMW data). While this method is certainly able to produce accurate interpretations, it is very time-consuming and simply cannot keep up with the vast number of SAR images generated.

While deep learning is indeed well-suited to this kind of high-volume data processing task, previous efforts have relied on manually annotated SAR images for training, which are generally only available in specific regions and only describe SIC in discrete 10–20% concentration bands, which leads to coarse and biased interpretations. In [6], SIC information generated automatically from PMW data was used to generate concentration labels corresponding to a set of SAR images. These image-label pairs were then used to train a deep neural network based on the DenseNet architecture [7].

It is worth discussing some of the challenges, design choices, and outcomes presented in [6], as they were a major influence on the path taken in this investigation. First, the size of SAR images (around 10000 × 10000 pixels) is prohibitively large, and cannot be used directly in neural networks. This was addressed by breaking each image into smaller patches, which were each processed individually. Second, the difference in resolution between the SAR images and the existing SIC data necessitated the use of an interpolation and resampling scheme to generate a concentration label corresponding to the footprint of each SAR image. Third, not all of the available image patches were used for training – only those that contained a representative range of SIC values were included in the training dataset. In terms of neural network architecture, the popular DenseNet structure was used with simple modifications made to the output to allow for single value regression, rather than classification for which the architecture is already well known. Last, this approach presented evidence that models trained in this way were able to predict SIC features at a higher resolution than the data with which they were trained and also suggested that some models would inherit the flaws and inaccuracies of the PMW data used for training, whereas others were able to avoid these flaws and actually produce more accurate predictions than the data with which they were trained [6].

Much of the novelty in the original DenseNet structure was in the way it utilized skip connections to aid in gradient propagation during training, making larger networks more efficient to train [7]. One of the most well-known architectures to also leverage skip connections is U-Net [8], which was introduced one year prior to DenseNet. Where DenseNet was designed for classification, U-Net was designed for image segmentation tasks, and its encoder-decoder type structure is naturally suited to producing outputs of a similar dimension to the input. This is particularly interesting in this context, where the improved resolution of SAR images is perhaps one of the most important factors in trying to generate higher spatial resolution SIC maps. Additionally, the original U-Net implementation also used sample weighting maps in the training process, which served to emphasize particular features and caused the network to learn those features above others [8]. This method of training manipulation by error-weighting presents some interesting opportunities in terms of allowing the training process to account for known inaccuracies in the training data and may play an important role in allowing learned models to avoid inheriting the flaws present in the models used to generate training data.

10.1.3 Major contributions

The work presented here is part of a more extensive project with the intention of presenting the most interesting outcomes in a more concise manner. The full report and more details about the project can be found on GitHub. For the purposes of this chapter, here are the major contributions that will be discussed.

1. a fully cloud-based workflow for machine learning research and prototyping
2. building interactive tools in Google Collaboratory for ease of use
3. identification of an error in the Copernicus image data processor
4. insight into the benefits of multi-stage/mixed-data training schemes
5. evaluation of existing concentration estimates using in situ observations.

10.1.4 Chapter structure

The development process in this investigation can be broken into three broad sections:

1. data pre-processing
2. neural network development and
3. serving targeted predictions.

The full details of each of these sections can be found in the project GitHub repository, but for the purpose of this chapter, the discussion will cover an overview of the data processing pipeline, some discussion around the filtering, batching, and validation processes, followed by an overview of the machine learning models and custom objective function used for this investigation. Some of the major contributions are then discussed in more detail, justifying their significance in this context. Finally, some brief conclusions are presented, as well as recommendations for continued work in this direction.

10.2 Data pipeline design

Every stage of the processing pipeline, machine learning development, and model evaluation described in this chapter was built online using Google Colaboratory running Python 3. This presents a number of useful advantages, such as free access to GPU hardware accelerators for machine learning development, but is not without its challenges, such as limited runtime and issues around storage virtualization. Even so, the speed increase offered by the use of GPUs compared to CPUs is so significant that Colaboratory was an obvious choice in the absence of any other available hardware accelerators. Much of the focus during development was to produce an end-to-end pipeline capable of demonstrating that free cloud platforms can, in fact, be used for more comprehensive tasks. The solutions and workarounds presented here can be easily adapted to other projects and will hopefully make free cloud computing more accessible for prototyping and research. The specific Python modules used at each stage will be mentioned as they are used, but, for now, it is worth mentioning that all image processing was done using *OpenCV*; *Matplotlib* was used to display all results, and *Keras* was used for all machine learning development using the *TensorFlow* backend.

One of the complications that arise when trying to work with machine learning using an online development environment such as Colaboratory is the need to access large datasets for training each model. While premium services such as Google Cloud or AWS EC2 are highly customizable, free services such as Colaboratory always spin up as a fresh instance, and any previous work will have been lost. To make the most of this free service, a Google Drive folder was mounted as a remote storage device to each new instance. All files were then saved into this remote drive so that work could continue from one instance to the next. This method of data management has some significant speed limitations when dealing with larger datasets, which had to be accommodated in the processing pipeline.

10.3 Data processing and validation

10.3.1 Creating a labeled dataset

In supervised machine learning, much of the challenge often lies in finding or creating suitable datasets on which to train the machine learning models. In this case, the objective was to generate concentration maps from SAR images. Training data was generated from an existing PMW-derived SIC model. The resulting dataset contained image–label pairs, where the images are resampled dual-polarization SAR measurements and the labels are the concentration and uncertainty patches corresponding to the SAR image footprint for the same day. In this investigation, SENTINEL-1 SAR data was chosen, as it has good coverage over polar regions and is freely available for research use. The PMW-derived SIC product chosen is published by the Ocean and Sea Ice Satelite Application Facility (OSI SAF) and designated OSI-401-b. This product has global coverage, meaning that its concentration maps cover both polar regions fully. Since each individual SAR image has a much smaller coverage, a resampling and interpolation algorithm was devised, which could generate concentration and uncertainty patches from the global SIC chart that matched the location and resolution of the SAR image. This process of automatically generating label data was crucial for this methodology, as training was not limited to a handful of manually interpreted samples as has often been the case in previous works.

10.3.2 Selective filtering

Previous works using SAR images for machine learning have usually selected a handful of images with which to work, and, in doing so, it is ensured that the data being learned is in some way representative of the property of interest. However, in this experimental setup, *all* SAR images from the region of interest were downloaded. This meant that many of the images contained entirely 0% concentration (open water) or 100% concentration (consolidated ice or land). These images are of no interest, since they do not contain any variation in ice concentration and will skew the training process in favor of single-value concentration patch estimations, which is undesirable.

The solution proposed in this investigation is to filter the dataset prior to training and retain only images that contain interesting variations in ice concentration such as at the ice edge where the concentration values in a single image span most of the 0–100% range. A histogram was plotted for a subset of the labeled images, which is shown in Figure 10.1. Pixel-wise variance in the concentration patch label was used to quantify the spread of concentration values within each image, and a threshold was used to filter the dataset based on the concentration variance – images with a suitably high variance were marked as acceptable, whereas the others were excluded from all further use.

10.3.3 Batching and augmentation

Keras provides a number of helpful tools for batch training on image data, and the *ImageDataGenerator* class is particularly powerful when it comes to feeding batches into a neural network. All tests were conducted both with and without data augmentation (random flips and rotations) to understand the efficacy of the augmentation process, and a 5×5 median filter was also applied to all concentration and uncertainty patches. This filter helped to smooth out the rather grainy patch labels, while preserving edges, as shown in Figure 10.2. The granularity in the unfiltered

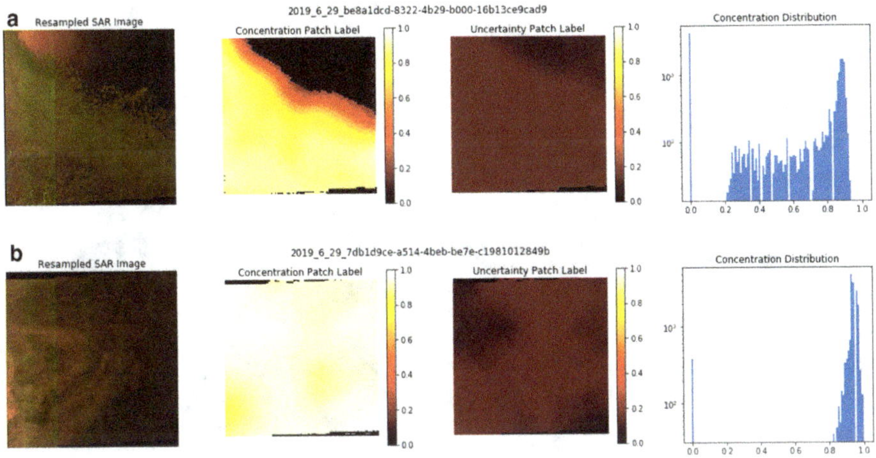

Figure 10.1 *Pixel-wise concentration distribution. In this figure, the resampled SAR image, concentration patch label, and uncertainty patch label are shown for two representative samples in the dataset. The histogram on the right shows the distribution of concentration values present in each concentration patch label. The first sample (top) is clearly more 'interesting' in terms of concentration features, since it includes a section of ice edge, whereas the second sample (bottom) contains almost no variation, with most of the concentration values being greater than 80%. A variance threshold was used to filter all samples, ensuring only 'interesting' samples were included.*

patches is due in part to the poorer 10 km resolution offered by the OSI SAF concentration chart used, and the median filter applied had the desired effect of smoothing the patch without blurring.

Google Collaboratory's storage virtualization imposes a significant bottleneck when trying to access so many files individually. To reduce the number of file access

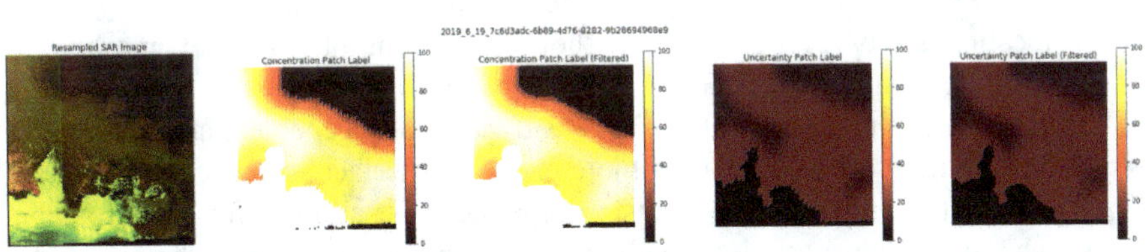

Figure 10.2 *Median filter (5 × 5) applied to concentration and uncertainty labels. The median filter does well at smoothing the patch labels without blurring the image and also does not introduce any concentration values that were not already present in the vicinity of each pixel, to begin with.*

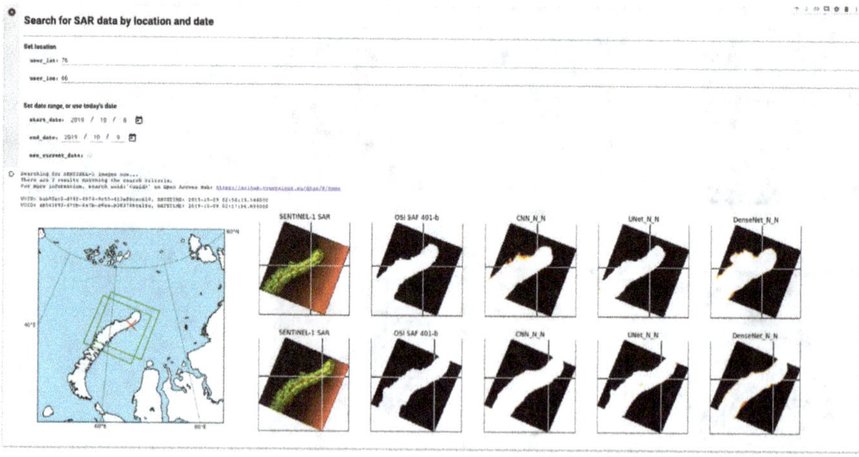

Figure 10.3 *This search interface allows a user to find and compare SAR images with concentration estimates from a variety of models side-by-side, with additional context provided by the map on the left-hand side*

calls and directory jumps during training, the newly labeled dataset was processed into batches, and each batch was saved as a single file. With a batch size of 16, this reduced the number of file access calls by a factor of 48, which successfully avoided starving the GPU while training. Recalling batches of input and label data like this was facilitated by extending the *Keras Sequence* class.

10.3.4 Evaluation interfaces

While existing SIC chart products, such as OSI-401-b, tend to focus on achieving global coverage in a single chart, the aim of this investigation was to produce a more focused concentration estimation for specific regions on demand. This kind of product is much more useful when providing close navigational support for shipping and research activities. A boilerplate search interface was developed in Google Colaboratory using *Forms* allowing a user to search for SAR images covering a particular location and date range. A screenshot of the Form interface and an arbitrary result set is shown in Figure 10.3.

The map plot shown in Figure 10.3 was dynamically generated at runtime and is centered on the specified location of interest. This interface was a very useful tool when developing the data pre-processing pipeline, as it allowed for specific edge cases to be visually inspected and validated. It also provided a very direct way to visually compare the concentration estimations generated by each model. Lastly, this interface offers a proof of concept for the intended workflow in the context of serving targeted predictions for close navigational support.

A secondary objective for this investigation was to compare estimates made by the existing numerical SIC models with a set of recent in situ observations.

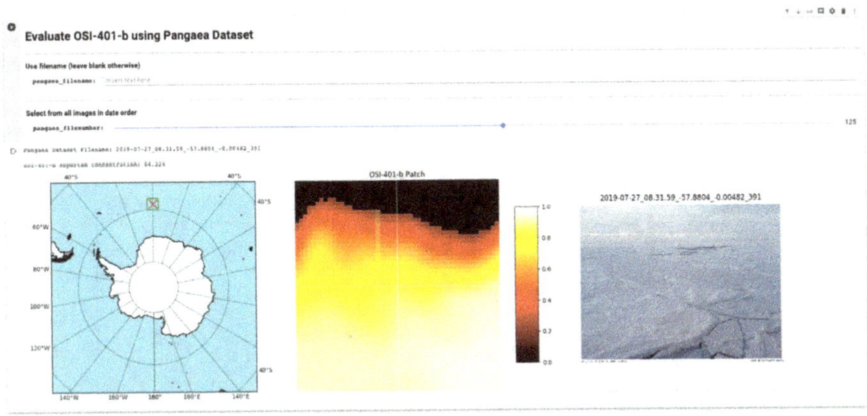

Figure 10.4 *This comparison interface shows images taken in the MIZ and compares them with the PMW-derived estimates for the same day and location, with the exact location indicated with crosshairs*

A 2019 research expedition to the MIZ in the Southern Ocean collected a large set of geotagged images of the sea ice conditions [9]. This kind of comparison is ideal for understanding the validity of the existing models in the Southern Ocean where extensive in situ observation data is lacking. Figure 10.4 shows a screenshot of the interface used to compare OSI-401-b predictions with in situ observations of sea ice conditions.

10.3.5 Pass direction error

The search interface shown in the previous section rotated each SAR image so that it appears aligned with the global reference frame. However, after implementing this rotation and footprint overlay process, some images still appeared to be incorrectly rotated. Extensive troubleshooting revealed that the cause was an inconsistent mislabeling of some SAR products' 'pass direction' attribute, which caused the image to be pre-processed incorrectly by Copernicus.

While troubleshooting this issue, it was noticed that the product *OData* 'footprint' attribute (a set of points defining the footprint polygon) follows a predictable structure: the order in which the points are listed is always top-left, top-right, bottom-right, and bottom-left (in the satellite's frame of reference). This is depicted in Figure 10.5.

This insight allows the actual pass direction to be determined without relying on the attribute reported by either *OData* or *OpenSearch*, which are both unreliable. One way to do this is to compare the latitudes of points 0 and 3. If the latitude of point 0 is greater than that of point 3, then the satellite was north facing at acquisition time, and the product is therefore ASCENDING. If not, then the product is DESCENDING. This derived pass direction was then compared to the reported *OData* attribute. If the two agree, then the pass direction attribute is correct, and

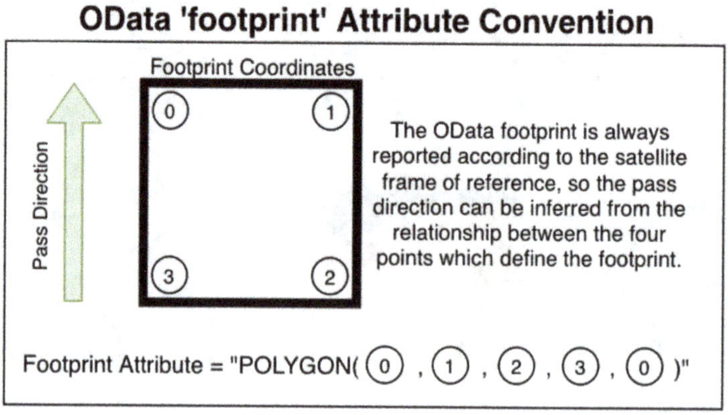

Figure 10.5 *A diagram explaining the OData 'footprint' structure. This can be leveraged to determine the satellite's actual orientation at acquisition time, without relying on the 'pass direction' attribute.*

the Quicklook image has been correctly pre-processed before being downloaded. If not, then the product was incorrectly labeled, and therefore the Quicklook image was incorrectly pre-processed before being downloaded. This erroneous processing could then be easily corrected, and further processing can continue as usual.

This information was reported to the Copernicus team for review and has been acknowledged as a known issue. It has been labeled as a low priority issue since most research activity uses the full SAR measurement, not the Quicklook image, which is not processed in the same way. This inquiry has been noted as the first time this issue has been reported externally, and the Copernicus team has stated that they intend to implement a fix in the next processor revision.

10.4 Architectures and training schemes

10.4.1 Summary of models used

Three different neural network architectures were used in this investigation, with the intention of comparing the performance and complexity of each. Hyperparameters were iteratively tuned until the mean absolute error on unseen test data was below 10%. All neural network models were developed using *Keras*.

Before investigating any exotic model architectures, a relatively straightforward sequential Fully Convolutional Neural Network (FCNN) was implemented to act as a baseline with which to compare the predictions of more complex models implemented later. Each layer was composed of a convolutional layer with $n \times (3 \times 3)$ filter kernels and ReLU activation functions, followed by a dropout layer with an adjustable dropout percentage. A simple two-layer version of this architecture is shown in Figure 10.6. The convolution operators were padded so as to maintain the same image resolution throughout the network, which meant that the output layer

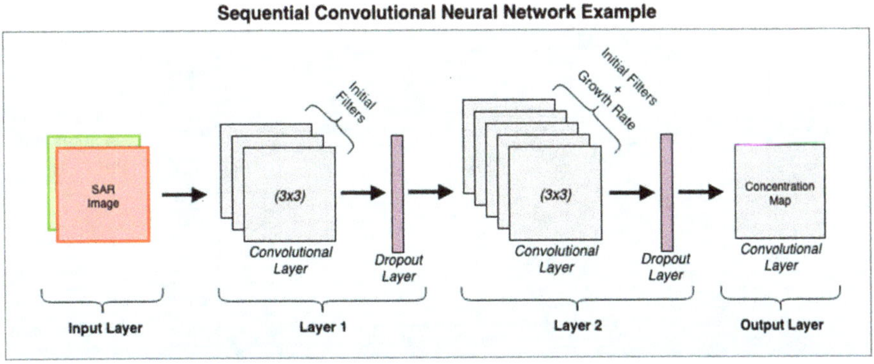

Figure 10.6 *An illustration of the sequential FCNN architecture. This example contains two hidden layers and illustrates how the number of convolutional filters is increased at each layer.*

was simply a single convolutional filter with a Sigmoid activation so that all pixel values fall in the range $(0, 1)$ representing the estimated concentration percentage.

The U-Net architecture was introduced in 2015 and has been successful in many image-segmentation tasks [8]. Since segmentation in this context is little more than pixel-wise classification, this architecture can be easily modified for concentration mapping, which is essentially pixel-wise regression. By modifying the final output layer, all the advantages of this network structure are retained, and very little design work is required. Figure 10.7 shows the U-Net architecture as described in the original paper. A *Keras* implementation of the U-Net architecture found on GitHub was used for the bulk of the model definition [10].

Building on the success of skip connections as demonstrated in U-Net, another unique architecture was introduced where convolutional layers were densely connected to one another, called DenseNet [7]. The network was originally intended for use on classification problems, and Figure 10.8 describes how the dense blocks (containing densely connected convolutional layers) are arranged sequentially with some transition operators in between each block. A *Keras* implementation of this architecture was used as a template for this investigation, found on GitHub [11]. Besides the changes to convolution padding and the removal of max pooling, the models' weight decay hyperparameter was also set to zero. The purpose of this hyperparameter is to penalize excessively large weights in any of the model layers, in order to prevent overfitting to particular training examples, and overreliance on particular feature sets. However, this interfered with the training process by diluting the training loss value. So instead of minimizing the objective function by estimating more accurate concentration maps, the model training process simply minimized the layer weights, leading to very poor visual performance despite the training loss following the expected exponential decay trajectory.

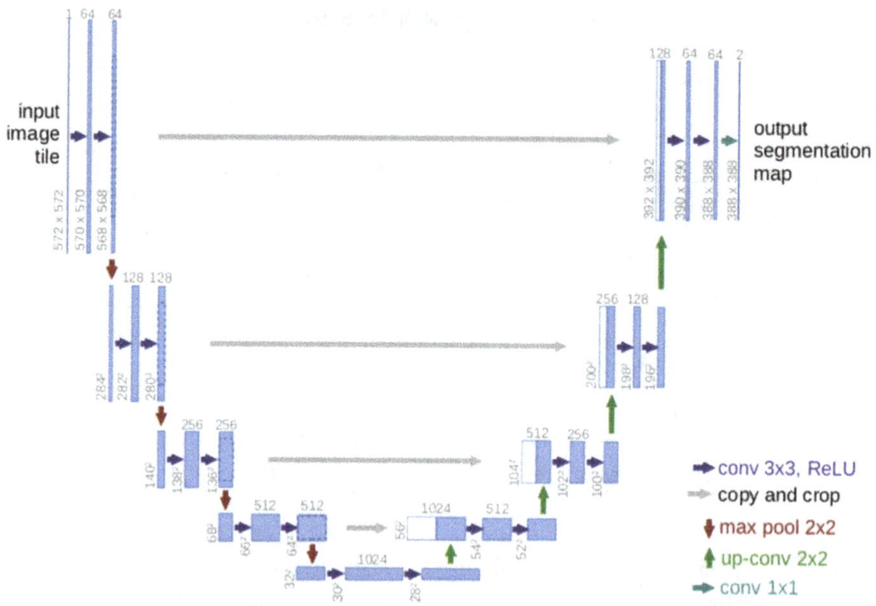

Figure 10.7 *A diagram illustrating the original U-Net architecture. Max pooling layers were removed so as to preserve the image dimension throughout the network [8].*

For each of the model architectures described above, a number of variants were trained and evaluated. Each structure was iteratively tuned over a number of generations until the estimation mean absolute error on unseen data was acceptable. The final configuration of each model is listed in Table 10.1.

10.4.2 Objective function

One of the primary objectives of this investigation was to train a model on data from an existing numerical model, without inheriting all the inaccuracies of that model. The proposed solution presented in this section was to incorporate the OSI-401-b uncertainty field into the objective function calculation so that the model would not be heavily penalized for estimation errors if the concentration label had a high

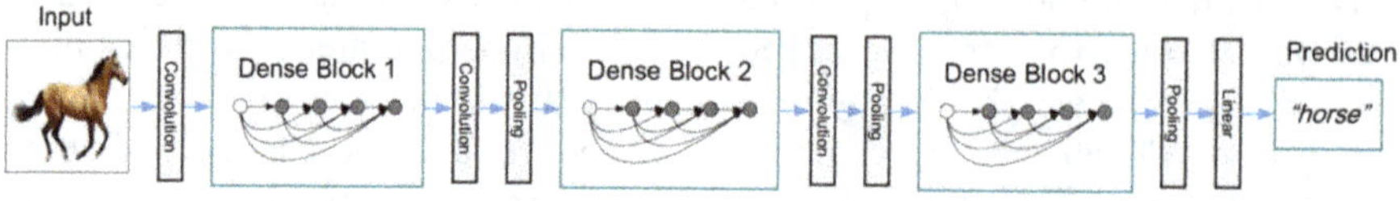

Figure 10.8 *A diagram illustrating the original DenseNet architecture. Although originally designed for classification, the structure was easily modified to predict the concentration maps needed in this investigation [7].*

Table 10.1 Summary of neural network models used in the investigation: details and important differences

Base architecture	Structure	Description
CNN	Layers: 10 Initial filters: 32 Growth rate: 32 Dropout rate: 0.2 Parameters: 3 043 937	A straightforward sequential CNN, where each layer is composed of a convolutional layer followed by a dropout layer.
U-Net	Layers: 4 Initial filters: 128 Growth rate: ×2 Dropout rate: 0.5 Parameters: 124 108 545	Based on the successful image segmentation network, this model used skip connections to feed low-level features forward to layers near the output.
DenseNet	Dense blocks: 4 Dense Layers: 8 Growth rate: 8 Dropout rate: 0.2 Parameters: 186 459	Based on the novel classification model architecture, this model was modified to retain the same image dimension throughout the network.

degree of uncertainty. The intention was for the neural network model to learn from features where the existing numerical models are certain. With sufficiently large datasets, the model would be exposed to enough features with a high level of certainty for the overall training process to be successful.

Each model was compiled to use a custom loss function, which first separated the label into its two layers so that the concentration and uncertainty patches could be used independently. The loss function then computed the pixel-wise prediction error and scaled it by the pixel-wise 'certainty'. Figure 10.9 shows the uncertainty weighted mean absolute error loss function Python definition.

The reasons for defining a custom loss function like this are so that multiple pieces of information can be utilized in determining the loss for each sample, and so that the loss can be constructed as desired for any application. In this case, the estimation error is scaled by the pixel-wise certainty, which prevents the model from

```python
def uncertainty_weighted_MAE(y_true, y_pred):
  # y_true contains the concentration label, and the uncertainty as two two channels
  y_target    = y_true[:, :, :, :1]
  uncertainty = y_true[:, :, :, 1:]
  # Calculate the weighted MAE
  loss = K.abs(y_pred - y_target)
  loss = loss * (K.ones_like(loss) - uncertainty) # Scale the error by the 'certainty' of the label
  return K.mean(loss, axis=-1)
```

Figure 10.9 Python definition for the uncertainty weighted mean absolute error function used as the objective function for all models in this investigation. The label passed to the objective function contains both the concentration and uncertainty layers.

Table 10.2 Summary of results obtained from each training strategy for each model type under investigation. Multiple sets of training data indicate that the model was partially trained on the first set and then trained to completion using the second set.

Model name	**Training data**	**Test data**	**Weighted MAE (Train, Test)**	**MAE (Train, Test)**
CNN_N_N	North	North	(0.0640, 0.0697)	(0.0727, 0.0793)
CNN_N_N_A	North_A	North	(0.0674, 0.0792)	(0.0763, 0.0897)
CNN_S_S	South	South	(0.0579, 0.0660)	(0.0652, 0.0748)
CNN_S_S_A	South_A	South	(0.0624, 0.0755)	(0.0703, 0.0855)
CNN_NS_S	North + South	South	(0.0561, 0.0631)	(0.0631, 0.0715)
CNN_NS_S_A	North_A + South_A	South	(0.0597, 0.0742)	(0.0675, 0.0842)
UNet_N_N	North	North	(0.0145, 0.0386)	(0.0166, 0.0437)
UNet_N_N_A	North_A	North	(0.0174, 0.0468)	(0.0200, 0.0530)
UNet_S_S	South	South	(0.0150, 0.0324)	(0.0170, 0.0369)
UNet_S_S_A	South_A	South	(0.0218, 0.0693)	(0.0248, 0.0786)
UNet_NS_S	North + South	South	(0.0165, 0.0325)	(0.0187, 0.0370)
UNet_NS_S_A	North_A + South_A	South	(0.0198, 0.0477)	(0.0225, 0.0541)
DenseNet_N_N	North	North	(0.0603, 0.0797)	(0.0683, 0.0904)
DenseNet_N_N_A	North_A	North	(0.0645, 0.124)	(0.0730, 0.138)
DenseNet_S_S	South	South	(0.0585, 0.120)	(0.0658, 0.130)
DenseNet_S_S_A	South_A	South	(0.0632, 0.130)	(0.0709, 0.141)
DenseNet_NS_S	North + South	South	(0.0581, 0.0813)	(0.0654, 0.0897)
DenseNet_NS_S_A	North_A + South_A	South	(0.0633, 0.0830)	(0.0711, 0.0919)

learning uncertain details from the existing numerical models as readily as it would without any kind of uncertainty compensation.

10.4.3 Training and evaluation

Each model type was compiled using the same custom objective function and then trained with various combinations of datasets. Table 10.2 shows a full list of all models, according to the following naming convention:

Model_TrainingData_TestData[_DataAugmentation]

The most interesting results came from using mixed-data training strategies, where the models were partially trained on Northern hemisphere data, and then trained to completion using Southern hemisphere data. These models trained for the same number of epochs as the rest, but exhibited some interesting behavior in their loss trajectories, which is discussed in more detail later on.

Model performance was compared using pure and uncertainty-weighted MAE and MSE. These metrics were tracked throughout the training process, which allows for analysis of loss trajectories as shown in Figure 10.10, which has the potential to offer more insight than simply looking at the final loss values as shown in Table 10.2. This discussion is continued in Section 10.5.

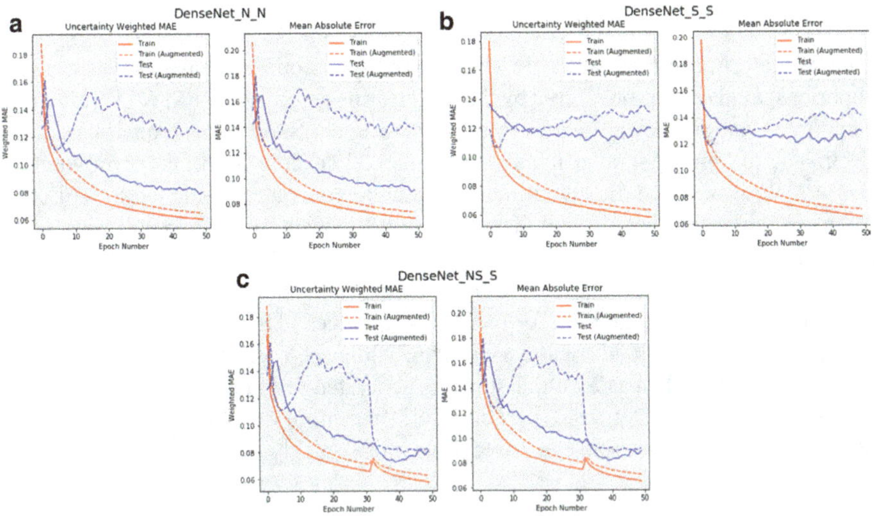

Figure 10.10 *Loss trajectories for DenseNet, covering all training strategies. The divergence of test results from the training curves is noteworthy and warrants further discussion.*

For a more subjective analysis, the search interface was used to serve concentration estimations on demand and could be configured to show any combination of model results side-by-side for visual comparison.

10.5 Results and discussion

10.5.1 Performance metrics and loss trajectories

Model performance during and after the training procedure is presented in two ways in this section. Models are compared primarily based on their final performance metrics, but the loss trajectories throughout the training process are also used to gain some insight into each model's behavior. Both the weighted MAE objective function and a pure MAE metric were tracked during all training procedures. Table 10.2 shows all combinations of model architecture and training strategy used in this investigation and presents the final performance metrics achieved by each model at the end of 50 epochs. It is worth noting that in all cases the weighted MAE scores are less than the pure MAE for each strategy, which confirms that the objective function is indeed penalizing the estimation error less in the presence of uncertainty.

Observing the progression of loss metrics throughout the training process can offer some insight into model behavior and may inform decisions when tuning hyperparameters. A particularly interesting set of DenseNet loss trajectories is included in Figure 10.10, which shows a number of features, all relating to the divergence of test results from the decreasing training error curve. The significance of this behavior is discussed in more detail later on.

10.5.2 Spatio-temporal resolution

One of the primary motivators for choosing to focus on SAR images was the far superior spatial resolution offered by these platforms compared to PMW instruments. This investigation showed that even after the original SAR measurement has been decimated to a much smaller image size, the pixel spacing is still an improvement over PMW data. Since the neural network architectures were designed to replicate the input dimension in their estimates, all of the concentration estimates presented here offer a resolution better than the original data with which they were trained.

The temporal resolution offered by SAR instruments is also an improvement. In many cases, when searching for recent SAR images of a particular location, two images would be returned for the same day, while the most recent PMW-derived estimate was more than a day old. Estimates generated from much more recent SAR data present a clear advantage, particularly in the more dynamic Southern Ocean MIZ where sea ice conditions tend to evolve more quickly.

10.5.3 Advantages of learned models

The reasons for pursuing a learning approach to generating SIC estimates were based on the hypothesis that a sufficiently well-designed learning structure would be able to overcome the inaccuracies inherent in the PMW-derived models used for training, even if these inaccuracies are not fully understood. This is similar to how the learned models were able to learn higher-resolution features than were present in any one sample of the training data. This is a non-trivial claim, but the results have shown positive signs indicating that the underlying theory has promise. Some of the more prominent advantages are discussed in detail here.

10.5.3.1 Arbitrary loss definition

Perhaps the most useful result of this investigation in the context of trying to create reliable new models using less reliable existing models is the implementation of a certainty-weighted objective function. In the limited scope of this investigation, the weighting relied solely on the uncertainty attribute published with the OSI-401-b concentration product. This theory can be applied to any arbitrary definition of 'certainty' in future research.

10.5.3.2 Data augmentation

Data augmentation is a commonly used tool in machine learning. It is particularly useful when dataset size is limited, or to avoid overfitting and in turn help the model generalize better to unseen test data. However, the loss trajectory results presented in the previous section suggest the opposite. In all cases where the use of augmented data was compared with an otherwise identical neural network and training strategy, the test results were worse when the model was trained with augmented data. Put simply, the use of augmented data decreased model generalization performance. One possible explanation for this effect is that there may be so little difference between training data and test data that 'overfitting' in training actually yields better test results. However, it should be noted that this is merely conjecture, and any further investigation into this result is out of scope.

10.5.3.3 Multi-stage training

It is not uncommon to schedule or dynamically adjust hyperparameters at key points during a model training process, and this is often referred to as multi-stage training. However, in this investigation, this staging approach was used to change the entire dataset being used. This has a similar effect to transfer learning, although transfer learning tends to be applied in a cross-domain fashion, whereas in this case both datasets were of SIC. The rationale for this hybrid training strategy was to allow the model to learn most of its feature associations from Northern hemisphere data, which is known to be more accurate and reliable, before completing the training process with Southern hemisphere data in order to help the model generalize to the new environment. The most interesting observation as a result of this is shown in Figure 10.10, specifically the DenseNet_NS_S loss trajectory. While training on Northern hemisphere data for the first 32 epochs the train and test scores diverge, which is likely due to overfitting and the fact that the DenseNet model used had far fewer parameters compared to the other two model types (see Table 10.1). However, almost as soon as the source of training data was switched to the Southern hemisphere after 32 epochs, the test scores rapidly converged with the training scores. As can be seen in Figure 10.10, this allowed the DenseNet_NS_S strategy to achieve a 4% reduction in weighted MAE on test data compared to the DenseNet_S_S, which both used the same Southern hemisphere test data (see Table 10.2). This presents another interesting result demonstrating the effects of mixed-data training strategies like this, which could certainly be investigated further in future research.

10.5.4 Reliability of existing models

As a secondary objective, this investigation aimed to offer some insight into the reliability of the PMW-based OSI-401-b SIC product used for generating all label information. The comparison tool introduced in Figure 10.4 allowed for a number of recent in situ observations to be retrospectively compared to the concentration estimates for the same day and location. This comparison revealed significant errors in the concentration estimates within the Southern Ocean MIZ, which illustrates beyond doubt that the naïve application in the Southern hemisphere of estimation models designed and calibrated for the Northern hemisphere will not yield reliable results.

10.6 Conclusions and future work

The problem of SIC estimation is clearly far from solved, although a number of interesting results have been presented over the course of this investigation. Satellite instrument data sources have been under-utilized in the past, but developments in deep learning, particularly in the field of image processing, have led to promising results in this context. The outcome of this investigation is a detailed implementation of an end-to-end workflow for estimating concentration maps from SAR images, based on three neural network models trained on data from existing PMW concentration estimation models. The neural network development process was a primary focus of this investigation, and many of the results led to insights on the effects of various data processing techniques used in the course of development, and also pointed to the potential strengths of mixed-data training strategies. Care was taken throughout the development process to validate results and procedures, thus ensuring that the proposed workflow is scalable and robust. The estimation models developed can be improved in many ways but have already shown signs of being able to learn higher resolution features than are contained in any single training sample, and the use of a novel weighted objective function allows the training process to account for known uncertainty within the training data. A comparison tool was presented, which allows existing concentration estimates to be compared with in situ observations and has offered more evidence asserting that the use of Arctic models in the Southern Ocean MIZ will not yield reliable results.

For any future research conducted in this direction, a number of recommendations have already been made, such as defining a more effective 'certainty' metric to be used in the objective function. For example, if one were to generate a 'bias map' using a series of in situ observations for a given region that quantifies any discrepancies between estimates and reality, this bias map could be incorporated into the objective function in a similar way and may allow the learning process to correct for errors in the training data. One could also investigate the effects of data augmentation in this context, which have appeared to behave counterintuitively in these results or propose more effective ways to utilize the benefits of multi-stage and mixed-data training strategies shown here. Additionally, one should consider alternative satellite data sources that may have been overlooked and also consider scaling up the neural network input dimension to make better use of the high-resolution SAR data, which is so freely available.

References

[1] Alberello A., Bennetts L., Heil P., *et al.* 'Drift of pancake ice floes in the winter Antarctic marginal ice zone during polar cyclones'. *Journal of Geophysical Research: Oceans*. 2020, vol. 125(3),e2019JC015418.

[2] Kwok R., Pang S.S., Kacimi S. 'Sea ice drift in the southern Ocean: regional patterns, variability, and trends'. *Elementa: Science of the Anthropocene*. 2017, vol. 5(32), pp. 1–16.

[3] Vichi M., Eayrs C., Alberello A., *et al.* 'Effects of an explosive polar cyclone crossing the Antarctic marginal ice zone'. *Geophysical Research Letters*. 2019, vol. 46(11), pp. 5948–58.

[4] Notz D. 'Challenges in simulating sea ice in earth system models'. *Wiley Interdisciplinary Reviews: Climate Change*. 2012, vol. 3(6), pp. 509–26.

[5] Häkkinen S., Cavalieri D.J. 'Sea ice drift and its relationship to altimetry-derived Ocean currents in the Labrador sea'. *Geophysical Research Letters*. 2005, vol. 32(11), pp. 1–4.
[6] Cooke C.L.V., Scott K.A. 'Estimating sea ice concentration from SAR: training convolutional neural networks with passive microwave data'. *IEEE Transactions on Geoscience and Remote Sensing*. 2019, vol. 57(7), pp. 4735–47.
[7] Huang G., Liu Z., van der Maaten L. Densely connected convolutional networks. Proceedings of the IEEE conference on computer vision and pattern recognition; 2017. pp. 4700–8.
[8] Ronneberger O., Fischer P., Brox T. U-Net: convolutional networks for biomedical image segmentation. International Conference on Medical image computing and computer-assisted intervention, Springer; 2015. pp. 234–41.
[9] de Vos M., Ramjukadh C.-L., Liesker C.G., de Villiers M., Lyttle C.T. Southern Ocean Marginal Ice Zone Photographs from a 2019 Winter Research Cruise (SA Agulhas II). 2019. Available from https://doi.pangaea.de/10.1594/PANGAEA.905497 [Accessed 5 Oct 2019].
[10] Żak K. 'keras-unet'. 2019. Available from https://github.com/karolzak/keras-unet [Accessed 5 Oct 2019].
[11] seasonyc. 'densenet'. 2018. Available from https://github.com/seasonyc/densenet [Accessed 5 Oct 2019].

Chapter 11

Cognitive radar for ground moving target imaging

Samuele Gelli[1], Elisa Giusti[1], and Marco Martorella[2]

In future decades there will be a growing demand for multi-function radar systems operating in resource-constrained and congested and contested spectrum environment. Due to the complexity of several missions, radar systems are required to perform multiple functions simultaneously, such as surveillance, clutter/interference cancellation, moving target detection (MTD), imaging and recognition. This leads to the development of multichannel, multi-static, multi-frequency and multi-polarization, in a single word, multi-dimensional, radar systems which enrich the received signal information content and increase the degrees of freedom that can be used to perform the above-mentioned radar tasks [1, 2]. Such systems need to share the spectrum with other concurrent and contrasting systems and, therefore, must be able to adapt to the environmental conditions. Moreover, to meet demanding requirements and to reduce costs, size and weight, modern radar systems should share the same hardware and software with other systems [3]. In light of recent developments in the matter of cognitive radar, recent radar demonstrators and prototypes have started to embrace this new technology as it provides the best performances in dynamically changing environments. As it happens in the human brain, as well as in other mammals, a cognitive radar system learns about the environment through the interaction with it and acts consequently by transmitting optimal waveforms and adopting optimal signal processing on receive [4, 5]. This process can be considered the core of a cognitive system. The concept of cognition has been introduced and extended to radar systems for the first time by Simon Haykin in 2006 [6]. One of Haykin's citations is emblematic of the concept of cognitive systems:

A cognitive system continuously learns about the environment through experience gained from interactions with the environment, continually updates the receiver with relevant information on the environment, makes adjustments to the transmitted signal in an effective and robust manner. The whole radar system constitutes

[1]CNIT Radar and Surveillance System (RSS), National Laboratory, Italy
[2]Department of Information Engineering, University of Pisa, Italy

a dynamic closed feedback loop encompassing the transmitter, environment, and receiver

Consequently, when the external environmental changes, the system must be able to sense it and automatically change the rules and parameters that define the way the system operates, consequently updating the memory to take external changes into account [6]. An appropriate technology that supports a cognitive architecture has yet to be developed even if, at least at present, it is tricky to conceive a system that perfectly emulates the human brain's ability to interact with the external environment. However, technology has made a step towards software-defined radar systems, thus allowing for key radar parameters to dynamically vary over a range of possible values [7–9]. Therefore, it makes sense to start thinking about a radar architecture that is capable of autonomously managing different parameters (transmitted power, Tx and Rx channels, polarization, bandwidth, pulse repetition frequency (PRF), etc.) in response to external changes. The concept formulated by Haykin may be interpreted in the following terms: a cognitive radar can be thought of as a system where transmit and receiver parameters can be dynamically changed over time in response to the observed scene with the aim to optimize performances according to a specific mission and the environmental conditions. The ability to change system parameters appropriately and promptly is given by the system's ability to perceive and code environmental changes through specific indicators. This coding mechanism necessarily implies the use of a memory, which keeps track of past actions and observations [10–12]. In the radar framework, adequate signal processing is needed that transforms the information contained in the received signal into actions towards the transmitter and receiver. In this perspective, rules can be defined for a cognitive system to implement decisions in order to fully adapt to the external environment. Recently, in response to the ever-increasingly complex missions and operational environments where modern radars are employed and the inability of traditional adaptive methods to address rapidly varying interference environments, the Defense Advanced Research Projects Agency (DARPA) has been pioneering the development of the first-ever real-time knowledge-aided (KA) adaptive radar architecture [13]. The goal of DARPA's program was to demonstrate that a high-performance embedded computing architecture was capable of integrating high-fidelity environmental knowledge (i.e., priors) into the most computationally demanding subsystem of a modern radar: the adaptive space-time beamformer. In this chapter, we will be taking this as a starting point and will propose a practical implementation of a cognitive radar system for the detection and imaging of non-cooperative moving target in the presence of a complex environment. Real data acquired with a multichannel airborne Synthetic Aperture Radar (SAR) system will be used to verify the effectiveness of a scaled version of the proposed architecture.

11.1 State of the art

This section discusses the state of the art regarding the detection and imaging of moving targets embedded in heterogeneous environments. SAR is a well established technology, dated back to the early 1950s, which allows for electromagnetic images of the observed area to be obtained with accuracy. In the last decades, airborne and spaceborne SAR systems have been heavily used in applications such as 2D/3D mapping of the Earth's surface, climate change monitoring and for security-related applications [14–16]. From this point of view, high-resolution imaging of non-cooperative moving targets become of particular interest [17]. Due to the lack of knowledge between the SAR platform and target's own motion, coherent processing of the target received echoes is not allowed. This leads to a target that appears blurred and misplaced in the SAR image. Inverse SAR (ISAR) has been suggested as an alternative way to look at the problem of high-resolution imaging of moving targets [18]. ISAR geometry is strongly related to the SAR geometry since with a proper choice of the reference system it is possible to change one to another. The real substantial difference is in the cooperation of targets present in the observed scene. As a matter of fact, in ISAR systems the target own motion is exploited to synthesize a large aperture antenna and the relative motion between radar and moving targets is estimated and compensated during the image formation process. In recent years, a solution based on ISAR processing has been proposed for high-resolution imaging of moving target in SAR images [18, 19].

In order to apply ISAR processing targets must be detected first. Ground MTD is not a trivial step since ground clutter return completely masks target return making the detection particularly challenging. The earliest proposed techniques concerning single-channel SAR systems were based on Doppler discrimination between moving targets and static scene clutter [20]. The main shortcoming of these techniques is the need for a high PRF. When in fact, the PRF is much larger than the Doppler clutter bandwidth the moving targets will appear at Doppler coordinates free of clutter. However, high PRF assumption causes both a reduction in SAR swath width and an increase in the size of the data to be processed. Moreover, these techniques, and in general all single-channel techniques, fail in the presence of slow-moving targets since the target Doppler frequency completely falls in the static Doppler region. With the development of multichannel imaging systems, more powerful techniques making use of more than one channel have been proposed [21, 22]. Multichannel systems have the capabilities of both acquiring simultaneous echoes of the same scene from different direction of arrival and echoes at different PRF. These space-time samples can be used in different (MTD methods to obtain better filtering performance. Principal MTD methods for multichannel SAR systems are Displaced Phase Centre Array [23, 24], Alog Track Interferometry [25–27], Time-frequency Transforms [28, 29] and Space-Time Adaptive Processing (STAP) that has been proven to offer the best MTD result for SAR [21, 30–33]. STAP processing has been extensively studied in the last few decades for clutter suppression, while in recent years an innovative processing chain that jointly combines different formulations of conventional STAP processing, termed Space Doppler Adaptive Processing (SDAP) and ISAR to obtain high-resolution images of non-cooperative moving targets in SAR images has been proposed [34].

In SDAP, such as in STAP, it is mandatory to correctly estimate the statistical properties of the clutter in order to obtain good enough clutter cancellation performance. The clutter statistical properties are included in the clutter covariance matrix which must therefore be correctly estimated. As a consequence, training data used in the clutter covariance matrix estimation must refer to target free range cells in order to avoid the target self-cancellation [35]. The majority of STAP techniques assumed that the training data are Independent and Identically Distributed (IID), that is the statistical property of clutter in each training data is the same of the cell under test. Under this hypothesis, the convergence issue of the clutter covariance matrix is only related to the amount of training data used in the covariance matrix estimation. IID hypothesis is equivalent to the restrictive homogeneous clutter environmental assumptions. This assumption is violated in some real scenarios due, for example, to the presence of Internal Clutter Motion effects and, more in general, in the presence of temporal and spatial clutter variations [36, 37]. As a matter of fact, heterogeneous clutter and jammers can seriously degrade the clutter covariance matrix estimation and then the SDAP clutter suppression performance. Moreover, a minimum number of training cells is mandatory to suppress clutter properly. The Reed-Mallett-Brennan (RMB) rule [38] guarantees a performance loss of 3dB with respect to the optimum case in which the clutter covariance matrix is perfectly known. For a typical observation time and radar PRF, the required number of training data satisfying the RMB rule results very high [39]. As a consequence the number of training cells and the clutter homogeneity, i.e. clutter with the same statistical properties of the clutter to be cancelled, are the two basic requirements for accurate clutter suppression. A sub-optimum SDAP approach [34] can be exploited to reduce the required number of training data imposed by the RMB rule. However, distortions in the final image can occur when the number of cells is much lower than the limit imposed by the RMB rule [40]. Moreover, the training data used for clutter covariance matrix estimation are conventionally chosen close to the cell under test. However, this choice does not avoid the presence of clutter heterogeneity. For example, the presence of structure or jammers leads to a wrong estimation of the clutter to be cancelled. In past years the DARPA has been pioneering the development of a KA adaptive radar architecture that integrates environmental knowledge into the adaptive space-time beamformer [13]. A knowledge-based SDAP (KB-SDAP) approach has also been investigated in [40–42]. Differently, a cognitive radar can be thought of as a system in which the software parameters can be changed over time in response to the observed scene change with the aim to optimize radar performances given limited resources and environmental constraints. Specifically, as a system that optimally chooses the range cells used for the covariance matrix estimation, based on the a priori knowledge about the environment. The radar, therefore, has to reason about what is being observed and has to take decisions about how to use its limited resources in the best possible way to improve its performance [40, 41]. A rule-based cognitive architecture is addressed as a way to design a cognitive radar that has to manage its resources dynamically in a complex and changing scenario. Specifically, the system must be able to think about the heterogeneous environment it is observing in order to select a training area as homogeneous as possible with the area under test. The system must also be able to understand if the training data is sufficient or not and if so, it must be able to better manage its resources to obtain the best possible clutter cancellation. Finally, the system must be able to evaluate the result obtained in clutter suppression and targets imaging and then undertake the necessary corrections in order to improve it.

11.2 Rule-based cognitive architecture

In a rule-based cognitive architecture, a set of rules can be made that define how the radar system reacts to environmental changes [41, 43]. More specifically, to constrain the problem to applications of interest, we will consider an airborne multi-dimensional radar system (e.g. multi-frequency, multichannel, multi-polarisation, etc.) for GMT Img applications. Typically, any of such systems is conceived to deal with limited resources, environmental disturbances like noise, clutter and in-band interferences and platform constraints like the maximum payload which results in hardware/software restraints, e.g. antenna element size and positioning, maximum number of channels, maximum transmitted power, overall system power consumption and computational power. The concept idea behind the word *cognitive* refers to a radar system that intelligently manages its multiple, but constrained, resources and its signal processing algorithms based on evolving rules and criteria in order to achieve optimal performances. In the considered case, performances will be assessed in terms of target detection in the presence of a complex and dynamic environment. However, the proposed rule-based architecture can fit more complex missions that require not only clutter suppression and target detection but also, for example, automatic target classification and recognition. The high-level architecture which emphasises the connection among all system components, i.e. radar transmitter, radar receiver, environment and cognitive sub-block, is shown in Figure 11.1. To deal with the GMT Img task, the signal processing requires three hierarchical layers: SAR image formation and segmentation (*L1*), training data selection and clutter covariance matrix estimation (*L2*) and, finally, clutter suppression and target imaging (*L3*).

A description of the main rule-based architecture macro components follows [41, 43]:

- **Transmitter and receiver**: The transmitter acts on the transmitted waveform over time according to the environmental changes to optimise performance through an “intelligent” signal processing on receive. The concept of cognition on receive is based on a learning process that is enabled through a continuous interaction between the system and the environment.
- **Signal processing block**: Process the received echoes according to the radar mission and cooperate with the cognitive block to obtain the best performance that meets the needs of the radar mission.
- **The cognitive block**: The information produced by the signal processing block are used to drive the transmitting parameters. This process is based on a comparison through some performance indicators between past actions and actual results, which ensures that the system learns from its past actions. The cognitive

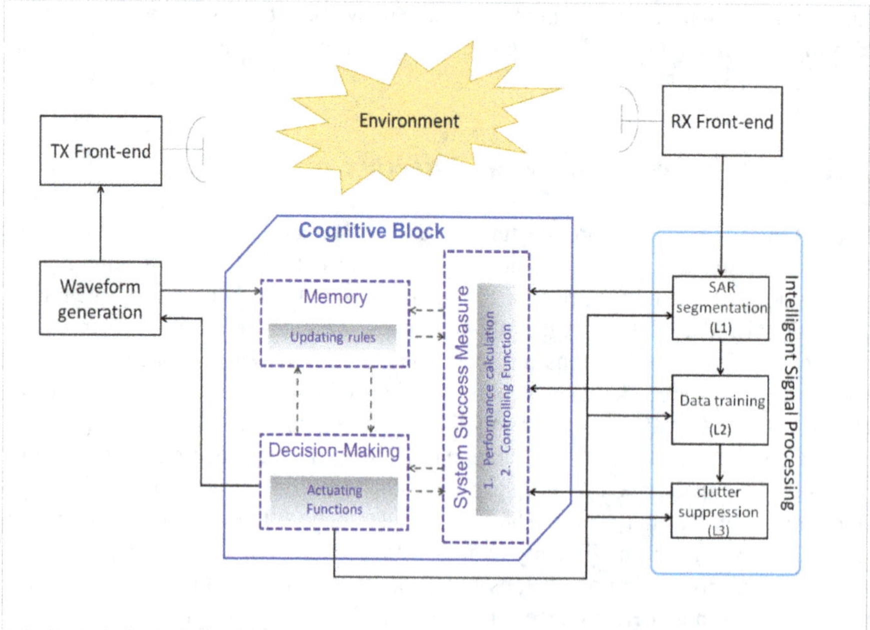

Figure 11.1 Rule-based cognitive radar architecture

block includes a *System Success Measure*, *Memory* and *Decision-Making*. The first one defines the rules, i.e. the controlling functions that represent external changes and that are based on performance indicators. The controlling functions drive, through the actuating functions, the system's response to adjust the transmitting parameters. The memory keeps track of all the changes in the observed environment and is used by the system to learn from its past actions. Finally, the decision-making block makes, according to the actuating function's output, adjustments on the transmitting parameters to best respond to environmental changes.

Summarising, radar returns are fed into the receiver, which processes them in order to extract target and clutter information. For each of the three processing layers i.e. SAR image segmentation, training data selection/clutter covariance matrix estimation and clutter suppression/target imaging, the mission success is measured through some defined metrics (i.e. performance indicators). Therefore, the controlling functions, which depend on such metrics, give the system the ability to take decisions according to the current results and the previous system state (memory). The decision embodies the action to be taken in order to improve the system performance according to the mission priority. This process is carried out by means of evolving rules that act on reconfigurable system parameters both at transmitter and receiver levels. The cognitive block shown in Figure 11.1, is further detailed in the block scheme as represented in Figure 11.2.

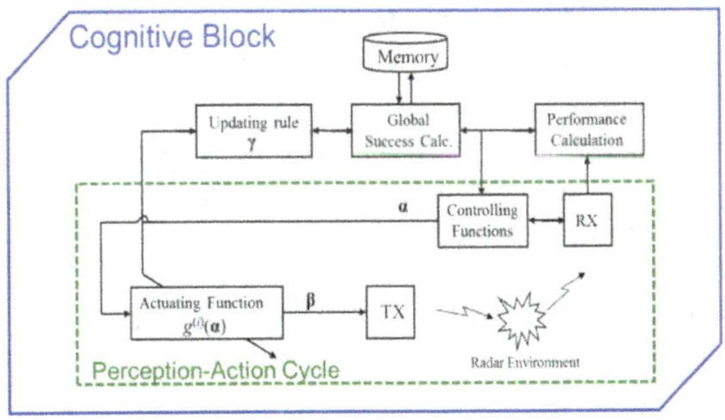

Figure 11.2 Detailed rule-based cognitive architecture

The perception-action cycle modifies the system parameters through the actuating function outputs β, which implement the decision-making algorithm. This function is driven by some controlling functions, namely α, that represent a compressed measure of the information contained in the received signal. Such functions may change with respect to the specific mission. One problem to be solved is how the system is able to update the actuating function and, as a consequence, how to change its parameters in order to maximize the system performance in a given environment.

The *updating rule* block accounts for this purpose. Specifically, the updating rule block provides the weights, namely γ, that define the actuating function. The γ parameters are modified in accordance with the feedback given by the *global success calculation* block, which compares the measured performance indicators (output of the performance calculation block) with those stored in the system memory. If the measured performance is better than those stored in memory, this is updated with both the actual weights and the actual performance values; otherwise, the weights are modified through some rules aiming at performance improvement driven by mission priorities and resource availability. To accomplish the GMT Img objective, the architecture behaviour has been characterised through the following elements:

- **Input parameters**: namely the signal backscattered from the environment (e.g., target, clutter, interference). These elements are not dependent on the model but affect it.
- **Output parameters** β: namely the system reconfigurable parameters (e.g., transmitted frequency f_0, instantaneous bandwidth B, transmitted power, P_t), which embodies the effect of the system behaviour on the input

$$\beta = [\beta_1, \beta_2, ..., \beta_N] \tag{11.1}$$

The output parameters are bounded by lower and upper limits given by hardware constraints and/or minimum desired performance. For instance, the instantaneous bandwidth has an upper limit that is imposed by the receiver bandwidth and a lower limit that is imposed by the minimum required resolution. Such bounds may also be imposed by spectrum regulations. The power has an upper limit that is due to the available power resources on board the platform and a lower one that depends on the desired range coverage. Furthermore, the choice of frequency mainly depends on the external environment and, in particular, on the clutter type. Considering that longer wavelength, and therefore lower frequencies, have a higher foliage penetration capability, higher frequencies can be used to detect the moving target in case of bare soil or low vegetation. Conversely, lower frequencies would be preferred in the presence of targets concealed under foliage.

- **Performance indicators** are strictly connected to the system performances in terms of the ability to remove the clutter and to obtain high-resolution images of moving targets.
- **Controlling functions** α enable a link from the receiver to the transmitter so as to inform it about the received signal processing results. The controlling functions are functions of the performance indicators.

$$\alpha = [\alpha_1(\mathbf{k}), \alpha_2(\mathbf{k}), \ldots, \alpha_M(\mathbf{k})] \tag{11.2}$$

where $\mathbf{k}$ indicates a vector containing the performance indexes. Each controlling function depends on a set of performance indexes as will be detailed in the following.

- **Actuating functions** are implemented through a set of rules to update the system reconfigurable parameters. They can be formalised as follows:

$$\beta = g[\alpha] \tag{11.3}$$

The new set of the system parameters, namely β, represents the ith system state. The previous set of actuating function, $\beta(i-1)$, is updated with the current one, $\beta(i)$, depending on the new set of controlling function $\alpha(i)$, which in turn depends on the current system memory and the current performance indicators, $\mathbf{k}(i)$. The memory is updated almost continuously according to specific controlling functions that have been defined to interact with (writing to or reading from) the memory.

In the following sections, the controlling functions and the actuating functions will refer to the ith system state. The cognitive design of each processing algorithm is established first. This includes the definition of the performance indicators, the use of the system memory, the definition of the controlling functions and finally the actuating rules. The following theoretical design refers to a multiband-multichannel system that is able to manage the operative frequency, some waveform parameters (instantaneous bandwidth and transmitted energy) and the number of receiving channels in a certain coherent processing interval (CPI). For the sake of simplicity, a radar system composed of a single RF transmitting channel and multiple receivers is here considered.

11.3 A case study: multichannel multi-band SAR for GMT Img

The aim of this section is to define the cognitive design of the GMT Img algorithm. The performance indicators will not only concern the transmission parameters, such as the transmitted bandwidth or the carrier frequency, but also operations of reading from and writing in the system memory. To accomplish this task, the performance indicators should be able to measure the capability of the SDAP processing to filter out the clutter and the interferences from the received signal. It is worth pointing out that the considered indicators cannot be directly related to the presence of targets in the scene since their presence is not guaranteed. However, it is clear that the performance indicators must be, in some way, connected to the system detection performance and the waveform parameters. The selected indicators that will be used in the signal processing steps (see Figure 11.1) are reported below and will be better analysed in the following sub-sections [41, 43]:

- SAR image segmentation accuracy
- Minimum number of training cells
- Improvement factor (IF)
- SDAP filter null position in the Doppler domain (DN)
- Doppler null bandwidth (DNB).

The processing algorithms, namely SAR image formation and segmentation, training data selection and clutter covariance matrix estimation and clutter suppression and target imaging, will be detailed in the following sub-sections. For each of them the performance indicators and the respective control functions will be defined. It is worth pointing out that each algorithm makes use of different signal processing techniques, which are detailed, for the sake of easier readability, in Appendix.

11.3.1 SAR image formation and segmentation

The first step is to focus on the received SAR data. This processing can be considered as a two-dimensional compression of the backscattered signals. The key differences between the SAR processing algorithms are related to the way the range cell migration compensation (RCMC) and azimuth compression are performed [44]. The effect of the range cell migration is related to the changes in the distance between a particular target and the moving SAR platform. As a result, the backscattered signal migrates through adjacent range cells during the SAR platform motion (and signal acquisition). Thus, the range profile of any point target on the ground will be concentrated along a so-called migration curve. The straightening of the range migration curve is called RCMC and it should be performed for each range. In this work, the range doppler algorithm (RDA) is considered. For sake of clarity the block diagram of the RDA is shown in Figure 11.3. A brief description of the main RDA steps is provided below. Further details can be found in [44].

- Range compression is performed with a fast convolution when the data are in the azimuth domain. In other words, a range FFT is performed followed by a range matched filter multiply, and finally a range IFFT, to complete the range compression.

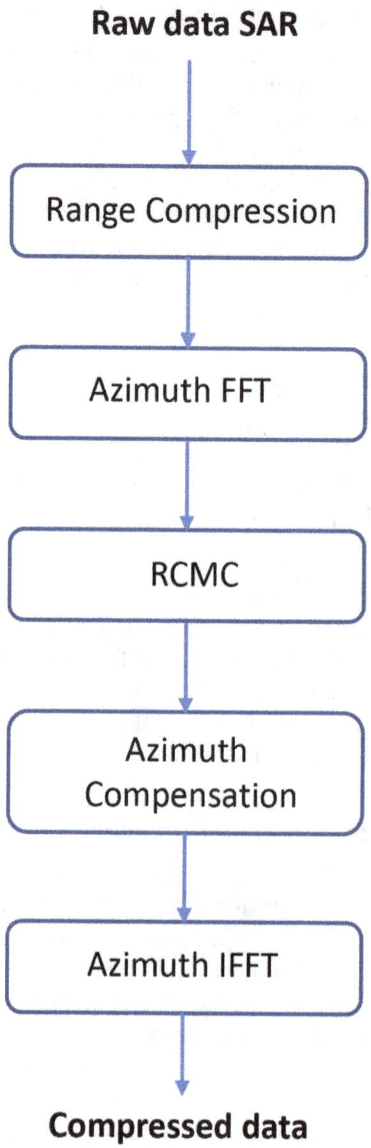

Figure 11.3 Functional block diagram of the basic range Doppler algorithm

- An azimuth FFT transforms the data into the range DN where Doppler centroid estimation and most of the subsequent operation are performed.
- RCMC is performed in the range DN where multiple target trajectories at the same range are transformed into one single trajectory. RCMC straightens the trajectories out in such a way that they run in parallel to the azimuth frequency axis.

- Azimuth compression i.e. matched filtering can be performed as a frequency domain matched filter multiply at each range gate.
- Azimuth FFT finally transforms the data back to the time domain, resulting in a compressed complex image.

It is worth pointing out that the main assumption of the RDA processing is that the two-dimensional phase history function is separable and, therefore, it can be represented as the product of two one-dimensional functions in the range and time domain. One refers to the range migration function and the other refers to the phase function. The problem is that this assumption does not hold in some cases, for example for large values of the squint angle. The existing azimuth and range phase cross-coupling effect influences the aperture synthesis performance and degrades the SAR image quality. However, in [45] it was shown that such cross-coupling effect can be eliminated by the application of an additional phase compensation in the two-dimensional frequency domain.

After SAR image formation, automatic image segmentation is applied to detect and separate different clutter types such as terrain, black top, grass, structures [46–48]. This, in turn, allows for the selection of a homogeneous clutter area i.e. with the same statistical properties as the one under test. It is also worth pointing out that one of the main drawbacks in SAR images is the speckle. Speckle is a signal-dependent granular noise inherent of all active coherent imaging systems that visually degrade the appearance of images. Speckle may severely diminish the performances of automated scene analysis and information extraction techniques. For these reasons, a speckle reduction (despeckling) is of crucial importance for a number of applications including SAR image segmentation. Such processing should be carefully designed to avoid losing useful information, such as local backscatter mean, point targets, linear features and textures. In this work, one of the most relevant speckle filters has been considered, namely the Lee filter [49]. Segmentation is a basic technique of digital image processing that aims at improving an image for subsequent analysis such as the classification of land cover. A SAR image segmentation can be obtained by looking at texture diversity. The texture is an essential parameter for land classification tasks. Texture originates from a scale-dependent spatial variability and is observed as a fluctuation of grey levels in SAR images. One of the tools that has received much attention in recent years for texture analysis is the Wavelet transform (WT) since it has the ability to examine a signal at different scales [46]. A two-dimensional wavelet transform (2D-WT), which is a separable filter bank in row and column directions, decomposes an image in four sub-images [47]. Iterative decomposition of approximated images forms a pyramidal image tree in which the filtered version of each sub-images is down-sampled by a factor of two. This typically refers to a dyadic decomposition and is shown in Figure 11.4.

A dyadic transform has no redundancy and, as a consequence, is quite attractive for some applications. However, it is not the best way to carry out a texture analysis since the decomposition is not translational invariant. For this reason overcomplete wavelet decompositions, which omit the downsampling, are preferred when extracting local texture features from SAR images. In Figure 11.4, the *LL* image is obtained

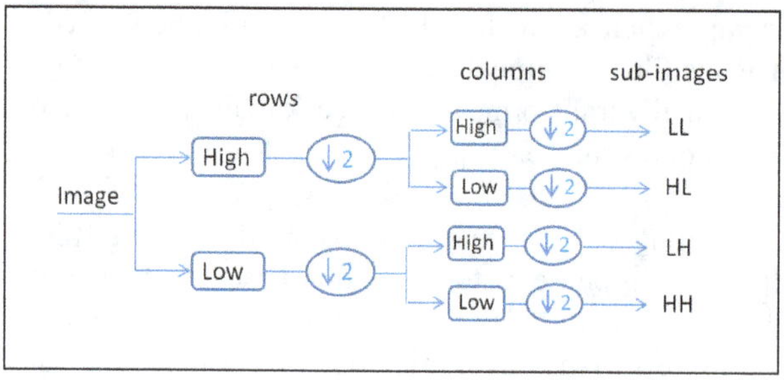

Figure 11.4 2D wavelet decomposition

by low-pass filtering in both the row and column directions while the detail images, i.e. *LH*, *HL* and *HH*, contains high-frequency components. It is clear that in the case of the overcomplete transform, the filtered images have the same size as the previous images. As explained in [47, 48], the energy l_1 *norm* of each sub-image can be a suitable feature for the representation of the texture because it indicates dominant spatial-frequency channels in the original image. Therefore, after decomposing a square local area in the image by an overcomplete Wavelet, the energy of each sub-images is calculated and assigned as a component of a feature vector relative to the central pixel of the area. All the *LL* sub-images obtained during the decomposition are low-pass sub-images that approximate the original image with a strong correlation among each other. As a consequence, it is preferable to add the energy of the *LL* sub-image only at the first level into the feature vector. Consequently when the square area is decomposed into M level, the feature vector has a dimension of $3M+1$ and can be represented as [46]:

$$\mathbf{v}_f = (e_{LL-1}, e_{LH-1}, e_{HL-1}, e_{HH-1}, \ldots, e_{LH-M}, e_{HL-M}, e_{HH-M})^T \in C^{1\times 3M+1} \tag{11.4}$$

where e_{LL-1} denotes the energy of the *LL* sub-image at the first level of Wavelet decomposition. The distance between the feature vector of a pixel and the template vector, relatively to a particular texture, leads to a supervised classification. In this case the classification is based on the concept of minimum distance, e.g. Euclidean and Mahalanobis. A template vector that represents a particular texture can be obtained by averaging the feature vectors of pixels in a training area with known characteristics. It is worth pointing out that a feature vector needs to be standardized, i.e. it must be transformed into a zero-mean and unitary energy vector. It is worth highlighting that several methods can be considered for image segmentation. For example, neural networks can be used for semantic segmentation on optical Google images of the observed area. Since optical images are easily retrievable for the majority of the Earth surface, they can be used as prior information. However, a large number of images are needed to train the neural network. Image segmentation can also be applied directly to the observed SAR scene by exploiting, as a priori information, the texture of the SAR image. Differently from neural networks, no training set is required. Anyway, texture analysis for accurate image segmentation is not a trivial step since textures depend on the system geometry. In the proposed processing chain, an image segmentation based on texture analysis will be followed. It is worth pointing out that the processing chain does not depend on the segmentation algorithm. It is therefore possible to use another segmentation technique while maintaining the same architecture. As previously stated, the clutter suppression performance is related to the training data used for the clutter covariance estimation. Such training data should be as homogeneous as possible for the clutter to be suppressed effectively. A predefined set of feature vectors can be stored in the system memory. Such vectors may be representative of different classes of clutter, which can be used as a priori information to perform image segmentation. The feature vectors identify some predetermined classes such as land, grass, urban and so on. In this context, a performance index is needed to assess the effectiveness of image segmentation [41]. The index can be defined as the ratio between the number of unclassified pixels P_{NC} and the number of classified pixels P_C as in (11.5)

$$Cl = \frac{\#P_{NC}}{\#P_C} \tag{11.5}$$

For a more accurate evaluation of the image segmentation, the area of interest can be divided into sub-blocks and the performance index Cl can be evaluated for each sub-block i.e. Cl_b. Starting from the evaluation of the image segmentation, it is possible to define a controlling function that aims to improve the image segmentation as follows [43]:

$$\alpha_{M_1,b} = \begin{cases} 0 & Cl_b \leq \lambda_c \\ 1 & Cl_b > \lambda_c \end{cases} \tag{11.6}$$

where λ_c is a predefined threshold on the maximum amount of non-segmented pixels within a sub-block. If the index does exceed the threshold, the considered sub-block is assumed to be correctly segmented. Otherwise, a new type of clutter may be present in the image that is not stored in the system memory. As a consequence, the corresponding feature vector can be extracted by exploiting the WT from the considered sub-block and stored in the system memory. When at least one $\alpha_{M_1,b}$ is equal to 1, the segmentation algorithm retrains itself including the new feature vectors in the training set. Through the performance index $Cl^{b,}$ the system is able to identify new clutter types that may be present in an observed area and therefore store the corresponding new feature vector in the memory. In other words, the system is able to learn from the observed scene by collecting new information to improve the image segmentation. According to the dominant clutter type in the observed area, the system is also able to choose the operating frequency that best suits the purpose of the mission. As an example, in the case of forest clutter, the L-band is best suited because of its penetrating property, whilst in the case of sea clutter, the X-band outperforms the L-band. The controlling function that directly controls the amount of energy to be assigned to each available frequency can be defined as follows:

$$\alpha_{d_1} = \frac{\mathbf{p} \otimes \mathbf{w}}{\sum_k \mathbf{p}_k \otimes \mathbf{w}_k} \tag{11.7}$$

where **p** vector contains the number of the pixels for each clutter class, **w** is a vector that contains the priority assigned, depending on the specific mission, to each clutter class and where ⊗ is the Hadamard product. In order to account for the properties of the illuminated scene, let Ω_β be the set of classes that exhibits the best performance in the β bands. For example, sea and ground clutter (not forested areas) could be the classes that exhibit the best performance in high-frequency bands (e.g. X-band). In this case, the amount of energy assigned to a particular frequency band can be expressed as follows:

$$E_\beta = \sum\nolimits_{k \in \Omega_\beta} \alpha_{d_1}[k] P_T CPI \tag{11.8}$$

where P_T is the total amount of power available at the transmitter. In case of a radar system able to transmit both high and low frequencies, (11.7) suggests how to split the overall power on board between high and low frequencies based on the observed scenario. The CPI can be used to enhance the Signal-to-Noise Ratio (SNR) at each frequency band. Although the use of longer CPI can help to enhance the SNR, it is however constrained by the SAR acquisition mode (e.g. stripmap or spotlight) and the SAR image formation algorithm, for example. Therefore, E_β is constrained by the minimum value of energy required to reach a desired value of SNR. It is worth pointing out that the controlling function in (11.6) does not influence the transmitter parameters but affects the system memory. The controlling functions acting on the system memory are indicated with the subscript M while the others that act on the detection are indicated with subscript d.

11.3.2 Training data selection and clutter covariance matrix estimation

After image segmentation, a homogeneous clutter area with respect to the clutter to be suppressed can be selected for the clutter covariance matrix estimation. However, some outliers can also be present in the considered training set. For example, unintentional temporally sporadic electromagnetic interference or intentional blanking interference in the main channel of the system could cause outliers [13]. Another common source that causes a severe degradation in clutter suppression performance is due to moving targets located in training range cells. In order to deal with this issue, a non-homogeneity detector (NHD) based on the generalized inner product (GIP) metric can be exploited by the system to screen and excise non-homogeneous data from the training set [50, 51]. For the sake of clarity, some details about GIP test can be found in the Appendix of the present chapter. It is worth pointing out that a minimum number of training cells is mandatory to effectively suppress stationary clutter. More specifically, the RMB rule [38] guarantees a performance loss of about 3 dB with respect to the optimum case if an adequate number of training cells is used. Such a number is specified by the RMB rule, as expressed in (11.9).

$$N_r = 2MP \tag{11.9}$$

where M represents the number of slow-time samples and P is the number of channels dedicated to the clutter suppression. If the number of selected cells is lower than that imposed by the RMB rule, the clutter suppression performances would produce a SNR loss greater than $3dB$ and this would be declared insufficient. The corresponding controlling function is defined as follows:

$$\alpha_{d,2} = \begin{cases} 0 & N_r \geq 2MP \\ 1 & N_r < 2MP \end{cases} \tag{11.10}$$

In the case where the number of cells is not enough, i.e. $\alpha_{d,2} = 1$, the system tries to improve the performance by applying the Bayesian SDAP approach. The Bayesian SDAP algorithm combines the clutter covariance matrix stored in the system memory with the estimated one [52, 53]. More details about the Bayesian approach can be found in this chapter's Appendix.

11.3.3 Clutter suppression and target imaging

After the training data selection has been completed, the clutter covariance matrix can be estimated and the SDAP filter can be applied for clutter suppression and MTD. The attenuation factor (AF), can be considered as a performance index to assess the capability of SDAP to remove both clutter and interference [40, 41]. This performance index can be defined as the ratio between the multichannel SAR image power before and after clutter suppression as follows:

$$\frac{1}{IF} = \frac{SCNR_{ns}}{SCNR_s} \approx \frac{P_{slc+n}}{P_{cl+n}} = AF \tag{11.11}$$

in which the signal to clutter plus noise ratios can be expressed as:

$$\begin{aligned} SCNR_{ns} &= \frac{P_{tot}}{P_{cl+n}} - 1 \\ SCNR_s &= \frac{P_{tot}}{P_{scl+n}} - 1 \end{aligned} \tag{11.12}$$

where P_{scl+n} is the residual clutter plus noise power after SDAP processing and $P_{tot} = P_{sn} + P_{cl+n}$ is the total power of the received signal in which P_{sn} is the power of the useful signal while P_{cl+n} is the clutter plus noise power. It is worth mentioning that the AF is not to be confused with the Clutter Attenuation (CA), which is defined as the ratio between the clutter power before and after the filter application. The IF is a parameter that describes the behaviour of the SDAP filter and is therefore the figure of merit of the filter. The IF defines the improvement in the SNR introduced by the SDAP filter as the Doppler frequency, or in an equivalent manner the radial velocity, of the target changes. Hence the more $1/IF$ is close to 0 the better the clutter is suppressed. Let λ_{IF} be a predefined set of thresholds that defines the level of clutter suppression required by the user. Threshold values must be specified for the different radial velocities of the target of interest. For sake of clarity of notation, the dependence of IF on the target Doppler frequency is omitted. The controlling function on SDAP performance can be defined as:

$$\alpha_{d,3} = \begin{cases} 0 & \frac{1}{IF} \leq \lambda_{IF} \\ -1 & \frac{1}{IF} > \lambda_{IF} \end{cases} \tag{11.13}$$

According to (11.13), the controlling function $\alpha_{d,3}$ produces a flag that indicates that the SDAP algorithm is not working as expected. To improve the SDAP performance, the system may ask to reduce the bandwidth because it suspects the presence of an in-band interference or may ask for additional spatial channels to enhance the AF. When $\alpha_{d,3} = -1$, the SDAP filter is declared ineffective and the following procedure is carried out to understand whether this result is due to the presence of strong interference or whether the number of spatial channels is insufficient. Then, the overall system analyses the spectrum through conventional spectrum sensing technique in order to identify the presence of interferences in the signal bandwidth. A specific flag, FSS will be set to $FSS = 1$ when an interference is detected and to $FSS = 0$ otherwise. The system suggests to change the signal bandwidth or spatial channels depending both on the current system state and the presence of an interference. This is expressed through the controlling functions $\alpha_{d,4}$ and $\alpha_{d,5}$ reported in (11.14) and (11.15), respectively.

$$\alpha_{d,4} = \begin{cases} 0 \ \ \alpha_{d,3} = 0 \\ +1 \ \ \text{otherwise } and\ \alpha_{d,2} = 1, FSS = 0 \\ -1 \ \ \text{otherwise } x\ and\ FSS = 1 \end{cases} \tag{11.14}$$

The second line of (11.14) addresses the case where no interference is present and the Bayesian SDAP does not perform well. In fact the number of training cell is not adequate i.e. $\alpha_{d,2} = 1$ and then the Bayesian SDAP approach is applied. Since the bandwidth affects the range resolution (the greater the bandwidth the more cells in the range are available), in this case the system asks for more bandwidth assuming that the bandwidth used by the system in its current state is less than the maximum available bandwidth i.e. $B^{(i)}_{max}$. In case of $B^{(i)} = B_{max}$, the system declares a failure and restart the system starting from the standard configuration for that particular scenario. On the contrary, if an in-band interference is present, i.e. $FSS = 1$, the system tries to reduce the bandwidth.

$$\alpha_{d,5} = \begin{cases} 0 \ \ \alpha_{d,3} = 0 \\ 0 \ \ \text{otherwise } and\ \alpha_{d,2} = 1\ FSS = 0 \\ +1 \ \ \text{otherwise } and\ \alpha_{d,2} = 0, FSS = 0 \end{cases} \tag{11.15}$$

At the beginning of the mission the number of spatial channels to use may depend on the clutter type. When $\alpha_{d,5} = 1$ the system switch to the maximum number of available channels unless the Bayesian has been used. In fact, if the Bayesian has been used (second line of (11.14)) it means that the number of available range cells is not enough and using a higher number of spatial channels would require a higher number of training cells. In this case, the system will ask for more bandwidth. Similarly, when no interference is detected, the system automatically uses the maximum available instantaneous bandwidth. An overall block diagram of the SDAP cognitive architecture is depicted in Figure 11.5

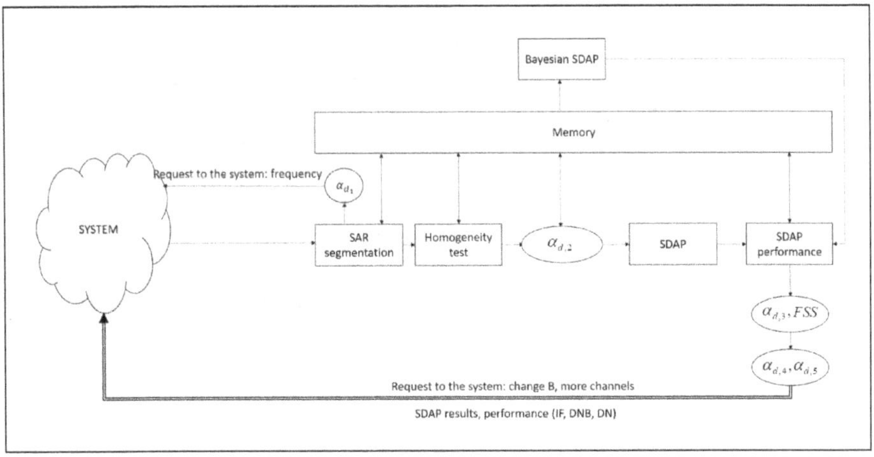

Figure 11.5 *Block scheme of the SDAP cognitive architecture*

11.4 Overall cognitive radar system concept

This section summarises the way the overall system takes the decisions based on the controlling functions defined so far. The system reconfigurable parameters are modified through the actuating function that depends on the overall controlling functions. These functions are obtained through a linear combination of the control functions defined for each processing block. The weights that define the linear combination depend both on the mission and on the observed scene [41, 43, 54]. In this specific case the mission concerns only the clutter suppression and non-cooperative ground MTD. However, the mission can be extended to other tasks and the system parameters can be modified according to other goals such as high-resolution imaging and target classification. In this case, there will be more control functions related to the different tasks that must be weighed with different priorities according to the purpose of the mission. In what follows, it is assumed that the system can be reconfigured in terms of operating frequency, bandwidth and number of channels.

11.4.1 Operative frequency selection

The theoretical framework of the cognitive system described so far refers to a hypothetical radar system able to switch the transmitted frequency within a large set of bands, e.g. L-, C-, X-, Ka-band or even more. Such capability may allow for the same system to handle different missions, e.g. detection of target concealed under foliage and very high resolution imaging of moving targets on bare soil or at sea. Some preliminary experiments have been conducted in this direction to show the benefit of using a system able to transmit in multiple frequency bands [55]. Although these systems are difficult to realise, the rapid progress growth of technology and data processing, due to the increasingly higher sampling rate of receivers and the ever-lower cost of electronics, pushes on the research effort in this direction. In this framework, the system has to manage its resources optimally but without wasting resources that can be useful for other applications. In this framework, the selection of the operative frequency is guided by the SAR segmentation block. In particular, the SAR segmentation block analyses the clutter scene and then produces an input to the system. As an example, if the scene contains a forest with thick canopies then the system would automatically select a low-frequency band, L-band or C-band. Otherwise, if the scene is mostly covered by sea or ground with either little or no vegetation, the system would select higher frequencies, e.g. X-band, Ku-band, Ka-band. Let Ω_L and Ω_X be the set of clutter types for which it is preferable to use, respectively, a low-frequency or a high-frequency band. Then, the energy to assign to each band can be derived by applying (11.8), which is reported hereinafter for the sake of clarity:

$$E_\beta = \sum_{k \in \Omega_\beta} \alpha_{d_1}[k] P_T CPI \tag{11.16}$$

The system will then choose the highest frequency band within the chosen set of frequencies that guarantees the necessary SNR to detect the target. In a more complex system, the choice of frequency could also be guided by the objective of obtaining well-focused images of the target.

11.4.2 Instantaneous bandwidth selection

Others controlling functions steer (increase/decrease ΔB) the amount of bandwidth for each frequency channel, according to the following rule, which acts directly as an actuating function [43, 54]:

$$B_\beta^{(i+1)} = \begin{cases} B^{(i)} & \alpha_B = 0 \\ B_{max} & \alpha_B = +1 \\ B^{(i)} - \Delta B & \alpha_B = -1 \end{cases} \tag{11.17}$$

where α_B defines the overall controlling function. In this case the goal of the system is clutter suppression and therefore a single control function that increases/decreases the amount of the bandwidth, i.e. α_B can be defined. However, if imaging and target classification are also considered, other controlling functions may also be present. In general the overall controlling function can be defined as follows:

$$\alpha_B = \gamma_1 \alpha_{d,4} + \gamma_n \alpha_n \tag{11.18}$$

where α_n is a generic controlling function. The weights γ are positive real numbers lower or equal to 1 such that their sum is equal to 1. In this case, having only one controlling function $\gamma_1 = 1$. The instantaneous bandwidth in each frequency channel is modified according to the sign of α_B. Specifically, when $\alpha_B = +1$ then $B_\beta^{(i+1)} = B_{max}$, where B_{max} is the maximum instantaneous transmitted bandwidth. Otherwise, when $\alpha_B = -1$ it means that the signal bandwidth needs to be decreased of ΔB to exclude the interfering signal. Finally, when $\alpha_B = 0$ the transmitted signal bandwidth remains unchanged, because $\alpha_B = 0$ means that all the processing steps perform effectively with the current signal bandwidth. The weights γ_1 and more in general γ_n are stored in the memory and depend on the specific scenario of interest and particularly on the clutter type, for example, forest, ground clutter, maritime, urban or mixed clutter scenario. Depending on the observed scenario, more emphasis should be given to the detection of moving targets (as for urban or forest scenario where the clutter may be high and may mask some low detectable targets) for others instead more emphasis can be given to the imaging and classification as for example in the case of maritime or ground clutter where the clutter is typically not high enough to mask moving targets.

11.4.3 System channel management

SDAP algorithms may require a higher number of receiving channels to suppress ground clutter and to detect moving targets. As mentioned before, a radar system with a single transmitter and multiple receivers N_c is assumed. The radar can transmit a single frequency band at a certain CPI but can change the transmitted frequency band among different CPIs. In a more general case, a polarimetric system can be considered. The polarimetric channels could be exploited to recognise the target. Therefore, the same transmitter can switch between H and V polarisation between two adjacent pulses. Under this hypothesis, two channels for each spatial receiver, one acquiring H and the other acquiring V polarisation, are required to have both spatial and full-pol acquisition. For example in case of $N_c = 6$ these can be used for a single pol-SDAP using up to six channels or a full-pol SDAP using up to three channels. In this case, the overall controlling function can be defined as [41, 43]:

$$\alpha_{N_c} = \gamma_1 \alpha_{d,5} - \gamma_2 \alpha_c \tag{11.19}$$

where α_c is a hypothetical controlling function that concerns the use of a fully polarimetric acquisition. For example, when a system is asked to perform target recognition, and not only target detection, the use of polarimetry can significantly improve the recognition rate. Therefore, the system needs to decide how to use the available receiving channels. When $\alpha_{N_c} = 0$, it means that both $\alpha_{d,5}$ and α_c are equal to 0 and that the number of actual receiving channels is enough for both detection and recognition. It is possible to have $\alpha_{N_c} < 0$ when $\alpha_{d,5} = 0$ and $\alpha_c = 1$ or when both are equal to 1 but $\gamma_1 < \gamma_2$. In both cases, however, more importance should be given to the classifier. The other possible combination is $\alpha_{N_c} > 0$ when $\alpha_{d,5} = 1$ and $\alpha_c = 0$ or both are equal to 1 but $\gamma_1 > \gamma_2$. In both cases, more importance should be given to the SDAP processing. Then the actuating function may be mathematically formulated as in the following equations, where N_c^i is the current combination of spatial and polarimetric channels and the symbol $\times$ is the multiplication operator.

$$N_{c,\beta}^{i+1} = N_c^i \quad \text{when} \quad \alpha_{N_c} = 0 \tag{11.20}$$

Equation (11.20) states that when $\alpha_{N_c} = 0$ the combination of the spatial and polarimetric channels at the next system status $i+1$, will remain unchanged. When instead $\alpha_{N_c} < 0$ the system switched on the full-pol mode while the number of spatial channels remains unchanged or decreases depending on its values in the current system status, N_c^i

$$N_{c,\beta}^{i+1} = \begin{cases} N_{cs}^{i} \times 2 \text{ if } \alpha_{N_c} < 0 \text{ and } N_{cs}^{i} < \frac{N_{c,max}}{2} \\ \frac{N_{c,max}}{2} \times 2 \text{ if } \alpha_{N_c} < 0 \text{ and } N_{cs}^{i} \geq \frac{N_{c,max}}{2} \end{cases} \quad (11.21)$$

When $\alpha_{N_c} > 0$ the number of spatial channels used for SDAP need to be increased. However, if the system is currently using full-pol mode and a number of receiving channels less than $\frac{N_{c,max}}{2}$, the system tries to maintain the full-pol mode and therefore increases the number of spatial channels to $\frac{N_{c,max}}{2}$. Otherwise, the full-pol mode is switched off and all the available channels are assigned to the SDAP algorithm.

$$N_{c,\beta}^{i+1} = \begin{cases} \frac{N_{c,max}}{2} \times 2 \text{ if } \alpha_{N_c} > 0 \text{ and } N_{cs}^{i} < \frac{N_{c,max}}{2} \\ \frac{N_{c,max}}{2} \times 1 \text{ if } \alpha_{N_c} > 0 \text{ and } N_{cs}^{i} \geq \frac{N_{c,max}}{2} \end{cases} \quad (11.22)$$

It is worth pointing out that in the considered case, the system does not have polarimetric channels so the number of channels will be regulated exclusively by the controlling function $\alpha_{d,5}$, and then γ_1 in (11.19) is equal to one.

11.5 Preliminary results – scaled use case, real dataset

In this section the results obtained by analysing a real dataset will be shown. In particular, the cognitive radar imaging system defined in the previous sections has been tested on a multichannel airborne platform for GMT Img in a heterogeneous environment. The system is a single polarization X-band SAR system, therefore it appears to be a scaled system compared to that considered theoretically. However, it is still possible to test the rule-based cognitive architecture proposed in this chapter. The measured campaign took place on 18 July 2018 near the Teuge airport (NL). The radar system mounted on the airborne platform is a FMCW SAR system with four receiving channels. The acquisition and radar parameters are summarised in Table 11.1, while details about the functioning of the FMCW system can be found in [56]. Results for each block of the proposed processing chain are shown in the following of this section.

Table 11.1 Acquisition parameters

Parameter	Value
Carrier frequency f_0	9.9 GHz
PRF	2.9 kHz
Tx bandwidth	600 MHz
ADC sampling frequency	25 MHz
Platform velocity	45 m/s
Incident angle	55°
Antenna beamwidth	$\theta_{el} = 20°$, $\theta_{az} = 20°$
Acquisition time	0.6 s
Platform altitude	996 m
Baseline	0.08 m
Numbers of Rx channels	4

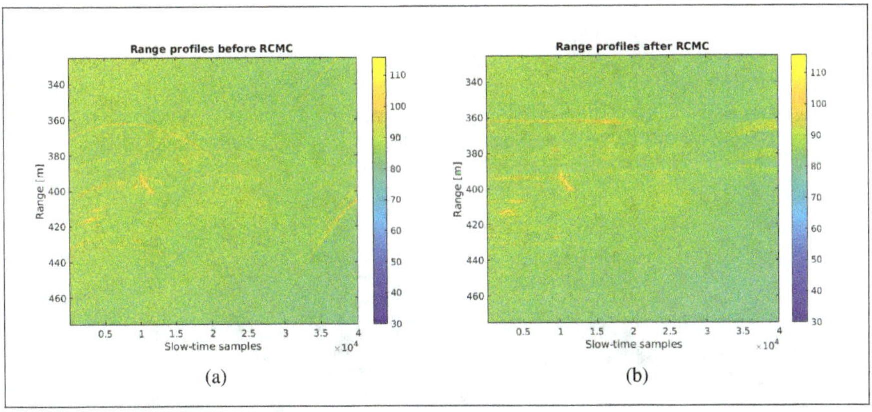

Figure 11.6 *Range profiles of the considered SAR image (a) range profiles before RCMC, (b) range profiles after RCMC*

11.5.1 SAR image formation

The RDA algorithm is used for SAR image formation. As known, due to the platform motion, a range migration occurs during the CPI. Range profiles considered for a particular synthetic aperture are shown in Figure 11.6a. As it is possible to see the range profiles must be realigned, then RCMC is mandatory to obtain the SAR image of the illuminated area. In real acquisition, factors that may be cause a misalignment between the true position of the SAR platform and the position measured by the Inertial Measurement Unit (IMU) can be present, some of which difficult to compensate. The range profiles after RCMC step are shown in Figure 11.6b. The radar platform (e.g. aircraft), may undergo roll, pitch and yaw motions during the CPI. As it is possible to see in Figure 11.6b, a rough range realignment is obtained. However, a residual range migration caused by an imperfect compensation of the platform motion is still present. After RCMC, the SAR image related to a synthetic aperture can be obtained by applying azimuth compression. Combining the SAR images related to adjacent synthetic apertures, the SAR image of the whole observed area during the acquisition can be formed. The synthetic aperture is expressed as follows:

$$L \simeq R_0\theta_{az} = 607\text{m} \tag{11.23}$$

where $R_0 = h/cos(\theta_{inc}) = 1740$m is the slant range distance between the radar platform and the centre of the scene while $h \simeq 1000$m is the platform altitude and θ_{inc} is the incidence angle shown in Table 11.1. The observation time and the achieved cross-range resolution are expressed as follows:

$$\begin{aligned} T_{obs} &= \frac{L}{v_p} = 13.4\text{s} \\ \delta_{az} &= \frac{cR_0}{2f_0 v_p T_{obs}} = 0.045\text{m} \end{aligned} \tag{11.24}$$

where v_p is the platform velocity and f_0 is the carrier frequency. The SAR image of the whole observed area is shown in Figure 11.7 while the corresponding optical

Figure 11.7 *Range Doppler SAR image of the Teuge Airport (NL). The box highlighted in red is the area that will take as the area under test.*

Google image is depicted in Figure 11.8. Although a perfect range migration has not been obtained due to the misalignment between the IMU and the true platform position, a quite accurate SAR image is obtained. In this case only a despeckle Lee filter is applied and no contrast stretching is considered.

11.5.2 SAR image segmentation

The segmentation of the SAR image is achieved by 2D-WT, which decomposes the image into four sub-images. As previously said, the energy of each sub-image can be used as a suitable feature for image segmentation. Then, a 4 × 4 area around a pixel

Figure 11.8 *Optical Google image of the Teuge Airport (NL)*

Figure 11.9 Feature extraction for six different classes of clutter in the SAR image relative to the aerodrome de Spa La Sauvenière (NL)

is decomposed through the 2D-WT and the feature vector that includes the energy of each corresponding sub-image is obtained. Based on the minimum Euclidean distance between the considered feature vector and a set of template vectors representing different classes of clutter, the pixel under test is classified. The set of template vectors are stored in the system memory and it represents the a priori information. As shown in Figure 11.9, the a priori information can be obtained by considering a SAR image of an area different from the one observed in which the main classes of clutter i.e. grass, terrain, road and structure are present. In this case the selected area is close to the aerodrome de Spa La Sauvenière (NL). Different clutter types are selected with different boxes and subsequently decomposed through the 2D-WT. A representation of the a priori information stored in the system memory is shown in Figure 11.10, in which the mean value of the energy for each class is depicted. It is worth pointing out that the mean values of the energy remain approximately the same regardless of the SAR image considered. A particular crop of the SAR image, highlighted in the red box in Figure 11.7, is considered. A result of the image segmentation using a priori information is shown in Figure 11.11, in which the pixels that are not assigned to any class of clutter are represented in yellow. In the segmented SAR image, it is however possible to recognise the airport runway and some structures adjacent to the runway. Based on the evaluation of the segmented image, the system may try to detect additional classes present in the image in order to improve the a priori information that is contained in the memory.

The segmented image includes many unknown pixels because the corresponding feature vectors do not have an adequate Euclidean distance to be associated with any classes stored in memory. After image segmentation, the system splits the image into sub-blocks and calculates for each of them the number of unclassified

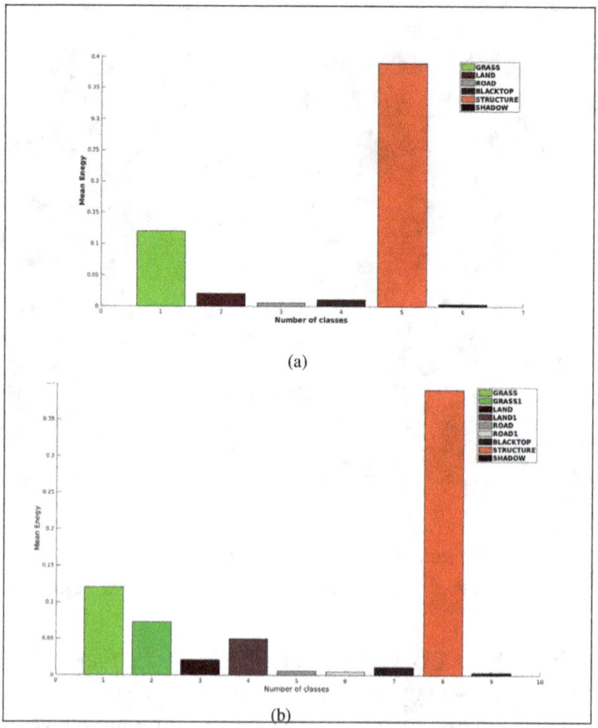

Figure 11.10 *A representation of the content of the system memory (a) initial content in which only a priori information is present. (b) The contents of the memory after the image has been divided into sub-blocks: New clutter classes have been detected.*

pixel and the number of classified pixels in order to evaluate the performance index expressed in (11.5), i.e. $Cl = \frac{\#P_{NC}}{\#P_C}$. The considered sub-blocks are shown with a red line in Figure 11.12. Each block in which the ratio Cl exceed a predefined threshold is considered not well segmented. In this case, an additional type of clutter not stored in the system memory may be present. As a consequence, the corresponding feature vector can be extracted through the 2D-WT and stored in the system memory. The sub-blocks that exceed the predetermined threshold are highlighted in blue in Figure 11.12. According to the controlling function in (11.6), i.e. $\sum_b \alpha_{M_1,b} \geq 1$, the system memory is updated as shown in Figure 11.10, where a new class of clutter are detected. Specifically, a class with mean energy similar to the grass is detected and also two other classes with energy compared to the land and to the road are present. After increasing the a priori information, a new segmentation is performed. The image after the new segmentation is depicted in Figure 11.13. For this image the controlling function $\alpha_{M_1,b}$ is 0 for each sub-blocks, i.e. $\sum_b \alpha_{M_1,b} = 0$. Then the content of the memory is not updated again and a new segmentation is not

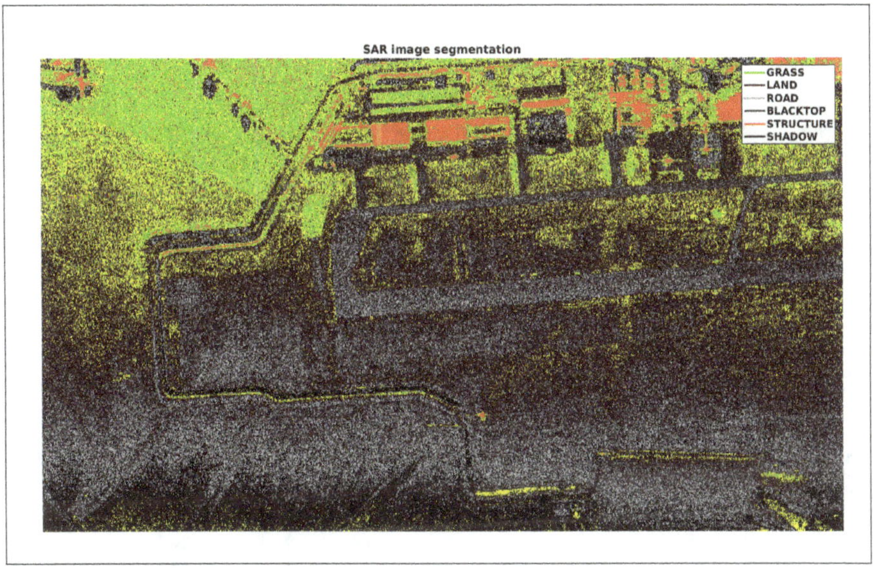

Figure 11.11 SAR image segmentation using a priori information

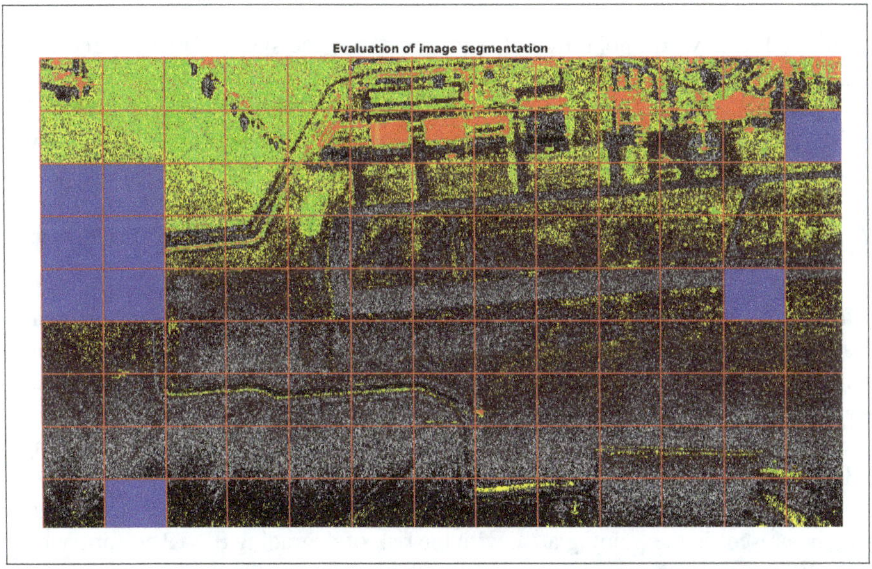

Figure 11.12 Performance evaluation of the SAR image after image segmentation: SAR is divided into sub-blocks to detect the presence of additional classes not stored in the system memory. The sub-blocks containing new classes are highlighted in blue.

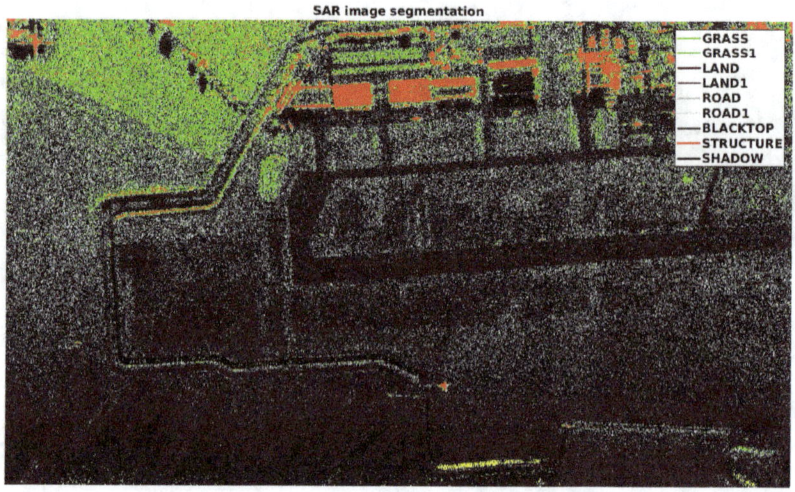

Figure 11.13 *Segmented SAR image after a new segmentation step in which the memory has been updated with new textures belonging to new classes of clutter*

performed. It is worth pointing out that in this case the system is an X-band SAR system and, therefore, it is not possible to choose other types of operating frequency as in (11.8), i.e. $E_\beta = E_X$. However, the ground clutter (bare soil) is dominant in the observed area, which makes the X-band a preferred candidate with respect to lower frequencies.

11.5.3 Training data selection and clutter covariance matrix estimation

After image segmentation, the training area on which to estimate the covariance matrix is no longer selected randomly by the system. Instead, this is carefully chosen in order to select a homogeneous clutter. The structures and the airport runway, which have clutter statistics that are different from the area under test, can be avoided, as they would lead to an incorrect estimate of the covariance matrix. Range cells that range from 455 m to 480 m are selected for clutter covariance matrix estimation as shown in Figure 11.14. However, some outliers such as moving targets may be present in the training area, with the risk of degrading SDAP performances. The GIP test has been considered to screen and excise non-homogeneous data that are still present in the selected training area [50, 57]. In this work, a dummy target with a radial velocity of 3 m/s has been included in the training cells around 468m in order to validate the effectiveness of the GIP test. The result of the application of the GIP test is shown in Figure 11.15, where the red line represents the GIP

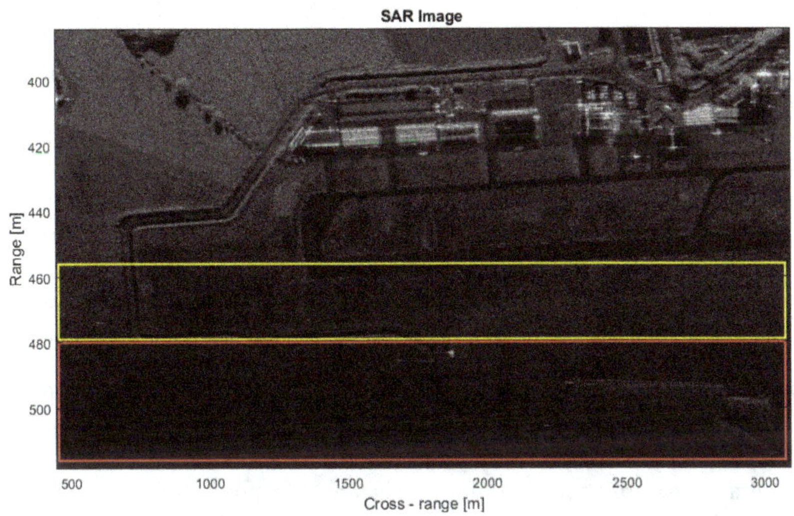

Figure 11.14 *SAR image under test in which the training area is included in the yellow box, while the area on which to apply SDAP is included in the red box*

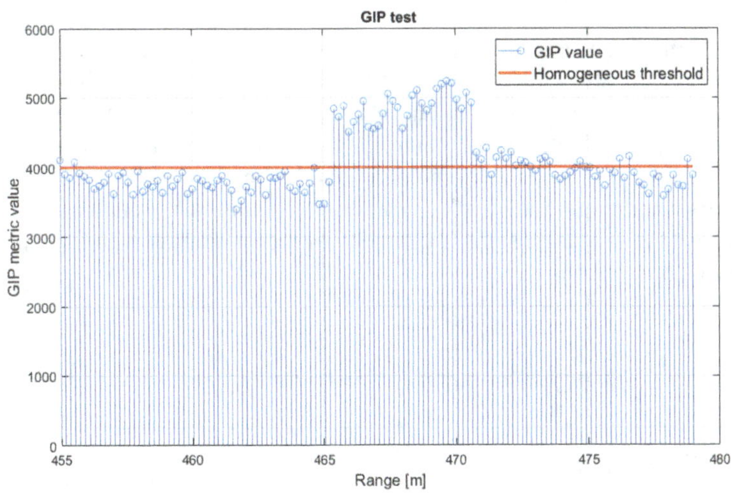

Figure 11.15 *Result of the GIP test applied on the training data set*

homogeneity threshold (more details can be found in Appendix). As it is possible to see, the dummy target is well detected and then the corresponding heterogeneous training cells can be excluded from the clutter covariance matrix estimation. Since the number of slow-time samples is very high (see (11.25)), the required training range cells is not available in order to fulfil the RMB rule i.e. $N_r = 2MP$ (see (11.9)). As in [34], a sub-optimum SDAP approach is exploited to reduce the number of required training data to $N_r = 2LP$, where P is the number of system channels and, for the considered case, $L = 10$. For the sake of clarity, more detail can be found in Appendix. As a consequence of the application of the sub-optimal SDAP processing, the number of required data can be reduced to $N_{r_{required}} = 80$. By considering a range resolution of $\delta_r = \frac{c}{2B} = 0.25$m (see Table 11.1), the initially selected range cells interval (from 455 m to 480 m) corresponds to $N_r = 100$. However, because the presence of the interfering target in the training area reduces the number of available range cells to $N_r = 72$. Since $N_{rr_{required}}$, the number of training cell is not sufficient, which leads the system to setting the controlling function $\alpha_{d,2} = 1$. Because of this flag, as previously explained, the system moves on and attempts to improve the STAP performances by applying the Bayesian SDAP approach, which consists of combining the estimated clutter covariance matrix with a clutter covariance matrix stored in the system memory.

11.5.4 Clutter suppression and target imaging

After selecting the training cells, the clutter covariance matrix can be estimated and the clutter can be suppressed via SDAP processing [34, 39]. It is worth pointing out that SDAP, similarly to STAP, processing is computationally burdensome. As a matter of fact, a long synthetic aperture (see (11.23)) makes it necessary to process a large number of samples:

$$N_{samp} = T_{obs}PRF \approx 39\,000 \tag{11.25}$$

To make this analysis viable on a general-purpose PC, a reduced number of samples $\left(N_{samp} \approx 2\,000\right)$ will be considered here. It is clear that a reduced number of samples degrades the SAR azimuth resolution. The operating parameters and the threshold λ_{IF} exploited by the controlling functions $\alpha_{d,3}$ at ith system status are shown in Table 11.2. It is worth pointing out that the IF threshold is defined by the user for several target radial velocity intervals. In this case a clutter rejection level with respect to the target signal of 25 dB was set for a target with a radial velocity between 4 m/s and 5 m/s.

Table 11.2 System parameters and thresholds at i*th status*

Carrier frequency f_0	9.9 GHz
Bandwidth B	$B^i = B_{max} = 600$ MHz
Channels	$N^i_{ch} = N_{chmax} = 4$
Level of clutter suppression	$\lambda^i_{IF} = -25$dB

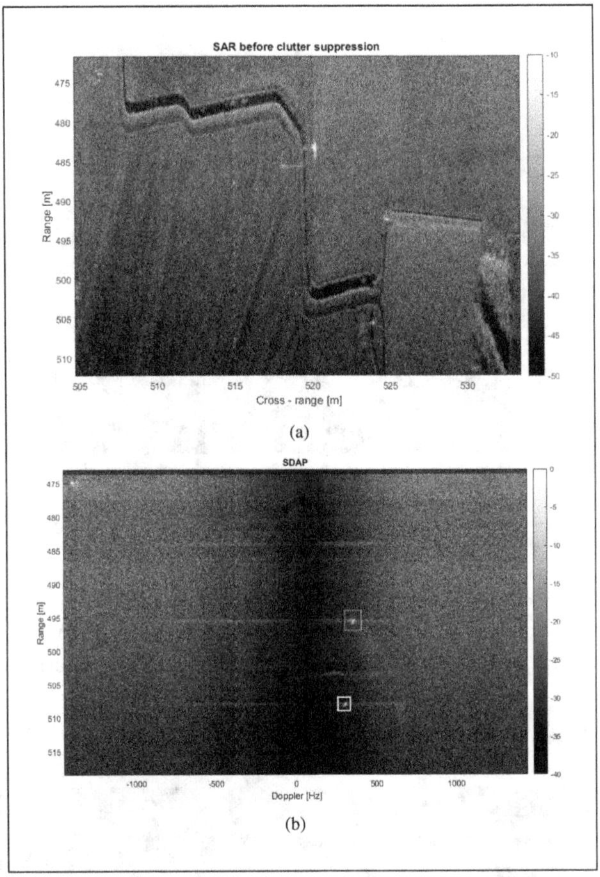

Figure 11.16 *Result of the cognitive SDAP processing. (a) Original SAR image of the area under test before clutter suppression, (b) SAR image after clutter suppression through SDAP processing.*

Results relative to the area under test (highlighted in the red box of Figure 11.7) before and after SDAP processing are shown in Figure 11.16. The optical google image of the selected area is shown in Figure 11.17. The clutter seems to be suppressed effectively and two moving targets, highlighted by the blue and yellow boxes, are detected. Detected targets are also superimposed on the original SAR image as in Figure 11.18. It is worth pointing out that the points having a high backscatter, highlighted with yellow circles, corresponds to a structure close to the airport runway. As shown in Figure 11.7b, these structures are completely suppressed by means of SDAP and only moving targets are detected. These structures are also visible in the optical Google image and are depicted in Figure 11.19. It is worth pointing out that without a cognitive approach, the selection of the training area remains completely random and then even the SDAP performance will be random.

Figure 11.17 Optical image of the area under test

Results obtained after clutter suppression in the case of a random training area selection is shown in Figure 11.20. As it is possible to see, non-cooperative moving targets are not well detected since the training data are not properly selected. As a

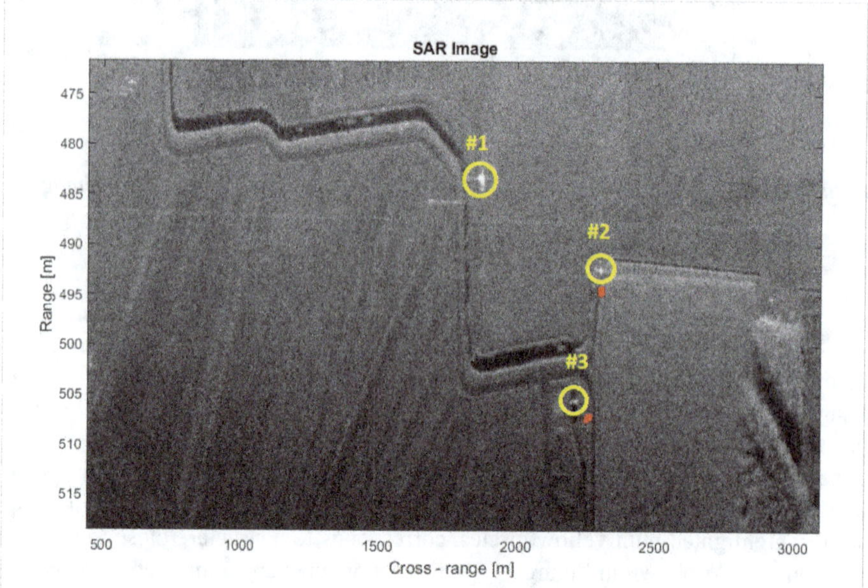

Figure 11.18 The detected non-cooperative moving targets are superimposed with red points on the original SAR image while the structures are highlighted with yellow circles

Figure 11.19 Structures present in the optical Google image relative to the area where KB-SDAP is applied. (a) First structure, (b) second structure, (c) third structure.

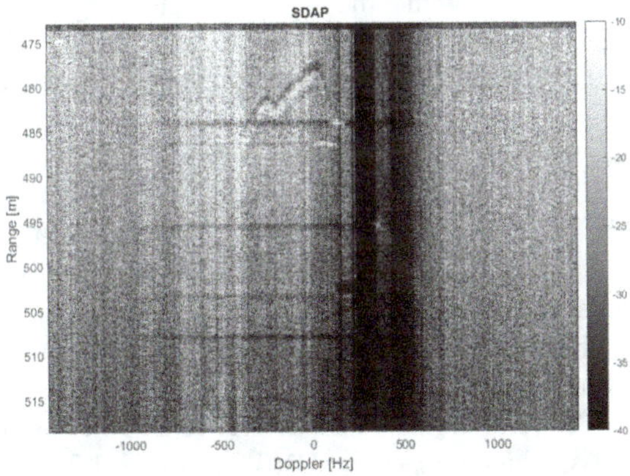

Figure 11.20 SDAP in the case where the training area is non-accurately selected

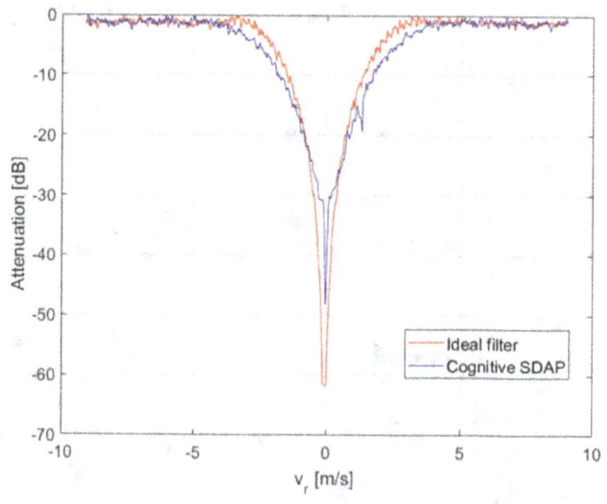

Figure 11.21 Filter comparison between cognitive SDAP filter and an ideal one

matter of fact, some structures fall within the training area, making the estimated clutter non-homogeneous and different from the clutter to be suppressed.

The performances in terms of clutter suppression are evaluated by the system in term of the AF, as reported in (11.11) and by comparing the CA as a function of radial velocity. As discussed previously, AF can be considered as a performance index to evaluate the SDAP capability in terms of clutter suppression. The more the AF is close to 0 the better the clutter is suppressed. The controlling function $\alpha_{d,3}$ compares the AF with a predefined threshold, i.e. λ_{IF}^{i}. In this case the obtained AF is equal to $AF \approx -31$dB and then $\alpha_{d,3} = 0$. The corresponding SDAP filter is shown in Figure 11.21 with the blue trend. A comparison with a reference Doppler profile can be obtained. The reference Doppler profile depends on the minimum radial target velocity to detect. Assuming that the minimum radial target velocity to detect is 4 m/s, the reference Doppler profile can be obtained by using the minimum number of channels that satisfy the requirement in ideal conditions (ideal conditions mean homogeneous clutter and a correctly estimated covariance matrix). The red line in Figure 11.21 represents the clairvoyant SDAP filter. The DNB is defined as the ratio between the notch bandwidth of the actual and the reference Doppler profiles and can be calculated in several ways. For example, it can be defined as the bandwidth computed at −10 dB with respect to the notch. The DNB is linked to the estimation of the clutter covariance matrix and the number of the receiving spatial channels. The notch bandwidth for the ideal filter corresponds to $NB_{ideal} = 2.8$m/s while the notch bandwidth of the SDAP filter corresponds to $NB_{SDAP} = 3.48$m/s. This means that the SDAP filter is very similar to the reference Doppler profile, i.e. the ratio between the two NBs, which defines the DNB is, $DNB = 1.24$. In this case study, the controlling functions $\alpha_{d,3}$ is equal to 0, which means that the SDAP suppresses the clutter effectively and no improvement is required, therefore the system requires neither an increase nor a decrease in bandwidth, nor an increase in the number of

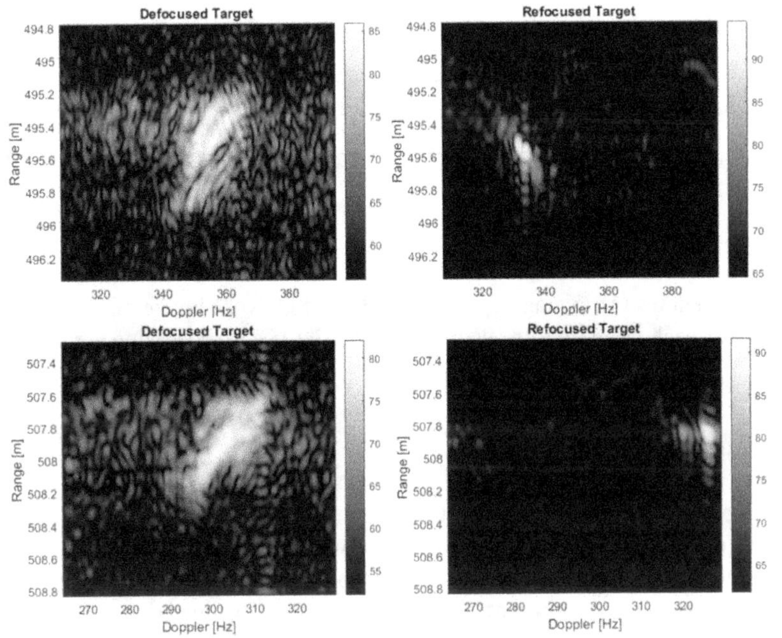

Figure 11.22 Refocusing through ISAR processing of Target 2 (a) before ISAR, (b) after ISAR

channels, i.e. $\alpha_{d,4} = 0$, $\alpha_{d,5} = 0$. In the case where the clutter is not suppressed effectively (see, for example, Figure 11.20), the system declares a failure since $B^i = B_{max}$ and $N^i_{ch} = N_{chmax}$ and restarts the system starting from the standard configuration for that particular scenario.

It is finally worth pointing out that for this dataset the ground truth is not available and the clutter suppression performance cannot be assessed directly. In fact, it is not possible to know whether the detected targets are moving or not. However, as previously described, ISAR processing can be exploited in this sense. Since each target has its own motion, to effectively apply ISAR processing, each target must be separated from both the contributions of the static scene and the other targets. In Figure 11.22a,b, the detected targets depicted in the blue and yellow box (see Figure 11.16) are shown before and after ISAR refocusing, respectively. The improvement in the image focus is quite evident also from a visual point of view, which means that the detected targets are most likely moving targets. Moreover, the improvement in the image focus can be also evaluated by exploiting the Image Contrast (IC) value, which is calculated based on the image before and after the application of ISAR processing. For the sake of completeness, the definition of IC is shown below [18]:

Table 11.3 IC before and after ISAR processing

	IC_b	IC_a
Target 1	3.14	10.32
Target 2	0.21	4.76

$$\mathrm{IC} = \frac{\sqrt{E\{I - E[I]\}}}{E\{I\}} \tag{11.26}$$

where $E\{.\}$ denotes the average operation and I is the ISAR image magnitude. The IC results for each target before (IC_b) and after (IC_a) ISAR processing are shown in Table 11.3.

11.6 Conclusion

STAP has been extensively studied in the last decades as an effective tool for GMTI since it offers the best performances when compared to other single-channel and multichannel techniques. However, for STAP to be applied effectively, the statistical properties of clutter and other disturbances must be estimated accurately. This requires an adequate number of training cells to be available. In real operational cases, the presence of highly heterogeneous scenarios often limits the number of range training range cells and, therefore, the applicability of STAP techniques. This chapter has shown some techniques that are able to solve this problem by designing a multi-function radar system that is able to operate in a complex and non-stationary environment for ground MTD and imaging. The cognitive paradigm has been considered and applied to improve the STAP algorithm's ability to remove clutter and improve the detection of moving targets. In particular, a rule-based cognitive architecture has been proposed as a way to design a cognitive radar system that has to dynamically manage its limited resources to maximise GMTI and GMTImg performances. The proposed architecture bases its functioning on the definition of a set of performance indicators and rules, which drive actions to be taken. Performance and rules have been defined by taking into account specific radar functions, namely, target detection and imaging. In particular, performance indicators measure the success of the mission while rules act on the system reconfigurable parameters. Each action is based both on measures of success and past experiences, which are store in the system memory. As a consequence, general rules are defined in such a way that they select the most appropriate set of waveform parameters. Such rules can be updated depending on the information contained in the system memory and the sensed environment, with the aim to improve ground MTD and imaging. A small case study referring to a cognitive multichannel SAR system for target detection and imaging has been considered and shown as an example. Even though a single and scaled case study has been considered, results show that the system is able to autonomously make decision about the best strategies to maximise clutter cancellation performances. These results are intended to show the benefits of using the proposed method in a resource-constrained and heterogeneous environmental scenario. The results show that, with a careful selection of the training data, it is possible to significantly improve the STAP performances in terms of clutter suppression and therefore in terms of target detection and imaging. Results clearly show that the better the selection of training cells, the better the clutter suppression is. Therefore, when the system is able to segment the observed area by discriminating the homogeneous clutter regions from the heterogeneous ones, the clutter cancellation performance improves. The segmentation of the SAR scenario is therefore a sensitive task that needs to be investigated in more details. It is worth pointing out that the proposed architecture does not depend on the segmentation algorithm. It is therefore possible to exploit other segmentation techniques while maintaining the same architecture. Another sensitive assignment to be taken into account concerns the setting of the predefined threshold that defines the required level of clutter suppression and that assesses the effectiveness of the STAP filter. These thresholds can be initially set to a value that is pre-estimated by using simulated data or previously acquired data and may be updated subsequently by the system to adapt to the dynamically changing environment. Recommendations to improve the proposed architecture in future work are related to the use of deep learning algorithms. As these need quite large amounts of data to be trained, a solution may be viable when using optical images to perform ground segmentation and then applied to co-registered SAR images. More generally, it is possible to think of an architecture as a black box based entirely on deep learning algorithms where the reconfigurable parameters of the system and the parameters that characterise the observed scenario are provided as input and the new parameters to be used in transmission are directly given as outputs. In this case, however, the proposed rule-based architecture would no longer remain the same as the proposed one. It should also be pointed out that black-box approaches have the disadvantage to not provide any insights about why a decision is made rather than another one, whereas, rule-based provide have a more constrained and structured architecture that is controllable by design. Hybrid architectures may also be envisioned that may still provide both benefits and potentially outperform current solutions.

Appendix

In this appendix, some aspects of signal processing present in the chapter are better detailed. In particular, the SDAP, the Bayesian approach and the GIP test as NHD test are addressed.

Space Doppler adaptive processing: signal model

In this section the signal model for multichannel SAR system will be provided. An extension of this model for a two-dimensional array can be found in [39]. Consider the geometry shown in Figure 11.23. in which a multichannel SAR with P channels aligned along the fight path is mounted on an airborne platform. The centre of the

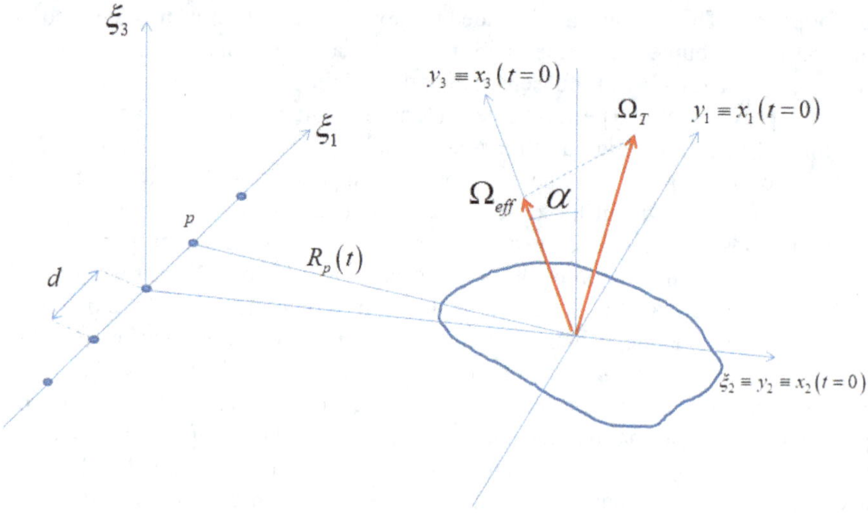

Figure 11.23 Multichannel SAR geometry

reference system of T_ξ is in the phase centre of the transmitter and is oriented so that the axis ξ_2 is parallel with the radar Line of Sight (LoS) while the axes ξ_1 corresponds to the baseline. The rotation vector applied to the reference point is expressed as Ω_T and its projection on a plane orthogonal to the LoS represents the effective rotation vector Ω_{eff}. The reference system T_x is embedded on the reference point on the scene and the axes x_3 is oriented along the Ω_{eff} while the axes x_2 is oriented along the LoS. The reference system T_y is fixed in time and its orientation coincides with T_x for $t = 0$.

The received signal at p_{th} channel can be expressed as follows:

$$S_{R_p}(f, m) = \sum_k \sigma_k e^{-j4\pi \frac{f}{c} R_p\left(x_1^{(k)}, x_2^{(k)}, m\right)} W(f, m) \tag{11.27}$$

in which $p = \left[-\frac{P}{2}, \ldots, \frac{P}{2} - 1\right]$, f is the frequency and $m = \left[-\frac{M}{2}, \ldots, \frac{M}{2} - 1\right]$ is the discrete slow-time instant, i.e. $(t = mT_R)$, where T_R is the pulse repetition interval (PRI). In (11.27), $W(f, m)$ is the function that defines the region in the Fourier domain where the signal is defined, σ_k is the reflectivity function of the k_{th} scatterer and $R_p\left(x_1^{(k)}, x_2^{(k)}, m\right)$ is the distance between the p_{th} channel and the generic scatterer k. The distance term can be expressed as below:

$$R_p\left(x_1^{(k)}, x_2^{(k)}, m\right) = \sqrt{(R_0 + x_2^{(k)})^2 + (v_p m T_R + pd - x_1^{(k)})^2} \tag{11.28}$$

where R_0 is the distance between the array centre and the reference point on the scene, v_p is the platform velocity and d is the baseline. Under straight iso-range approximation [18] and under the assumption that the array dimension is much smaller than the radar-target distance, (11.28) can be approximated as:

$$R_p\left(x_1^{(k)}, x_2^{(k)}, m\right) = R_0^{(p)} + \mathbf{x}_k(m)_{losx}^{(p)}(m) \tag{11.29}$$

In (11.29), $R_0^{(p)}$ represents the distance between the pth channel and the reference point on the scene while $\mathbf{x}_k$ is the kth scatterer position varying in time and the $i_{losx}^{(p)}(m)$ is the LoS relative to the pth channel with respect to the T_x reference system. Under the assumption of constant rotation vector during the observation time, i.e. $\Omega_T(m) \approx \Omega_T$ and considering the scatterer position $\mathbf{x}_k(m) = \mathbf{y}_k + \mathbf{c}mT_R$ [39], the inner product in (11.29) can be expressed as follows:

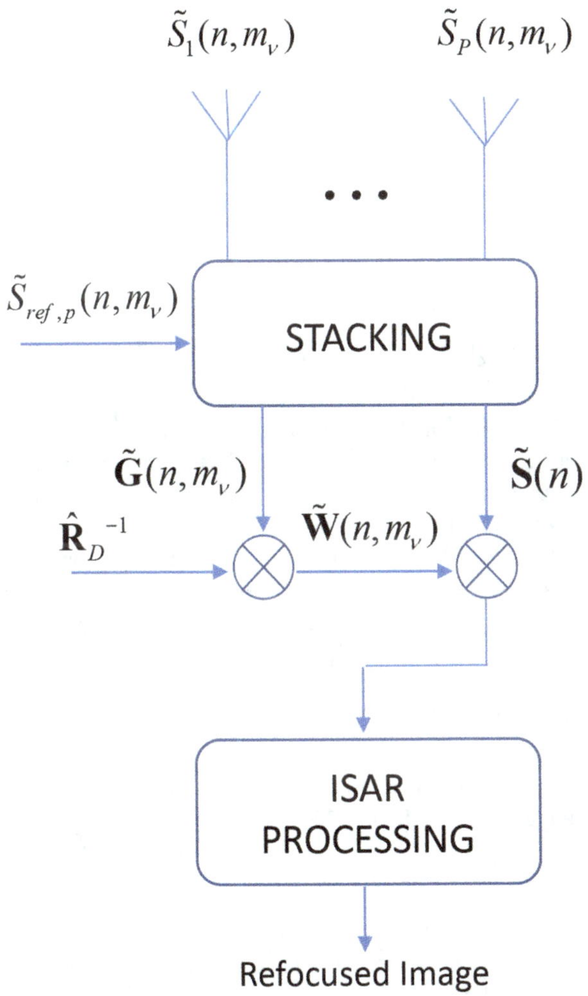

Figure 11.24 Optimum SDAP ISAR functional block

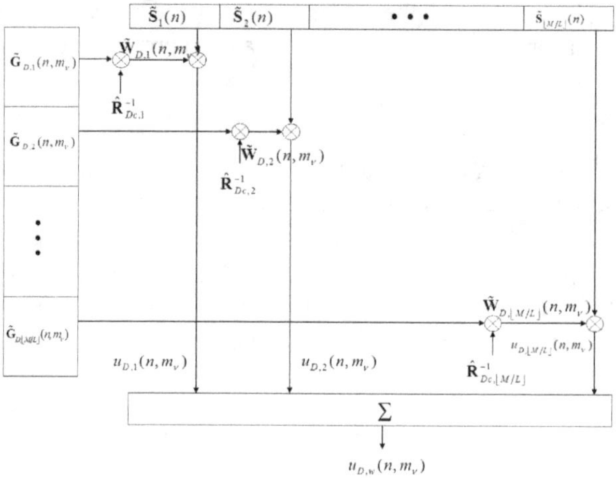

Figure 11.25 Sub-optimum SDAP functional block

$$\mathbf{x}_k(m) \cdot i_{losx}^{(p)}(m) = K_{0,k}^{(p)} + K_{1,k}^{(p)} m T_R \tag{11.30}$$

where

$$K_{0,k}^{(p)} = y_2^{(k)} - \frac{d}{R_0}\left[y_1^{(k)} p\cos\alpha - y_3^{(k)} p\sin\alpha\right] \tag{11.31}$$

$$K_{1,k}^{(p)} = c_2^{(k)} - \frac{d}{R_0}\left[c_1^{(k)} p\cos\alpha - c_3^{(k)} p\sin\alpha\right] \tag{11.32}$$

in which $c_1^{(k)}$, $c_2^{(k)} = \Omega_{eff} y_1^{(k)}$, and $c_3^{(k)}$ are the components of the vector $\mathbf{c} = \Omega_T \times \mathbf{y}_k$. It is worth pointing out that the angle α identifying the effective rotation vector orientation in the plane perpendicular to the radar LoS, y_2. The Multichannel Range Doppler image is obtained by summing the P complex images obtained by processing the signals received by the single p_{th} channels. The complex Point Spread Function relative of the generic p_{th} channel relative to a single scatterer located at $\mathbf{y}_k$ can be obtained after $2D - IFT$ of the compensated signal and can be expressed in the time-delay/DN, i.e. (τ, ν) as [17]

$$I^{(p)}(\tau,\nu) = BT_{obs}\sigma_k e^{j2\pi f_0 \frac{2}{c} K_{0,k}^{(p)}} sinc\left[T_{obs}\left(\nu + \frac{2f_0}{c} K_{1,k}^{(p)}\right)\right] sinc\left[\left(\tau - \frac{2f}{c} K_{0,k}^{(p)}\right)\right] \tag{11.33}$$

where f_0 is the central frequency. After a scaling operation from the time-delay/DN to range/cross-range domain [18], (11.33) can be reasonably approximated as [19]

$$I^{(p)}(y_1,y_2) = BT_{obs}\sigma_k e^{j2\pi f_0 K_{0,k}^{(p)}} sinc\left[\frac{2f_0}{c}\Omega_{eff}T_{obs}\left(y_1 + y_1^{(k)}\right)\right] sinc\left[\frac{2B}{c}\left(y_2 - y_2^{(k)}\right)\right] \tag{11.34}$$

It is worth pointing out that under the assumption of array size much smaller than radar-target distance, the terms present in (11.31) and (11.32) can be approximated, respectively, as $K_{0,k}^{(p)} = y_2^{(k)}$ and $K_{1,k}^{(p)} = c_2^{(k)} = \Omega_{eff} y_1$ only within the sinc function. By looking at (11.34), it is evident that the P images are aligned in phase only for the focusing point. Then, in order to coherently sum the P images, the phase term should be kept as low as possible. This condition determines the maximum extent of the SAR image. The condition can be empirically determined by imposing an upper bound to the maximum phase difference among the images as follows:

$$\frac{4\pi}{\lambda}(P-1)\frac{d}{R_0}\left(y_1 cos\alpha - y_3 sin\alpha\right) \leq \frac{\pi}{8} \tag{11.35}$$

where the term $\frac{\pi}{8}$ has been considered as the maximum phase difference allowed to consider all the contribution aligned in phase. it is quite clear that (11.35) depends on the array size and the target scene size, $\left(y_1 cos\alpha - y_3 sin\alpha\right)$, in the ξ_1 dimension. Moreover, this method can still be applied when the target size is larger than the bound expressed in (11.35) but in this case the whole region must be split into smaller sub-regions and the focusing step must be applied separately to each sub-region. In order to better understand the distortion introduced when the requirement in (11.35) is not satisfied, the attenuation term can be evaluated. Considering the summation among all the P channels and set $\alpha = 0$ to simplify the notation

$$I\left(y_1, y_2\right) = \sum_{p=-P/2}^{P/2-1} I^{(p)}\left(y_1, y_2\right) = I^{(0)}(y_1, y_2)\sum_{p=-P/2}^{P/2-1} e^{j\frac{4\pi}{\lambda}\frac{d}{R_0}y_1 p} \tag{11.36}$$

After some mathematical manipulations (11.36) becomes

$$I\left(y_1, y_2\right) = I^{(0)}(y_1, y_2)e^{\frac{j2\pi Pd}{2\lambda R_0 y_1}} J_{y_1} \tag{11.37}$$

where the term $J_{y_1} = \frac{sin\left(\frac{2\pi d y_1 P}{\lambda R_0}\right)}{sin\left(\frac{2\pi d y_1}{\lambda R_0}\right)}$ produces a distortion in the amplitude of the image due to the fact that the cross-range region is far away from the focusing point and the image does not sum coherently. The array size and the target scene size have been carefully selected in the system requirements specification present in the previous technical report.

Space Doppler adaptive processing: SDAP-ISAR processing

The availability of more than one receiving channel located along the flight direction allows the exploitation of STAP to suppress the undesired clutter return. This can significantly improve the detection performance and allows the ISAR from SAR processing to be used to obtain high-resolution images of non-cooperative moving targets. ISAR from SAR processing has already been described in [19]. STAP processing has been extensively studied in the last few decades for clutter suppression and target detection [21, 30–33], but its use for both clutter suppression and moving target imaging has been addressed only in recent years [34]. In [34], a different version of STAP processing termed SDAP directly derived from the RDA used in ISAR processing is proposed. Joint SDAP-ISAR processing addresses the problem of clutter suppression from an imaging point of view and combined ISAR processing in order to refocus the moving target after the suppression of the clutter. This algorithm is here recalled. First, it is worth pointing out that in an actual system both the variables in the signal domain, i.e. in the frequency/slow-time domain (f, m), and the image domain, i.e. in the time-delay/Doppler (τ, ν), are discrete:

$$\begin{aligned} f &= f_0 + n\Delta f \\ t &= mT_r \\ \tau &= n_\tau \Delta\tau \\ \nu &= m_\nu \Delta\nu \end{aligned} \tag{11.38}$$

where $n = \left[-\frac{N}{2}, \ldots, \frac{N}{2} - 1\right]$, $n_\tau = \left[-\frac{N_\tau}{2}, \ldots, \frac{N_\tau}{2} - 1\right]$ in which N and N_τ are, respectively, the samples along the frequency and time-delay domain and where $m = \left[-\frac{M}{2}, \ldots, \frac{M}{2} - 1\right]$, $m_\nu = \left[-\frac{M_\nu}{2}, \ldots, \frac{M_\nu}{2} - 1\right]$ in which M and M_ν are, respectively, the samples along the slow-time and the DN. Since the relationship between frequency/slow-time domain and time delay/DN is only a Fourier transform, the sampling interval can be expressed as

$$\begin{aligned} \Delta\tau &= \frac{1}{N\Delta f} = \frac{1}{B} \\ \Delta\nu &= \frac{1}{MT_R} = \frac{1}{T_{obs}} \\ N_\tau &= N \\ M_\nu &= M \end{aligned} \tag{11.39}$$

The fully discrete signals are then defined as follows

$$S_{R,p}(n, m) = S_{R,p}\left(f_0 + n\Delta f, mT_R\right) \tag{11.40}$$

$$I^{(p)}\left(n_\tau, m_\nu\right) = I^{(p)}\left(n_\tau \Delta\tau, m_\nu \Delta\nu\right) \tag{11.41}$$

As known, high-range resolution can be obtained by transmitting wideband signals and matched filtering the received signal. High resolution in azimuth can be achieved by applying the RDA [17]. Considering the received signal on p_{th} channel from the generic scatterer k. The range Doppler processing can be expressed as

$$u_D^{(p)}\left(n, m_\nu\right) = DTF_m\left\{S_{R,p}(n, m) S_{ref,p}(n, m)\right\} \tag{11.42}$$

where DFT_m is the discrete Fourier transform along the slow-time domain and $S_{ref,p}(n, m)$ is the reference signal that performs the motion compensation. Equation (11.42) can be expressed as a convolution in the DN between the received and the reference signal both expressed in the frequency/Doppler domain:

$$u_D^{(p)}\left(n, m_\nu\right) = \tilde{S}_{R,p}\left(f, \nu\right) \otimes_\nu \tilde{S}_{ref,p}\left(f, -\nu\right) \tag{11.43}$$

in which

$$\begin{aligned} \tilde{S}_{R,p}\left(f, \nu\right) &= DFT_m\left\{S_{R,p}(f, m)\right\} \\ \tilde{S}_{ref,p}\left(f, \nu\right) &= DFT_m\left\{S_{ref,p}(f, m)\right\} \end{aligned} \tag{11.44}$$

are the Fourier transform of the received and the reference signal. It is clear from (11.43) that the range Doppler processing can be seen as a matched filter in the frequency DN. This processing can be easily extended to a multichannel scenario and as a consequence the multichannel range Doppler processing can be expressed as follows:

$$u_D\left(n, m_\nu\right) = \sum_{p=1}^{P} u_D^{(p)}\left(n, m_\nu\right) = \sum_{p=1}^{P} \tilde{S}_{R,p}\left(f, \nu\right) \otimes_\nu \tilde{S}_{ref,p}\left(f, -\nu\right) \tag{11.45}$$

Equation (11.45) can be expressed more compactly in a vectorial form after a stacking operation first along the channel dimension, i.e. (11.46) and (11.47) and after along the Doppler dimension, i.e. (11.48) and (11.49)

$$\tilde{\mathbf{S}}(n, m_\nu) = \frac{1}{P}\left[\tilde{S}_{R,1}(n, m_\nu), \tilde{S}_{R,2}(n, m_\nu), .., \tilde{S}_{R,P}(n, m_\nu)\right]^T \varepsilon C^{P\times 1} \tag{11.46}$$

$$\tilde{\mathbf{S}}_{ref}(n, m_\nu) = \frac{1}{P}\left[\tilde{S}_{ref,1}(n, m_\nu), \tilde{S}_{ref,2}(n, m_\nu), .., \tilde{S}_{ref,P}(n, m_\nu)\right]^T \varepsilon C^{P\times 1} \tag{11.47}$$

$$\tilde{\mathbf{S}}(n) = \left[\tilde{\mathbf{S}}^T(n, 0), \tilde{\mathbf{S}}^T(n, 1), .., \tilde{\mathbf{S}}^T(n, (M-1))\right]^T \varepsilon C^{MP\times 1} \tag{11.48}$$

$$\tilde{\mathbf{G}}_D(n, m_\nu) = \left[\tilde{\mathbf{S}}_{ref}^T(n, m_\nu), ..., \tilde{\mathbf{S}}_{ref}^T(n, (m_\nu - (M-1)))\right]^T \varepsilon C^{MP\times 1} \tag{11.49}$$

The output of the Doppler matched filter is then expressed as

$$u_D\left(n, m_\nu\right) = \tilde{\mathbf{G}}_D(n, m_\nu)^H \tilde{\mathbf{S}}(n, m_\nu) \tag{11.50}$$

It is worth pointing out that in order to obtain the multichannel image formation a sum along the spatial dimension must be performed. In order to avoid distortions in the resulting image, the sum must be performed coherently leading to a limitation in the array and the imaging area size as previously described. The maximum array and area size allow for the distortion in the image to be avoided are evaluated in the previous technical report. At this point, to obtain the final image in the time-delay/DN a Fourier transform along the frequency dimension must be performed. Optimum SDAP involves the replacement of the space/Doppler reference vector $\mathbf{G}_D(n, m_\nu)$ with a weight vector:

$$u_D\left(n, m_\nu\right) = \tilde{\mathbf{W}}_D(n, m_\nu)^H \tilde{\mathbf{S}}(n) \tag{11.51}$$

The optimum weight vector can be selected with different criteria. In this case the objective is in target detection than in the signal-to-interference-plus-noise ratio (SINR) maximization. The expression for the optimum filter [21, 22] in the SDAP case can be defined as follows:

$$\tilde{\mathbf{W}}_D(n, m_\nu) = \gamma \hat{\mathbf{R}}_D^{-1} \tilde{\mathbf{G}}_D(n, m_\nu) \tag{11.52}$$

where γ is a constant that not affect the SINR while $\gamma \hat{\mathbf{R}}_D$ is the estimation of the clutter plus noise covariance matrix and it is obtained from a set of target-free training data taken along the range dimension:

$$\hat{\mathbf{R}}_D = \frac{1}{N_r} \sum_{nr=1}^{N_r-1} \tilde{\mathbf{Z}}(n_r) \tilde{\mathbf{Z}}(n_r)^H \varepsilon C^{MP\times MP} \tag{11.53}$$

The vector $\tilde{\mathbf{Z}}(n_r)$ define the target-free vector that is clearly defined in the space/DN. It is worth highlighting that if the target contribution is present in the training data, the adaptive filter considers the useful signal of the target as clutter leading to a target cancellation. Optimum processing defined in (11.51), performs both clutter suppression and image formation by means of RD processing. It is clear that SDAP is intimately linked to the RD image formation algorithm and as a consequence all the RD assumption, i.e. the straight iso-range approximation and the constant aspect angle variation is mandatory in order to apply this type of processing. In order to obtain a well-focused image of a non-cooperative moving target present in the observed area, the reference vector $\mathbf{G}_D(n, m_\nu)$ must compensate both the platform and the target's own motion. This is not a realistic situation since the target is not cooperative and the relative motion between the platform and the target is not a priori known. Therefore, SDAP performs clutter suppression and platform motion compensation leading to a well detected but unfocused target. At the output of SDAP, ISAR processing can be applied in order to refocus moving target by compensating the platform and moving target relative motion. ISAR processing is described in [17, 18]. The overall functional block processing is shown in Figure 11.24.

Another issue to be addressed concerns the number of training data necessary to obtain a correct estimate of the clutter covariance matrix. In order to obtain a performance loss of 3 dB with respect to the optimal case, a number of $N_r = 2MP$ target-free training data are needed. This is the well-known RMB rule [38]. It can be easily demonstrated that for typical values of PRF and observation time, the number of required training data that satisfies the RMB rule is very high. Since training data are taken in the range dimension and considering standard range resolution, a very wide area of homogeneous clutter is required to obtain a properly estimation of covariance matrix. Obviously this is an unrealistic situation since clutter is in general heterogeneous and stays homogeneous only in a small range of areas. A sub-optimum SDAP processing to overcome this issue and reduce the dimensionality of clutter covariance matrix to be estimated is presented in next subsection.

Sub-optimum approach

The reduction in the dimensionality of the clutter covariance matrix can be obtained by splitting the filtering operation in (11.52) in sub-windows as shown in Figure 11.25. this operation consists of slitting the Doppler dimension in sub-windows of length L and applying a reduced dimension SDAP on each sub-block. As shown later, the reduction in the estimated covariance matrix is given by the length of the choose sub-window. It is important to note that if only the Doppler matched filtering operation is considered, i.e. $\tilde{\mathbf{W}}(n, m_\nu = \tilde{\mathbf{G}}_D(n, m_\nu)$ any loss is expected. However, when the clutter mitigation is performed, i.e. $\tilde{\mathbf{W}}(n, m_\nu = \gamma \hat{\mathbf{R}}_D^{-1} \tilde{\mathbf{G}}_D(n, m_\nu)$, the windowing operation involves an approximation of the clutter covariance matrix that consists of considering only a set of blocks of the overall covariance matrix aligned along the principal diagonal that leads to some losses in the final image. The approximated matrix considered in the sub-optimum processing is shown in Figure 11.25. The sub-optimum processing can be formalized as follows. Let $i = 1, \ldots \frac{M}{L}$ denotes the window index and L the window length, the output of the SDAP filter in a generic i_{th} sub-window is:

$$u_{D,i}\left(n, m_\nu\right) = \tilde{\mathbf{W}}_{D,i}(n, m_\nu)^H \tilde{\mathbf{S}}_i(n) \tag{11.54}$$

where $\tilde{\mathbf{W}}_{D,i}(n, m_\nu)$ and $\tilde{\mathbf{S}}_i(n)$ are the weight vector and the signal vector with respect to the sub-block. The global output of the sub-optimum processing is the sum of the SDAP filter applied on each block:

$$u_{D,w}(n, m_\nu) = \sum_i u_{D,i}(n, m_\nu) \tag{11.55}$$

The weight vector relative to the *i*th sub-block can be expressed as follows:

$$\tilde{\mathbf{W}}_{D,i}(n, m_\nu) = \hat{\mathbf{R}}_{D,i}^{-1}\tilde{\mathbf{G}}_{D,i}(n, m_\nu) \tag{11.56}$$

where

$$\tilde{\boldsymbol{S}}_i(n) = \begin{bmatrix} \tilde{\mathbf{S}}(n,(i-1)L) \\ \tilde{\mathbf{S}}(n,(i-1)L+1) \\ \vdots \\ \tilde{\mathbf{S}}(n, iL-1) \end{bmatrix} \in C^{LP\times 1} \tag{11.57}$$

and

$$\tilde{\mathbf{G}}_i(n, m_\nu) = \begin{bmatrix} \tilde{\mathbf{S}}_{ref}(n, m_\nu-(i-1)L) \\ \tilde{\mathbf{S}}_{ref}(n, m_\nu-(i-1)L+1) \\ \vdots \\ \tilde{\mathbf{S}}_{ref}(n, m_\nu-(iL-1)) \end{bmatrix} \in C^{LP\times 1} \tag{11.58}$$

The clutter covariance matrix in the i_{th} sub-block, $\hat{\mathbf{R}}_{D,i}$ is obtained as:

$$\hat{\mathbf{R}}_{D,i} = \frac{1}{N_r}\sum_{nr=1}^{N_r-1}\tilde{\mathbf{Z}}_i(n_r)\tilde{\mathbf{Z}}_i(n_r)^H \varepsilon C^{LP\times L} \tag{11.59}$$

Since the clutter covariance matrix to be estimated is reduced, a reduced number of training range cells are needed. In this case, only a $N_r = 2LP$ training data are needed in order to obtain a performance loss of 3 dB. This windowing approximation may induce some losses caused by an approximated covariance matrix.

Non-homogeneity detector: the generalized inner product test

As described previously, STAP or SDAP processing has been proposed as a means for improving the detection performance of airborne surveillance sensors. A very important, practical issue in fielding a STAP-based system concerns accurately estimating the interference covariance matrix and then computing an improved adaptive weight vector. Non-homogeneous interference, a feature of real-world airborne radar, complicates interference covariance matrix estimation and adaptive weight formulation. In past years, the problem of non-homogeneous interference was introduced by Wang in [58], while the problems of the applicability of the STAP in the event that a non-homogeneous interference was present in the training data were discussed in [59] and demonstrated using MCARM data in [60]. In order to improve STAP performance, potential non-homogeneity detection methods to excise non-homogeneous interference from adaptive weight computation based on ranking and selection [61], the inner product [59], and the sample matrix inversion test statistic [38] were developed. In this section the GIP Test is recalled. The GIP Test [50] can be used to assess the similarity between the unknown covariance matrix of a received signal vector for a particular range cell and a test covariance matrix that can be derived from a priori knowledge about clutter environment. In order to obtain the highest number of homogeneous training data, a sensible approach is to screen and excise non-homogeneous training data from the clutter covariance matrix estimation using the STAP filter. The GIP can be expressed as follows:

$$\mathbf{z}_i = \mathbf{X}_i^H \mathbf{R}_C^{-1} \mathbf{X}_i \tag{11.60}$$

where $\mathbf{X}_i \varepsilon C^{MP\times 1}$ represents the space-time snapshot relative to a particular range cell under test (M is the number of slow-time pulses while P is the number of the channels) while $\mathbf{R}_C$ is the test covariance matrix estimated using a set of homogeneous training data that can be located in the observed scenario by using a priori knowledge. This matrix can be also obtained from a clutter model derived from a priori knowledge. By defining the whitening filter output as the vector $\tilde{\mathbf{X}}_i = \mathbf{R}_C^{-1/2}\mathbf{X}_i \varepsilon C^{MP\times 1}$, the GIP test expressed in (11.60) can be interpreted as the measures of the power in the received signal vector $\mathbf{X}_i$ after being whitened by $\mathbf{R}_C^{-1}$:

$$\mathbf{z}_i = \tilde{\mathbf{X}}_i^H \tilde{\mathbf{X}}_i \tag{11.61}$$

The covariance matrix of the whitened signal vector can be then expressed as follows:

$$\tilde{\mathbf{R}}_i = E\left[\tilde{\mathbf{X}}_i \tilde{\mathbf{X}}_i^H\right] = \mathbf{R}_C^{-1/2}\mathbf{R}_i \mathbf{R}_C^{-1/2} \tag{11.62}$$

where $\mathbf{R}_i$ represents the true unknown covariance matrix of the received signal vector $\mathbf{X}_i$. If the covariance matrix of the received signal vector relative to the range cell under test is equal to the test covariance matrix of the desired homogeneous clutter environment, i.e. $\mathbf{R}_i \approx \mathbf{R}_C$, the covariance matrix of the received signal vector after GIP test is equal to an identity matrix, i.e. $\tilde{\mathbf{R}}_i \approx \mathbf{I}_{MP}$, and then $E\left[\mathbf{z}_i\right] \approx MP$. Instead, when the true covariance matrix of $\mathbf{X}_i$ deviates from the desired homogeneous clutter model, i.e. $\mathbf{R}_i \neq \mathbf{R}_C$, the expected value of the GIP $\mathbf{z}_i$ deviates from the product MP, i.e. $E\left[\mathbf{z}_i\right] \neq MP$. The GIP test can be used to assess the similarity between the unknown covariance matrix of the space-time snapshot relative to a range cell under test and a test covariance matrix derived from a priori knowledge. In an equivalent way, it is possible to consider homogeneous the received signal vectors with $\mathbf{z}_i$ equal to the GIP value (MP) and non-homogeneous the received signal vectors with a $\mathbf{z}_i$ that significantly varying from the GIP value MP. Therefore, the (11.60) can be used as the basis for evaluating and excising heterogeneous training data. The test covariance matrix is assumed to be derived from a perfect a priori knowledge about clutter model. It is clear that with no perfect knowledge about the clutter environment some distortions in the model appear that degrades the performance of STAP/SDAP.

Bayesian approach

As explained in Section 11.3.2, the estimation of clutter covariance matrix in STAP processing is crucial for target detection. STAP techniques derive data-dependent weighting by the calculation applied to the training data (also called secondary data), which leads to the estimation of clutter covariance matrix. The presumption of statistically IID secondary data set is common when STAP processing is considered. Under the IID hypothesis, the maximum likelihood estimation of the covariance matrix converge to the exact one as the number of training data increase. In real scenarios, heterogeneous environment degrades the quality of the secondary data set and then diminishing the performance of STAP through covariance matrix estimation error. In the past years several methods that integrate a priori knowledge in STAP processing with the aim of improving STAP performance are developed. Bayesian filtering belongs to this category. Bayesian filtering is suitable for its general applicability in many adaptive sensor signal processing systems and in this case, the prior knowledge is directly used in the non-stationary adaptive clutter filtering. In general, the prior information can be used in two principal ways: indirectly, in order to select useful training data (e.g. by merging prior information with the criterion based on complex metric such as NHD) or directly to form the filter. In the second case, the prior information can be summarised in a a priori matrix that is used to form the STAP filter. Coloured Loading (CL) technique is a common algorithm that allows to incorporate the a priori matrix into the detection scheme. In [52], the Bayesian approach is undertaken, which also leads to the CL technique. The samples of the estimated covariance matrix that derive from the sum of the scalar product between L Gaussian IID training sample obey to a Wishar distribution of L degree, as shown in [52]. If a prior estimate of the covariance matrix $\hat{\mathbf{R}}_0$ is noted, it is reasonable to assume that this initial matrix is Wishar distributed too. Such estimate can be derived from a priori SAR image information or can result from a physical-based model. For example, $\hat{\mathbf{R}}_0$ could be derived from a physical scattering model of the terrain in which the IID training data vector, present in the summation, represents the clutter patch steering vector weighted by the estimated reflectivity:

$$\hat{\mathbf{R}}_0 = \frac{1}{N_c} \sum_{i=1}^{Nc} G_i v_i v_i^H \tag{11.63}$$

where N_c and G_i represent the number of clutter patches and the power in each patch, respectively. As demonstrated in [52], the Bayesian estimate of the covariance matrix (i.e. a maximum a posteriori estimate) is given by

$$\hat{\mathbf{R}} = \frac{1}{L_0+L_1} \left(L_0 \hat{\mathbf{R}}_0 + L_1 \hat{\mathbf{R}}_1 \right) \tag{11.64}$$

where the previously defined matrix $\hat{\mathbf{R}}_0$ is based on L_0 training data sample while $\hat{\mathbf{R}}_1$, i.e. the ML estimate of the clutter covariance matrix, is based on L_1 sample and also have a Wishard distribution of L_1 degree. As it is possible to see from (11.64), the a posteriori covariance estimate is formed as a weighted sum of the a priori covariance model and current estimate with weighting factors that are proportional to the amount of training data sample. A generalization of (11.64) can be expressed as follows:

$$\hat{\mathbf{R}} = \alpha\hat{\mathbf{R}}_0 + \beta\hat{\mathbf{R}}_1 \tag{11.65}$$

A practical advantage to express the (11.64) in the form of (11.65) is, for example, that in the case when the training data used in the estimation of $\hat{\mathbf{R}}_0$ lose relevance (this problem is addressed as the problem of stale weights), the same matrix effectively contains much less information. A simple method to take into account this event is, therefore, to set the weighting factor α as

$$\alpha = e^{-\gamma t}L_0 \tag{11.66}$$

where t represents the time elapsed since the prior covariance estimate was formed and γ is a scalar constant. Another good method to set this weighting parameter is to choose α adaptively so as to maximally whiten the observed interference data:

$$min_{\{\alpha\}}\mathbf{Z}_L(\alpha) \tag{11.67}$$

where

$$\mathbf{Z}_L(\alpha) = \left\| \sum_i \mathbf{y}_i\mathbf{y}_i^H - \mathbf{I} \right\| \tag{11.68}$$

$$\mathbf{y}_i = \left(\alpha\hat{\mathbf{R}}_0 + \beta\hat{\mathbf{R}}_1\right)^{-1/2}\mathbf{x}_i \tag{11.69}$$

in which $\mathbf{x}_i$ represents the space-time snapshot for a particular range cell, while the term $\left(\alpha\hat{\mathbf{R}}_0 + \beta\hat{\mathbf{R}}_1\right)^{-1/2}$ corresponds to the whitening matrix for a particular value of α. It can also be noted that the vector $\mathbf{y}_i$ is a residue vector and then the major advantage of this method is to remove quasi deterministic trends in the data so that the resulting residue vector is less non-stationary. The choice of the weighting factors remains, however, a delicate issue. In [52], this choice is discussed in more detail. In particular, the weight of the a priori matrix is derived in order to optimise the whitening filter for the observed interference.

References

[1] Wicks M. 'Spectrum crowding and cognitive radar'. *2010 2nd International Workshop on Cognitive Information Processing*; 2010. pp. 452–7.

[2] Griffiths H., Cohen L., Watts S., *et al.* 'Radar spectrum engineering and management: technical and regulatory issues'. *Proceedings of the IEEE*. 2015, vol. 103(1), pp. 85–102.

[3] Hassanien A., Amin M.G., Zhang Y.D., Ahmad F. 'Dual-function radar-communications: information embedding using sidelobe control and waveform diversity'. *IEEE Transactions on Signal Processing*. 2016, vol. 64(8), pp. 2168–81.

[4] Blunt S.D., Mokole E.L. 'Overview of radar waveform diversity'. *IEEE Aerospace and Electronic Systems Magazine*. 2016, vol. 31(11), pp. 2–42.

[5] Bell K.L., Baker C.J., Smith G.E., *et al.* 'Cognitive radar framework for target detection and tracking'. *IEEE Journal of Selected Topics in Signal Processing*. 2015, vol. 9(8), pp. 1427–39.

[6] Haykin S. 'Cognitive radar: a way of the future'. *IEEE Signal Processing Magazine*. 2006, vol. 23(1), pp. 30–40.

[7] Guerci J.R. 'Next generation intelligent radar'. 2007 IEEE Radar Conference; 2007. pp. 7–10.

[8] Haykin S., Xue Y., Setoodeh P. 'Cognitive radar: step toward bridging the gap between neuroscience and engineering'. *Proceedings of the IEEE*. 2012, vol. 100(11), pp. 3102–30.

[9] Haykin S., Zia A., Arasaratnam I. 'Cognitive tracking radar'. 2010 IEEE Radar Conference; 2010. pp. 1467–70.

[10] Horne C., Ritchie M., Griffiths H. 'Proposed ontology for cognitive radar systems'. *IET Radar, Sonar & Navigation*. 2018, vol. 12(12), pp. 1363–70.

[11] Smith G.E., Cammenga Z., Mitchell A., *et al.* 'Experiments with cognitive radar in CAMSAP'. *2015 IEEE 6th International Workshop on Computational Advances in Multi-Sensor Adaptive Processing*. 2015, pp. 293–6.

[12] Mitchell A.E., Smith G.E., Bell K.L. 'Fully adaptive radar cost function design'. 2018 IEEE Radar Conference; 2018. pp. 1301–6.

[13] Guerci J.R., Baranoski E.J. 'Knowledge-aided adaptive radar at DARPA: an overview'. *IEEE Signal Processing Magazine*. 2006, vol. 23(1), pp. 41–50.

[14] Ulaby F.T., Moore R.K., Fung A.K. 'Microwave Remote Sensing: Radar Remote Sensing and Surface Scattering and Emission Theory'. *Remote Sensing. Addison-Wesley Publishing Company*. 1981.

[15] Carrara W.G., Goodman R.S., Majewski R.M. *Spotlight Synthetic Aperture Radar: Signal Processing Algorithms*. Artech House, Incorporated. 1995. Available from http://books.google.it/books?id=uztiQgAACAAJ.

[16] Curlander J.C. *Synthetic Aperture Radar: Systems and Signal Processing*. Wiley; 1991.

[17] Berizzi F., Martorella M., Giusti E. *Radar Imaging for Maritime Observation. Signal and Image Processing of Earth Observations*. Taylor & Francis. 2016. Available from https://books.google.it/books?id=3kaNMQEACAAJ.

[18] Chen V.C., Martorella M. 'Inverse synthetic aperture radar imaging; principles, algorithms and applications'. Institution of Engineering and Technology. 2014. Available from https://books.google.it/books?id=xWmABAAAQBAJ.

[19] Martorella M., Giusti E., Berizzi F., *et al.* 'ISAR based techniques for refocusing non-cooperative targets in SAR images'. *Radar, Sonar Navigation, IET*. 2012, vol. 6(5), pp. 332–40.

[20] Raney R. 'Synthetic aperture imaging radar and moving targets'. *IEEE Transactions on Aerospace and Electronic Systems*. 1971, vol. AES-7(3), pp. 499–505.

[21] Klemm R, of Engineering I, Technology. *Principles of Space-Time Adaptive Processing*, 3rd Edition. IET Radar, Sonar, Navigation and Avionics Series. 2006. Available from http://books.google.it/books?id=z6ICP3PtYi0C.

[22] Rosenberg L. *Multichannel Synthetic Aperture Radar*. Australia: University of Adelaide; 2007.

[23] Dickey J.F.R., Labitt M., Staudaher F.M. 'Development of airborne moving target radar for long range surveillance'. *Aerospace and Electronic Systems, IEEE Transactions on*. 1991, vol. 27(6), pp. 959–72.

[24] Shnitkin H. 'Joint stars phased array radar antenna'. *Aerospace and Electronics Conference, 1994. NAECON 1994. Proceedings of the IEEE 1994 National*. 1994, vol. 1, pp. 142–50.

[25] Pascazio V., Schirinzi G., Farina A. 'Moving target detection by along-track interferometry'. *IEEE Transactions on Geoscience and Remote Sensing Symposium, 2001. IGARSS '01*. 2001, vol. 7, pp. 3024–6.

[26] Gierull C. *Moving Target Detection by Along-track SAR Interferometry*. DREO Technical Report; 2002.

[27] Bao M., Bruning C., Alpers W. 'Simulation of ocean waves imaging by an along-track interferometric synthetic aperture radar'. *Geoscience and Remote Sensing, IEEE Transactions on*. 1997, vol. 35(3), pp. 618–31.

[28] Cohen L. 'Time-frequency distributions-a review'. *Proceedings of the IEEE*. 1989, vol. 77(7), pp. 941–81.

[29] Rieck W. ''Time-frequency distribution of multichannel SAR-data for autofocusing of moving targets'. *Radar 97*1997. pp.. 224–8.

[30] Guerci J.R. *Space-Time Adaptive Processing for Radar*. Artech House. 2003. Available from http://books.google.it/books?id=cmdI8pvOdqAC.

[31] Ward J. *Space-Time Adaptive Processing for Airborne Radar*. MIT Lincoln Laboratory. 1994. Available from http://www.dtic.mil/cgi-bin/GetTRDoc?Location=U2&doc=GetTRDoc.pdf&AD=ADA293032.

[32] Ender J.H.G. 'Space-time processing for multichannel synthetic aperture radar'. *Electronics Communication Engineering Journal*. 1999, vol. 11(1), pp. 29–38.

[33] Ender J.H.G. 'Space-time adaptive processing for synthetic aperture radar'. *IEE Colloquium on Space-Time Adaptive Processing*; 1998. pp. 611–8.

[34] Bacci A., Martorella M., Gray D.A., *et al.* 'Space-Doppler adaptive processing for radar imaging of moving targets masked by ground clutter'. *IET Radar, Sonar Navigation*. 2015, vol. 9(6), pp. 712–26.

[35] Melvin W.L. 'A STAP overview'. *IEEE Aerospace and Electronic Systems Magazine*. 2004, vol. 19(1), pp. 19–35.

[36] Techau P.M., Bergin J.S., Guerci J.R. 'Effects of internal clutter motion on STAP in a heterogeneous environment'. *Proceedings of the 2001 IEEE radar Conference*. 2001, pp. 204–9.

[37] Melvin W.L. 'Space-time adaptive radar performance in heterogeneous clutter'. *IEEE Transactions on Aerospace and Electronic Systems*. 2000, vol. 36(2), pp. 621–33.

[38] Reed I.S., Mallett J.D., Brennan L.E. 'Rapid convergence rate in adaptive arrays'. *IEEE Transactions on Aerospace and Electronic Systems*. 1974, vol. AES-10(6), pp. 853–63.

[39] Bacci A. 'Optimal space time adaptive processing for multichannel inverse synthetic aperture radar imaging'. *University of Pisa, Italy & University of Adelaide*. Australia; 2014. Available from http://hdl.handle.net/2440/86480.

[40] Gelli S., Bacci A., Giusti E., *et al*. 'Effectiveness of knowledge-based STAP in ground targets detection with real dataset'. *International Conference on Radar Systems (Radar 2017)*; 2017. pp. 1–5.

[41] Gelli S., Martorella M.A. 'A cognitive architecture for space time adaptive processing'.. 2020 IEEE International Radar Conference (RADAR); 2020. pp. 7–12.

[42] Bacci A., Gelli S., Giusti E., *et al*. 'Knowledge based space Doppler adaptive processing for imaging of moving targets in strong ground clutter'. *International Conference on Radar Systems (Radar 2017)*; 2017. pp. 1–5.

[43] Giusti E., Saverino A.L., Martorella M., *et al*. 'A rule-based cognitive radar design for target detection and imaging'. *IEEE Aerospace and Electronic Systems Magazine*. 2020, vol. 35(6), pp. 34–44.

[44] Cumming I.G., Wong F.H. *Digital Processing of Synthetic Aperture Radar Data: Algorithms and Implementation*. Artech House. 2005. Available from https://books.google.it/books?id=e9xxQgAACAAJ.

[45] Gorovyi I., Bezvesolniy O., Vavriv D.M. 'Modifications of range-Doppler algorithm for compensation of SAR platform motion instabilities'. *International Journal of Electronics and Telecommunications*. 2014, vol. 09(60), pp. 225–31.

[46] Fukuda S., Hirosawa H. 'A wavelet-based texture feature set applied to classification of multifrequency polarimetric SAR images'. *IEEE Transactions on Geoscience and Remote Sensing*. 1999, vol. 37(5), pp. 2282–6.

[47] Fukuda S., Hirosawa H. 'Suppression of speckle in synthetic aperture radar images using wavelet'. *International Journal of Remote Sensing*. 1998, vol. 19(3), pp. 507–19.

[48] Chang T., Kuo C.J. 'A wavelet transform approach to texture analysis'. *1992 IEEE International Conference on Acoustics, Speech, and Signal Processing*. 1992, vol. 4, pp. 661–4.

[49] Lee J.S. 'Digital image enhancement and noise filtering by use of local statistics'. *IEEE Transactions on Pattern Analysis and Machine Intelligence*. 1980, vol. 2(2), pp. 165–8.

[50] Melvin W.L., Wicks M.C. 'Improving practical space-time adaptive radar'. *Proceedings of the 1997 IEEE National Radar Conference*. 1997, pp. 48–53.

[51] Adve R.S., Hale T.B., Wicks M.C. 'Transform domain localized processing using measured steering vectors and non-homogeneity detection'. *Proceedings of the 1999 IEEE Radar Conference. Radar into the Next Millennium*; 1999. pp. 285–90.

[52] Bergin J.S., Teixeira C.M., Techau P.M., *et al*. 'Improved clutter mitigation performance using knowledge-aided space-time adaptive processing'.

IEEE Transactions on Aerospace and Electronic Systems. 2006, vol. 42(3), pp. 997–1009.

[53] Bergin J.S., Teixeira C.M., Techau P.M., *et al*. 'Stap with knowledge-aided data pre-whitening'. *Proceedings of the 2004 IEEE Radar Conference*. 2004, pp. 289–94.

[54] Giusti E., Bacci A., Stinco P., *et al*. 'Cognitive multichannel ISAR imaging for maritime coastal surveillance and ground border control' in CAMSAP'. *2015 IEEE 6th International Workshop on Computational Advances in Multi-Sensor Adaptive Processing*; 2015. pp. 277–80.

[55] Baqu R., Ruault du Plessis O., Castet N., *et al*. 'SETHI / RAMSES-NG: new performances of the flexible multi-spectral airborne remote sensing research platform". *2017 European Radar Conference (EURAD)*; 2017. pp. 191–4.

[56] Meta A., de Wit J.J.M., Hoogeboom P. 'Development of a high resolution airborne millimeter wave FM-CW SAR'. *First European Radar Conference, 2004 EURAD*. 2004, pp. 209–12.

[57] Melvin W.L., Wicks M.C., Brown R.D. 'Assessment of multichannel airborne radar measurements for analysis and design of space-time processing architectures and algorithms'. *Proceedings of the 1996 IEEE National Radar Conference*. 1996, pp. 130–5.

[58] Wang H., Cai L. 'On adaptive spatial-temporal processing for airborne surveillance radar systems'. *IEEE Transactions on Aerospace and Electronic Systems*. 1994, vol. 30(3), pp. 660–70.

[59] Marshall D. 'Evaluation of STAP training strategies with mountaintop data'. MIT Lincoln Lab, TR MTP-5, Bedford, MA, February; 1996.

[60] Melvin W.L., Wicks M.C., Brown R.D. 'Assessment of multichannel airborne radar measurements for analysis and design of space-time processing architectures and algorithms'. *Proceedings of the 1996 IEEE National Radar Conference*. 1996, pp. 130–5.

[61] Wicks M.C. 'Application of ranking and selection theory to radar signal processing'. *Rome Laboratory Internal Tech Memo, unpublished*. 1993.

Chapter 12

GAN4SAR: generative adversarial networks for synthetic aperture radar imaging of targets signature

Luc Vignaud[1]

This chapter describes the generation of Synthetic Aperture Radar (SAR) images of man-made targets using modern deep learning methods called Generative Adversarial Networks (GANs). These relatively new techniques have shown impressive results in the optical image field, enabling, for instance, the generation of very convincing *fake* images, i.e. images that look very much like the ones the generative algorithm has been trained on. Thus, we wish to investigate how these methods can be used or adapted for similar data generation in radar imaging. We follow the progressive development path of GANs and explore some applications for SAR data augmentation and image domains transformations, while monitoring their expected performances for Automatic/Assisted Target Recognition (AsTR) purposes.

12.1 Introduction

Many radar applications require the fine knowledge of targets signatures. However, compared to the optical field, databanks of real radar measurements are rather scarce as data acquisition on real targets needs expensive field trials and observation on wide angular extents due to the anisotropic behaviour of targets scattering centres. Large progress have been made in the simulation of targets radar signatures, but still, the complexity of the ElectroMagnetic (EM) interaction often makes them not enough representatives of the actual real measurements. Therefore, there is a real need for radar data augmentation but with a strong constraint on the quality of their generation.

Meanwhile in the optical domain, the joint availability of large image databases and the development of deep learning networks has revolutionized large-scale

[1]The French Aerospace Lab, ONERA, France

classification tasks, boosted synthetic data generation and facilitated the translation between several image domains. For instance, exciting breakthroughs have emerged in image super-resolution, art style transfer or generation of high-resolution facial photos of "fake celebrities", to name a few.

Our objective is to investigate how modern generative networks, such as deep learning GANs, may be used to generate useful and trustworthy radar images. This new area is only starting to get attention from the radar community [1–3]. Among the large number of GAN-based methods proposed in optics, we need to select those potentially applicable to radar images generation. Moreover, efforts have to be made on measuring the quality of the augmented data according to the physics of the electromagnetic interaction with the target, such as its angular dependency. Since this cannot be properly performed by a simple visual inspection, we need to infer this quality from reference classifiers.

We apply the chosen methods on radar images coming from the publicly available X-Band Moving and Stationary Target Acquisition and Recognition (MSTAR) and SAMPLE databases ([4–9]), which we briefly describe in Section 12.2.

In Section 12.3, we introduce a reference classifier built on a Convolutional Neural Network (CNN) and derive the performance analysis tools that we will use throughout this chapter. We also present our general implementation framework using the MATLAB® Deep Learning toolbox.

We then describe in Section 12.4 the basic GANs architectures and derive some specific adaptations for radar signatures generation. We start with a simple introduction to GANs architecture in Section 12.4.1. In Section 12.4.2, we show how augmented images can be generated from scratch using a noise vector as input and real radar images as reference. In Section 12.4.3, we present methods to condition a GAN in order to get control over the generated target class and orientation. Finally, in Section 12.4.4, we see how a GAN may learn the translation between two image domains such as simulated images to real ones. Whenever feasible, we propose methods to measure the realism and representativeness of the various GAN generations as well as their potential generalization capabilities, either for interpolation between known examples or for extrapolation, which requires going beyond the space of known training examples.

12.2 MSTAR and SAMPLE databases

The MSTAR was a joint Defense Advanced Research Projects Agency and Air Force Research Laboratory program [4–6]. This famous dataset consists of X-band SAR images with one square foot resolution measured in spotlight mode. These data have been acquired around 1995 at X band by Sandia National Laboratory SAR sensor platform at fixed incidences (15° and 17°) and for 360 azimuth orientations each degree. The various targets on the ground include wheel and track vehicles, some with variants and articulations. For decades, MSTAR has been the reference database for testing various Automatic Target Recognition (ATR) machine learning methods, from template-based algorithms to CNNs leading to

Figure 12.1 *Optical images of the MSTAR (ten targets) database*

the achievement of excellent target classification results [7]. Here we use the public ten targets version with no variant. Photographs of the targets are shown in Figure 12.1.

The SAMPLE database is a recently released collection of simulated SAR signatures of ground targets paired with MSTAR image chips [8, 9]. The simulations have been computed using asymptotic ray-tracing techniques based on 3D CAD models carefully built to match ground truth photographs taken on the acquisition site (see Figures 12.2 and 12.3). The public version contains ten targets observed at fixed incidences (15° to 17°) for a limited 70° span of azimuth orientations each degree. Photographs of the targets are shown in Figure 12.2.

The two databases have five identical targets in common sharing roughly the same image chips over the SAMPLE limited angular observation span.

Since our primary focus is the target signature, we extract the centred 64 × 64 pixels target region of interest from image chips and apply a fixed threshold to mitigate inhomogeneities of the SAR background clutter, target shadow and speckle noise that could affect generation results. Interesting studies on targets surrounding influence on ATR and targets signature segmentation methods can be found in [9–11].

Figure 12.2 *Optical images of the sample database*

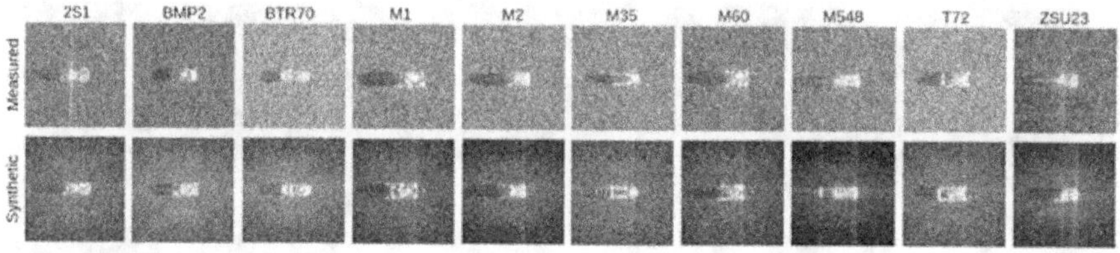

Figure 12.3 *Measured (top) and synthetic (bottom) SAR image from each target class within the SAMPLE database*

12.3 Implementation choices and reference classifier

We have chosen to compute the various networks architectures using MATLAB Deep Learning Toolbox. MATLAB has introduced the possibility to work with GANs in its 2019b release (September 2019). This long-awaited feature now gives access to user-defined training loops and loss functions.

Implementations of the various networks presented Section 12.4 and corresponding to [12] to [13] can be found in Yui Chun Leung GitHub repository: https://github.com/zcemycl/Matlab-GAN [Accessed 1 May 2020].

Computations were made on a Dell Alienware Intel Core i9 with NVIDIA GeForce GTX 1080 8Go.

Networks are trained and tested with log magnitude 64 × 64 pixels cropped images chips from MSTAR or SAMPLE datasets using a fixed threshold above underlying clutter.

When needed, we will use a generic CNN as a reference classifier. Its detailed structure is presented in Figure 12.4. It is made of four blocks (stride two convolution + batchnormalization + relu) followed by a fully connected softmax classification layer.

This network reaches over 93 per cent mean accuracy on the classical MSTAR dataset trained at 17° incidence and tested at 15°. Accuracy on the SAMPLE dataset trained on simulated images and tested on real ones is around 30 per cent; this low result is coherent with previous similar tests reported in [8] and [14].

We do not expect this basic classifier to be optimal and compete with dedicated optimized ATR ones, but rather to provide a standardized performance evaluation tool in order to monitor the quality of the generated images when compared to the real ones.

Note that, unfortunately, we cannot count on the classical training data augmentation improvement used in optics due to the strong dependency of radar signatures to their observation aspect angle: no manipulation such as rotation, mirroring, or brightness change are allowed for SAR images [9]. This remark also applies to the various GAN trainings discussed in Section 12.4.

	Name	Type	Activations
1	ImagesIn 64x64x1 images	Image Input	64×64×1
2	conv1 32 4x4x1 convolutions with stride [2 2] an	Convolution	31×31×32
3	relu1 Leaky ReLU with scale 0.2	Leaky ReLU	31×31×32
4	conv2 64 4x4x32 convolutions with stride [2 2] a	Convolution	14×14×64
5	bn2 Batch normalization with 64 channels	Batch Normalization	14×14×64
6	relu2 Leaky ReLU with scale 0.2	Leaky ReLU	14×14×64
7	conv3 128 4x4x64 convolutions with stride [2 2]	Convolution	6×6×128
8	bn3 Batch normalization with 128 channels	Batch Normalization	6×6×128
9	relu3 Leaky ReLU with scale 0.2	Leaky ReLU	6×6×128
10	conv4 256 4x4x128 convolutions with stride [2 2]	Convolution	2×2×256
11	bn4 Batch normalization with 256 channels	Batch Normalization	2×2×256
12	relu4 Leaky ReLU with scale 0.2	Leaky ReLU	2×2×256
13	fc 10 fully connected layer	Fully Connected	1×1×10
14	softmax softmax	Softmax	1×1×10
15	classOut crossentropyex	Classification Output	-

Figure 12.4 Detailed structure of the reference CNN classifier

12.4 Generative adversarial networks

12.4.1 Introduction

GANs were introduced by Ian Goodfellow in [12], which led to numerous publications exploiting the new ideas brought by this 2014 paper. Indeed, Yann Le Cun, a well-known father of modern deep learning methods, described GAN as 'the most interesting idea in the last 10 years in Machine Learning'!

GANs rely on a *zero-sum game* between two networks: a Generator and a Discriminator. The objective of the Generator is to minimize the ability of the Discriminator to be correct on fake samples, while the Discriminator's goal is to minimize its classification error between real and fake samples. The two networks are trained jointly with opposite goals at the same time. So, at each iteration of the training process, the weights of the generative network are updated in order to

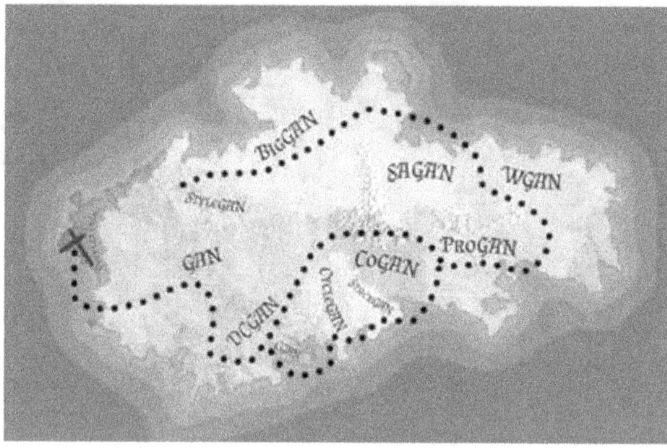

Figure 12.5 'The GAN Island' (https://blog.floydhub.com/gans-story-so-far/)

increase the classification error, whereas the weights of the discriminative network are updated to decrease this error.

This technique learns to generate new data with the same statistics as a given training set. GANs are thus able to model complex, high dimensional distributions as, for instance, images. For example, a GAN trained on photographs can generate new photographs that look 'authentic' to human observers or at least that have many realistic characteristics. Fake samples are specifically generated to fool the Discriminator which is basically a binary classifier. We may then expect these samples to be challenging for any classifier. Usage of the generated images will depend on the intended application, but one may already think to use this data augmentation to better train a classifier either to recognize class variants or confusers, especially in the context of unbalanced datasets.

Initial 'Vanilla' GAN implementation (presented Section 12.4.2) has led to many variants since its introduction. Most of these adaptations have been named by adding a prefix to the GAN acronym, forming a large family of GAN-based algorithms as illustrated in Figure 12.5 by a fancy *GAN island* drawing. The underlying objective of all these methods remains roughly the same, leaving the main differences to loss functions definitions and networks architectures, especially concerning their inputs and outputs. A recent and interesting review of the wide family of GANs is available in [15].

However simple this competitive process may appear, GANs are plagued with difficulties, especially in their training stability known as *Nash equilibrium*. Even with many useful tweaks [16], GANs often fail to train or experience mode collapse, where the Generator produces only a few distinct images. They often need patience and a fairly large number of tries to optimize the various training parameters.

Contrary to other networks that just stop when a minimum loss is reached, the construction of GANs makes them unable to decide by themselves when to end the training and use the generated products. That is the reason, whenever feasible, we

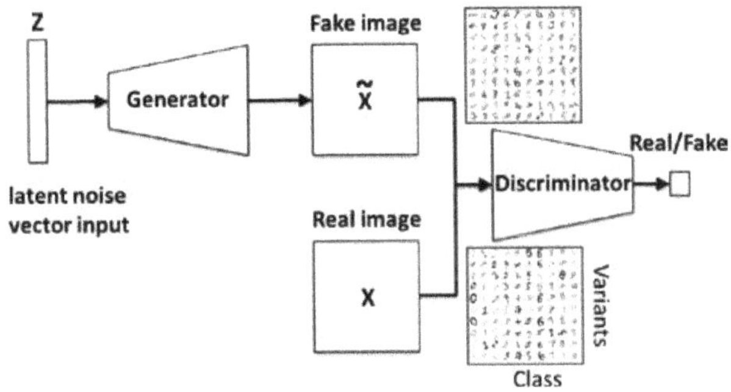

Figure 12.6 *Vanilla GAN architecture and example of 100 fake and real images on MNIST dataset*

have decided to add auxiliary reference classifiers to monitor their production quality along their training.

12.4.2 'Vanilla' GAN

The original GAN architecture described in [12] is presented in Figure 12.6, along with an application example on MNIST handwritten digit database. The input of the Generator, often named 'latent space', is a simple noise vector z. Outputs of the Generator are the *fake* images $G(z) = \tilde{X}$. These $G(z) = \tilde{X}$ *fake* images are presented, among *real* images X, as inputs to the Discriminator. The Discriminator has a single binary output, giving either a *real* or *fake* image result.

The Generator goal is to make $D(G(z)) = real$, while the Discriminator aims at $\left[D(G(z)) = fake \,\&\, D(X) = real\right]$. The Generator tries to replicate the *real* images probability distribution. The GAN uses loss functions that reflect the distance between the distribution of the data generated by the GAN and the distribution of the real data. The original minimax loss described in [12] monitors the cross-entropy between the real and generated distributions. Since, several alternative loss functions have been proposed including the simple Mean Square Error (MSE) [17] or the Earth Mover Distance in Wasserstein GAN [18].

The training reference set is divided into small batches that are used by GAN networks upon desired convergence. Once the GAN has been trained for several epochs,[2] its Discriminator part is disregarded and the Generator is simply fed with noise vectors to generate new images that look like real from the Discriminator's final state perspective.

The GAN strategy may be applied to any relevant couple of Generator/Discriminator networks, shallow [12] or deep [19].

For our experiment, we use two CNNs detailed in Figure 12.7. Note that the Discriminator is almost identical to the reference classifier defined in Section 12.3 (see Figure 12.4), except for its binary output. The Generator uses several transposed convolutions with stride 2 to gradually up-sample an input latent vector of

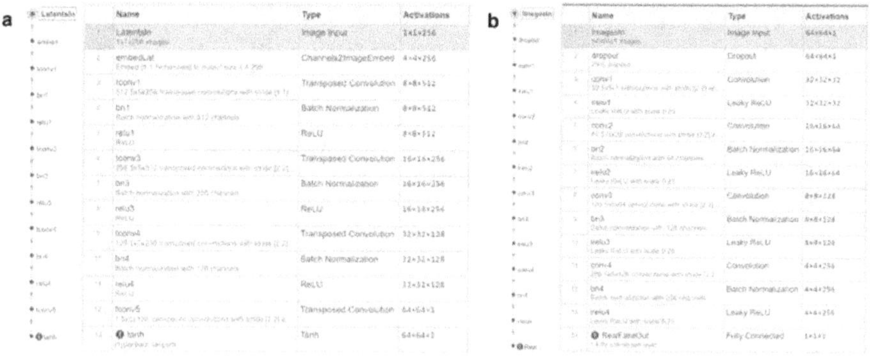

Figure 12.7 *Detailed structure of Vanilla GAN Generator (left) and Discriminator networks (right)*

length 256 to a 64 × 64 image. The latest activation layer is traditionally a *tanh* to control the final image dynamics.

We then apply this algorithm to the MSTAR dataset using all available real SAR images in batches of 128 images and random vectors with length 256 as latent inputs.

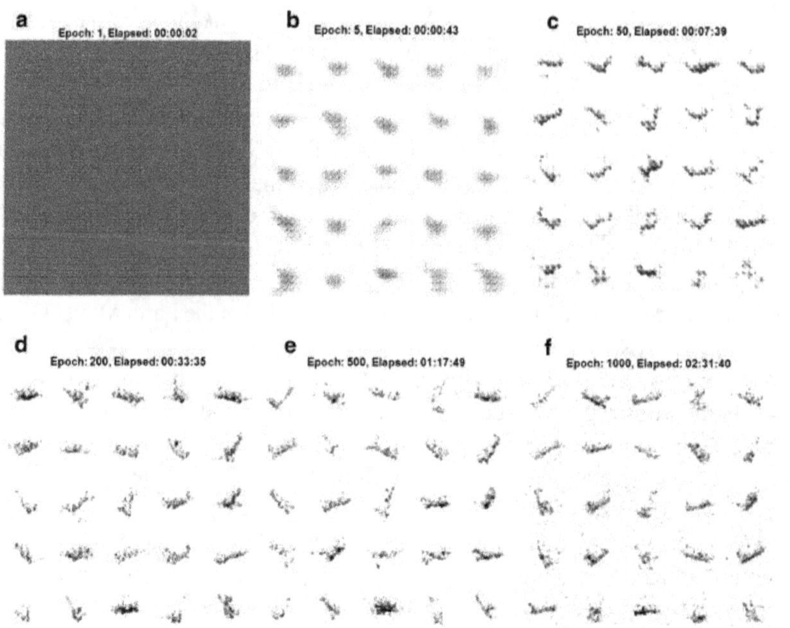

Figure 12.8 *Examples of 25 GAN-generated images along training epochs for Vanilla GAN trained on real MSTAR images*

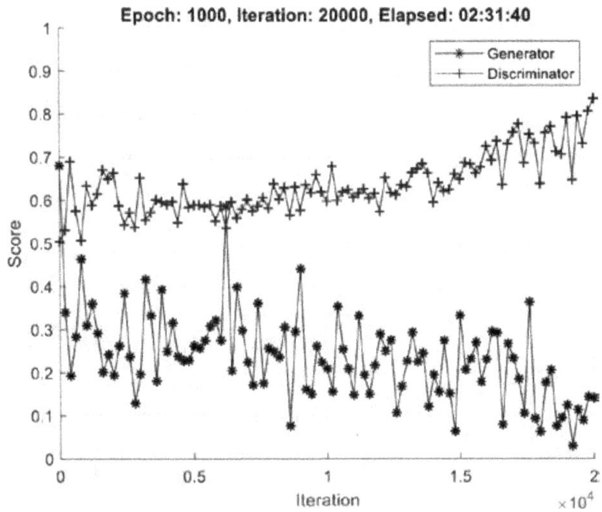

Figure 12.9 Discriminator and generator scores during Vanilla GAN training on MSTAR images

Figure 12.8 shows the progression of 25 validation examples of generated images through training iterations, starting from noise at first iteration and progressively converging to plausible images of different targets and orientations. Figure 12.9 shows the relatively good balance of the Generator and Discriminator scores along training, with a small divergence after 750 epochs indicating a possible lack of future innovation in the generated samples.

However, we notice that there is no fine control over the data being generated: we cannot choose the generated target class or its observed orientation. Neither are we able to directly measure the quality of the reconstructed images in any other way than a visual correspondence or by trying to find the closest neighbour to a real reference image. This lack of labelling hinders us from using the generated images for future supervised classification tasks, other than populating confusers' classes.

In Figure 12.8 we have kept the same input random latent vector through time for each validation image. We observe a tendency to the specialization of each latent towards a certain target class and orientation. This is an indication that a posteriori clustering in the input latent space could probably help to guide the generation but at a somehow prohibitive cost.

A trivial solution to regain control over the generated target class is to build a class-specific GAN by feeding its training with SAR images coming only from a chosen class or group of classes. Of course, this requires separated networks for each desired class, which is obviously sub-optimal when the number of classes increases. Moreover, we cannot get the GAN to specialize in both target class and orientation parameters: there will not be enough data to feed its training because of the natural angular dispersivity of SAR images. We always have to remember that, unlike optical images, radar images are very sensitive to their observation point of view.

12.4.3 Conditioned GAN

A limitation of the Vanilla GAN model is that it may only generate a random image from the reference domain it has been trained on. As seen in Section 12.4.2, there is a relationship between points in the latent space to the generated images, but this relationship is complex and hard to map. GAN variants have been proposed to better guide the data generation process by conditioning the model on additional information.

In our case, such conditioning can be based on class labels and/or target orientation, for instance. However, any additional information that is correlated with the input images may be used to condition the GAN. Moreover, authors of [20] have observed that synthetic optical images generated by Vanilla GAN were of low quality when trained on datasets with high variability, which is obviously relevant for radar images. Several conditioned GAN architectures have shown to actually improve the GAN performances: improvement may come in the form of more stable and faster training and generated images that have better quality. Among them, let us cite:

- Deep convolutional GANs [19]
- Auxiliary classifier GAN [21]
- Conditional GAN [22]
- Information Maximizing Generative Adversarial Nets [23].

These various GANs mostly differ in the way conditions are used by both their Generator and Discriminator and the impact of the conditioning loss added to the conventional GAN loss. Figure 12.10 summarizes the various ways to add condition(s) to a GAN:

- at generator input by conditioning/augmenting the noise latent vector
- at discriminator input by merging/embedding the image with the condition

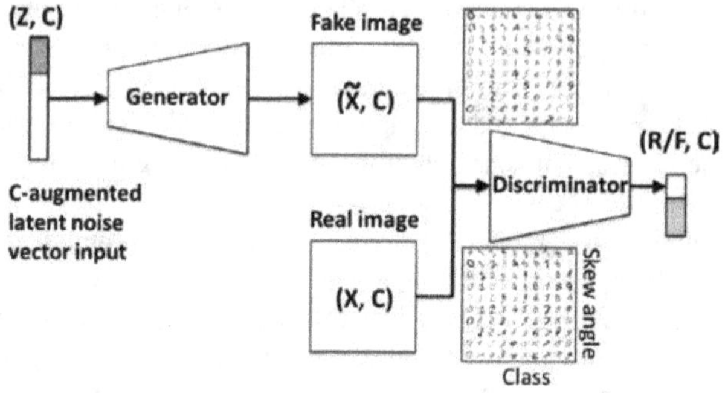

Figure 12.10 GAN with condition architecture and example of 100 fake and real images on MNIST dataset with class label and skew angle conditions

- at discriminator output by augmenting the binary decision with extra condition outputs.

Conditioning the Generator input is mandatory if we need to target the salient structured semantic features of the data distribution and have some control over the data generation once training is done. The latent vector is not totally random anymore, enabling a semi deterministic partition of the latent space.

It may not be easy to condition the Discriminator at its input, especially when generated data and condition(s) are different by nature or size, and thus requires a non-trivial embedding. The special case where the condition is an image is discussed separately in Section 12.4.4.

At Discriminator output, GAN losses have then to be adapted to take into account the following condition(s):

- Generator goal is to make $\left[D\left(G\left(z\right)\right) = real \,\&\, D\left(G\left(z\right)\right)\ meets\ C\right]$, where C stands for the condition (loss).
- Discriminator goal is $\left[D\left(G\left(z\right)\right) = Fake \,\&\, \left[D\left(X\right) = Real \,\&\, meets\ C\right]\right]$. Note that the Discriminator has no interest in fulfilling the condition for fake inputs.

The added condition loss is adapted to the nature of condition variables: for instance, *softmax* cross-entropy for discrete ones such as classes or MSE for continuous values such as angles.

The structure of a GAN fully conditioned on class labels and orientation angles is presented in Figure 12.11. This architecture only differs from the Vanilla GAN (Figure 12.7) in that adding conditions inputs (class and angle) that are merged and concatenated, respectively, with the latent embedding for the Generator and with images for the Discriminator, and two extra outputs (inferred class and angle) to the Discriminator which are used to compute the GAN conditional loss terms.

We have trained such a conditioned GAN on the MSTAR dataset. Figure 12.12 shows the generation of a subset for ten classes and ten orientations with the corresponding real images used for training. The GAN Generator has successfully learned to generate plausible SAR images for each class and orientation couple, despite the rather low cardinality of the reference dataset. If we focus on one specific signature (Figure 12.13), we see that we can obtain a variety of generated signatures bearing similar features as the real ones (main shape, orientation and size, for instance), without being a perfect copy of them. It is visually impossible to distinguish the real image from the *fake* ones.

We are now able to generate augmented images of reference classes for any orientation. However, are these trustworthy? To check this point, we introduce a pair of independent reference Classifiers (see Section 12.3) to monitor the GAN performance generation all along its training.

- The first Classifier is trained only once on the MSTAR real images and regularly tested on an image set produced by the Generator in its running state of training. The classification accuracy gives us an indication of the ability of the

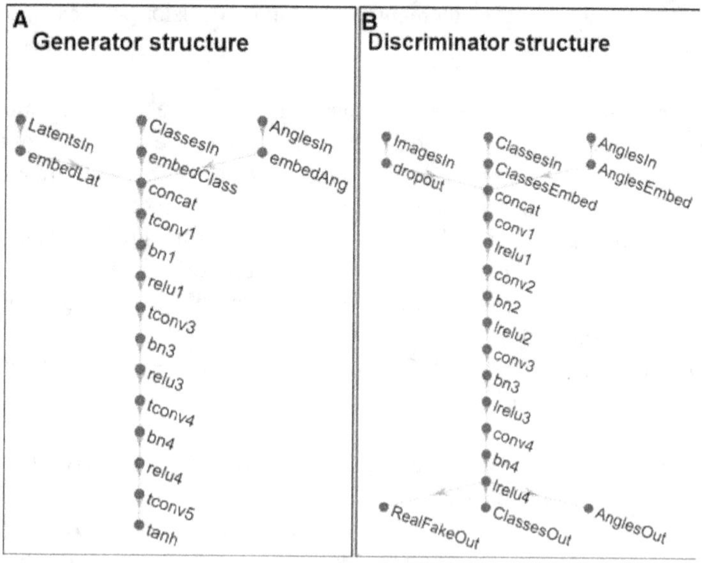

Figure 12.11 *Structure of a fully conditioned GAN (left) Generator and (right) Discriminator*

GAN to produce images that are consistent with their intended condition class label. It is related to the realism of generation but not a final proof, as the generator could also produce class correct but unrealistic *spoof* samples in this closed world classification with no confuser rejection.

- The second Classifier is trained at regular steps during the GAN training phase using the running production of generated images and tested on real images. The classification accuracy is an insight into the representativeness of the GAN generation regarding the real image population.

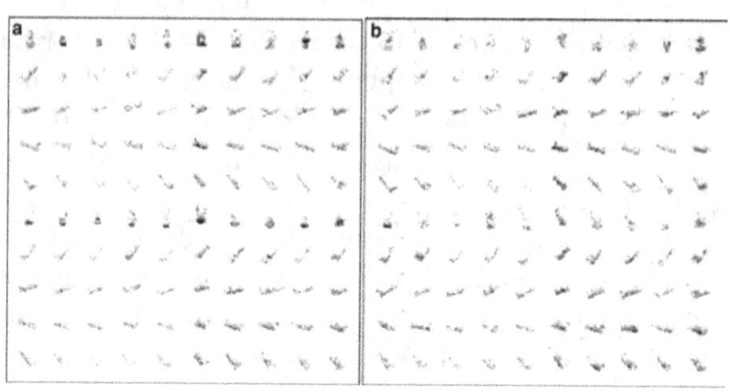

Figure 12.12 *(left) Real images used for training (one class per column and one orientation per line); (right) output of GAN conditioned on class and orientation*

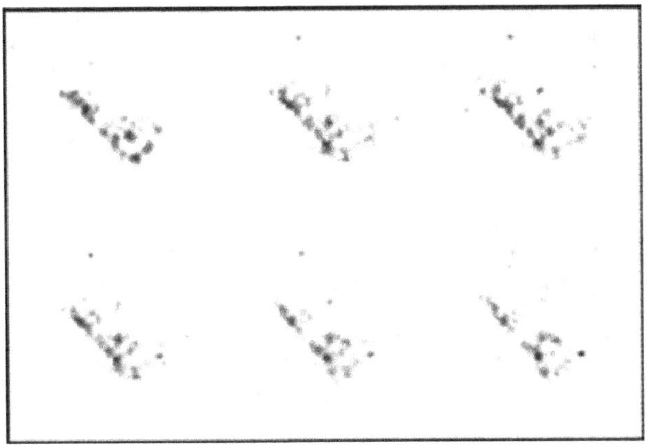

Figure 12.13 Five GAN-generated images with conditioning on one class and one orientation, and the corresponding real image (top left!)

Combination of both accuracies is a reasonable measure of GAN performance throughout its training. Results of such monitoring, with accuracies transformed to [0, 1] scores, are shown in Figure 12.14 along with the Generator and Discriminator scores (as in Figure 12.9). We see that the accuracy of the first Classifier climbs rapidly after the first 100 epochs and stabilizes around 95 per cent, whereas the second Classifier accuracy has a smoother ascent reaching a plateau around 80 per cent.

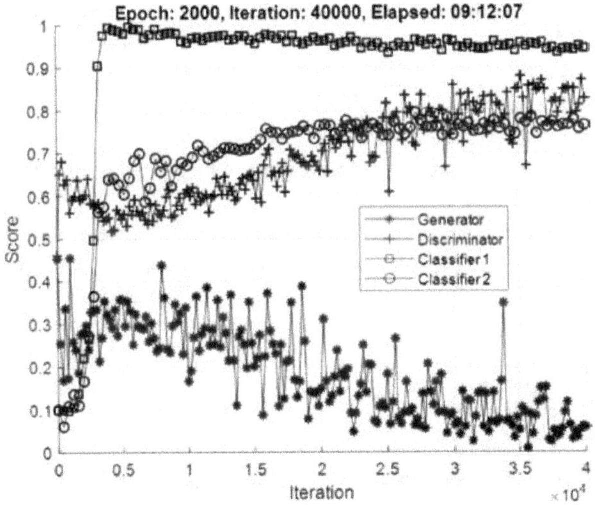

Figure 12.14 History of scores of GAN Generator (star) Discriminator (plus) and accuracies of auxiliary Classifier 1 (train = real and test = fakes] (square) and Classifier 2 [train = fakes and test = real] (circle)

Meanwhile, the Generator score degradation confirms the GAN innovation decay. This shows that the final quality of the generated images is quite good. However, some improvements would of course be needed to use them as full substitutes for real images.

Next, we test the GAN interpolation capabilities. We retrain the GAN on a subset of MSTAR images keeping only those standing on a regular 5° azimuth grid (*real train set @5°*), all other images being kept out of the GAN training (*real test set @1°*). We keep the first 2 Classifiers introduced previously to monitor the training performance of the under-sampled train set. We introduce two extra Classifiers:

- The third Classifier is trained once on the *real test set @1°* and regularly tested on a set produced by the Generator in its running state on the full span of possible azimuth condition angles (*fake set @1°*). The classification accuracy informs us of the GAN possibility to produce class consistent images interpolated within the trained angular grid.
- The fourth Classifier is trained at regular steps during the GAN training phase using the running production of generated images on all angles (*fake set @1°*) and tested on the *real test set @1°*. The classification accuracy represents the GAN capability to generate representative interpolated real images.

These two accuracies help to qualify the GAN interpolation capabilities on the angular condition.

Accuracies of the four Classifiers are presented in Figure 12.15. As expected, we find similar performances for the first two Classifiers as previously stated. The final accuracy of Classifiers 3 and 4 is around 90 per cent and 65 per cent, respectively; it is an indication that the GAN bears some interesting interpolation capabilities, even if not perfect.

In a separate test, we have verified that no accuracy gain was brought when augmenting the under-sampled real train set with the interpolated GAN-generated images, which is not surprising as no new information on targets signature is meant to be brought by the GAN.

A further limitation lies in the fact that the conditioned GAN architecture cannot be used to extrapolate the generation to unknown classes, i.e. different from the reference ones used for training.

12.4.4 *Image-conditioned GAN*

12.4.4.1 Pix2Pix GAN

As mentioned in Section 12.4.3, a GAN can also be conditioned by images and used for image-to-image translation tasks. This idea was first introduced by Pix2Pix GAN [17] using paired images. This GAN architecture (see Figure 12.16) is able to learn the mapping from an input image to a paired output image. The novelty lies in the way the input image is first fed to an Encoder network that compresses the information into a pseudo latent space that is no longer randomly chosen. The Encoder acts as a translator of input images into the latent space, forming a kind of latent space

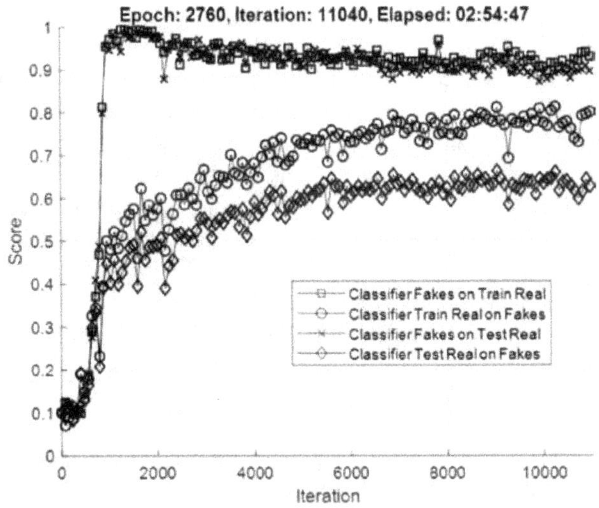

Figure 12.15 *Accuracies of auxiliary Classifier 1 [train = real @ 5° and test = fakes @ 5°] (square), Classifier 2 [train = fakes @ 5° and test = real @ 5°] (circle), Classifier 3 [train = real @ 1° and test = fakes @ 1°] (cross) and Classifier 4 [train = fakes @ 1° and test = real @ 1°] (diamond)*

discriminator so to speak. As a matter of fact, the Encoder + Generator network becomes the new Generator of the image-conditioned GAN. The latent vector may even disappear from the *bottleneck* of this structure if no condition is used inside; in that case, it is worth noticing the similarity with auto-encoder or Unet networks. A nice feature is that this new Generator needs only images as input, which means we will be able to test it on any image once trained on a given training set.

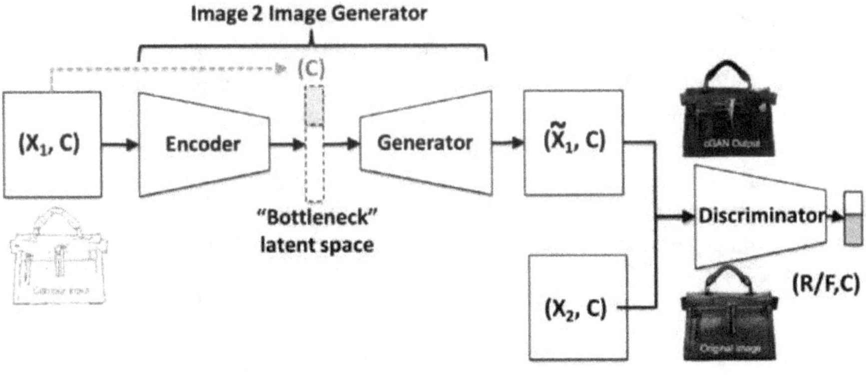

Figure 12.16 *Image-conditioned GAN architecture (C is empty in Pix2Pix GAN) and illustration of image-generation from sketch*

The generator is trained via the classical adversarial loss (see Section 12.4.2), which encourages the generator to generate plausible images in the target domain. The generator is also updated by adding the mean absolute error L1 loss measured between the generated image and the expected output image. This additional loss encourages the generator model to create more plausible translations of the source image. The Pix2PixGAN has been demonstrated on a range of image-to-image translation tasks such as converting maps to satellite photographs, black and white photographs to colour, photographs to paintings and vice versa, sketches of products to product photographs, etc.

In the original Pix2Pix GAN architecture, the input image is the only condition. But extra conditions C can be added to images pairs, e.g. sharing the same class or orientation (see Figure 12.16).

Conditioning may then occur:

- at generator input with the input image C, of course, barely with C as it may require a complicated embedding between C and X_1 and will limit possible extrapolations outside the trained condition space
- at generator bottleneck, as it may be easier than at its input but this is not used in practice to our knowledge
- at Discriminator input by merging the image with the condition. In Pix2Pix GAN, the generated *fake* image $\tilde{X}_1$ is concatenated with the *real* reference image X_2 in the channel space (*colour*); the reference X_2 is also concatenated with itself when sent to the Discriminator. The discriminator is then provided with both a source image and a target image, and it must determine whether the target is a plausible transformation of the source image or not.
- at Discriminator output by augmenting the binary decision with extra condition outputs as previously seen in Section 12.4.3.

We have adapted this image-conditioned GAN to the translation of EM simulations to real SAR measurements on the SAMPLE dataset. The Discriminator remains unchanged (see Figure 12.11), except for the input concatenation of two images in their channel dimension. The Generator structure is shown in Figure 12.17. Note that its Encoder is indeed very similar to the Discriminator structure.

Figures 12.18–12.20 show the result of such training on a subset of ten targets for five orientations, with a focus example on a specific signature (Figure 12.21). The training reaches a perfect 99 per cent accuracy for performance monitoring both Classifiers 1 and 2 (as defined in Section 12.4.3) in only a few hundred epochs. This means that the GAN Generator has successfully learned a mapping to translate simulated images to plausible SAR images for each class and orientation couple.

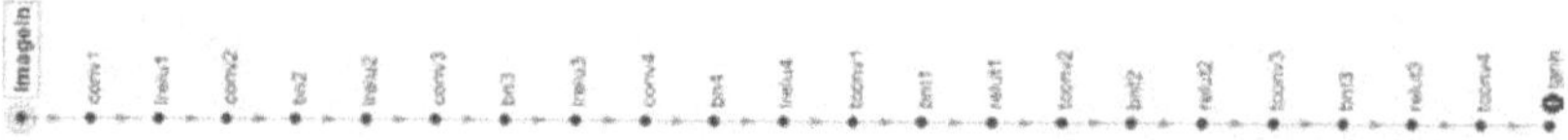

Figure 12.17 Generator structure of the image-conditioned GAN

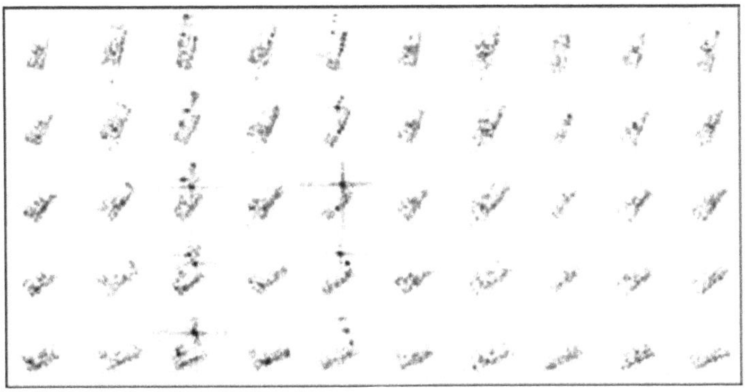

Figure 12.18 Input-simulated SAMPLE images (ten classes and five angles shown)

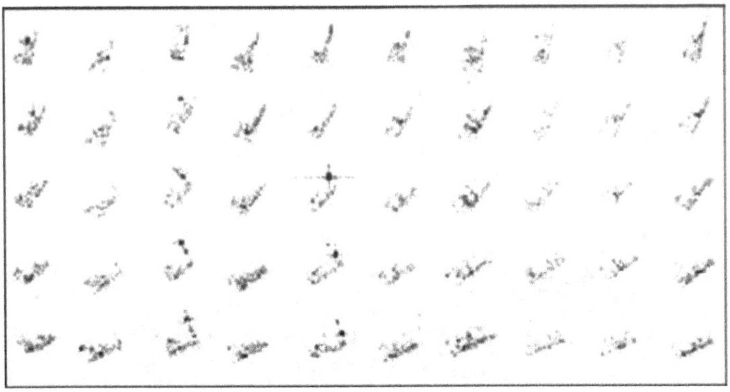

Figure 12.19 Real reference SAMPLE images (ten classes and five angles shown)

We are now theoretically able to translate simulated images into pseudo-real SAR images for any available reference class and any orientation. It is interesting to note that we have lost the ability to generate multiple plausible samples for each condition input as seen in Section 12.4.3 with conditioned GAN. This feature can be naïvely implemented by adding some noise at the Generator bottleneck, but a more stable solution has been proposed in BicycleGAN [24] by introducing two loops between images and latent spaces.

To test the potential interpolation quality of the image-conditioned GAN architecture, we run a similar experiment as we did in Section 12.4.3 by limiting the training to only 5° azimuth spaced images and comparing the generated images produced by the trained image-conditioned GAN on unseen simulations to the reference real images, also kept unknown from the GAN. Reference Classifiers accuracies along the training are reported in Figure 12.22. The lowest final accuracy of the four Classifiers is 87 per cent, showing a very valuable interpolation property for

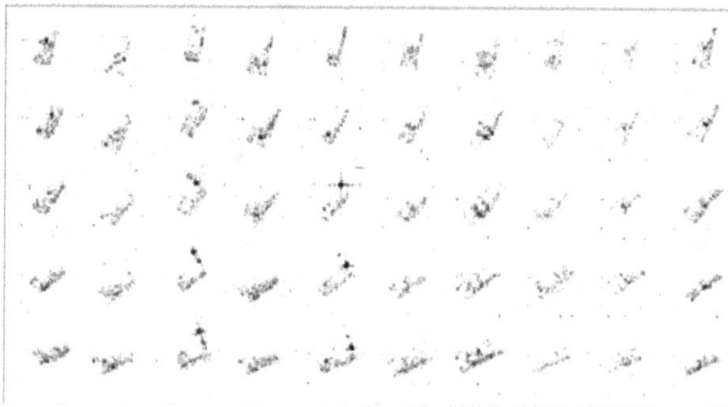

Figure 12.20 *Image-conditioned GAN Generator output of Figure 12.18 input set*

this architecture. This proves a real data augmentation gain that could save real data measurements.

Furthermore, this neural architecture can also be used to produce generated images of unknown classes, i.e. different from references used for training. Figure 12.23 shows the comparison of generated images formed by feeding the Generator, trained on only five classes, with new simulation images (i.e. from five classes that have never been shown to the GAN), compared to the expected ones (i.e. real measurements paired with input simulations and never used for training). We observe that the generated images show at least a coherent overall shape and orientation with real images. In particular, we notice that the GAN has learned to minimize the side lobes caused by trihedral returns whose contribution is often exaggerated in the simulations images. This proves that this kind of network has some capability to extrapolate the image generation to unknown classes, therefore, emulating a kind of *simulation refiner function.*

Still, the extrapolation quality measured by the Classifiers 3 and 4 remains low, cumulating at around 40 per cent accuracy.[3] This may be explained by the relatively low number of classes used for training. Further experiments should be conducted with more classes to better assess this property.

Figure 12.21 *(left) Input image (simulations); (centre) reference image (real measurements); (right) GAN Generator output*

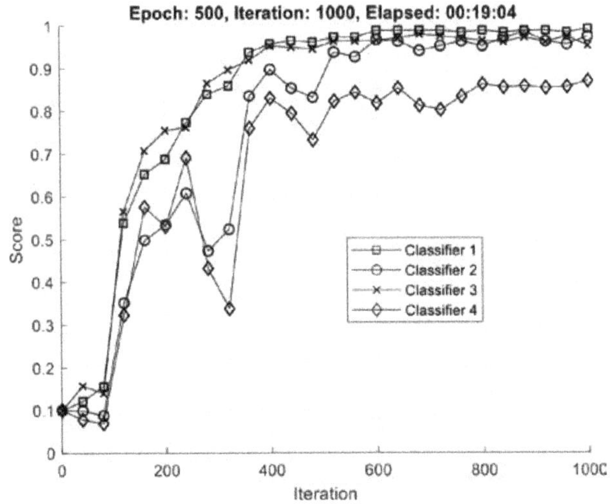

Figure 12.22 *Image-conditioned GAN accuracies of auxiliary Classifier 1 [train=real @ 5° and test = fakes @ 5°] (square), Classifier 2 [train= fakes @ 5° and test= real @ 5°] (circle), Classifier 3 [train=real @ 1° and test=fakes @ 1°] (cross) and Classifier 4 [train= fakes @ 1° and test= real @ 1°] (diamond)*

12.4.4.2 CycleGAN

Training a model for image-to-image translation typically requires a large dataset of paired examples. These datasets can be difficult and expensive to prepare. A later extension of Pix2Pix GAN to unpaired images was proposed in [13] using Cycle-Consistent Adversarial Networks (CycleGANs). A similar scheme was simultaneously studied in CoGAN [25]. The models are trained in an unsupervised manner using a collection of images from the source and target domain that do not need to be

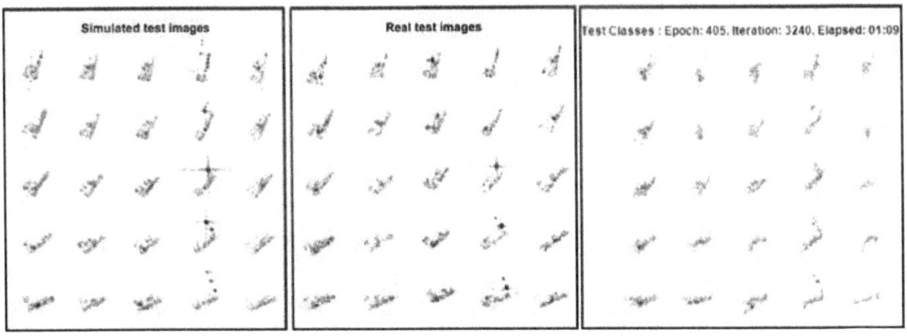

Figure 12.23 *(left) Input simulations (none from classes used for training); (centre) expected images: real measurements (not used for training); (right) image-conditioned GAN-generated images*

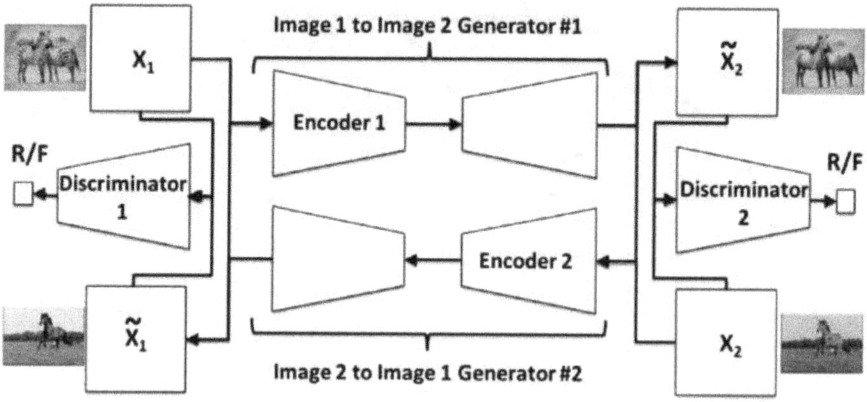

Figure 12.24 *Unpaired image-conditioned GAN (CycleGAN) architecture and illustration of zebra to horse mapping*

related in any way. This simple technique is powerful, achieving visually impressive results on a range of application domains, most notably translating between two texture domains, see famous zebra to horse example: https://www.youtube.com/watch?v=9reHvktowLY. [Accessed 1 May 2020].

The CycleGAN involves the simultaneous training of two Generator models and two Discriminator models (see Figure 12.24); it looks like a two-way Pix2Pix GAN. Generator #1 takes images from the first domain as input and outputs images for the second domain and Generator #2 takes images from the second domain as input and generates images for the first domain. Discriminator models are then used to determine how plausible the generated images are on each side and update the Generator models accordingly.

This extension alone might be enough to generate plausible images in each domain but not sufficient to generate translations of the input images. Thus, the CycleGAN uses an additional extension to the architecture called *cycle consistency*. The underlying idea is that an image output by the first Generator could be used as input to the second Generator and the output of the second Generator should match the original image. The reverse is also true; that is, an output from the second Generator can be sent as input to the first Generator and the result should match the input to the second Generator. The CycleGAN encourages *cycle consistency* by adding an additional loss to measure the difference between the generated output of the second Generator and the original image, and the reverse. This acts as a regularization of the generator models, guiding the image-generation process in the new domain toward image translation.

A conditioned version of CycleGAN, named Augmented CycleGAN, was logically proposed later in [26]. It is worth noticing that, with such architecture, the two image domains may theoretically have their own independent condition space.

When applied to the translation between simulated images and real ones using the SAMPLE dataset, we have not observed any significant performance change

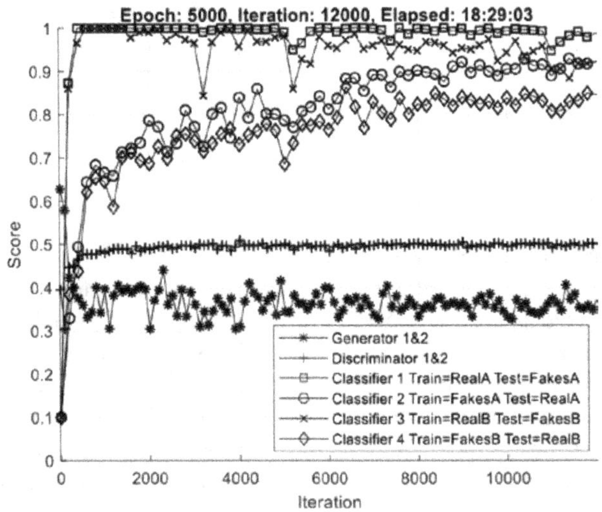

Figure 12.25 *Scores of CycleGAN Generators (star), Discriminators (plus) and accuracies of auxiliary Classifier 1 [train=real A and test = fakes A] (square), Classifier 2 [train = fakes A and test = real A] (circle), Classifier 3 [train = real B and test = fakes B] (cross) and Classifier 4 [train = fakes B and test = real B] (diamond)*

compared to the image-conditioned GAN previously discussed in Section 12.4.4.1, as expected using already paired images.

But the main advantage brought by this kind of GAN architecture is that the two image domains do not need to be strictly paired anymore. Therefore, we randomly divide the SAMPLE dataset into two separate parts (subsets A & B), with equal cardinality but with unpaired simulation and real images. We only use the first part (A subset) for training the GAN and use our four auxiliary Classifiers as performance indicators (Figure 12.25). Stability of the Discriminators and Generators scores indicates a very smooth training. Training was stopped at 5 000 epochs[4] with the following respective accuracy results for the four auxiliary Classifiers: 98, 93, 97 and 86 per cent.

This indicates a very accurate unpaired mapping, especially on the real test images (B subset) considering that

- They have been kept out of the GAN training.
- Half of the class and orientation conditions have never been 'activated' before testing since the GAN has not been trained on the corresponding simulated images (B subset).
- Classifier 4 was trained only on the generated images produced by the captive simulated images (B subset) and not by the whole SAMPLE simulated set (A+B).

This powerful tool opens a wide range of applications. For instance, we can use such mapping to populate a real measurements campaign facing occasional bad image dynamics, radar saturations or unfortunate interferences, accidental target occlusions or shadowing, etc.

12.5 Conclusions

We have presented a progressive methodology to generate SAR images of ground vehicles using various GAN architectures adapted from the literature. These algorithms were trained and tested on SAR image chips from the MSTAR and SAMPLE datasets. Networks were implemented with MATLAB Deep Learning Toolbox. We have shown that this class of machine learning algorithms bears potentially good performances in the generation of controlled and trustworthy simulated SAR signatures as well as a capability to perform image translation, interpolation and extrapolation between image domains. We have proposed an original method based on auxiliary reference classifiers to study the realism and usefulness of generated images and derived these performances for different GAN-based architectures.

There is still a lot of work to investigate in more detail the architecture, stability and optimization of GAN structures as well as their numerous training parameters and testing strategies. Yet, the natural variance of targets and clutter in radar images, including the complex interaction of targets with their surrounding background, will have to be taken into account to generate convincing SAR scenes in scenarios with more extended operational conditions.

As SAR resolution power increase, so will the image size; we know, from the optical domain experience, that this puts an extra challenge on the quality of GAN generations. StackedGAN [27, 28] and Multi-resolution GANs such as ProGAN and SinGAN [29, 30] seem promising ways to handle this problem.

We have only explored a small part of the *GAN Island* shown in Figure 12.5; however, we can be sure that multi-domain data augmentation studies, including radar, will benefit from the future improvements brought by this active machine learning field. We can already imagine building amazing mappings, not limited to image translations but, for instance, from 3D CAD models to complex radar signals and vice-versa!

References

[1] Guo J., Lei B., Ding C., Zhang Y. 'Synthetic aperture radar image synthesis by using generative adversarial nets'. *IEEE Geoscience and Remote Sensing Letters*. 2017, vol. 14(7), pp. 1111–5.

[2] Lewis B., Liu J., Wong A. 'Generative adversarial networks for SAR image realism'. Proceedings of SPIE 10647, Algorithms for Synthetic Aperture Radar Imagery XXV; Orlando, USA; April 2018.

[3] Lewis B., De Guchy O., Sebastian J., Kaminski J. 'Realistic SAR data augmentation using machine learning techniques'. Proceedings of SPIE 10987,

Algorithms for Synthetic Aperture Radar Imagery XXVI; Orlando, USA; May 2019.

[4] Moving and Stationary Target Acquisition Recognition (MSTAR). *Program Review*. Denver, CO; Nov. 1996. Available from https://www.sdms.afrl.af.mil/index.php?collection=mstar.

[5] Keydel E.R., Lee S.W., Moore J.T. 'MSTAR extended operating conditions. A tutorial'. Proceedings of SPIE, Vol. 2757, Algorithms for Synthetic Aperture Radar Imagery III; Orlando, USA; 1996. pp. 228–42.

[6] Ross T.D., Mossing J. 'MSTAR evaluation methodology'. Proceedings of SPIE, Vol. 3721, Algorithms for SAR Imagery VI; Orlando, USA; 1999. p. 705.

[7] He H., Wang S., Yang D., Wang S. 'SAR target recognition and unsupervised detection based on convolutional neural network'. Proceedings of the 2017 Chinese Automation Congress (CAC); Jinan, China; 2017. pp. 435–8.

[8] Lewis B., Scarnati T., Sudkamp E., Nehrbass J., Rosencrantz S., Zelnio E. 'A SAR database for ATR development: synthetic and measured paired labeled experiment (SAMPLE)'. Proceedings of SPIE Algorithms for Synthetic Aperture Radar XXVI; Orlando, USA; 2019.

[9] Scarnati T., Lewis B. 'A deep learning approach to the synthetic and measured paired labelled experiment (SAMPLE) challenge problem'. Proceedings of SPIE Volume 10987, Algorithms for Synthetic Aperture Radar Imagery XXVI; Orlando, USA; 2019.

[10] Schumacher R., Rosenbach Kh. 'ATR of battlefield targets by SAR – classification results using the public MSTAR dataset compared with a dataset by QinetiQ, UK'. *Proceedings of the RTO SET Symposium on Target Identification and Recognition Using RF Systems*; Oslo, Norway; 2004. pp. 31:1–31:12.

[11] Zhou F., Wang L., Bai X., Hui Y. 'SAR ATR of ground vehicles based on LM-BN-CNN'. *IEEE Transactions on Geoscience and Remote Sensing*. 2018, vol. 56(12), pp. 7282–93.

[12] Goodfellow I. 'Generative adversarial nets'. Advances in Neural Information Processing Systems 27 NIPS; 2014.

[13] Zhu J.-Y., Park T., Isola P., Efros A.A. 'Unpaired image-to-image translation using cycle-consistent adversarial networks'. IEEE International Conference on Computer Vision; 2017.

[14] Cha M., Majumdar A., Kung H., Barber J. 'Improving SAR automatic target recognition using simulated images under deep residual refinements'. IEEE International Conference on Acoustics, Speech and Signal Processing; 2018.

[15] Gui J., Sun Z., Wen Y., Tao D., Ye J. *A review on generative adversarial networks: algorithms, theory, and applications [online]*. 2020. Available from https://arxiv.org/pdf/2001.06937 [Accessed 1 May 2020].

[16] Soumith/ganhacks. *How to train a GAN? Tips and tricks to make GANs work [online]*. 2020. Available from https://github.com/soumith/ganhacks [Accessed 1 May 2020].

[17] Isola P., Zhu J.-Y., Zhou T., Efros A.A. *Image-to-image translation with conditional adversarial networks [online]*. 2016. Available from https://arxiv.org/pdf/1703.10593.pdf [Accessed 1 May 2020].

[18] Arjovsky M., Chintala S., Bottou L. *Wasserstein GAN [online]*. 2017. Available from https://arxiv.org/abs/1701.07875 [Accessed 1 May 2020].

[19] Radford A., Metz L., Chintala S. *Unsupervised representation learning with deep convolutional generative adversarial networks [online]*. 2016. Available from https://arxiv.org/pdf/1511.06434v2.pdf [Accessed 1 May 2020].

[20] Denton E.L., Chintala S., Szlam A., Fergus R. 'Deep generative image models using a Laplacian pyramid of adversarial networks'. *Advances in Neural Information Processing Systems 28*. 2015, pp. 1486–94.

[21] Odena A., Olah C., Shlens J. *Conditional image synthesis with auxiliary classifier GANs [online]*. 2016. Available from https://arxiv.org/pdf/1411.1784v1.pdf [Accessed 1 May 2020].

[22] Mirza M., Osindero S. *Conditional generative adversarial [online]*. 2014. Available from https://arxiv.org/pdf/1411.1784 [Accessed 1 May 2020].

[23] Chen X., Duan Y., Houthooft R., Schulman J., Sutskever I., Abbeel P. InfoGAN: Interpretable Representation Learning by Information Maximizing Generative Adversarial Nets, Conf. Neural Information Processing Systems (NIPS). 2016. Available from https://arxiv.org/pdf/1606.03657 [Accessed 1 May 2020].

[24] Zhu J.Y. *Toward multimodal image-to-image translation [online]*. 2016. Available from https://arxiv.org/abs/1711.11586 [Accessed 1 May 2020].

[25] Liu M.Y., Tuzel O. *Coupled generative adversarial networks [online]*. 2016. Available from https://arxiv.org/abs/1606.07536 [Accessed 1 May 2020].

[26] Almahairi A., Rajeswar S., Sordoni A., Bachman P., Courville A. *Augmented CycleGAN: learning many-to-many mappings from unpaired data [online]*. 2018. Available from https://arxiv.org/abs/1802.10151 [Accessed 1 May 2020].

[27] Huang X., Li Y., Poursaeed O., Hopcroft J., Belongie S. *Stacked generative adversarial networks [online]*. 2016. Available from https://arxiv.org/abs/1612.04357 [Accessed 1 May 2020].

[28] Zhang H. *StackGAN++: realistic image synthesis with stacked generative adversarial networks [online]*. 2017. Available from https://arxiv.org/abs/1710.10916 [Accessed 1 May 2020].

[29] Karras T., Aila T., Laine S., Lehtinen J. *Progressive growing of GANs for improved quality, stability, and variation [online]*. 2017. Available from https://arxiv.org/abs/1710.10196 [Accessed 1 May 2020].

[30] Shahal T.R., Dekel T., Michaeli T. 'SinGAN: learning a generative model from a single natural image'. ICCV; 2019.

Chapter 13
Micro-Doppler

Christoph Neumann[1] and Tobias Brosch[1]

Micro-Doppler is the designation for the spectral distribution of – mainly moving – targets in radar echo signals, derived from moving parts and vibrations of the target. Micro-Doppler analysis requires a sufficiently high Doppler resolution inside the radar, enabled by appropriate radar modes with appropriate high pulse repetition frequency (PRF) and coherent signal processing. Some of the most powerful automatic target recognition (ATR) techniques are based on micro-Doppler analysis.

In this chapter, the fundamentals and the physical background of micro-Doppler signatures and their dependencies on radar modes of operation and target types are explained and illustrated by some examples.

A mature classical ATR system based on feature extraction from micro-Doppler-Signatures and a classifier stage is presented, as well as modern artificial intelligence (AI) methods using deep learning (DL) principles to evaluate the spectral information without the need for feature definition. The capabilities of micro-Doppler analysis in combination with modern AI systems for threat assessments like drone attacks are presented, addressing also the challenges related to the database collection, processing needs and unknown target classes. The comparison of different approaches shows the benefit and potential drawbacks depending on the operational task.

13.1 Physical background

The Doppler effect is well known from acoustics and electromagnetic wave propagation and describes the frequency shift of a signal observed by a receiver moving relative to the signal source.

$$f = f_0 \cdot \left(1 + \frac{\Delta v}{c}\right) \tag{13.1}$$

[1]Hensoldt Sensors AG, Germany

where f_0 is the frequency of the transmitted signal, Δv is the relative velocity between transmitter and receiver, and c is the velocity of the wave propagation, i.e. the velocity of light for electromagnetic waves or – much stronger medium dependent – the velocity of sound for acoustic waves. The formula (13.1) is an approximation for cases where the relative movement Δv is small compared to the wave propagation velocity c and does not consider relativistic effects on electromagnetic waves. For usual measurements in free space or the earth's atmosphere, making use of this formula is appropriate in many cases. Each radar, making use of Doppler effects, calculates this way.

For active monostatic radars, the first "observer" of the Doppler effect is the object where the transmit signal is reflected. These reflections build the new source of the electromagnetic waves received by the radar, so the Doppler effect applies twice, leading to

$$f_{rx} = f_{tx} \cdot \left(1 + \frac{\Delta v}{c}\right)^2 = f_{tx} + \frac{2\Delta v}{\lambda} + O\left(\frac{1}{c^2}\right) \tag{13.2}$$

At the right side of this formula, the term f_{tx}/c is substituted by the corresponding wavelength λ. The last summand in the order of $\left(\frac{1}{c^2}\right)$ can be neglected, so the definition of the Doppler shift in non-relativistic approximation results in the deviation from the transmit frequency as

$$f_D = \frac{2\Delta v}{\lambda}. \tag{13.3}$$

The velocity difference Δv is the velocity component along the line-of-sight (LoS) from the radar to the reflecting object and can be calculated by the scalar product of the two ownship velocity vectors of radar and object. So it is obvious that Δv is zero when radar and object are moving in orthogonal directions, and Δv is maximised when they are moving along the LoS. Furthermore, the magnitude of the Doppler is directly proportional to the radar transmit frequency (i.e. $\sim 1/\lambda$), so the "sensitivity" to this effect increases with the radar frequency. An X-band radar operating at 9.4 GHz is ten times less sensitive than a W-band radar operating at 94 GHz. This means that the W-band radar needs only a tenth of the observation time required for the X-band radar to obtain a given Doppler velocity resolution – see more about this in the chapter "Radar Waveform Considerations".

13.2 Extension to bistatic radars

In "bistatic" radars, the location of the receiver Rx is remote from the location of the transmitter Tx, so a modified measurement geometry applies.

Assuming that the locations of Tx and Rx are stationary and using the definition according to Figure 13.1:

$$\begin{aligned} \beta &:= \text{angle}\ (\text{Tx} \rightarrow \text{Tgt} \rightarrow \text{Rx}) \\ \varphi &:= \text{angle}\ (\text{Tx} - \text{Tgt},\ \text{v}_{\text{own,Tgt}}) + \beta/2 \end{aligned} \tag{13.4}$$

This results in the angular symmetry

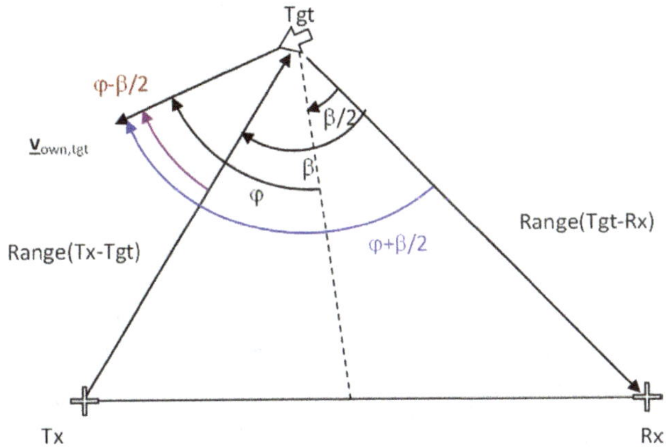

Figure 13.1 Definition of the "bistatic Doppler"

$$\begin{aligned} &\text{angle}\,(\text{Tx}-\text{Tgt},\ \text{V}_{\text{own,Tgt}}) = \varphi - \beta/2 \\ &\text{angle}\,(\text{Rx}-\text{Tgt},\ \text{V}_{\text{own,Tgt}}) = \varphi + \beta/2 \end{aligned} \tag{13.5}$$

with "Tx", "Rx", and "Tgt" naming the position of the transmitter, receiver, and target and with the target ownship velocity vector "$v_{own,Tgt}$". Using this angular definition, the bistatic Doppler formula can be expressed as

$$\begin{aligned} f_{D,bi} &= \frac{1}{\lambda}\cdot\frac{d}{dt}\left[\text{Range}\left(\text{Tx}-\text{Tgt}\right)+\text{Range}\left(\text{Rx}-\text{Tgt}\right)\right] \\ &= \frac{1}{\lambda}\cdot\left|\underline{v}_{own,Tgt}\right|\cdot\left(\cos\left(\phi-\frac{\beta}{2}\right)+\cos\left(\phi+\frac{\beta}{2}\right)\right) \\ &= \frac{2\cdot\left|\underline{v}_{own,Tgt}\right|}{\lambda}\cdot\cos\left(\phi\right)\cdot\cos\left(\frac{\beta}{2}\right) \end{aligned} \tag{13.6}$$

In comparison with (13.3) and considering that Δv from (13.3) equals $|\underline{v}_{\text{own,Tgt}}|\cos(\varphi)$ in this case (i.e. stationary Tx and Rx positions), the bistatic Doppler equals the "monostatic" Doppler, multiplied by the geometry factor $\cos(\beta/2)$, where β is the angle (Tx→Tgt→Rx):

$$f_{D,bi} = f_{D,mono}\cdot\cos\left(\frac{\beta}{2}\right) \tag{13.7}$$

For $\beta = 0°$ (i.e. position Tx equals position Rx, the "monostatic" case), $f_{D,bi}$ from equation (13.7) equals $f_{D,mono}$ from equation (13.3).

For $\beta = 180°$ (i.e. target is placed at the line Tx – Rx), $f_{D,bi}$ – i.e. the Doppler shift observed at the receiver Rx – equals to zero, regardless of the target's speed.

If $\varphi = 0°$ or $\varphi = 180°$ (target is moving orthogonally to the ellipsoid line defined by the focal points "Rx" and "Tx" and with "Tgt" positioned on this line, see Figure 13.2), the absolute value of $f_{D,bi}$ is maximised, resulting in $f_{D,bi} = 2|\underline{v}_{own,Tgt}|/\lambda\ \cos(\beta/2)$.

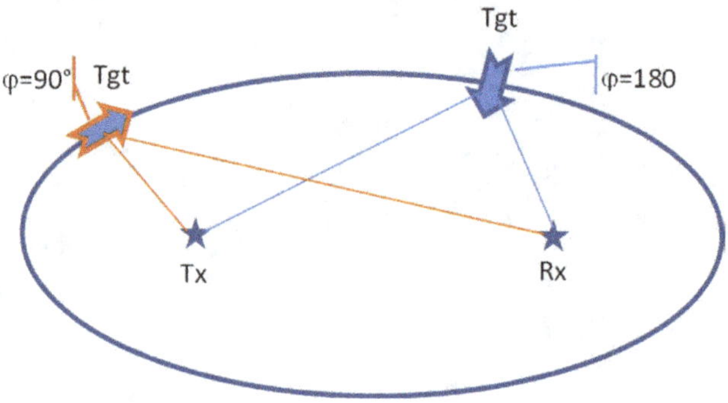

Figure 13.2 Measurement geometry for the "bistatic Doppler"

13.3 Radar waveform and digital sampling considerations

As Equation (13.3) implies, the radar receives a frequency shifted answer of a moving object even from a single pulse, which is spread in frequency according to the velocity distribution of the reflecting components of the object in the direction of the looking angle. Consider the answer of a pulsed waveform with transmit (Tx) and receive (Rx) periods as shown in Figure 13.3.

The Doppler shift of moving objects is normally much too small to detect it from a single pulse, since it is several orders of magnitude below the carrier frequency and the radar sampling at receive is too coarse to resolve this. As an example, at X-band (10 GHz) an object moving at a speed of 100 m/s relative to the radar produces a Doppler shift of 2 × 100 m/s/0.03 m = 667 Hz, which is more than 6 orders of magnitude below the carrier frequency of 10 GHz.

Consider again the X-band radar and, this time, an object moving at only 10 m/s. The range measurements shall be unambiguous up to 150 km, so it is appropriate to select a PRF of 1 000 Hz, so the radar gets N echoes from the target with time distance intervals (pulse repetition interval, PRI = 1/PRF) of 1 ms. The object is moving 0.01 m towards the radar within each PRI, and this is 1/3 of the radar wavelength. So the phase increment from pulse to pulse is equal to two times 2/3 × π = 4/3 × π because the wave travels two times the range distance.

In general, the phase difference from pulse to pulse is

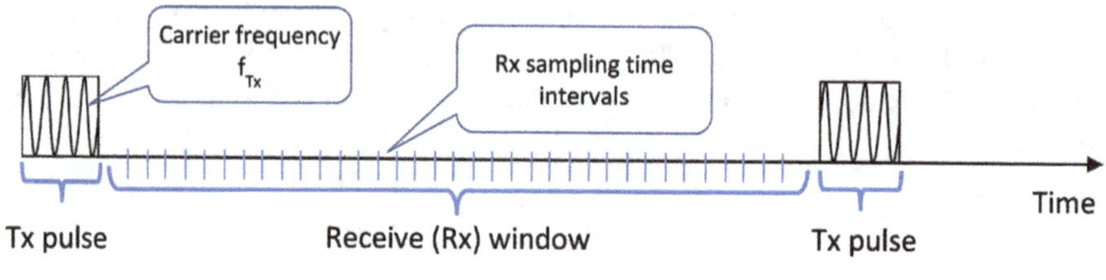

Figure 13.3 Transmit/receive scheme of pulsed radar

$$\frac{\Delta\phi}{2\pi} = \frac{2\cdot\Delta v\cdot \text{PRI}}{\lambda} = \frac{f_D}{\text{PRF}} \tag{13.8}$$

Due to the discrete pulse repetition, the radar will not observe a phase difference of 2π or multiples of 2π, only the fractional phase difference within the interval $[0, 2\pi]$. Therefore, the PRF defines the unambiguous range for direct velocity measurements, At a given PRF, the maximum unambiguous velocity range is

$$\Delta v_{max} = \Delta v\left(f_D = \text{PRF}\right) = \frac{\lambda\cdot\text{PRF}}{2} \tag{13.9}$$

Any velocity exceeding Δv_{max} is measured as residue modulo Δv_{max}:

$$\Delta v_{true} = k \cdot \Delta v_{max} + \Delta v_{observed}, k \in Z \tag{13.10}$$

In order to select an appropriate PRI for a specific measurement scenario, it is necessary to find a trade-off between desired unambiguous range and unambiguous velocity interval, taking into account the radar wavelength as a major impact factor.

In the scenario of the example above, Δv_{max} equals 15 m/s, which means that from the received and sampled signal it cannot distinguish between object velocities of 10 m/s, 25 m/s, −5 m/s, and so on. For aircraft target detection and recognition, it is necessary to cover a much larger unambiguous velocity interval, so the selected PRFs are typically in the order of some 100 kHz. For ground targets, normally PRFs of some few kHz are sufficient. If the radar transmit frequency is lowered, the PRF can be lowered in a proportional manner.

While the unambiguous velocity range (equation 13.10) is independent of the number of successive pulses N (see Figure 13.4), the velocity resolution and measurement accuracy strongly depend on N. Coherent integration over the received echoes from N pulses can improve the velocity resolution as well as the signal-to-noise ratio (SNR) by a factor of N but does not extend the unambiguous velocity measurement range, which is defined according to equation (13.9) independently of N. To overcome the "ambiguous problem" (i.e. range extent vs. velocity extent), many radars use "PRF staggering" modes of operation [a].

After collecting the echoes from N pulses (N is typically some hundreds or even some thousands, depending on the radar approach), the favourite method to perform

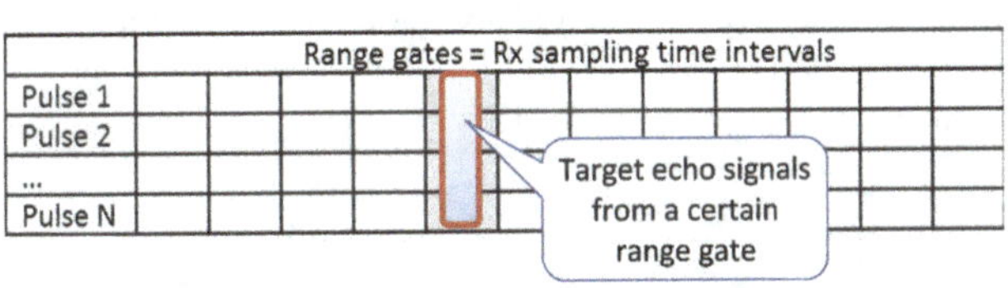

Figure 13.4 Pulses over range gates as "receive matrix"

[a]The PRF-staggering is changing the PRF from one CPI ("burst") to the next in order to get echo answers from the same target at different velocity ambiguities. Assuming that the target can be detected each time (or sufficiently often), the differences in the detection addresses w.r.t. the coordinates "range" and "velocity", caused by the different PRF values, are used to resolve the ambiguities in these coordinates.

a coherent integration is to perform a transformation into the spectral domain, for each range gate in parallel, e.g. by a fast Fourier transform (FFT), but also any other frequency analysis algorithm can be used to transform the signal to the spectral domain. In this way, a "velocity filter bank" is established, dividing the unambiguous velocity interval Δv_{max} into N fractions δv:

$$\delta v = \frac{\Delta v_{max}}{N} = \frac{\lambda}{2 \cdot T_{int}} \tag{13.11}$$

where the coherent integration time T_{int} equals N/PRF = $N \times$ PRI.

Of course, there are also limitations for the minimum number of pulses N respective the shortest time period T_{int} necessary to get sufficient spectral resolution. These are:

- The maximum coherent integration time must not exceed the target illumination time (limited for scanning pencil-beam radars).
- N must be low enough that for the maximum assumed target velocity (projection onto the LoS), during the time T_{int} the target range must not change more than a fraction of the range resolution cell size.

The first of these two points is obvious; the second point means that the target echo signal must stay in the same range gate (shadowed and marked in Figure 13.4), which is also obviously necessary to get the Doppler processing gain. Range walk compensation algorithms might be applied to the matrix data shown in Figure 13.4 to extend the limit for N in this case.

13.4 Impact of object vibrations on the Doppler spectrum

A single tone vibration is defined by its acceleration function

$$a(t) = g_{vib} \cdot \sin\left(2\pi \cdot f_{vib} \cdot t\right) \tag{13.12}$$

where g_{vib} is measured in m/s [1] and f_{vib} is measured in Hz.

The resulting velocity $v(t)$ and range deviation $r(t)$ are the integrals

$$v(t) = \int a(t)\, dt = -\frac{g_{vib}}{2pi \cdot f_{vib}} \cdot \cos\left(2\pi \cdot f_{vib} \cdot t\right) \tag{13.13}$$

$$r(t) = \int v(t)\, dt = -\frac{g_{vib}}{\left(2pi \cdot f_{vib}\right)^2} \cdot \sin\left(2\pi \cdot f_{vib} \cdot t\right) \tag{13.14}$$

where $r(t)$ describes the movement of the skin of a vibrating object relative to the centre of gravity of this object. In reality, this movement is much more complicated, but for general considerations of the impact on the Doppler spectrum, equation (13.13) is sufficient. We see from this formula that the amplitude decreases with the square of the vibration frequency, so high vibration frequencies may not be visible in the Doppler spectrum since their amplitudes may produce signal deviations with amplitudes below the sensor noise floor.

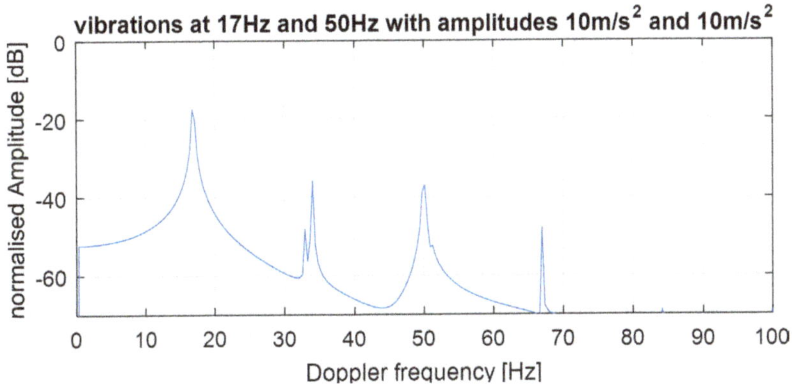

Figure 13.5 Impact of vibrations to the Doppler spectrum (example)

The following example simulates an object vibrating at frequencies 17 Hz and 50 Hz, both with the acceleration g_{vib} = 10 m/s [1], which is quite strong for a real target object. The resulting range deviation amplitude is 0.88 mm at 17 Hz and 0.1 mm at 50 Hz. For an X-band radar, this results in a sinusoidal phase variation of about 20° at 17 Hz and only about 2° at 50 Hz. Figure 13.5 shows also some intermodulation products appearing. To assess the signal power relative to the radar cross section (RCS) value of the target, zero dB corresponds to the skin echo power level of the object.

The power level of vibration signals in the Doppler spectrum depends also on the radar centre frequency, since $r(t)$ leads to the wavelength-dependent phase variation $4\pi/\lambda \cdot r(t)$. So the power will increase proportional to the radar centre frequency and the distance of the spectral position from the main skin echo depends on the vibration frequency (equation 13.14) and the radar frequency (equation 13.3) but not on the target velocity.

13.5 Impact of amplitude variations on the Doppler spectrum

Amplitude variations appear when the reflectivity of moving, e.g. rotating, parts of an object depends on the angular orientation w.r.t. the incident radar beam, which is normally the case. Especially at 90° incident condition, a rotor blade may reflect some order of magnitudes higher than at orientations some degrees away from this incidence condition.

Consider a two-propeller aircraft with a propeller diameter of 2 m and a blade tip speed of 250 m/s. This gives an angular frequency of 125 rad/s or a modulation frequency of about 20 Hz per blade or 40 Hz for a two-blade propeller. Figure 13.6 shows the appropriate Doppler spectrum for a one-engine aircraft and a two-engine aircraft.

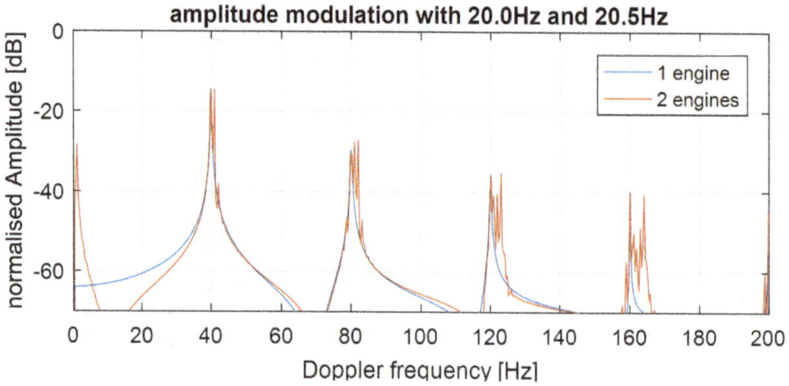

Figure 13.6 Impact of vibrations on the Doppler spectrum (example)

Note that

- In opposite to vibrations, the spectral appearance is independent of the radar frequency respective of the radar wavelength.
- The spectral amplitude depends on the reflectivity of the rotating parts only and, therefore, on the aspect angle of the radar illumination.
- The amplitude modulations produce a series of harmonics of the base frequency with magnitudes decreasing according to the maxima of the Bessel functions of appropriate order.

13.6 Geometrical considerations and spectral appearance

The velocity measurement described above is fine to detect targets away from clutter but does not help to classify or identify targets as long as the target itself has no characteristic movement components that distinguish it from other objects. However, many objects show such specific characteristics and the spectral appearance of them after Doppler processing is called the "micro-Doppler" signature.

Some examples of object movements producing "micro-Doppler" signatures are given in Sections 13.6.1–13.6.6.

13.6.1 Walking human

Characteristic movements of arms and legs with velocities both higher and lower as the body movement leads to according Doppler shifts which overlay in the spectral domain, i.e. to periodical spectral widening and narrowing with a period equal to the the movement sequence. Figure 13.7 shows the characteristic movement of a walking person. Figure 13.8 the appropriate Micro-Doppler spectrum of real measurement.

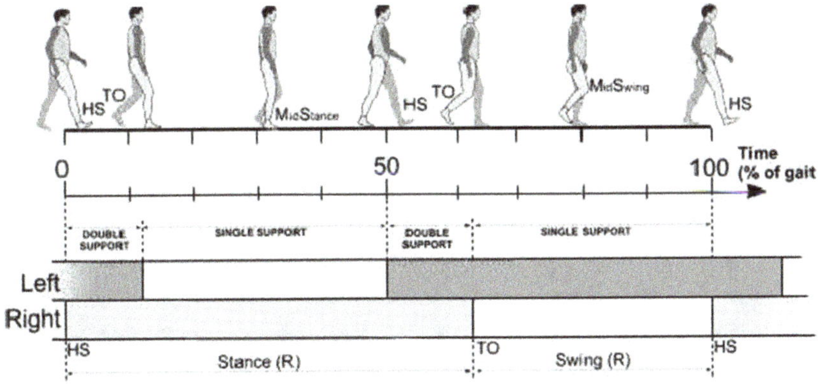

Figure 13.7 *Human walking (picture from [2])*

13.6.2 Wind turbine

Wind turbines show a quite clear and geometrically easy comprehensible velocity distribution, which is reflected in the Doppler spectrum analysis. Figure 13.9 shows the velocity distribution along the three blades of a wind turbine.

Figures 13.10 and 13.11 show the appropriate spectral distribution, resulting from a RCS simulation of a wind turbine. The radar is operating at S-band (2.8 GHz) with a PRF of 2 kHz. In Figure 13.10, the wind turbine blades are rotating with 12.5 rounds per minute (rpm), resulting in Doppler shifts all within the unambiguous

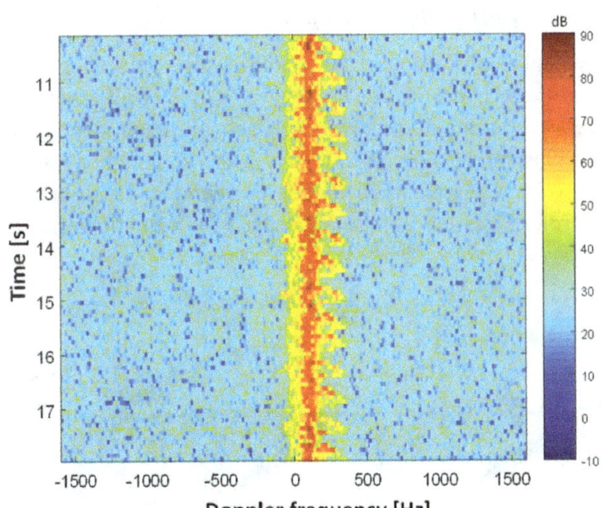

Figure 13.8 *Doppler signature of human walking. Source: HENSOLDT Sensors GmbH. Radar: X-Band, PRF ~3 kHz, coherent integration time ~80 ms (256 pulses).*

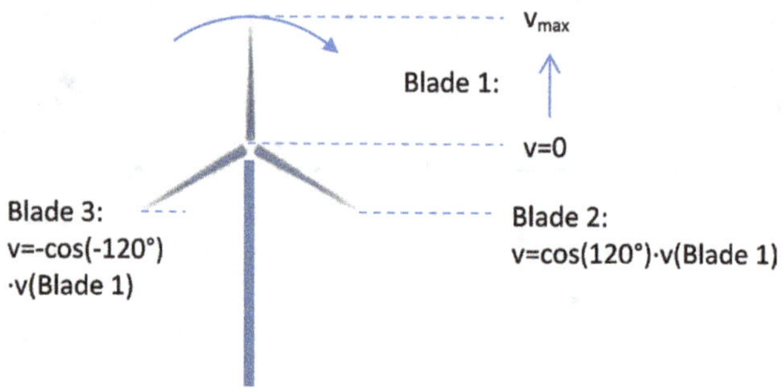

Figure 13.9 Velocity distribution of wind turbine blades

measurement bandwidth of 2 kHz. In Figure 13.11, the rotation speed is doubled to 25 rpm, so that aliasing effects appear.

13.6.3 Jet engines

Jet engine modulation (JEM) enables modern fighter aircraft radars to identify the engine type of illuminated aircraft engines and hence the aircraft type itself. The waveform used aliasing effects in the Doppler spectrum and to maximise spectral resolution has typically some hundred kHz PRF; the processing integrates several thousand pulse echoes signals coherently. Figure 13.12 explains the appearance of JEM lines (coloured in red) placed along the Doppler axis surrounding the target position (coloured in blue):

13.6.4 Wheeled and tracked vehicles

At the top of a wheel or track mounted on a vehicle, the velocity is twice the vehicle velocity, while at the ground, it is zero. The velocities of all other parts are within these boundaries. Vibrations of the tracks can be significant, depending on the kind of terrain. Vibrations can also be observed from the vehicle body and may be characteristic of the vehicle engine. The felloes of vehicles – also of civil cars – can produce quite strong Doppler signatures, caused by reflections from the edges of

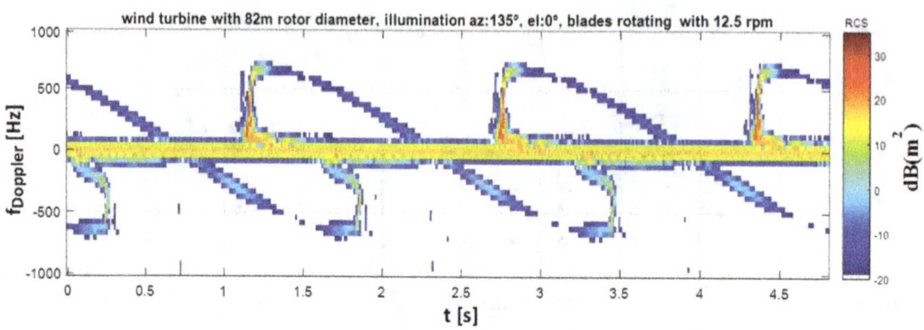

Figure 13.10 Spectral distribution of wind turbine echoes for a 2.8 GHz radar

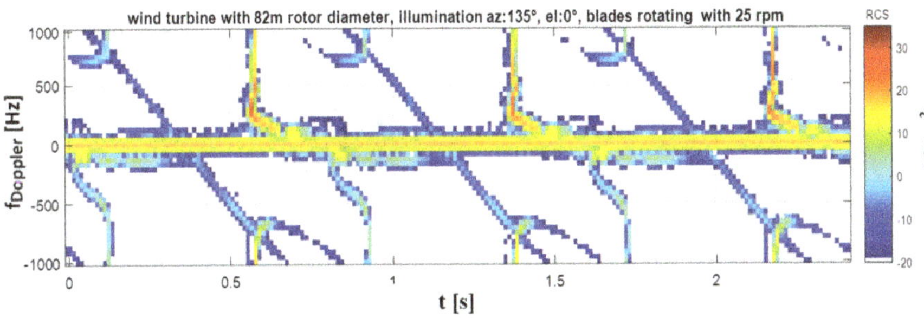

Figure 13.11 *Aliasing effects shown for wind turbine echoes*

spokes holes. In Figure 13.13, the mechanical reasons for the raising Doppler shifts are illustrated.

The two examples in Figures 13.14 and 13.15 show radar measurements over a time of about 25 seconds. The Doppler spectra show the main skin echo of the object (visible as the strongest red line) as well as the micro-Doppler signatures varying over movement state, consisting of a mixture of vibrations and amplitude modulations from wheels or tracks. It is obvious that a classification of the object type requires a great variety of such signatures, respective moving states reflected by the characteristic Doppler spectra to be considered.

From the colour-coded dB scaling, it can also be seen that the power level of most of the characteristic micro-Doppler spectrum is 30–60 dB below that of the main skin echo signal, so it is important for the radar to be designed for low noise power and high instantaneous dynamic range; otherwise, the micro-Doppler spectrum will be hidden by the receive noise floor.

13.6.5 Helicopters

The micro-Doppler spectrum of a helicopter is characterised by the body skin echo signal indicating the helicopter motion (which can be also zero if the helicopter is hovering) and multiple echoes from the main rotor and the tail rotor (or whatever rotors instead to enable steering, e.g. a second main rotor). Depending on the illumination aspect angle, the rotor blades or only the rotor heads may produce significant echo signals. Also, the amplitude modulation caused by blade passing may or may

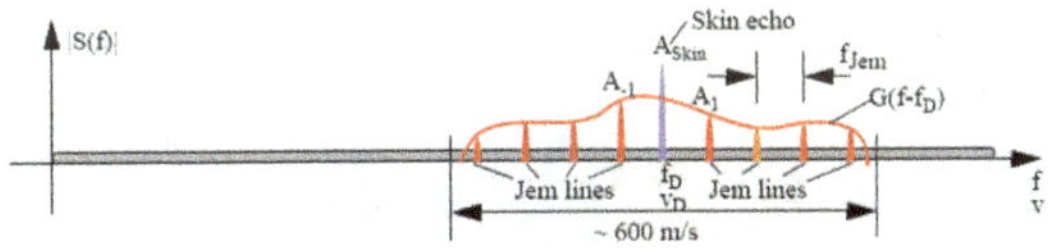

$$S_{Jem}(f) = W(f) \times A_{Skin}\delta_0(f - f_D) + W(f) \times P(f - f_D)$$

$$P(f) = \sum_{i=-N_{Jem}}^{N_{Jem}} A_i\delta_0(f - if_{Jem})$$

Figure 13.12 *JEM principles. Legend: f=Doppler frequency, |S(f)| = spectral amplitude, Askin at f=fD is the skin echo signal of a target with LoS velocity vD, A-1 and A1 are the first order JEM lines (basic harmonics), fJEM is the JEM line frequency distance, G(f-fD) is the characteristic "envelope" of the JEM spectrum.*

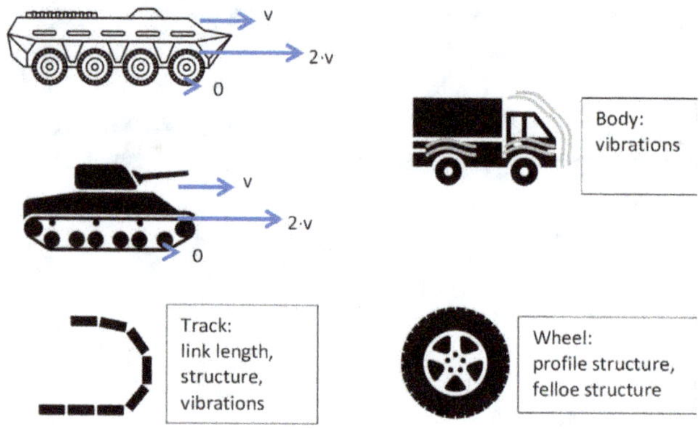

Figure 13.13 Wheeled and tracked vehicles: components with influence on micro-Doppler. Source: HENSOLDT Sensors GmbH. Radar: X-Band, PRF ~3 kHz, coherent integration time ~80 ms (256 pulses).

not be observable. Vibrations of the body skin and the rotor heads are also present. If all this is observable, i.e. the micro-Doppler signals produced by the helicopter echo is above the receive noise level, a typical helicopter signature looks like Figure 13.16.

Typically, the main rotor spins with 250–600 rpm, depending on the rotor size (ensuring the tips speed remain in the subsonic region). The blade passing frequency (BPF) is then two to five times higher, according to the number of blades, so we expect BPF values between 8 Hz and 50 Hz. Since the Doppler frequency resolution in the measurement shown in Figure 13.16 is only about 30 Hz (the FFT grid

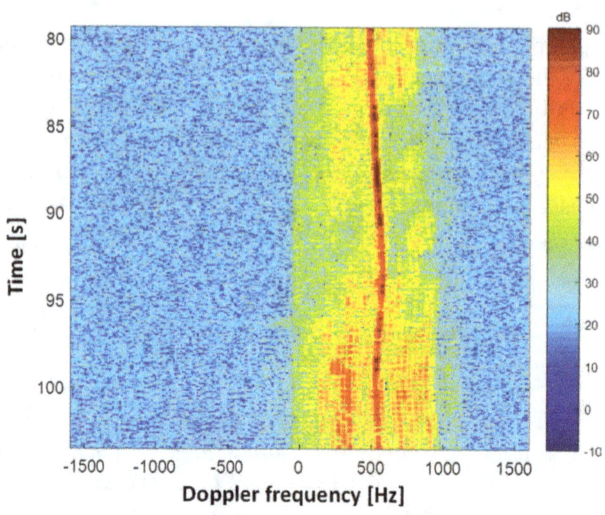

Figure 13.14 Micro-Doppler spectrum of a civil car

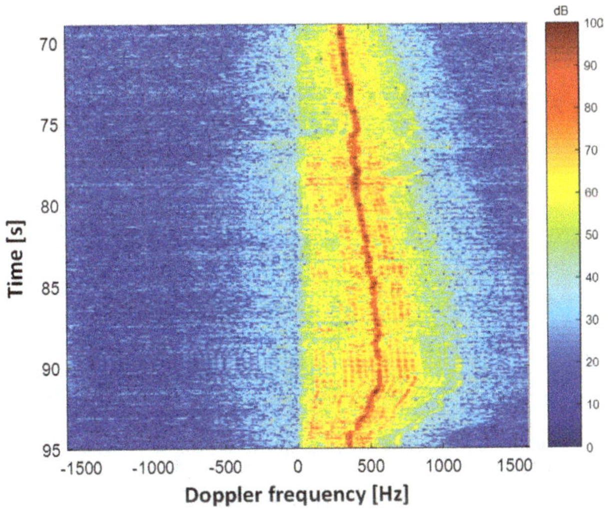

Figure 13.15 *Micro-Doppler spectrum of a tracked vehicle*

spacing is 12.5 Hz), most of the spectral lines visible in this figure derive from Doppler movement, not from blade passing or vibrations. This is also observable by the changes in the spectral appearance when the LoS velocity of the helicopter changes, resulting in a spectral widening proportional to the speed change occurring.

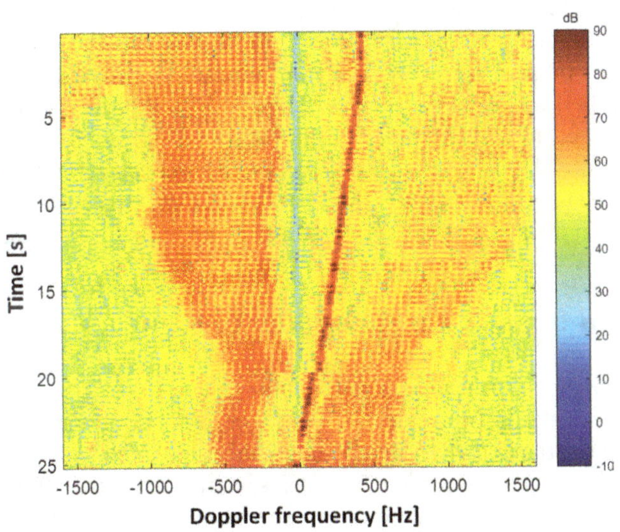

Figure 13.16 *Micro-Doppler spectrum of a Helicopter. Source: HENSOLDT Sensors GmbH. Radar: X-Band, PRF ~3 kHz, coherent integration time ~80 ms (256 pulses).*

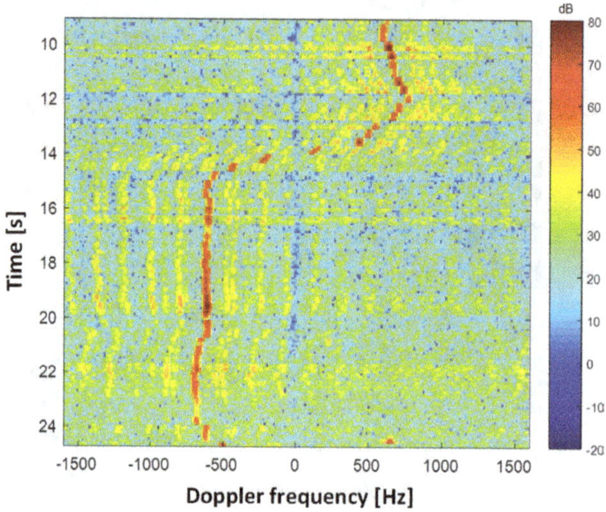

Figure 13.17 Micro-Doppler spectrum of a quadcopter drone. Source: HENSOLDT Sensors GmbH. Radar: X-Band, PRF ~3 kHz, coherent integration time ~80 ms (256 pulses).

13.6.6 Drones

The rotor speed of small drones ("small" defines the weight categories from less than 1 kg to about 10 kg) exceeds those of helicopters by a factor of 10 to 20; rotor spin frequencies between 1 000 rpm and 8 000 rpm can be expected, with four to eight times two-blade propellers present. This results in BPFs between 30 Hz and 260 Hz for each propeller. The example given in Figure 13.17 shows a frequency spacing of about 150 Hz, according to a propeller spin rate of 4 500 rpm. As characteristic of the amplitude modulation, this frequency spacing does not change with target velocity changes. The target velocity is proportional to the main skin echo signal position, clearly visible in the figure as the strongest (red coloured) reflection line.

13.7 Micro-Doppler-based classifier concept

The micro-Doppler spectrum evaluation allows to recognise and discriminate objects with characteristic movement of the object components and/or characteristic vibrations and rotational parts as described in the examples above.

Viewing the pictures of micro-Doppler spectrum examples (Figures 13.14–13.17), the classifier normally has only one row out of those 2-dimensional images available for spectral analysis, i.e. one "time sample" (one coherent processing interval (CPI)). At maximum, a very few rows (spectra from CPIs) can be used to feed into a classifier system, according to the illumination time of a scanning radar system. A typical micro-Doppler signature is shown in Figure 13.18.

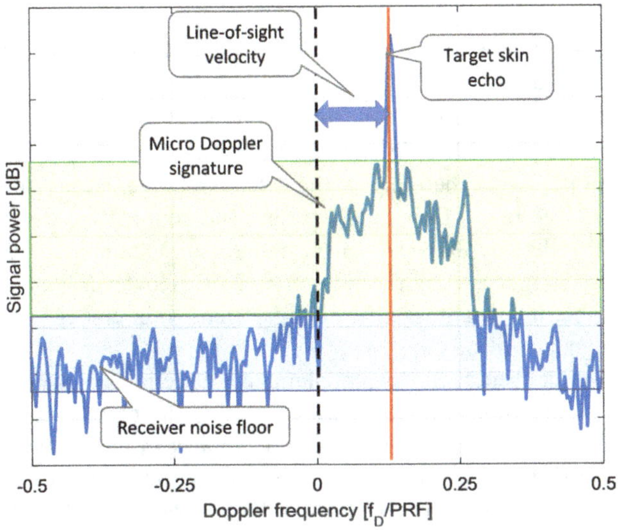

Figure 13.18 Micro-Doppler signature analysis

The physical understanding of the content of micro-Doppler signatures is important to formulate motion invariant[b] features for target classification or feed a network forming such features automatically.

It is important to notice that the frequency spacing of micro-Doppler signatures caused by target motion will scale with the projection of this velocity onto the LoS, while signatures caused by vibration and rotor blade reflections do not (see the geometrical considerations above). The noise floor level is normally known from internal radar control measurements and so can be determined independently from the actual Doppler spectrum.

However, there are some drawbacks of the classification approach via micro-Doppler analysis. These are:

- Sensitive to shadowing effects
  The LoS might be shadowed, e.g. by bushes, roadside ditches, or other objects, so that the wheels or tracks of vehicles or the legs of persons are not illuminated by the radar and so do not produce contributions to the Doppler spectrum.
- Requires high SNR ratio.

[b]"Motion invariant" means motion components not being the characteristic of the target object, like absolute value or the sign of the LoS velocity, which mainly depends on the looking angle.

As seen from the different examples above, the main contribution to the Doppler spectra between the different target classes has power levels of typically 30 dB and down to 60 dB below of the main signal, normally given by the object skin echo signal, depending on the target characteristics and illumination angle. Reliable object detections require about 15 dB contrast in the signal-to-noise level. Therefore, at the detection range, it may not be possible to recognise the micro-Doppler spectrum. The contrast increases with range reduction by the fourth power of the range, so at half the detection range, the contrast is increased by an additional 12 dB. However, this calculation is not applicable in signal-to-clutter configurations: rain clutter might cover the Doppler signature also.

- Partly low effectiveness for tangential moving directions

  The classical Doppler effect is proportional to the projection of the relative velocity to the LoS between radar and target object. If the motion directions of radar platform and target object are orthogonal to each other, the velocity component is zero. All motion contributions of wheels or tracks of a vehicle are also zero in this case. Only vibrations and movements causing amplitude modulation (like blade passing) are still present then.

A good classifier system should take measurement conditions into account as far as possible and indicate limitations in classification performance together with the classification result.

A classifier system implemented in a radar also has to comply with the radar search and track tasks, which normally do not allow to spend much time for target illumination, since a wide angular area has to be observed within a small so-called revisit time. As a typical example, if the radar has to observe 120° in azimuth within two seconds, using a radar beam with an azimuthal width of 6°, the average time on target can be only 100 ms. Intelligent beam scanning options allow to spend more time on "interesting" targets and less time on the other objects, where "interesting" can be defined by rules based on the target range and approaching velocity. So the time on target for classification purposes is in the order of 100 ms, ranging from below 50 ms up to very few hundred milliseconds, depending on the radar type and task.

The examples given in Figure 13.19 show *one observation* per object type for several objects (noted at the left), each consisting of three successive coherent integration times of 80 ms resulting three Doppler spectra from the 240 ms total observation time.

For target classification, some appropriate features to discriminate between target classes can be derived from these examples. A selection of them is listed in Table 13.1. Details of feature definition techniques can be found in [3] and [1].

A general guideline for feature definition is to formulate them independently from the sign or absolute value of the velocity, since velocity-dependent features would represent the measurement geometry, not only the target characteristics. The spacing of frequency lines caused by wheel or track movement should be scaled by the velocity for the same reason. The spacing of frequency lines caused by vibrations

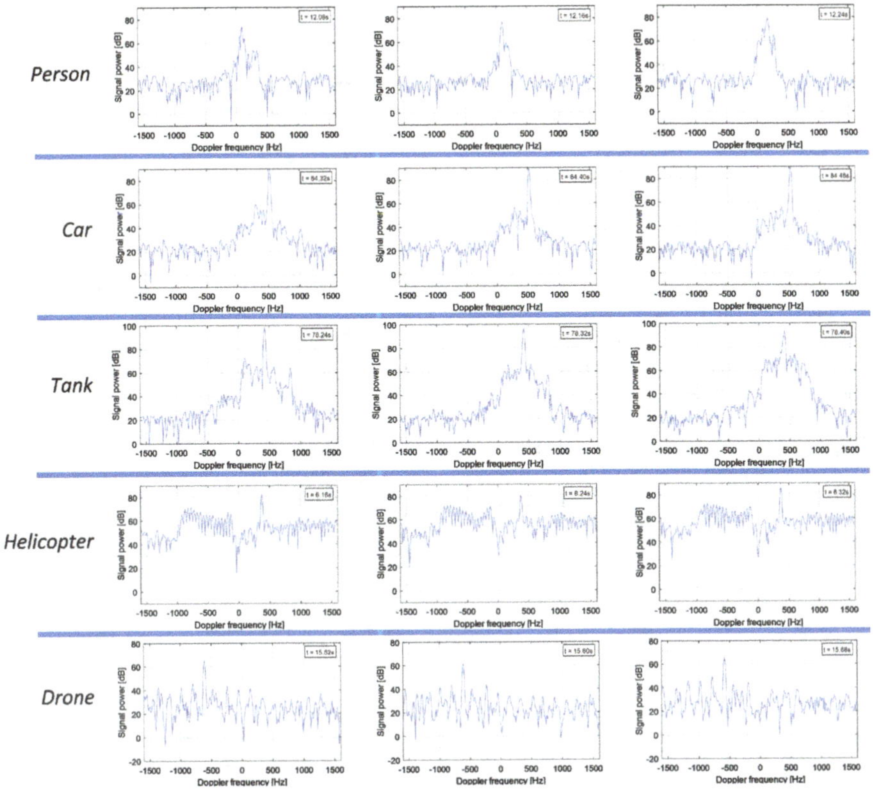

Figure 13.19 Micro-Doppler signatures for target recognition. Source: HENSOLDT Sensors GmbH. Radar: X-Band, PRF ~3 kHz

or blade passing should *not* be scaled by the velocity, hence they are due to amplitude modulation. Normally, the origin of spectral lines is not known a priori, so it is a good idea to use both scaled and unscaled spacing calculations as features.

The absolute power of spectral components cannot contribute to target separation features either. This power varies with the target range and position inside the radar beam. However, after normalisation of these values (i.e. range and target position relative to the antenna beam position), it can be useful to derive minimum and maximum assumable radar cross sections from the absolute power to compare it with a target characterising database.

Further features can be extracted, based on target track observation when signatures and/or extracted features from signatures are observable over a longer time section than possible during a single target illumination. Also the target's velocity can be extracted only based on target tracks, not on single target detections, as the Doppler spectrum measures only the fraction of the LoS component of the target velocity modulo PRF (see Section 13.3).

Table 13.1 Features extracted from micro-Doppler for classification purposes

Feature	**Class separation capability**	**Mathematical algorithm (e.g.)**	**Remark**
Shape (e.g. width) of the "main skin echo"	Persons↔Other objects	Central moments	
Shape and spectral extension of the micro-Doppler signature	All classes	Moments, autocorrelation	To be normalised to main skin echo velocity
Harmonics analysis within the micro-Doppler signature	Vehicles↔ Helicopters↔Drones	Cepstrum analysis, autocorrelation, target model correlation	Dependent or independent of target velocity
BPF	Helicopter and drone sub-types, Helicopters↔ Drones	Autocorrelation, target model correlation	Independent of target velocity, sufficient high PRF required
Change of spectral appearance over target illumination time	Persons or animals↔Man-made objects	Short-time FFTs (sub-CPI), cross-correlation	Mainly to recognise irregularities in movements

Feature extraction requires permanent testing of the feature's effectiveness, done by classical measures like Mahalanobis distance or preliminary classifier systems, in all cases, based on a statistical process using the whole database. Feature extraction from the complex numbers representing the micro-Doppler signature should result in a significant data reduction from the spectral representation to the feature representation of the signature.

The performance of a classifier system depends mainly and strongly on the database ensemble, i.e. on how good the database captures the real world, but also on the selected classifier type. Indeed, today [4], modern classifiers using a DL approach can outperform the classification performance of "classical" neuronal networks or other classifier types by a significant margin, still depending on a good database as a prerequisite.

In the radar world, collecting a good database is one of the most expensive expenditures during the development of a statistically based classifier system like a neuronal network, since it requires real radar data measurement of all kinds of possible targets under all possible measurement conditions. Some of these conditions can be modelled based on other real measurements, while most of the target objects, micro-Doppler signatures cannot be modelled up to now, at least not with realistic results.

This is because the micro-Doppler spectrum is an overlay of many components of the very complex movement of all the object parts contributing with individual and non-stationary radar cross sections causing Doppler shifts, vibrations and other frequency and amplitude modulations in the echo answer.

Modelling is helpful, however, when the classifier is not statistically based but knowledge-based, e.g. when the algorithm is evaluating certain main frequencies and frequency distances as in the case of JEM-based target recognition. Nevertheless, in the following section, a statistically based classifier is presented dealing with "manually" defined features extracted from micro-Doppler signatures as well as automatically generated features by a neuronal network.

The database can – and for DL approaches must – be increased by data augmentation methods. Applicable methods for micro-Doppler signatures are:

- Change SNR

  SNR can be lowered by adding random noise – distributed like the receiver noise, i.e. Gauss distribution in both real and complex parts of the signal. If the target range is used for feature normalisation, this value has to be changed accordingly.
- Change target velocity

  The target LoS velocity can be changed by scaling the time axis, originally defined by the PRI steps. This is done by resampling the data in the time domain. Motion-based Doppler lines will scale accordingly as shown in Figure 13.20, but vibration or other amplitude modulation-based lines will then also scale in opposite to their behaviour at real movement changes. Therefore, this method should be limited to small amounts of changes.
- Change or add clutter signals

  Ground clutter of the earth surface appears at the Doppler zero position if the radar is stationary (green line in Figure 13.18) or at the Doppler position according to the radar platform movement. Rain clutter appears within a certain bandwidth around the ground clutter position (green-shaded areas in Figure 13.18). The Doppler band extension and the power level of the rain clutter echo depend on the wind conditions and the rain rate. For maritime targets, appropriate sea clutter models can be applied.

The database used for the classifier comparison shown below roughly contains ten thousand to several ten thousand labelled signatures per target class. "Labelled" means that these signatures are from known target types, so a label indicating the target class is attached to each signature. The labelling procedure was done semi-automatically, regarding all plots associated to a target track of a known object as belonging to the tracked object. Nevertheless, the labelling procedure is the most expensive topic within the ATR enabling chain. For the classifiers presented here, the target classes to be discriminated were:

- "Clutter" (detected objects of no interest in this approach, like birds, windmills)
- "Persons" (walking and running, single and multiple)

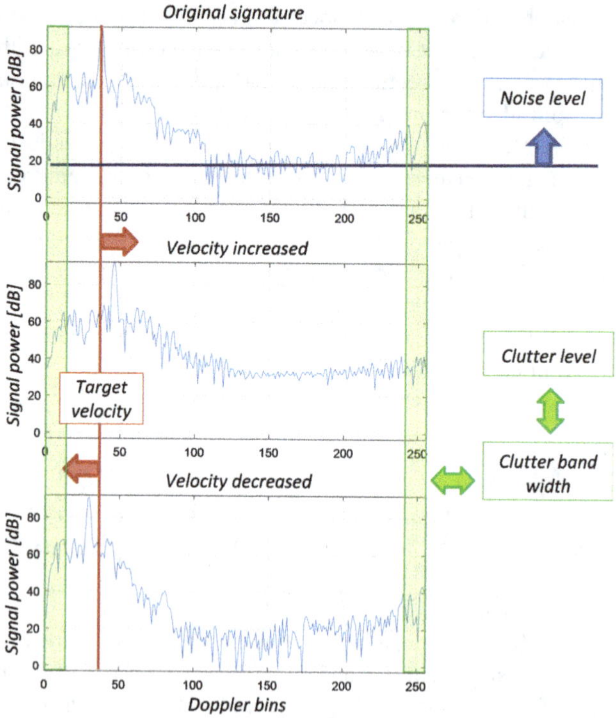

Figure 13.20 Data augmentation methods

- "Vehicles" (like cars, civil tracks, moving in all directions/with all velocities, single and multiple)
- "Wheeled armoured vehicles"
- "Tracked armoured vehicles"
- Drones ("small" and "very small").

Augmentation was done during the training process without saving the augmented signatures; the database was increased by augmentation by a factor of ten to fifty.

The "reference" classifier was a feature-based classical 3-nearest neighbour classifier, comparing each feature vector from actual measurements with target vector clusters from the training process, representing each defined target class by a number of more than one hundred clusters. This classifier was implemented to the same operational ground observation radar where all the measurements originate from. The advantages of this classifier type are:

- "white box" design
- easy to extend to new classes
- low memory consumption, also during training time

- compatible with self-supervised learning at runtime.

The classifier "competition" compares the results of this reference with those of other classifiers using state-of-the-art concepts, both feature-based and non-feature-based.

The feature-based classifiers are:

- decision tree
- "classical" 3-layer neuronal network
- Hidden Markov chain
- simple "dense" (convolutional) 5-layer neuronal network
- deep neural network with 12 layers.

Each of these classifiers could significantly improve the classification performance compared to the baseline nearest neighbour classifier accuracy of 84%. The decision tree achieved about 89.1% classification accuracy; the simple dense neural network reached 93.5% classification accuracy, while the dense neural network could gain 95.2% classification accuracy.

Instead, they use the entire spectrum in logarithmic scaling as shown in the different examples above and after applying a signal power normalisation.

In order to find the best performing architecture, a multitude of different deep neuronal networks (DNNs) was investigated by varying the number of layers, the number of (DNN-generated) features, the number of connections in deeper architectures, and the augmentation techniques. In addition, we investigated in modelling digesting complex-valued spectrums vs. models using only the absolute values. In sum, augmentation was vital to the success of the Fourier spectrum-based architecture. Architectures with even more parameters than the currently best performing ones with about 115 000 parameters did not improve the results. All these investigations did not focus on a further reduction of the number of parameters, which must definitely be investigated before implementing it into an operational system. The classifier types investigated are listed in Table 13.2.

13.8 Summary

The micro-Doppler spectrum is observable for all radars serving coherent signal processing over repetitive waveforms and ensuring target illumination times of a quite large number of such waveform repetitions.

The appearance of the target echo signal in the Doppler spectrum can vary extremely, depending on the aspect angle of illumination and target velocity, also for one distinct target object.

Physical background given by the formulas to transform the object movements to the expected spectral appearance is very helpful when designing a classifier system based on micro-Doppler. Also the assessment of the possible variations

Table 13.2 Comparison of micro-Doppler classification approaches

Feature-based classifiers	No. of parameters	Classifier accuracy (%))
Reference nearest neighbour	~50 000	~86.4
Decision tree	n.a.	~89.1
Simple dense neural network [5]	~6 000	~93.5
Deep neural network [6]	~47 000	95.2+/−0.1
Spectrum-based classifier		
DL network architecture 1 [6,7]	~468 000	91.5
DL network architecture 2 [7,8]	~595 000	93.9+/−0.1
DL network architecture 3 [7,8]	~115 000	95.0+/−0.2

The accuracy value is defined as the rate of true-positive decisions for all decisions. The deviation (if given) results from multiple training runs and reflects a 95% confidence interval.

in real-world measurements requires this background. The reference list contains sources for further information on all the topics addressed in this chapter.

13.9 Formula abbreviations

C	Velocity of light (m/s)
l	Wavelength (m)
f	Frequency (Hz)
f_D	Frequency deviation caused by the Doppler effect (Hz)
j	Phase (rad)
v	Velocity (m/s)

13.10 Terms and definitions

AI	Artificial intelligence
ATR	Automatic target recognition
Bistatic radar	Radar with transmit and receive antennas at different locations
BPF	Blade passing frequency (of rotors or turbines)
Burst	Series of equidistant pulses for coherent integration of the Rx signals
Clutter	Unwanted echo signals from other objects than targets
Carrier frequency	Mid-band frequency of the radar transmit signal (naming is common for signals with small modulation bandwidth)
Centre frequency	Mean frequency of a radar transmit signal (like carrier frequency, naming is common for waveforms with large modulation bandwidth)
CPI	Coherent processing interval, normally the sum of multiple PRIs

dB	Decibel
DL	Deep learning
DNN	Deep neuronal network
Dwell	Sequence of bursts over the defined PRI staggering
Main skin echo	Spectral component of the object's main reflection (normally reflection on the object body parts not moving against each other)
Micro-Doppler	Spectral Doppler signature of an object beside the main skin echo
Monostatic radar	Radar with transmit and receive antenna(s) at the same location
Range gates	Sequence of consecutive range resolution cell
FFT	Fast Fourier transform – one out of several methods to transform time series signals to frequency domain
Ownship velocity	Velocity of the target (or the radar) in an earth fixed Cartesian coordinate system
PRF	Pulse repetition frequency
PRG staggering	Changing the PRF from burst to burst in order to resolve ambiguities in range and velocity measurements
PRI	Pulse repetition interval = 1/PRF
RCS	Radar cross section, unit (m^2), is defined by the reflected signal power level from a metallised sphere skin of this size. The RCS of most structures depends on the radar wavelength.
rpm	Round per minute, 1 rpm ≈ 0.1 rad/s
Rx	Receive operation
Tx	Transmit operation

13.11 Radar band names

Name[3]	Frequency range (GHz)
L	1–2
S	2–4
C	4–8
X	8–12
Ku	12–18
K	18–27
Ka	27–40
W	80–110

[c]According to ITU: Recommendation ITU-R V.431-7: Nomenclature of the Frequency and Wavelength Bands Used in Telecommunications

References

[1] Duda R., Hart P., Stork G. *Pattern Classification*. 2nd ed. Wiley Interscience; 2001.

[2] Skolnik M. *Radar Handbook*. 2nd ed. New York: McGraw-Hill, Inc; 1990.

[3] Theodoridis S., Koutroumbas K. *Pattern Recognition*. 3rd ed. Academic Press Elsevier; 2006.

[4] Neumann C., Brosch T. 'Deep learning approach for radar applications'. *International Radar Symposium*. 2020.

[5] Fukushima K., Miyake S., Ito T. 'Neocognitron: a neural network model for a mechanism of visual pattern recognition'. *IEEE Transactions on Systems, Man, and Cybernetics*. 1983, vol. SMC-13(5), pp. 826–34.

[6] He K., Zhang X., Ren S., Sun J. *Deep residual learning for image recognition* [online]. 2015. Available from https://arxiv.org/abs/1512.03385 [Accessed 16 Jun, 2016].

[7] Ioffe S., Szegedy C. *Batch normalization: Accelerating deep network training by reducing internal covariate shift* [online]. 2015. Available from https://arxiv.org/pdf/1502.03167v3.pdf [Accessed 8 Nov, 2019].

[8] Zagoruyko S., Komodakis N. *Wide residual networks* [online]. 2017. Available from https://arxiv.org/pdf/1605.07146.pdf [Accessed 7 Sep, 2018].

Chapter 14
Kinematic and linguistic interpretation of human motion via RF signal analysis

Sevgi Z. Gurbuz[1] *and Evie Malaia*[2]

14.1 Introduction

The first demonstration of 24 GHz-phased-array transceiver (a.k.a "radar-on-a-chip") technology [1] in 2005, followed by increasingly miniaturized and higher frequency, millimeter wave (mmw) designs [2, 3], has made possible the use of low-cost, low-power Radio Frequency (RF) sensors in a wide range of commercial applications, including Internet of Things (IoT) and automotive sensing. More recently, the last five years have seen a dramatic increase in research in deep learning for RF signal classification, motivated by applications that require recognition of much finer human motions, such as activity and gesture recognition for smart environments, remote health monitoring, and intruder detection.

RF sensors have many unique advantages, such as being non-contact, operable in the dark and through-the-wall, and effective even over large distances. In the context of indoor monitoring, such as in a home or assisted living facility, RF sensors also provide the essential element of privacy: even if hacked, RF sensors cannot record faces, background scenery, or tell what someone is wearing. RF sensors use radio detection and ranging (radar) signal processing techniques to provide high-resolution measurements of distance, direction-of-arrival, velocity, and micro-Doppler signature, which has biometric properties exploitable for the recognition of specific individuals and gait patterns [4, 5]. Thus, RF sensors possess many qualities greatly desirable for human motion monitoring; yet, accurate and robust machine understanding of human motion has been precluded by several fundamental challenges, as summarized in Figure 14.1.

- ***Human motion has nearly unlimited diversity.*** It is typical for classification approaches to involve the recognition of a finite number of pre-defined activities, e.g. "walking," "running," and "jumping." However, human motion is not

[1]Department of Electrical and Computer Engineering, The University of Alabama, USA
[2]Department of Communication Disorders, The University of Alabama, USA

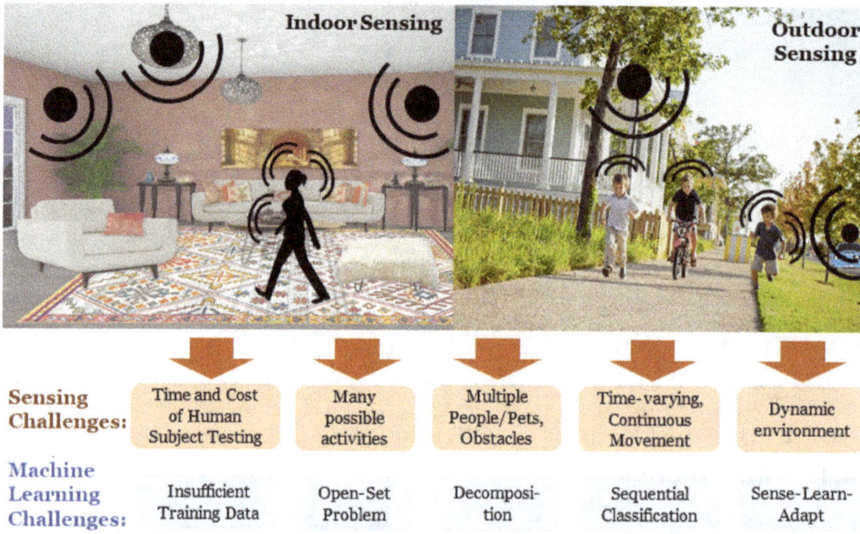

Figure 14.1 Sensing and machine learning challenges to RF signal classification

always so easily categorized, especially when intra-class diversity is considered. For example, over two dozen different types of falling have been described [6]. Broadly, grouping class descriptors may reduce the total number of classes but can also increase the number of misclassified data. Even so, it is highly likely to encounter motions other than the pre-defined classes, resulting in what is known as an *open-set problem*—the case when the test data involves a class that has not been included in the training dataset. Approaches for addressing this challenge include hierarchical classification [7, 8], adding an "unknown" target class [9], or conditioning the classes considered using sequential state models [10]. The design of deep neural networks (DNNs) that can generate novelty outside trained model classes remains an active area of research [11, 12].

- ***Physical factors can distort, block, or mask patterns seen in the RF data.*** RF sensors measure radial, not absolute velocity, so the path of motion and the angle of observation modulate the amplitude of micro-Doppler frequencies measured, resulting in mismatch with training data that is often acquired by participants moving along a straight line and reducing classification accuracy to just 5% in some cases [13]. Also, obstructions and the presence of other motion sources, such as other people, pets, or ceiling fans, can result in missing components or intertwined signatures. Additionally, other RF sources may cause interference, blocking part of the transmit band [14, 15].
- ***Human motion is time-varying, continuous stream of sequential data.*** Current approaches typically classify sequences of finite duration snapshots. This approach forces the RF data to conform to the structure of DNNs originally designed for computer vision applications and is not well suited to the

processing of sequential data. Moreover, natural motion does not instantly change from one class to another and, instead, has a period of transition that is constrained by biomechanics. While recurrent neural networks or their variants [16–18] have been proposed to better model the sequential dependencies in the data, kinematic transitions have not been adequately modeled and accurate temporal segmentation remains a challenge. There has been some recent work, however, in using state-models to capture physical dependencies [10] and Bi-Directional Long Short-Term Memory (Bi-LSTM) networks in conjunction with micro-Doppler and trilaterated range profiles from an RF sensor network [19] for sequential classification of continuous RF data streams.

- ***Environmental factors can also vary with time***. Life is dynamic: not just the person of interest but other vehicles, people, animals, and motion sources also move. Obstructions and masking are typically transient, not persistent, causes for degradation. Algorithms for human recognition should not be optimized for just one environment or scenario but be adaptable under real-life dynamic conditions.
- ***Insufficient training data***. The training of deep models requires a lot of data that spans all probable subject profiles, e.g. people of all heights, walking styles, and possible motions—a typically unfeasible task given the time and costs involved with human subject experiments. The low training sample support common in RF applications (less than several thousand samples) severely limits DNN depth and, hence, the accuracy and generalization of resulting models.

Overcoming these challenges requires the design of new DNN architectures and training strategies that are fully optimized to RF datasets. Deep learning has led to incredible advancements in computer vision and speech processing, however, inherently RF data are neither images nor acoustic time-series. RF datasets are incredibly rich and can measure range, velocity, and angle as a function of time. Through radar signal processing, 1D, 2D, three-dimensional (3D), and even 4D data representations can be generated. Current deep learning approaches focus on converting RF data into images that can be processed using computer vision-based approaches, and it has been only recently that investigation in broader representations and architectures has begun.

This chapter provides an overview of machine learning for RF signal classification, especially as applied to human motion recognition. Current innovations that integrate a knowledge-aided, physics-enabled approach to DNN design are discussed in the context of recognition of activities of daily living. A newly emerging application of RF sensing to the study of American Sign Language (ASL) is also presented.

14.2 RF Sensor phenomenology and data representations

The signal received by a radar is, in general, a time-delayed, frequency-shifted version of the transmitted signal. But the received radar backscatter, $s_r(t)$, consists not just of reflective signals from the subject of interest, $s_h(t)$. Reflections from other moving objects, persons, or animals, which we will henceforth refer to as parasitic Doppler returns, $s_p(t)$, since they obscure the desired returns, are also present. Backscatter from stationary objects, such as buildings, furniture, and trees, is known as clutter, $s_c(t)$. Thus, the total received signal for an RF sensor may be expressed as

$$s_r(t) = s_h(t) + s_p(t) + s_c(t) + w(t), \tag{14.1}$$

where $w(t)$ represents noise and sensor artifacts. Most modern RF transceivers are equipped with quadrature demodulators so that the received signal is the complex sum of in-phase (I) and quadrature (Q) components: $s_r(t) = I(t) + jQ(t)$.

While the structure of a vehicle may be represented by the scattering of a discrete point target, for a high resolution, linear frequency-modulated continuous wave (CW) radar, the human body behaves as a distributed target. Finite-element electromagnetic models [20] can be used to approximate the radar cross-section (RCS) of the human body; but for practical scenarios, the scattering from the entire human body, s_h, may be approximated using superposition of the returns from K body parts [21, 22] as

$$s_h[n] = \sum_{i=1}^{K} a_i exp\left\{ -j\frac{4\pi f_c}{c} R_{n,i} \right\}, \text{ for } a_i = \frac{\sqrt{G_x G_{rx}}\lambda \sqrt{P_t \sigma_i}}{(4\pi)^{3/2} R_i^2 \sqrt{L_s}\sqrt{L_a}}, \tag{14.2}$$

where $R_{n,i}$ is the range to the ith body part at time n; f_c is the transmit frequency; c is the speed of light; a_i is the square root of the power of the received signal; G_{tx} and G_{rx} are the gains of the transmit and receive antennas, respectively; λ and P_t are the wavelength and power of the transmitted signal, respectively; σ_i is the RCS of the ith body part; and L_s and G_{tx} are system and atmospheric losses, respectively.

The effect of target kinematics is predominantly reflected in the frequency modulations of the received signal. While the distance of the target to the radar does affect the received signal power, these variations are not significant for human subjects given typical body sizes. Rather, the amplitude of the received signal is primarily dependent upon the transmit frequency being used and RCS of target observed. Thus, at higher frequencies, such as 77 GHz, the received signal power from a person will be much lower than that of a radar at lower frequencies, such as 24 GHz or 5.8 GHz. Not just transmit frequency but also bandwidth, pulse repetition interval, observation duration, and aspect angle have been shown to affect the radar received signal and the performance of machine learning algorithms applied to the data [13].

A fundamental question is then how to best make accessible those signal properties that are most critical to classification. Although there has been work [23] proposing a direct classification of the 1D complex ($I + jQ$) time stream, it is more common for radar signal processing techniques to be applied that convert the data into 2D or 3D data representations, shown in Figure 14.2, that better reveal the two fundamental measurements made by a single aperture radar: slant range and radial velocity. The range, R, is computed from the round-trip travel time, Δt, between the sensor and object, while the radial velocity, v, can be found by using Fourier analysis to measure the Doppler shift, f_D, due to object motion:

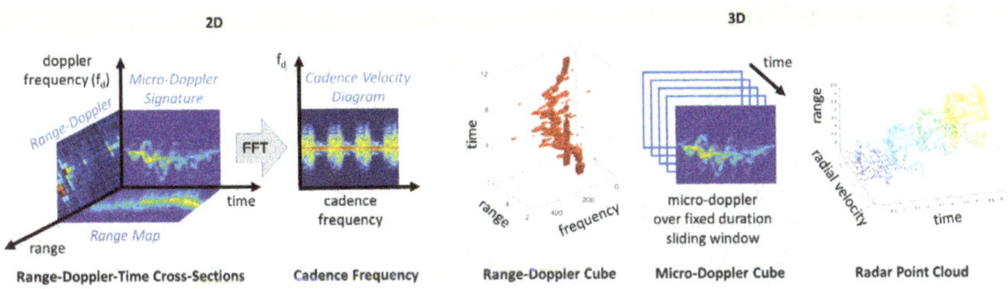

Figure 14.2 2D and 3D RF data representations

$$R = \frac{c\Delta t}{2} \text{ and } f_D = \frac{2vf_c}{c}. \tag{14.3}$$

14.2.1 2D representations

Four different types of 2D inputs can be generated from raw radar data: micro-Doppler signatures, range maps, range-Doppler (RD) images, and cadence velocity diagrams (CVDs). The range map shows the variation of target position with slow-time, while RD images show the relationship between target position and velocity. Micro-Doppler signatures reflect the variation of frequency with time and may be computed using a variety of time-frequency (TF) transforms. The CVD is computed by taking the Fourier transform at a given frequency bin across time [24].

TF analysis methods can be classified into linear transforms and quadratic time-frequency distributions (QTFDs) [25, 26]. The former includes the short-time Fourier transform (STFT) [25], Gabor transform [27], fractional Fourier transform [28], and wavelet transform [29]. The most commonly applied TF transform is the spectrogram, denoted by $S(t, \omega)$, which is the square modulus of the STFT, and a special case of a Cohen QTFD. It can be expressed in terms of the employed window function, $w(t)$, as

$$S(t, \omega) = \left| \int_{-\infty}^{\infty} w(t-u)x(u)e^{j\omega u}du \right|^2 \tag{14.4}$$

While reaping the benefits of linearity, STFTs must trade-off time and frequency resolution due to fixed length windows. Other TF transforms have been proposed, such as Gabor-Wigner transform, defined as the product of the STFT with the Wigner distribution, which removes cross-term interference while maintaining the clarity and resolution of the result [30]. QTFDs, on the other hand, aim at concentrating the signal power along the instantaneous frequency of each signal component. In the case of radar backscattering from a moving target, such concentration accurately reveals the target velocity, acceleration, and higher-order terms. The Cohen's class of QTFDs can suppress signal cross-terms and preserve auto-terms, leading to reduced interference distributions [31].

14.2.2 3D representations

The richness of RF data representations has led to not just the proposal of fusing results achieved with different 2D representations [32, 33] but also to 3D representations in terms of time, range, and frequency. The most common approach to forming a 3D data cube is to form a sequence of frames as a function of time that would effectively resemble video but for radar data. Both sequences of RD maps [17, 34–36] as well as sequences of micro-Doppler snapshots [18] have been applied to radar micro-Doppler classification of human activities and gestures. A RD map at a single instant of time is obtained by applying a Fourier transform at each range bin over a non-overlapping coherent processing interval of slow-time on the range map. Additionally, a radar point cloud [37] may be derived from RD data cubes by applying constant false alarm rage detection on the data, computing an iso-surface mesh from points with similar intensity [38], and then applying farthest point sampling to arrive at a 3D, discrete representation [39].

14.3 Deep learning for human activity recognition

Conventional machine learning involves the extraction of pre-defined or handcrafted features from the data. In supervised techniques, training data is used to define class boundaries based on the values of the handcrafted features, which are often developed based on domain knowledge or mathematical models and transforms. Thus, the procedure for computing features is fixed. In contrast, *data-driven* feature learning techniques modify or adapt the computation of features based on the specific training data acquired. Deep learning is a popular approach for data-driven learning in which network weights are optimized using training data to maximize resulting classification accuracy on test data.

The degree to which the resulting network can generalize greatly depends on how well the training data statistically represents all possible presentations of the class. In the case of human motion, this requires training data that spans different heights, speeds, gait styles, paths, and individual variations. Deep learning leverages massive amounts of data to avoid any dependence on domain knowledge. However, in radar applications, typically the amount of real data available is greatly limited and does not necessarily sufficiently represent all possible statistical variations within a class. This has limited the depth with which deep models have been constructed to 15–20 layers in contrast with DNNs used in computer vision, which can have as much as 152 layers [40]. In this section, various strategies for training deep models for human activity recognition with small amounts of data are presented, along with recent radar-specific techniques which leverage domain knowledge to improve generalization and classification accuracy.

14.3.1 Unsupervised pre-training versus transfer learning

Convolutional neural networks (CNNs) are a popular architecture that has advantages in image processing due to their use of convolutional filtering, which allows

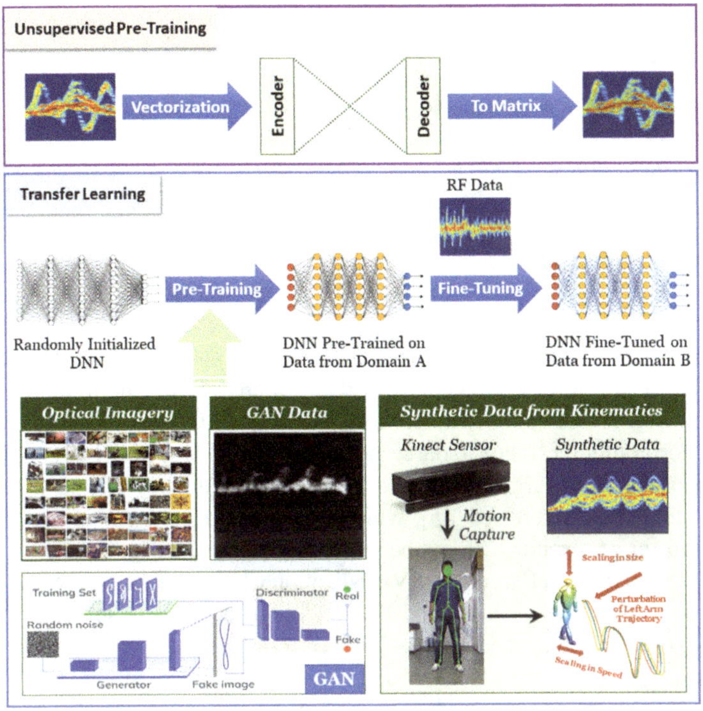

Figure 14.3 Techniques for training RF data when real sample support is low

the network to capture local spatial variations. The weights of a CNN are typically initialized with random values. However, the objective function of CNNs is highly non-convex, meaning that the parameter space of the model may contain many local minima. As a result, random initialization of weights may allow gradient-based optimization algorithms to converge to a local minimum that is in fact not optimal in a global sense [41].

An alternative way to train networks, which has been shown to yield greater accuracy with smaller amounts of training data, is *unsupervised pre-training*, illustrated in Figure 14.3. An encoder-decoder structure, such as used in autoencoders, can be used to greedily pre-train network weights [42] to learn an identity mapping in an unsupervised fashion. This approach has been shown to be effective in capturing the underlying variations of the data without any labels [43]. In particular, *convolutional autoencoders* (*CAEs*), which combine the benefits of convolutional filtering with unsupervised pre-training, have especially been shown to be effective in the classification of radar micro-Doppler signatures [44].

Another way of initializing network weights when there are very few real data samples available is *transfer learning*. The objective of transfer learning is to learn the weights of a network from a source domain, and then transfer those weights to the model for another "related" problem, or a target domain. As with unsupervised pre-training, transfer learning can be used for initializing the model

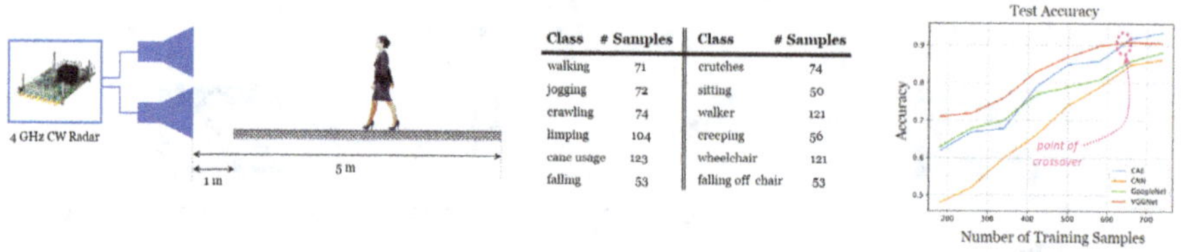

Class	# Samples	Class	# Samples
walking	71	crutches	74
jogging	72	sitting	50
crawling	74	walker	121
limping	104	creeping	56
cane usage	123	wheelchair	121
falling	53	falling off chair	53

Figure 14.4 Comparison of unsupervised pre-training and transfer learning from optical imagery

weights to values that bring the model closer to a global optimum. A small amount of real data from the target domain can then be applied to fine-tune the weights in a supervised fashion. Transfer learning also allows us to leverage very DNNs designed for computer vision (source domain) to initialize networks for learning models of radar data (target domain). There is an abundance of image databases available, including ImageNet [45], which contains over one million images, and can be used for pre-training. It has been popular to use DNNs that have achieved success in the ImageNet Large Scale Visualization Challenge, such as AlexNet [46], VGGnet [47], GoogLeNet [48], and ResNet [40], for transfer learning. Transfer learning has been used in radar micro-Doppler applications, such as human activity recognition [49–51], gesture recognition [52], and voice recognition [53].

In a recent study [54], the performance of unsupervised pre-training and transfer learning from optical imagery was compared on a 12-class human activity dataset acquired by a 4-GHz CW radar. The test setup and comparison are given in Figure 14.4, where the accuracy attained by a randomly initialized CNN, unsupervised pre-training (e.g. CAE), and two networks (VGGnet and GoogleNet) pre-trained on ImageNet is shown as a function of the amount of available training data. When hardly any training data is available, the best performance is given by VGGnet. But, beyond 600 samples, the accuracy attainable by VGGnet saturates and at roughly 650 samples is surpassed by that given by the CAE, which continues its linear upward trend.

Thus, if truly hardly any data is available, transfer learning from optical imagery can be a powerful technique for the classification of RF data; however, the difference in phenomenology between optical imagery and RF data limits potential gains. One potential remedy is to instead train the network on synthetic data, which is much easier to generate, in contrast with the high human costs of real data acquisition.

14.3.2 Synthetic RF data generation with motion capture

Accurate synthetic data generation relies on replication of not just the target motion, including position, velocity, and acceleration, but also the components of clutter, noise, and sensor artifacts. Equation 14.2 provides a point target approximation to target scattering and relies on superposition to compute the overall target response as a function of the time-varying ranges of each body part. The kinematic modeling of human activities, which would provide information on the trajectories of each body part, is a challenging problem, especially for human subjects due to their great potential diversity. Both kinematic models, such as the Boulic model [55], and motion capture (MOCAP) data [56–62] have been proposed for the animation of the human skeletal model. However, the Boulic model is only valid for walking and cannot take into account individual differences in gaits, since its only two input parameters are height and speed.

Using MOCAP data to derive the time-varying ranges of each body part avoids the limitations of the Boulic model. However, the acquisition of MOCAP still requires engaging test subjects to separately enact the motion for each sample. Thus, the practical amount of data that can be acquired is still limited by human time, effort, and costs—just as required for the acquisition of real RF measurements. *Data augmentation* can be applied to increase the number of training samples. Examples of data augmentation techniques from image processing include translation, rotation, cropping, and flipping; however, these methods have a corrupting effect on micro-Doppler signatures because they map to kinematically impossible human motions.

Alternatively, data augmentation may be applied to the underlying skeletal animation, so that the resulting micro-Doppler signatures remain consistent with physical constraints. Different body sizes can be created by scaling the dimensions of the skeleton, and different speeds generated by scaling the time axis of the animation. The statistical variations created by individualized gait styles can be emulated by adding perturbations to a parametric model of the time-varying joint trajectories. An overview of the diversification transformations applied to the skeletal model and resulting efficacy for network pre-training is given in Figure 14.5. A total of 55 MOCAP measurements of 5 test subjects performing 7 different human activities was used to generate a diversified MOCAP dataset of 32 000 samples [63]. The large amount of synthetic training data generated enabled the training of a residual neural network with 15 layers (DivNet-15)—much deeper than the three-layer CNN trainable with the four hundred seventy-four 4-GHzC-W radar samples available. DivNet-15 improved the attainable classification accuracy from 86% using a CNN to 97%.

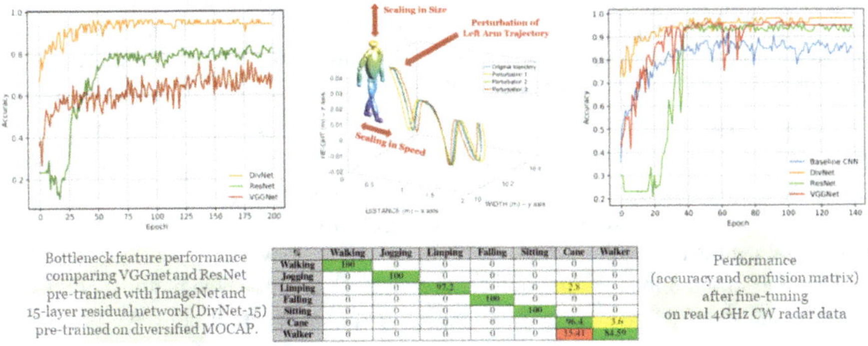

%	Walking	Jogging	Limping	Falling	Sitting	Cane	Walker
Walking	100	0	0	0	0	0	0
Jogging	0	100	0	0	0	0	0
Limping	0	0	97.2	0	0	2.8	0
Falling	0	0	0	100	0	0	0
Sitting	0	0	0	0	100	0	0
Cane	0	0	0	0	0	96.4	3.6
Walker	0	0	0	0	0	15.41	84.59

Figure 14.5 Classification of human activities using diversified MOCAP data for network initialization

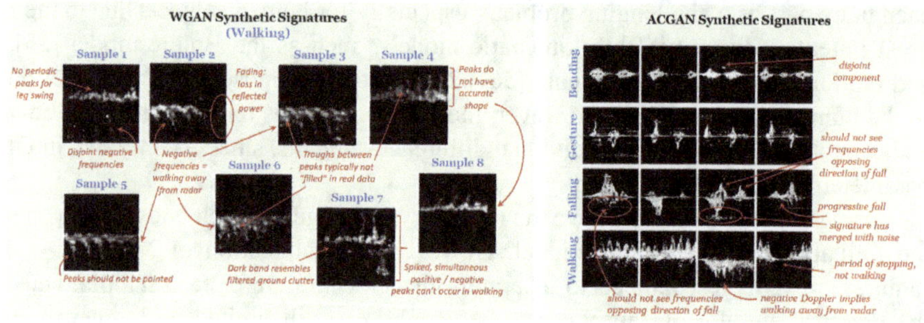

Figure 14.6 Kinematic fallacies of GAN synthesized micro-Doppler signatures

14.3.3 Kinematic considerations in adversarial learning

An alternative approach for generating synthetic data, which is gaining increased attention in the radar community, is adversarial learning. Generative adversarial networks (GANs) have been proposed for synthesizing realistic images in a variety of applications [64], including synthetic aperture radar [65] and micro-Doppler [9, 66–68]. In [69], a 4% increase in the classification accuracy of 50 running, walking, and jumping signatures was achieved using a GAN that augmented a training data set of 150 signatures with 1 000 GAN-generated signatures. The efficacy of different generative models to accurately synthesize radar micro-Doppler signatures varies. A recent study [70] shows, for example, that auxiliary conditional GANs (ACGANs) generate crisper signatures than conditional variational autoencoders (CVAEs).

For radar micro-Doppler applications, however, fidelity cannot be evaluated by considering image quality alone. A critical challenge is the possibility for generative models to provide misleading synthetic signatures. The characteristics of the micro-Doppler signature resulting from human activity are constrained by electromagnetic scattering and human kinematics. However, generative models have no knowledge of the physical constraints imposed by the human skeleton. It is possible, therefore, for GANs to generate synthetic samples that may appear visually similar but, in fact, correspond to impossible motions. This may be seen by observing the fallacies of the synthetic micro-Doppler signatures given in Figure 14.6, which have been generated using a Wasserstein GAN (WGAN) and ACGAN. The WGAN signatures exhibit properties that would never occur in any type of human motion, such as disjoint frequency components, fading, suppressed regions, and incorrect shapes, especially of the periodic components corresponding to leg swing and regions near the ground clutter line. Additionally, some signatures are generated reflected about the horizontal centerline, which would result the signature having negative frequencies and motion in the opposite direction as intended. The ACGAN signatures are more pristine and exhibit fewer fatal flaws but nevertheless continue to have problems with disconnected regions, reflected signatures, and additional artifacts, which can make the signatures appear similar to motion from a different class (e.g. progressive fall rather than an impulsive fall).

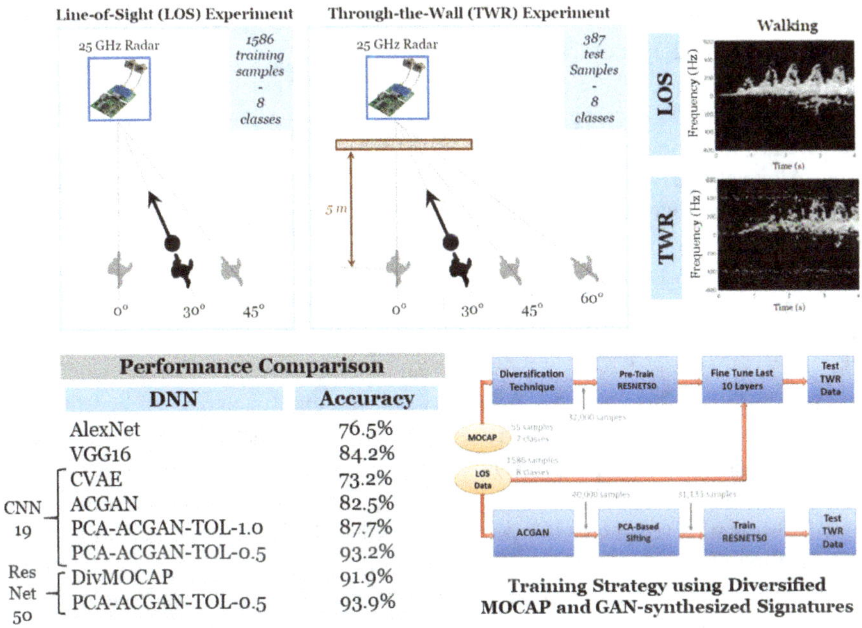

Figure 14.7 *Comparison of performance achieved by transfer learning with synthetic data generation approaches*

The presence of such fallacious signatures in the synthetic training database degrades the classifier performance. Consider an experiment in which a 24 CW GHz radar is used to observe 8 different human activities under 2 different environmental conditions: (1) radar observes motion with direct line-of-sight (LOS) and (2) with a plywood wall between the radar and target (TWR), as shown in Figure 14.7. The LOS data is used to drive an ACGAN to generate synthetic data for training, while the TWR data is used for fine-tuning and testing. Both datasets contain non-target-related artifacts in their signatures due to the RF sensor itself. If synthetic data were generated using MOCAP-based models of target kinematics alone, there would still be some mismatch between training and test data that would degrade test accuracy. The ability of GANs to replicate not just target components but clutter and sensor artifacts as well is a strength of the approach—but this benefit is somewhat offset by the degradation caused by kinematic inaccuracies in the signatures.

To eliminate such misleading signatures from the GAN-synthesized training database, principal component analysis (PCA) is used to determine the convex hull bounding measured micro-Doppler samples in feature space. This volume is expanded by an adjustable tolerance to allow from some diversity while excluding extremely deviant synthetic samples. By a process termed “kinematic sifting” [70], any synthetic data which falls outside the expanded convex hull for a given class is removed from the training database. This results in a large amount of

GAN-synthesized data which has been essentially quality checked to ensure somewhat accurate modeling of the target signature.

The efficacy of various approaches for pre-training a DNN for classification of the TWR dataset is compared in Figure 14.7. The first two approaches considered are pre-training on optical images from the ImageNet database and classifying using AlexNet and VGGnet. In the literature [44], VGGnet has consistently given good results on micro-Doppler data, a result that is reflected in the results for this experiment [71] as well. VGG16 yields 84.2% accuracy, a result that surpasses those achieved by GANs such as the CVAE and ACGAN without any kinematic considerations taken into account. This result highlights the degree to which the misleading GAN signatures can degrade potential performance. When PCA-based kinematic sifting is applied, the classification performance achieved with a 19-layer CNN now surpasses VGGnet, reaching 93.2% with the discarding of about 10% of the signatures. Changing the network to a deeper 50-layer residual network (ResNet-50) only incrementally improves performance to 93.9%. This shows how strongly performance is limited by the training data as opposed to the specific network being used. The use of diversified MOCAP with ResNet-50 (DivMOCAP) also surpasses the performance of VGGnet16, only slightly shy of PCA-sifted ACGAN, with an accuracy of 91.9%. The high performance attained with diversified MOCAP reinforces the significance of accurately representing the target signature, while the 2% difference can be attributed to the lack of clutter representation in the diversified MOCAP data.

14.3.3.1 Knowledge-aided neural network design

The challenge of generating signatures with good kinematic fidelity is a prime example of the importance of integrating domain knowledge into the design of neural networks. Deep learning was developed to automatically learn complex underlying relationships within the data by relying solely on the data itself. But in sensing problems, this means that the DNNs are also trying to learn aspects of the model that are actually well known due to physical principles. It is unnecessary to learn principles that are already well known in advance, and doing so not only increases complexity but also leaves the learning process completely unconstrained and results in avoidable errors. Just as humans rely on prior experience to temper our situational awareness, so should domain knowledge be integrated into the design of DNNs to boost performance and robustness.

For example, let us return to the problem of data synthesis with adversarial networks. Kinematic fidelity is greatly reliant on accurately capturing the motion of the lower body. Any type of gait abnormality, such as due to medical, foot injury, or neuromuscular disorder, is predominantly reflected in the kinematics of the legs, and, hence, the envelope of the micro-Doppler signature. To reinforce envelope features in the adversarial learning process, a multi-branch GAN (MBGAN) architecture has been recently proposed [72] that uses not just the micro-Doppler signature but also the envelope of the synthetic and real signatures, as shown in Figure 14.8. Visual comparison of the resulting signatures and distributions of features reveal

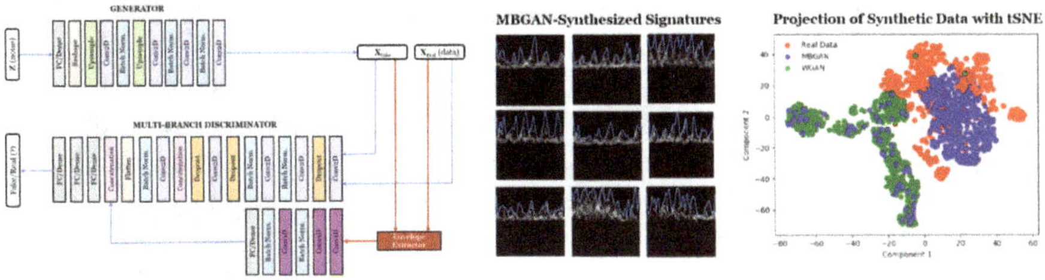

Figure 14.8 MBGAN architecture and resulting constrained envelopes in synthetic signatures generated

that the MBGAN signatures can better conform to the envelope than WGAN or ACGAN networks. Thus, integration of domain knowledge—namely, the insight that the time-varying envelope bounds kinematic components—was used to improve the accuracy of synthesized signatures.

14.4 Linguistic analysis of radar data for ASL recognition

Kinematics is just one form of domain knowledge that bounds RF human motion signatures. An emerging application of RF sensing is toward the understanding of sign languages, such as ASL, which use hands as primary articulators, and head/body motion and facial expressions (non-manuals) for communication. Although ASL is often likened to gesturing, it is not reducible to a sequence of handshapes and arm movements. ASL, like other sign languages, concurrently uses multiple channels (i.e. hand motion, facial actions/non-manuals, body posture changes) to convey linguistic information at the lexical, pragmatic, and syntactic levels [73]. Non-signers tend to be familiar with the fingerspelling component of sign language; this, however, constitutes only a small portion of sign language vocabulary. Instead, dynamic changes in handshape, hand orientation, and place of articulation constitute most signs, with alternations in motion patterns and scope contributing morphological, syntactic, and prosodic (e.g. emphasis) information.

There is often confusion with regard to the similarities and differences in gesture vs. sign language. Signing, as a gesture, does have some reliance on iconic forms. For example, temporal and spatial boundaries, which can be identified based on deceleration (and/or a stop) in signing articulators, can be interpreted semantically as event boundaries. However, if non-signers tend to make such interpretations sweepingly, on the basis of visual event segmentation heuristic [74], the meanings of individual signs within a language are much more subtle, and subject to the rules of grammar and prosody (suprasegmental envelope of motion) in a synthetic manner. The main difference between gesture and sign, however, is the amount of information per unit of time that each can convey. The multi-layered, multi-channel structure of sign language communication, in conjunction with rich and varied inventory of available handshapes and dynamic motions, allows for a much denser stream of information.

In this section, we focus on how the linguistic properties of language are manifested physically in ASL, and how these physical properties can be utilized as a part of domain knowledge in response to the sensing challenges as outlined above. We specifically focus on the applicability of domain knowledge to three problems: (1) separation of ASL from other activities; (2) use of machine learning to identify non-signers imitating ASL from native ASL users; and (3) recognition of up to 20 words in ASL.

14.4.1 Physical parameters of ASL signal

Figure 14.9 presents the velocity of the dominant hand during signing of a short connected narrative projected into the three orthogonal planes—X, Y, and Z—as recorded by 200 Hz motion capture system. The initial and final intervals of high-speed motion are transitional motions from neutral position (hands-down) to and from signing space in front of the signer. The physical space in which production occurs (signing space) is relatively small, such that most signs fall within an 80 cm vertical and 60 cm horizontal space in front of the signer [75].

Several important features of signing motion are readily apparent in this ground-truth motion data. First, the velocity of hands as articulators is more variable than that observed in everyday human activity [76]. The time-varying movement primarily occurs in the range of 0.5 to 1 m/s, although velocities of up to 4 m/s are attested in the literature [75]. The physical parameters of signing are determined by rapid hand motion and handshape changes contained within the signing space.

Thus, the temporal resolution of data collection is important. The parameters of temporal resolution depend on whether domain knowledge can be used as a back-end constraint for sign recognition. Native signers rely on predictive processing across linguistic levels: during sign language comprehension, they are visually sampling the input, while tracking possible grammatical structures, known vocabulary

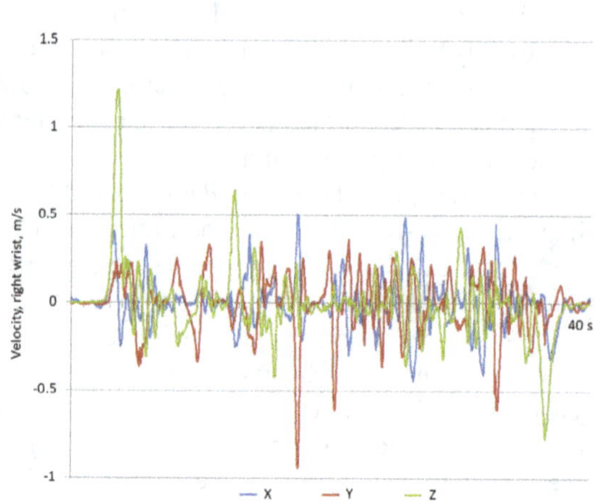

Figure 14.9 Velocity of the dominant hand along orthogonal axes

inventory, and phonotactic constraints (combinations of motion, place of articulation, and handshape sequences appropriate to each specific language), as well as physical constraints—e.g. that of the arm always being attached to the shoulder. All of these limit the space of "what I might see next," and reduce the sampling frequency needed for comprehension by the native signer. Predictive processing also compensates for obstructions at different viewing angles: proficient signers are able to reconstruct linguistic message from input that has been significantly downgraded in terms of spatial (e.g. one-handed WHISPER that uses only handshapes and motion, dispensing with the normal place of articulation) or temporal parameters (e.g. if the field of view between communicating signers is temporarily obstructed by someone else walking between them).

14.4.2 RF measurements of ASL

Current research in psycholinguistics of sign language comprehension indicates that dynamic properties of signs (e.g. dominant hand velocity, temporal contour of motion signature for manual and non-manual articulators) contribute crucial linguistic information to the meaning of signs as understood by native signers [77]. Although radar data does not provide for easy identification of hand shapes and place of articulation (i.e. static spatial features of signs), it does allow for improved measurements of the gross spatiotemporal dynamics of signing (i.e. shape change dynamics), combining information picked up from the moving hands with the information on other articulators (head and body).

Illustrative examples of the μD signatures for several ASL signs are shown in Figure 14.10. Each RF sensor is positioned so as to be directly facing the signer with a 0° aspect angle. Comparing the μD signatures for different sensors, it may be observed that as the transmit frequencies increase so does the peak Doppler shift. This result is consistent with expectations as the Doppler shift resulting from the motion of an object with velocity *vs* toward a radar is $f_D = 2v(f_c/c)$, where c is the speed of light and f_c is the center frequency of transmission. From the perspective of ASL recognition, the greater Doppler spread observed at higher frequencies also results in subtle, small micro-motions being revealed in the signature. Moreover, μD signatures obtained from the 77 GHz sensor appear to be crisper than especially those obtained with the lower frequency Xethru devices.

A number of important inferences about kinematics can be made from observations of the μD signatures:

- *The starting position of the articulators (hands) affects the initial and final frequency components measured.* In the case of our experiments, participants were seated and hands initially placed on the thighs. As signing requires raising the arms, typically with a motion away from the radar, toward the body, and at the level of the chest, this motion resulted in negative frequency spikes at the beginning of all samples.
- *For signs with uniform motion (i.e. where non-dominant hand mirrors the motion of the dominant hand, or where only the dominant hand is moving),*

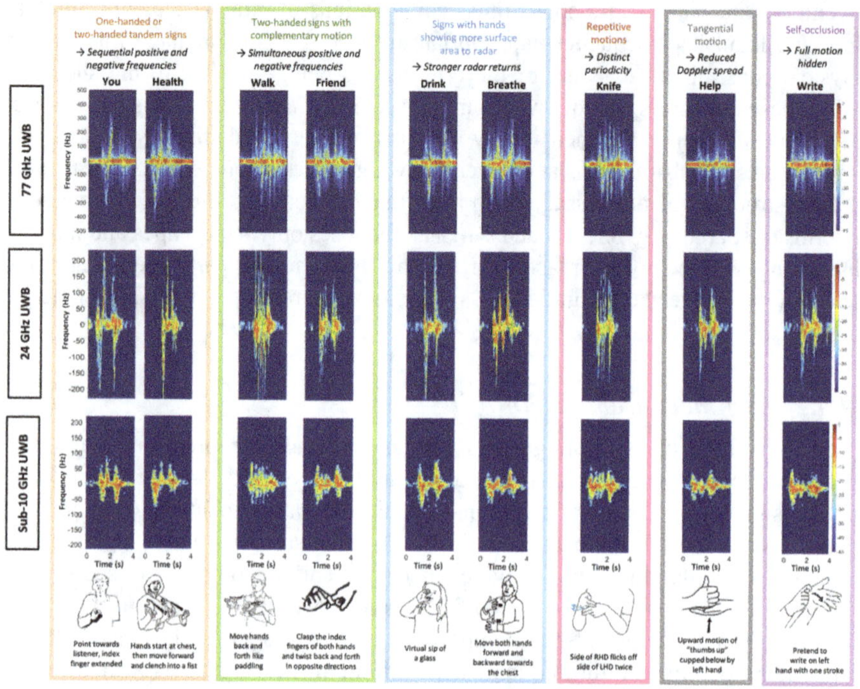

Figure 14.10 Micro-Doppler signatures of ASL signs

the temporal signature of the motion is evident from the μD signature. The signature has either a positive or a negative Doppler frequency at a given time, indicative of the articulator(s) moving toward the sensor, or away from it, respectively. Examples include the signs for YOU and HEALTH.

- *In signs where dominant and non-dominant hand motions are complementary, i.e. the two move in opposite directions, both positive and negative μD components are present simultaneously.* Examples include WALK and friend.
- *When more surface area of the hand(s) faces the radar LOS, the received power of the signal may be observed to be greater.* Examples include DRINK and BREATHE. Notice that the DRINK micro-Doppler signature has two vertical spikes; one due to raising the hand cupped, the second lowering the hand. The first peak has a greater intensity as the outside of the hand faces the radar. Once the virtual cup is lifted, the side of the hand faces the radar, and this dominates the return signature, but with lesser intensity.
- *In signs with reduplicated movements, the number of cycles can be counted from the μD signature.* Examples include KNIFE and WALK.
- *The effect of aspect angle between the LOS and direction of motion can be observed in the signatures.* For example, the sign for HELP involves primarily vertical movement, which is orthogonal to the radar LOS, and hence has a low Doppler frequency. As the hand moves higher, however, non-zero Doppler

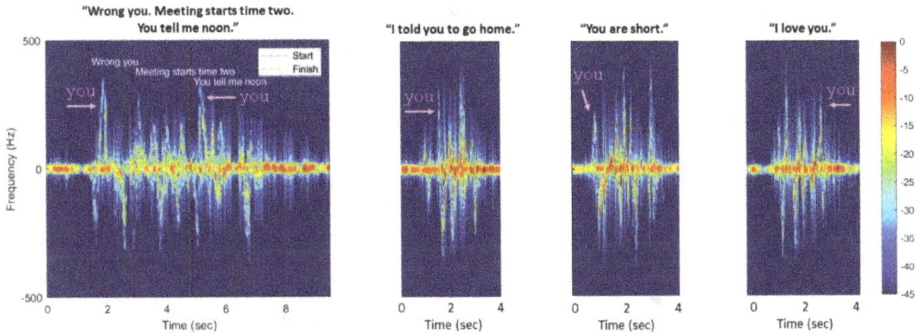

Figure 14.11 Coarticulation example: μD signature for YOU *in different positions within a sentence*

does appear, resulting in vertical streaks in the signature as the hands move upwards and downwards.

- *The effects of occlusion can be observed in some signatures.* Consider the sign for WRITE, which is expressed by holding the left hand and dragging the right hand (with fingers as though holding a pen) across the inside of the palm. Whereas the sliding motion in KNIFE is clearly visible (fingers graze top of the fingers in the left hand), in the sign for WRITE, the left hand shields the right hand swiping motion—at least partially—from the radar.

An important linguistic feature that is visible in the RF micro-Doppler signatures of ASL is *coarticulation*: the effect of the previous sign (more specifically, its place of articulation) influencing subsequent signs. In sign language, coarticulation is most obviously present in cases where the location from which the hand(s) move toward the place of articulation for the next sign affects both the total distance traveled, and the speed of the articulator motion. These kinematic features are directly visible in the micro-Doppler signature and result in signs appearing slightly differently in different sentences. In our experiments, the participant always began with hands on the knees; thus, when the same sign was used sentence-initially vs. mid-sentence, the hands did not start from the same position.

Consider, for example, the sentence "WRONG YOU. MEETING STARTS TIME TWO. YOU TELL ME NOON." depicted in Figure 14.11, which includes the sign for YOU twice. This sign is clearly visible in the micro-Doppler signature as the two peaks with the greatest positive velocity or micro-Doppler frequency. The extension and retraction of the hand are also clearly visible as two separate lines forming the sides of a narrow triangle. However, comparing the first and second occurrences, it may be observed that in the second occurrence the sign is entered rapidly, almost instantly reaching the peak of the μD signature, whereas the first occurrence is slower: the micro-Doppler has a finite slope, indicating that a longer duration was needed to reach peak speed. Note that this makes sense considering that more time is needed to travel a greater distance (from knees to chest level, arm extended). In contrast, when in a mid-sentence position, the hands are already chest level and the signer needs to just extend the hand forward. This observation is consistent with the greater slope exhibited by the first occurrence of YOU during the retraction of the hand. In this situation, very little time is needed for the hand to return to the chest (instead of the knees) and continue to the next sign. Hence, the slope is nearly vertical. Notice that in the sentence "YOU ARE SHORT." the shape of the sign YOU matches that observed in the aforementioned example, when YOU was the first sign in the sentence.

In other mid-sentence occurrences of YOU, exemplified by the μD signatures for "I TOLD YOU TO GO HOME" the burst is so narrow that the extension and retraction can no longer be distinguished. And in the final example "I LOVE YOU" the sign yields greater reflected power due to the prior sign involving both arms crossed over the chest (LOVE). As the arms uncross to permit the right hand to extend forward, strong reflections are generated from the arms crossed across the chest. The two arms uncrossing to permit the right hand to extend forward results in a stronger reflection signal than that observed in the previous instances of the sign YOU. Thus, some differences in the μD signature for YOU may be observed depending on the preceding and successive signs.

14.4.3 Native vs. imitation signing

In the domain of spoken language recognition, the analysis of a specific language's pronunciation is done on the data from native speakers of the language. It would not occur to anyone to develop French processing algorithms using the datasets of English speakers trying to pronounce individual French words. Yet, this is the approach frequently adopted for sign language processing research due to lack of data from native signers: non-signers are asked to attempt to copy signs made by native signers, and these data are then used for algorithm development. Perceptual data indicates that signers and non-signers differ drastically in their perception of rapidly changing visual stimuli. For example, when signers and non-signers viewed point-light displays of 'WRITING' [78], signers could identify 'STROKES (information-bearing portions of point-light movement) and 'TRANSITIONS' (movement of the point-light from the end of one meaningful portion to the beginning of another), while non-signers could not. In terms of production, sign language learners show much greater variability in spatiotemporal parameters of sign and sign sequence production [79], with variability decreasing as ASL language experience increases. Hilger *et al.* [79] also demonstrate that Deaf native signers' sign production (native production) is characterized by stability, which indicates (and requires) proficiency—and it can take learners of sign language at least three years to produce signs in a manner that is perceived as fluent by native signers [80]. In the process of learning to sign, L2 learners commit a large number of phonetic and semantic errors (i.e. errors in the use of appropriate handshape, motion, and places of articulation) where finer motor control is required. With regard to temporal parameters of signing, utterances, signs, and transitions between signs produced by learners of ASL are longer in duration as compared to native signing [81]. Appropriate timing of prosodic contours is, in fact, one of the features that can only be acquired through early exposure to, and use of sign language: for example, a study of early and late

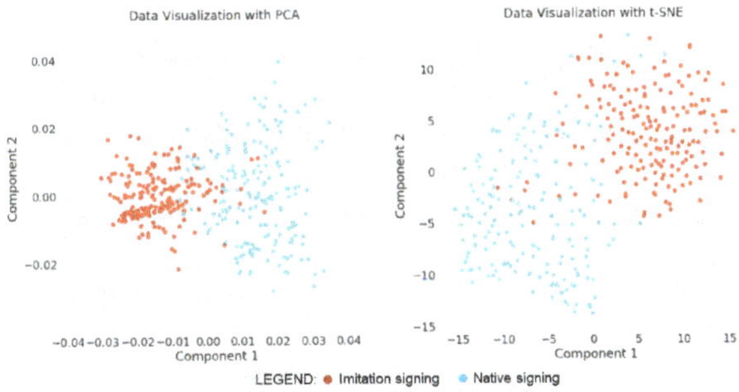

Figure 14.12 *Comparison of RF data from native ASL users and imitation signers using PCA and tSNE*

learners of Swiss-German Sign Language has indicated that phrase-structure sensitive change in signing dynamics (i.e. prosody) can be entirely absent in late learners [82]. Notably, appropriate (native-signer-like) application of prosodic, i.e. motion-envelope-related modifications is impossible without high stability of signing motor control—which, as noted above, is only achievable by proficient native signers.

Thus, copysign production by non-signers results in kinematic data that contain linguistic errors across phonological, semantic, syntactic, prosodic, and pragmatic levels, and, from the perspective of physical representation, are severely distorted in their spectrotemporal parameters. This means that the use of non-signer production data for training can lead to inappropriate capture of the underlying variations in the data and subsequent classification errors. However, due to the differences in kinematic parameters between native and non-native data, machine learning and quantitative analysis could potentially be used to evaluate signer proficiency, for example, in sign language learners. Consider, for example, visualizations of the feature space occupied by native versus imitation signing data. Both PCA and T-distributed Stochastic Neighbor Embedding (t-SNE) algorithms were applied to visualize and compare the feature spaces. The t-SNE algorithm converts similarities between data points into joint probabilities and tries to minimize the Kullback-Leibler divergence between the joint probabilities.

Figure 14.12 shows the feature space as represented by the first two principal components and t-SNE representations. To remove any bias due to sample size, 180 samples were randomly selected from the imitation dataset, an equal number to the native signing samples utilized. It is striking that although there is an overlap between the two datasets, the variance of the imitation signers' data is greater than that of the native ASL signers. Moreover, the centroids of each dataset are clearly shifted relative to each other. This indicates that imitation signing is biased relative to native ASL signing, and its distribution is not truly representative of native ASL signing.

The difference between imitation and native ASL data can be further quantified by using a support vector machine (SVM) classifier with radial basis function to explicitly classify imitation versus native signing data. The data for all 20 signs were combined into two sets: imitation data and native ASL data. To mitigate class imbalance between the datasets, the synthetic minority technique (SMOTE) [83, 84] was applied. SMOTE equalizes the class data by oversampling minority the classes with "synthetic" samples. First, the pixels of the 2D spectrogram are converted into a concatenated 1-D vector. Then, the synthetic samples are generated by taking the difference between the sample under consideration and its nearest neighbor. This difference is multiplied by a random number between 0 and 1 and added to the sample under consideration. This procedure can be generalized for k-nearest neighbors (kNNs) and *N*% oversampling. SVM was applied on the features generated via PCA using fivefold cross validation. We found that SVM could distinguish between native and imitation signers with a 99% accuracy.

These results underscore the need to test ASL recognition algorithms on native ASL data provided by deaf participants, rather than rely upon more easily acquired imitation signing data from students or other hearing participants via copysigning. Machine learning algorithms tested on imitation data are actually not being correctly assessed.

14.4.4 *Fractal complexity for discrimination of ASL from daily activities*

Human languages in general are characterized by complexity in production (auditory, and visual, i.e. print and signing) that are a result of interactions across linguistic subsystems at multiple levels of granularity, such as phonology and syntax [85, 86]. ASL signal thus contains a greater amount of information than non-communicative motion [76, 87]. This expectation can be verified in RF data through comparison of the fractal complexity of ASL with that of daily activities. Fractal complexity is calculated by first finding the power spectral density for each velocity. This is done by computing the Fourier transform of the velocity spectrum versus time. This results in a magnitude matrix $M(j,f)$, where (j,f) are velocity and frequency bins, respectively. The fitting parameter for the fractal complexity, $\beta(j)$, is related to the magnitude at each velocity bin j as

$$M(j,f) = \frac{a}{|f|^{\beta}}, \tag{14.5}$$

where a is an amplitude fitting variable [88]. Taking the logarithm of both sides, a linear expression may be found as

$$ln(M) = ln(a) - \beta ln|f| \tag{14.6}$$

A simple linear fit is then performed on $ln(f)$ versus $ln(M)$, where β relates to the slope and $ln(a)$ is the intercept on a log–log plot. If $M(j,f)$ is integrated over j, an overall velocity spectrum can be obtained, which, after fitting, then results in an overall fractal complexity parameter, $\bar{\beta}$. Note that $\bar{\beta}$ is inversely related to fractal complexity, so that a lower $\bar{\beta}$ implies greater information.

The fractal complexity for ASL was computed using the sentence sequences acquired under Experiment 2, which yielded a total of 30 samples of 8 seconds

Table 14.1 *Comparison of $\bar{\beta}$ values for ASL and daily activities using video and RF sensors*

		$\bar{\beta}$	
		ASL	**Daily activity**
Sensors	RF	0.86	0.93
	Video	0.16	0.48

duration each. Daily activity data was acquired from two hearing participants enacting eight different daily activities: (1) building legos, (2) playing chess, (3) cooking, (4) painting, (5) eating, (6) vacuuming, (7) folding laundry, and (8) ironing clothes. The test area was equipped with required props so that activities were conducted just as would be done were the participants at home. Data were acquired for a continuous duration of 10 minutes to ensure participants were moving about as naturally as possible. These data were then cropped into 8 seconds, non-overlapping segments to yield 30 samples per activity or a total of 240 samples. The average $\bar{\beta}$ values calculated for ASL and daily activities using video and RF sensors are given in Table 14.1. For both sensors, the $\bar{\beta}$ value for ASL is less than that for daily activities. Thus, both RF sensors and video show that ASL signing communicates greater information than daily activities; this feature of ASL signal can be used for discrimination between ASL and non-ASL signal.

14.5 ASL Recognition using RF μD signatures

In this section, the ability of RF sensors to distinguish various ASL signs is demonstrated through μD signature classification.

14.5.1 Handcrafted features

First, a wide range of features, drawn from four principal groupings, are extracted from the μD signatures:

14.5.1.1 Envelope features

Envelope features have been shown to be significant physical features [13, 89] of the μD signature as they describe the outline and peak response of the signature. In this work, the maximum, minimum, and mean values of upper and lower envelopes, as well as the difference between the average values of the upper and lower envelope are extracted using the percentile technique [90]. The time indices of the envelopes are found from the cumulative amplitude distribution of the columns as follows:

$$P(v,t) = \sum_{v=v_{min}}^{v} s(v,t) \Big/ \sum_{v=v_{min}}^{v_{max}} s(v,t) \tag{14.7}$$

where v is the radial velocity, t is the time, and s is the spectrogram image matrix storing the values for amplitude of the spectrogram. For each column, the value $P(U(t), t) = 0.97$ is used to calculate the upper envelope; $P(C(t), t) = 0.5$ is used to calculate the peak trajectory; and $P(L(t), t) = 0.03$ is used to calculate the lower envelope. The seven envelope features are computed from the upper and lower envelopes.

14.5.1.2 DCT coefficients

The discrete cosine transform (DCT) represents a given signal in terms of a sum of sinusoidal functions oscillating at different frequencies. We extracted 500 2D DCT coefficients [91] from each μD signature and vectorized them.

14.5.1.3 FWCC features

Frequency-warped cepstral coefficients (FWCC) are inspired from mel-frequency cepstral coefficients, common in speech processing, but whose filter bank is optimized to RF data using genetic algorithms [92–94].

FWCC features are found by first initializing a filter bank that contains M triangular filters $h_m(k)$. Then, the log-energy output of each filter, E_m, is computed by applying the filter on the spectrogram (S) as follows:

$$E_{(n,m)} = log\left(\sum_{k=0}^{N-1} S(n,k)h_m(k)\right) \tag{14.8}$$

for $n = 1, 2, \ldots T$, where N is the length of the filter and T is the time index of the spectrogram. We then compute the cepstral coefficients C by taking the DCT of the filter outputs:

$$C(j,n) = \sum_{m=1}^{M} E_{(n,m)} cos\left(j(m - \tfrac{1}{2})\tfrac{\pi}{M}\right) \tag{14.9}$$

for $j = 1, 2, \ldots J$, where J is the number of cepstral coefficients. The feature vector, C_i, for ith sample is then obtained by vectorizing the resulting matrix $C(j, n)$. It should be noticed that the resulting coefficients depend on the parameters that we used while creating the filters; namely, its starting, maximum, and endpoints. The filter bank itself is optimized using a genetic algorithm, where the length of a chromosome is equal to three times the total number of filters, $3M$, since we have three parameters to define a filter. The classification performance of a random forest classifier (RFC) with 100 decision trees is employed as the fitness function.

14.5.1.4 LPC coefficients

Linear predictive coding (LPC), a commonly used method in speech processing that has been proposed for μD [95], computes the coefficients of a forward linear predictor by minimizing the prediction error in a least squares sense. In this work, we extracted 100 LPC coefficients from each μD spectrogram by minimizing the residual error, $e[n]$ in

$$s[n] = \sum_{k=1}^{p} a_k s[n-k] + e[n], \tag{14.10}$$

where $\{a_k\}$ is the pth order linear predictor coefficients and s is the spectrogram matrix, first on the columns and then the rows.

14.5.2 Feature selection and classification results

Due to the curse of dimensionality, an effective subset of features is oftentimes more beneficial in classification than blindly using all possible features. While there are many available methods for feature selection, it has been shown [96] that for μD classification the minimum redundancy maximum relevance (mRMR) algorithm [97] yields the best performance without having any dependency on the specific classifier used, while also having better generalization capabilities. A total of 932 features are initially extracted for each RF sensor in the network: 7 envelope features, 500 DCT coefficients, 325 FWCC features, and 100 LPC coefficients. For all sensors, features were extracted from the spectrogram both with and without high-pass filtering (HPF). This is because while HPF removes stationary clutter, there is also the potential for important low-frequency information to be lost. Next, the mRMR algorithm was applied to select an optimal subset of features from each sensor. The number of features selected was varied between 20 and 250, while four different classifiers considered to evaluate performance: SVMs, kNNs, linear discriminant analysis using random subspace gradient boosting, and RFC. During classification, 75% of the data was used for training, while 25% was used for testing.

Figure 14.13(a) shows the selected features and accuracy obtained for classification of 20 ASL signs using RFC, as that was the classifier that offered the best overall performance. Notice that significant numbers of all types of features were selected, with the exception of LPC, which does not appear to be very effective towards ASL recognition. In contrast, features that reflect the decomposition of the signature into different frequency components, such as DCT and FWCC, are strongly preferred from an information-theoretic perspective.

Better performance is achieved at the higher transmit frequencies, with the accuracy of the 24 GHz sensor closely followed by that of the 77 GHz sensor. Notice that there is a great benefit to not using a HPF on 77 GHz data, for which the accuracy without the HPF is increased by 13% whereas other sensors benefit from the filtering. The ASL recognition accuracy can be greatly increased by fusing features from all inputs and performing a second round of feature selection with mRMR. This yields a classification accuracy of 72% for 20 ASL classes—about a 15%–30% improvement over the results obtained with just a single sensor. A classification accuracy of 95% is attained for the recognition of 5 ASL signs. The results obtained for all four classifiers as a function of the total number of signs discriminated is given in Figure 14.13(b), while the impact of HPF is compared in Figure 14.13(c).

14.6 Conclusion

This chapter provided an overview of challenges and techniques used in the application of machine learning for human motion recognition. Two specific types of human motions are considered: activity recognition and ASL recognition. In both

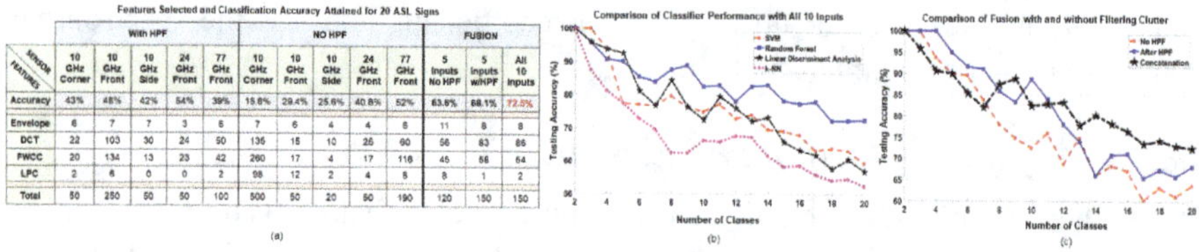

Features Selected and Classification Accuracy Attained for 20 ASL Signs

SENSOR / FEATURES	With HPF					NO HPF					FUSION		
	10 GHz Corner	10 GHz Front	10 GHz Side	24 GHz Front	77 GHz Front	10 GHz Corner	10 GHz Front	10 GHz Side	24 GHz Front	77 GHz Front	5 Inputs No HPF	5 Inputs w/HPF	All 10 Inputs
Accuracy	43%	48%	42%	54%	39%	15.6%	29.4%	25.6%	40.8%	52%	63.6%	68.1%	72.5%
Envelope	6	7	7	3	6	7	6	4	4	6	11	8	8
DCT	22	103	30	24	50	136	15	10	25	60	56	83	86
FWCC	20	134	13	23	42	260	17	4	17	116	45	58	54
LPC	2	6	0	0	2	98	12	2	4	8	8	1	2
Total	50	250	50	50	100	500	50	20	50	190	120	150	150

Figure 14.13 (a) Features selected by MRMR algorithm and resulting accuracy with RFC; (b) comparison of performance given by different classifiers; (c) comparison of performance achieved with and without filtering out clutter

cases, bringing a physics-aware perspective to the application of machine learning was seen to provide tangible improvements in performance. In the activity recognition case, integration of kinematics into adversarial learning was seen to improve the accuracy of synthetic data generated. In the case of sign language recognition, the linguistic information carried by visual parameters, such as movement dynamics and shape of articulators, was seen to be reflected in the RF data. This was evidenced by the observation of coarticulation, which changed the envelope profile of the micro-Doppler signature of the sign depending on the position in a sentence, and the ability of RF sensors to distinguish native from imitation signing. Thus, in both cases, it could be seen that the physics and context of the motion, be it kinematics or linguistics, was reflected in the data and critical to the development of better performing machine learning algorithms.

Acknowledgments

This work was funded in part by NSF Awards #1932547, #1931861, and #1734938. The authors thank their colleagues Dr. Darrin Griffin, Dr. Ali C. Gurbuz, and Dr. Chris Crawford for their contribution in acquisition and analysis of ASL data.

References

[1] Natarajan A., Komijani A., Hajimiri A. 'A fully integrated 24-GHz phased-array transmitter in CMOs'. *IEEE Journal of Solid-State Circuits*. 2005, vol. 40(12), pp. 2502–14.

[2] Natarajan A., Komijani A., Guan X., *et al*. 'A 77-GHz phased-array transceiver with on-chip antennas in silicon: transmitter and local LO-path phase shifting'. *IEEE Journal of Solid-State Circuits*. 2006, vol. 41(12), pp. 2807–19.

[3] Ku B., Schmalenberg P., Inac O., *et al*. 'A 77–81-GHz 16-element phased-array receiver with beam scanning for advanced automotive Radars'. *IEEE Transactions on Microwave Theory and Techniques*. 2014, vol. 62(11), pp. 2823–32.

[4] Silvious J., Clark J., Pizzillo T., *et al.* 'Micro-Doppler phenomenology of humans at UHF and Ku-band for biometric characterization'. *Proceedings of SPIE - The International Society for Optical Engineering*. 2009, vol. 05, p. 7308.

[5] Seifert A.-K., Grimmer M., Zoubir A.M. 'Doppler radar for the extraction of biomechanical parameters in gait analysis'. *IEEE Journal of Biomedical and Health Informatics*. 2020, vol. 25(2), pp. 547–58.

[6] Amin M.G., Zhang Y.D., Ahmad F., *et al.* 'Radar signal processing for elderly fall detection: the future for in-home monitoring'. *IEEE Signal Processing Magazine*. 2016, vol. 33(2), pp. 71–80.

[7] Seyfioğlu M.S., Serinöz A., Özbayoğlu A.M., Gürbüz S.Z. 'Feature diverse hierarchical classification of human gait with CW radar for assisted living'. International Conference on Radar Systems (Radar 2017); 2017. pp. 1–5.

[8] Li Y., Du L., Liu H. 'Hierarchical classification of moving vehicles based on empirical mode decomposition of Micro-Doppler signatures'. *IEEE Transactions on Geoscience and Remote Sensing*. 2013, vol. 51(5), pp. 3001–13.

[9] Yang X., Yin W., Zhang L. 'People counting based on CNN using IR-UWB radar'. 2017 IEEE/CIC International Conference on Communications; China; 2017. pp. 1–5.

[10] Amin M.G., Guendel R.G. 'Radar human motion recognition using motion states and two-way classifications'. 2020 IEEE International Radar Conference (RADAR); 2020. pp. 1046–51.

[11] Scheirer W.J., de Rezende Rocha A., Sapkota A., *et al.* 'Toward open set recognition'. *IEEE Transactions on Pattern Analysis and Machine Intelligence*. 2013, vol. 35(7), pp. 1757–72.

[12] Cherti M., Kégl B., Kazakçı A. 'Out-of-Class Novelty Generation: An Experimental Foundation'. 2017 IEEE 29th International Conference on Tools with Artificial Intelligence (ICTAI); 2017. pp. 1312–9.

[13] Gurbuz S.Z., Erol B., Cagliyan B., *et al.* 'Operational assessment and adaptive selection of micro-Doppler features'. *IET Radar, Sonar Navigation*. 2015, vol. 9(9), pp. 1196–204.

[14] Fioranelli F., Patel J., Gurbuz S.Z. 'Multistatic human micro-Doppler classification with degraded/jammed radar data'. 2019 IEEE Radar Conference (RadarConf); 2019. pp. 1–6.

[15] Patel J.S., Fioranelli F., Ritchie M., *et al.* 'Fusion of deep representations in Multistatic radar networks to counteract the presence of synthetic jamming'. *IEEE Sensors Journal*. 2019, vol. 19(15), pp. 6362–70.

[16] Klarenbeek G., Harmanny R.I.A., Cifola L. 'Multi-target human gait classification using LSTM recurrent neural networks applied to micro-Doppler'. 2017 European Radar Conference (EURAD); 2017. pp. 167–70.

[17] Wang S., Song J., Lien J. Interacting with soli: exploring Fine-Grained dynamic gesture recognition in the radio-frequency spectrum. Proc. of the 29th Annual Symposium on User Interface Software and Technology; New York, NY, USA: ACM; 2016. pp. 851–60.

[18] Zhang Z., Tian Z., Zhou M. 'Latern: dynamic continuous hand gesture recognition using FMCW radar sensor'. *IEEE Sensors Journal*. 2018, vol. 18(8), pp. 3278–89.
[19] Li H., Mehul A., Le Kernec J., Gurbuz S.Z., Fioranelli F. 'Sequential human gait classification with distributed radar sensor fusion'. *IEEE Sensors Journal*. 2021, vol. 21(6), pp. 7590–603.
[20] Dogaru M., Le C. '*Validation of Xpatch Computer Models for Human Body Radar Signature*. US Army Research Laboratory Technical Report, ARL-TR-4403'.
[21] Geisheimer J.L., Greneker E.F., Marshall W.S. 'High-resolution Doppler model of the human gait'. Proceedings of the AeroSense Conference; 2002.
[22] van Dorp P., Groen F.C.A. 'Human walking estimation with radar'. *IEE Proceedings - Radar, Sonar and Navigation*. 2003, vol. 150(5), pp. 356–65.
[23] Loukas C., Fioranelli F., Le Kernec J., Yang S. 'Activity Classification Using Raw Range and I Q Radar Data with Long Short Term Memory Layers'. 2018 IEEE 16th Intl Conf on Dependable, Autonomic and Secure Computing, 16th Intl Conf on Pervasive Intelligence and Computing, 4th Intl Conf on Big Data Intelligence and Computing and Cyber Science and Technology Congress(DASC/PiCom/DataCom/CyberSciTech); 2018. pp. 441–5.
[24] Otero M. 'Application of a continuous wave radar for human gait recognition'. Proceedings of SPIE; 2005.
[25] Cohen L. 'Time-frequency distributions – a review'. *Proceedings of the IEEE*. 1989, vol. 77(7), pp. 941–81.
[26] Chen V., Ling H. 'Time-frequency transforms for radar imaging and signal analysis'. *Artech House*. 2002.
[27] Tivive F.H.C., Phung S.L., Bouzerdoum A. 'Classification of micro-Doppler signatures of human motions using log-Gabor filters'. *IET Radar, Sonar Navigation*. 2015, vol. 9(9), pp. 1188–95.
[28] Almeida L.B. 'The fractional Fourier transform and time-frequency representations'. *IEEE Trans on Signal Processing*. 1994, vol. 42(11), pp. 3084–91.
[29] Daubechies I. 'The wavelet transform, time-frequency localization and signal analysis'. *IEEE Trans on Information Theory*. 1990, vol. 36(5), pp. 961–1005.
[30] Sridharan K., Thayaparan T., SivaSankaraSai S. 'Gabor-Wigner transform for micro-Doppler analysis'. 9th International Radar Symposium; India; 2013.
[31] Choi H., Williams W.J. 'Improved time-frequency representation of multicomponent signals using exponential kernels'. *IEEE Transactions on Acoustics, Speech, and Signal Processing*. 1989, vol. 37(6), pp. 862–71.
[32] Jokanovic B., Amin M., Erol B. 'Multiple joint-variable domains recognition of human motion'. 2017 IEEE Radar Conference; 2017. pp. 0948–52.
[33] Li Y., Peng Z., Pal R., *et al*. 'Potential active Shooter detection based on radar micro-Doppler and range-Doppler analysis using artificial neural network'. *IEEE Sensors Journal*. 2019, vol. 19(3), pp. 1052–63.
[34] Erol B., Amin M.G. 'Radar data cube processing for human activity recognition using multi subspace learning'. IEEE Transactions on Aerospace and Electronic Systems; 2019. pp. 1–1.

[35] He Y., Le Chevalier F., Yarovoy A.G. 'Range-Doppler processing for indoor human tracking by multistatic ultra-wideband radar'. 2012 13th International Radar Symposium; 2012. pp. 250–3.

[36] He Y. Human target tracking in multistatic ultra-wideband radar. PhD Dissertation; Delft University of Technology; 2014.

[37] Schumann O., Hahn M., Dickmann J. 'Semantic segmentation on radar point clouds'. 21st International Conference on Information Fusion; 2018. pp. 2179–86.

[38] He Y., Molchanov P., Sakamoto T., Aubry P., Le Chevalier F., Yarovoy A. 'Range-Doppler surface: a tool to analyse human target in ultra-wideband radar'. *IET Radar, Sonar & Navigation.* 2015, vol. 9(9), pp. 1240–50.

[39] Du H., Jin T., Song Y. 'A three-dimensional deep learning framework for human behavior analysis using Range-Doppler time points'. IEEE Geoscience and Remote Sensing Letters; 2019. pp. 1–5.

[40] He K., Zhang X., Ren S., Sun J. 'Deep residual learning for image recognition'. 2016 IEEE Conference on Computer Vision and Pattern Recognition, CVPR; Las Vegas, NV, USA, June 27–30; 2016. pp. 770–8.

[41] Bengio Y., Lecun Y. in Bottou L., Chapelle O., DeCoste D., Weston J. (eds.) *Scaling Learning Algorithms Towards AI.* MIT Press; 2007.

[42] Bengio Y., Lamblin P., Popovici D. *Greedy Layer-Wise Training of Deep Networks.* MIT Press; 2007.

[43] Erhan D., Courville A., Bengio Y. 'Why does unsupervised pre-training help deep learning?'(eds. *Proceedings of the Thirteenth International Conference on Artificial Intelligence and Statistics. vol. 9 of Proceedings of Machine Learning Research.* Chia Laguna Resort, Sardinia, Italy; 2010. pp. 201–8.

[44] Seyfioglu M.S., Ozbayoglu A.M., Gurbuz S.Z. 'Deep convolutional autoencoder for radar-based classification of similar aided and unaided human activities'. *IEEE Transactions on Aerospace and Electronic Systems.* 2018, vol. 54(4), pp. 1709–23.

[45] Russakovsky O., Deng J., Su H., *et al. Imagenet large scale visual recognition challenge [online].* CoRR. 2014;abs/1409.0575. Available from http://arxiv.org/abs/1409.0575 [Accessed 4 Aug 2021].

[46] Krizhevsky A., Sutskever I., Hinton G.E. 'Imagenet classification with deep convolutional neural networks'. Advances in Neural Information Processing Systems. 2012. Available from https://citeseerx.ist.psu.edu/viewdoc/summary?doi=10.1.1.299.205.

[47] Simonyan K., Zisserman A. Very deep convolutional networks for large-scale image recognition. 2014. Available from https://www.semanticscholar.org/paper/Very-Deep-Convolutional-Networks-for-Large-Scale-Simonyan-Zisserman/eb42cf88027de515750f230b23b1a057dc782108.

[48] Szegedy C., Liu W., Jia Y., *et al. Going deeper with convolutions [online].* CoRR. 2014;abs/1409.4842. Available from http://arxiv.org/abs/1409.4842.

[49] Park J., Javier R.J., Moon T., Kim Y. 'Micro-Doppler based classification of human aquatic activities via transfer learning of convolutional neural networks'. *Sensors.* 2016, vol. 16(12), p. 1990.

[50] Du H., Jin T., Song Y., *et al.* 'Efficient human activity classification via sparsity-driven transfer learning'. *IET Radar, Sonar & Navigation.* 2019, vol. 05, p. 13.

[51] Seyfioglu M.S., Erol B., Gurbuz S.Z., *et al.* 'DNN transfer learning from diversified micro-Doppler for motion classification'. *IEEE Transactions on Aerospace and Electronic Systems.* 2019, vol. 55(5), pp. 2164–80.

[52] Alnujaim I., Alali H., Khan F., *et al.* 'Hand gesture recognition using input impedance variation of two antennas with transfer learning'. *IEEE Sensors Journal.* 2018, vol. 18(10), pp. 4129–35.

[53] Khanna R., Oh D., Kim Y. 'Through-wall remote human voice recognition using Doppler radar with transfer learning'. *IEEE Sensors Journal.* 2019, vol. 19(12), pp. 4571–6.

[54] SeyfioÄŸlu MS., Gurbuz S.Z. 'Deep neural network initialization methods for micro-Doppler classification with low training sample support'. *IEEE Geoscience and Remote Sensing Letters.* 2017, vol. 14(12), pp. 2462–6.

[55] Boulic R., Thalmann N.M., Thalmann D. 'A global human walking model with real-time kinematic personification'. *The Visual Computer.* 1990, vol. 6(6), pp. 344–58.

[56] Blasch E., Majumder U., Minardi M. 'Radar signals dismount tracking for urban operations'. *Proc. SPIE 6235, Signal Processing, Sensor Fusion, and Target Recognition XV, 623504.* 2006, vol. 06.

[57] Majumder U., Minardi M., Blasch E., *et al.* 'Radar signals dismount data production'. *Proc. SPIE 6237, Algorithms for Synthetic Aperture Radar Imagery XIII, 623704 (17 May 2006).* 2006, vol. 05.

[58] Ram S., Ling H. 'Micro Doppler signature simulation of computer animated human and animal motions'. *2008 IEEE Antennas and Propagation Society International Symposium.* 2008, pp. 1–4.

[59] Ram S.S., Christianson C., Kim Y., *et al.* 'Simulation and analysis of human micro-dopplers in through-wall environments'. *IEEE Transactions on Geoscience and Remote Sensing.* 2010, vol. 48(4), pp. 2015–23.

[60] Erol B., Karabacak C., Gürbüz S.Z., Gürbüz A.C. 'Simulation of human micro-Doppler signatures with Kinect sensor'. 2014 IEEE Radar Conference; 2014. pp. 0863–8.

[61] Erol B., Karabacak C., Gurbuz S.Z. 'A kinect-based human micro-Doppler simulator'. *IEEE Aerospace and Electronic Systems Magazine.* 2015, vol. 30(5), pp. 6–17.

[62] Karabacak C., Gurbuz S.Z., Gurbuz A.C., *et al.* 'Knowledge exploitation for human micro-Doppler classification'. *IEEE Geoscience and Remote Sensing Letters.* 2015, vol. 12(10), pp. 2125–9.

[63] Seyfioglu M.S., Erol B., Gurbuz S.Z., Amin M.G., *et al.* 'DNN transfer learning from diversified micro-Doppler for motion classification'. *IEEE Transactions on Aerospace and Electronic Systems.* 2019, vol. 55(5), pp. 2164–80.

[64] Creswell A., White T., Dumoulin V., *et al.* 'Generative adversarial networks: an overview'. *IEEE Signal Processing Magazine.* 2018, vol. 35(1), pp. 53–65.

[65] Lewis B., Liu J., Wong A. 'Generative adversarial networks for SAR image realism'. Proceedings SPIE; 2018. p. 10647.

[66] Doherty H.G., Cifola L., Harmanny R.I.A., Fioranelli F. 'Unsupervised learning using generative Adversarial networks on micro-Doppler spectrograms'. 2019 16th European Radar Conference (EuRAD); 2019. pp. 197–200.

[67] Alnujaim I., Oh D., Kim Y. 'Generative Adversarial Networks to Augment Micro-Doppler Signatures for the Classification of Human Activity'. IGARSS 2019 - 2019 IEEE International Geoscience and Remote Sensing Symposium; 2019. pp. 9459–61.

[68] Erol B., Gurbuz S.Z., Amin M.G. *'GAN-based Synthetic Radar Micro-Doppler Augmentations for Improved Human Activity Recognition'*. 2019 IEEE Radar Conference (Radar Conf); 2019. pp. 1–5.

[69] Mi Y., Jing X., Mu J., x L., He Y. 'DCGAN-based scheme for radar spectrogram augmentation in human activity classification'. 2018 IEEE International Symposium on Antennas and Propagation USNC/URSI National Radio Science Meeting; 2018. pp. 1973–4.

[70] Erol B., Gurbuz S.Z., Amin M.G. 'Motion classification using kinematically sifted ACGAN-synthesized radar micro-Doppler signatures'. *IEEE Transactions on Aerospace and Electronic Systems*. 2020, vol. 56(4), pp. 3197–213.

[71] Erol B., Gurbuz S.Z., Amin M.G. 'Exploitation of motion capture data for improved synthetic micro-Doppler signature generation with adversarial learning.' in Ahmad F. (ed.). *Big Data II: Learning, Analytics, and Applications*. International Society for Optics and Photonics. SPIE; 2020. pp. 30–7. Available from https://doi.org/10.1117/12.2560118.

[72] Erol B., Gurbuz S.Z., Amin M.G. 'Synthesis of Micro-Doppler Signatures for Abnormal Gait using Multi-branch Discriminator with Embedded Kinematics'. 2020 IEEE International Radar Conference (RADAR); 2020. pp. 175–9.

[73] Blumenthal-Dramé A., Malaia E. 'Shared neural and cognitive mechanisms in action and language: the multiscale information transfer framework'. *Wiley Interdisciplinary Reviews: Cognitive Science*. 2019, vol. 10(2), p. e1484.

[74] Strickland B., Geraci C., Chemla E., *et al.* 'Event representations constrain the structure of language: sign language as a window into universally accessible linguistic biases'. *Proceedings of the National Academy of Sciences*. 2015, vol. 112(19), pp. 5968–73.

[75] Bosworth R.G., Wright C.E., Dobkins K.R. 'Analysis of the visual spatiotemporal properties of American sign language'. *Vision Research*. 2019, vol. 164, pp. 34–43.

[76] Malaia E., Borneman J.D., Wilbur R.B. 'Assessment of information content in visual signal: analysis of optical flow fractal complexity'. *Visual Cognition*. 2016, vol. 24(3), pp. 246–51.

[77] Malaia E., Wilbur R.B., Milković M. 'Kinematic parameters of signed verbs'. *Journal of Speech, Language, and Hearing Research*. 2013, vol. 56(5), pp. 1677–88.

[78] Klima E.S., Tzeng O.J., Fok Y., *et al.* 'From sign to script: effects of linguistic experience on perceptual categorization'. *Journal of Chinese Linguistics Monograph Series*. 1999, pp. 96–129.

[79] Hilger A.I., Loucks T.M.J., Quinto-Pozos D., Dye M.W.G., *et al.* 'Second language acquisition across modalities: production variability in adult L2 learners of American sign language'. *Second Language Research*. 2015, vol. 31(3), pp. 375–88.

[80] Beal J.S., Faniel K. 'Hearing L2 sign language learners: how do they perform on ASL phonological fluency?' *Sign Language Studies*. 2018, vol. 19(2), pp. 204–24.

[81] Cull A. *Production of Movement in Users of American Sign Language and Its Influence on Being Identified as "Non-Native"*. Gallaudet University; 2014.

[82] Braem P.B. 'Rhythmic temporal patterns in the signing of deaf early and late learners of Swiss German sign language'. *Language and Speech*. 1999, vol. 42(2–3), pp. 177–208.

[83] Chawla N.V., Bowyer K.W., Hall L.O., *et al.* 'SMOTE: synthetic minority over-sampling technique'. *Journal of Artificial intelligence research*. 2002, vol. 16, pp. 321–57.

[84] Bufler T.D., Narayanan R.M. 'Radar classification of indoor targets using support vector machines'. *IET Radar, Sonar Navigation*. 2016, vol. 10(8), pp. 1468–76.

[85] Coupé C., YM O., Dediu D., *et al.* 'Different languages, similar encoding efficiency: comparable information rates across the human communicative niche'. *Science Advances*. 2019, vol. 5(9), p. eaaw2594.

[86] Torre I.G., Luque B., Lacasa L., *et al.* 'On the physical origin of linguistic laws and lognormality in speech'. *Royal Society Open Science*. 2019, vol. 6(8), p. 191023.

[87] Borneman J.D., Malaia E., Wilbur R.B. 'Motion characterization using optical flow and fractal complexity'. *Journal of Electronic Imaging*. 2018, vol. 27(5),p. 051229.

[88] Borneman J.D., Malaia E., Wilbur R.B. 'Motion characterization using optical flow and fractal complexity'. *Journal of electronic imaging*. 2018, vol. 27(5),p. 1.

[89] Kim Y., Ha S., Kwon J. 'Human detection using Doppler radar based on physical characteristics of targets'. *IEEE Geoscience and Remote Sensing Letters*. 2015, vol. 12(2), pp. 289–93.

[90] Dorp P.V., Groen F.C.A. 'Feature-based human motion parameter estimation with radar'. *IET Radar, Sonar Navigation*. 2008, vol. 2(2), pp. 135–45.

[91] Molchanov P., Astola J., Egiazarian K. 'Ground moving target classification by using DCT coefficients extracted from micro-Doppler radar signatures and artificial neuron network'. 2011 Microwaves, Radar and Remote Sensing Symposium; 2011. pp. 173–6.

[92] Erol B., Gurbuz S.Z. 'Hyperbolically-warped cepstral coefficients for improved micro-Doppler classification'. 2016 IEEE Radar Conference (Radar Conf); 2016. pp. 1–6.

[93] Erol B., Amin M.G., Gurbuz S.Z. 'Automatic data-driven frequency-warped cepstral feature design for micro-Doppler classification'. *IEEE Transactions on Aerospace and Electronic Systems*. 2018, vol. 54(4), pp. 1724–38.

[94] Erol B., Seyfioglu M.S., Gurbuz S.Z., Amin M. 'Data-driven cepstral and neural learning of features for robust micro-Doppler classification' in Ranney K.I., Doerry A. (eds.). *Radar Sensor Technology XXII*. 10633. International Society for Optics and Photonics. SPIE; 2018. pp. 163–70. Available from https://doi.org/10.1117/12.2304396.

[95] Javier R.J., Youngwook Kim. 'Application of linear predictive coding for human activity classification based on micro-Doppler signatures'. *IEEE Geoscience and Remote Sensing Letters*. 2014, vol. 11(10), pp. 1831–4.

[96] Tekeli B., Gurbuz S.Z., Yuksel M. 'Information-theoretic feature selection for human micro-Doppler signature classification'. *IEEE Transactions on Geoscience and Remote Sensing*. 2016, vol. 54(5), pp. 2749–62.

[97] Peng H., Long F., Ding C. 'Feature selection based on mutual information criteria of max-dependency, max-relevance, and min-redundancy'. *IEEE Transactions on Pattern Analysis and Machine Intelligence*. 2005, vol. 27(8), pp. 1226–38.

Index

www.ingramcontent.com/pod-product-compliance
Lightning Source LLC
LaVergne TN
LVHW011551200926
848484LV00001B/303
* 9 7 8 1 8 3 9 5 3 1 8 8 0 *